STP 1027

Aquatic Toxicology and Hazard Assessment: 12th Volume

U. M. Cowgill and L. R. Williams, editors

ASTM
1916 Race Street
Philadelphia, PA 19103

ASTM: Publication Code Number (PCN): 04-010270-48
ISBN: 0-8031-1253-X
ISSN: 1040-306X

NOTE

The Society is not responsible, as a body,
for the statements and opinions
advanced in this publication.

Peer Review Policy

Each paper published in this volume was evaluated by three peer reviewers. The authors addressed all of the reviewers' comments to the satisfaction of both the technical editor(s) and the ASTM Committee on Publications.

The quality of the papers in this publication reflects not only the obvious efforts of the authors and the technical editor(s), but also the work of these peer reviewers. The ASTM Committee on Publications acknowledges with appreciation their dedication and contribution of time and effort on behalf of ASTM.

Printed in Ann Arbor, MI
July 1989

Foreword

This publication, *Aquatic Toxicology and Hazard Assessment: 12th Volume,* contains papers presented at the 12th Symposium on Aquatic Toxicology and Hazard Assessment, which was held 24–26 April 1988 in Sparks, Nevada. The symposium was sponsored by ASTM Committee E-47 on Biological Effects and Environmental Fate and its Subcommittee E47.01 on Aquatic Toxicology. U. M. Cowgill, Dow Chemical Co., presided as symposium chairperson and Llewellyn R. Williams, EPA/EMSL, served as cochairperson. Both served as coeditors of this publication.

Contents

Overview

During the past twelve years, the aquatic toxicology group (Subcommittee E47.01) of ASTM has sponsored an annual symposium for the major purpose of bringing together aquatic specialists from industry, government, and academe. The end result of these gatherings has been a debate on the merits of test development, animal and plant culture, nutrition and testing, and, last but not least, the lack of interagency harmony. The underlying intent of the 12th Symposium on Aquatic Toxicology and Hazard Assessment was to hold sessions devoted to updating all the various subfields of aquatic toxicology. Thus, the meeting began with a discussion of the benefits of interagency harmonization, which was largely devoted to the need for unity among the various regulatory bodies devoted to protecting the environment. Reviews of common modes of toxic action, target toxicant analysis, and field techniques offered a forum for discussion on advances that have occurred in these fields since the last deliberation on these topics some symposia ago. The problems associated with statistical interpretation of the results of microcosm testing occupied a full session. New approaches in sediment toxicity testing, the culturing and testing of new organisms, exclusively marine, and the never-ending association between nutrition and testing were brought up to date in several minisymposia. A session was devoted to quality assurance in ectotoxicity testing, which is represented in this symposium volume by a discussion of the New Jersey laboratory certification program. The symposium closed with a heavy attendance at a session on toxicity testing problems involving effluents.

These many minisymposia resulted in this volume, which records the events of this meeting in the form of 34 papers. An update of the diverse subfields of aquatic toxicology will serve as a summary useful to all aquatic specialists. It must be realized that the subjects these minisymposia covered are advancing quickly and, therefore, a review or update is advisable on an annual basis. This information is produced at such speed it would be difficult for one individual alone to be proficient in all areas. Thus, annual symposia of the present type serve a very useful purpose to aquatic specialists from industry, government, and academe by appraising them of the most recent advances in the many subfields of aquatic toxicology.

The symposium chairman is indebted to the minisymposia organizers and to their speakers, who made this symposium a success. The ASTM staff is gratefully acknowledged for their assistance in organizing this twelfth symposium and their efforts toward making its publication timely.

U. M. Cowgill

The Dow Chemical Co., Midland, MI 48674, symposium chairman and coeditor

Benefits of Interagency Harmonization of Methodology

William Horwitz[1]

Harmonization of Methodology: International Perspective

REFERENCE: Horwitz, W., **"Harmonization of Methodology: International Perspective,"** *Aquatic Toxicology and Hazard Assessment: 12th Volume, ASTM STP 1027,* U. M. Cowgill and L. R. Williams, Eds., American Society for Testing and Materials, Philadelphia, 1989, pp. 5–10.

ABSTRACT: Not long ago, most experimentation dealing with analytical methodology in the physical and biological sciences was conducted within a single laboratory. Method validation by other laboratories was not only considered unnecessary but also detrimental because, in the words of one commentator, "the results are too variable." Within the last two decades, however, largely as a result of the requirements of international environmental and food standards programs, it has become increasingly apparent that a collaborative interlaboratory study is the only way to estimate the systematic and random error characteristics of methods of analysis as they will be performed by the population of laboratories typical of those who will be using the method. To obtain a common basis for measuring the statistical characteristics of analytical methods, representatives of almost two dozen international organizations meeting in Geneva in May 1987 approved by consensus a protocol that will be useful for the design and interpretation of collaborative studies of chemical methods of analysis. Much of this protocol will also be useful for the study of biological measurement methods.

KEY WORDS: methods of analysis, accuracy, precision, interlaboratory studies, proficiency studies, collaborative studies, systematic error, random error, performance parameters

Biologists, chemists, toxicologists, and other scientists are constantly making and interpreting measurements without giving a second thought to the mechanics of the process. When investigating a phenomenon, whether it be the wavelength of light, the amount of copper in liver, or the growth of organisms in an aquatic environment, scientists have been taught to isolate the activity, maintain everything else constant, and then vary one factor at a time to see what happens to the response. If many things can be varied with little effect on the response, the system is said to be rugged, and a direct cause and effect relationship can be deduced when a variable does influence the system and the result produced. Sometimes several factors can be varied simultaneously, with mathematical statistics used to sort out the individual effects. This is the way things are done in the laboratory. As soon as we move away from the well-controlled environment of the laboratory into the unrestrained setting provided by nature, direct relationships seem to disintegrate and extraneous factors begin to perturb our observations. Sometimes we can identify the source of the disturbance and control that factor. But when numerous uncontrollable variables exist that have the ability to influence the results, the net effect is high variability in the observations.

About a century ago, when chemical analysis began to be used in the United States for the control of industrial processes, and particularly for the control of agricultural commodities such as fertilizers, animal feeds, and human foods, most of the substantial variability observed between results from different laboratories was ascribed to the methods of analysis. Steps were

[1]Scientific advisor, Food and Drug Administration HFF-7, Washington, DC 20204.

taken by the organizations with responsibility for the products to control this source of variability by specifying the method of analysis to be used. To this end, the Association of Official Analytical Chemists (AOAC) was established in 1884 to approve methods of chemical analysis to control products regulated by state and federal agencies, and the American Society for Testing and Materials (ASTM) was founded in 1898 to formulate standards for commercial commodities. On an international scale, the International Organization for Standardization (ISO) was organized in 1946.

Methods of analysis, the directions for conducting chemical measurements, are the most visible and consequently the most adjustable aspect of biological and chemical experimental systems. When dealing with the macrochemistry of analytes in the 1 to 100% concentration range, the effect of major external variables on the observed system was easily demonstrated because the inherent random and systematic variabilities of the measurements utilized in classical analytical chemistry were relatively small, of the order of 1 to 2% of the analytical value. Therefore, a perturbation from external sources resulting in a difference of several percent in the final value of an analyte was easily visible against a background of 1% or so of analytical noise. The analytes dealt with today, however, are present in the parts per million (ppm, 10^{-6}) and parts per billion (ppb, 10^{-9}) range, where analytical variabilities are of the order of magnitude of 10 to 20% of the analytical value, so the relative influence of variables affecting the system must be considerably greater than this magnitude for their effect to be noticeable above the analytical noise.

Errors

A method of analysis is basically a set of directions on how to conduct a specific biological, chemical, or physical measurement, which ultimately involves the determination of the amount or concentration of an analyte or property. The adequacy of the result is considered in terms of the difference (called systematic error) between the result found and the result known or assumed to be true as well as the degree of dispersion (called random error) of the multiple results among themselves. In discussing experimental measurements, the term "error" is not used in the ordinary sense of a mistake or blunder. Rather it is used in the statistical sense as a numerical measure of the difference between an experimental result and a true value. These experimental variabilities and errors of measurement can be classified into three major categories according to their source: sampling, within laboratory, and between laboratory.

Sampling Errors

Although sampling errors are often the greatest source of variability in environmental measurements, these errors are really beyond the control of the laboratory, unless the laboratory is responsible for the sampling operation. Regardless of who collects the samples, this source of variability must be considered individually for each lot or population to be examined. The magnitude of this source of variability is assessed by taking a number of random samples from the population and examining them individually. With experience, the amount of variability to be expected from similar sites or populations can be determined and used as a first approximation for calculating sample size in terms of number of units and their size. Such a protocol is called a statistical sampling plan and is intended to control the variability encountered when taking additional samples from the same population at different times.

Because each sampling situation is unique, the variability introduced by sampling will not be considered here. Sampling variabilities should always be kept in mind, however, when analytical variability appears to be excessive. Sometimes laboratories have been unfairly accused of producing unreliable results when, actually, the variability lies in the samples submitted to them. As pointed out by Youden [*1*], if the sampling error is greater than twice the analytical error, reduction in analytical error is of little avail in improving estimates of composition or properties. Emphasis must then be placed on the sampling aspects of the situation, not on the analytical aspects.

Analytical Errors

Analytical errors are inherent in any measurement system. They result from unavoidable differences which arise during the conduct of the measurement operations. There are numerous sources of these deviations, many of which occur without the knowledge of the operator. From the point of view of a single laboratory, it is assumed that the average of a series of measurements of the same analyte or property is the best estimate of that item. The variability of the individual within-laboratory measurements that compose an average provides an indication of the reliability of the estimate of that average. The within-laboratory variability can be broken down into smaller components such as the variability between analysts in the same laboratory, between runs and within runs, between days and within days, between calibration curves and within the same calibration curve, and between and within any perceived source of variability. However, many of these sources of variability overlap and some encompass others (nested) to form a complex error structure that can be very difficult to unravel. Many of these sources of variability are so small that they can be classified as "microprecisions." Sometimes a statistical analysis will indicate that a variable is significant, and at other times, with a different set of data, that it is not significant. Ascribing significance to such unstable factors is the source of many of the controversies involving the interpretation of biological data. All of these microprecisions can be encompassed in the single estimator "within-laboratory" standard deviation.

The situation changes dramatically when more laboratories enter the picture. Each laboratory produces a similar but individual within-laboratory variability and its own long-term average. A new source of variability, *between* laboratories, now emerges which did not exist previously. The *inter*laboratory variability presents a more complex problem than the *intra*laboratory variability because the between-laboratory variability includes the within-laboratory component. In fact, the between-laboratory component of the total variability is usually the largest fraction of the total analytical error. In biological work, this component can be so large that toxicologists, biologists, and microbiologists rarely perform interlaboratory experiments because, as one microbiologist put it, "the results are too variable." If this is the case, whose results are correct—those from my laboratory or from your laboratory? Actually, both results may be "correct" when there is no control of systematic differences in analytical methods between laboratories.

Organizations that provide standard methods of analysis, such as ASTM, AOAC, and ISO, have developed their own procedures, called the collaborative study, to predict how methods of analysis will operate in actual practice. Such a study consists of analyses by a number of laboratories of identical materials over the concentration and commodity range of interest. The results are analyzed statistically to extract the within-laboratory standard deviation, s_r, and the "pure" between-laboratory standard deviation, s_L, which when added together as variances give the overall between-laboratory standard deviation, $s_R = (s_r^2 + s_L^2)^{1/2}$. These parameters, even better expressed as relative standard deviations, RSD_r and RSD_R, are useful as summary statistics of previously performed collaborative studies. With further statistical calculations, the results are then used to predict the performance of methods when used in the future for the same analyte in the same matrix by similar laboratories. The maximum tolerable difference (with 95% confidence) between two individual readings from the same laboratory (r) or from two different laboratories (R) are the repeatability value and the reproducibility value, respectively:

$$\text{Repeatability value} = 2 \times 2^{1/2} \times s_r$$

$$\text{Reproducibility value} = 2 \times 2^{1/2} \times s_R$$

The same statements in terms of relative standard deviations or variances are obtained by substituting the appropriate RSD or s^2 for s.

In addition to the organizations which cover a number of commodity areas, many organizations cover specialized areas, such as water and wastewater, textiles, petroleum products, wines, cereal products, and oils and fats. Many of these organizations have overlapping interests with potential duplication of effort.

The conduct of a collaborative study involving numerous laboratories analyzing a number of identical blind test samples is an expensive undertaking. To avoid duplication of studies among the interested organizations, a working group of the International Union of Pure and Applied Chemistry (IUPAC) was established to harmonize the protocols of the various organizations for performing collaborative studies. The working group sponsored several international symposia over the past decade, culminating in a Workshop on Harmonization of Collaborative Analytical Studies, held in Geneva, Switzerland, at ISO headquarters in May of 1987. Representatives from about 20 international and national organizations interested in the subject agreed by consensus on minimum criteria for the design, conduct, and interpretation of collaborative studies. Adherence to these criteria will permit a collaborative study conducted by one group to serve for all organizations.

Minimum Criteria for Collaborative Studies

The representatives participating in the IUPAC harmonization workshop agreed on a definition of a collaborative study as a study to determine the performance characteristics of a method of analysis as distinct from a proficiency study to determine the performance of laboratories or analysts in conducting an analysis, or from a certification study to obtain a "true" or reference value for an analyte in, or property of, a material. Many individuals and organizations had used these terms, as well as such undefined terms as *round-robin, ring test,* and *intercalibration scheme,* interchangeably for any interlaboratory experiment involving more than one laboratory. A collaborative study, as agreed by the workshop, requires the analysis by at least eight laboratories of at least five materials utilizing either blind (that is, unknown to the analyst) replicates or a split level design. A split level consists of two materials of slightly different analyte concentration but sufficiently close together that they can be considered as having the same variance. When it is impossible to meet the above conditions, the study may be smaller, for example, five laboratories or three materials, but at the sacrifice of confidence in the reliability of the estimated parameters.

An important recommendation, dealing with aspects preliminary to a collaborative study, is to conduct interlaboratory trials only on methods that have received thorough testing within a single laboratory. A method that does not perform satisfactorily within a laboratory will perform less well when subjected to the abuse of several laboratories. Single-laboratory testing of a method includes analyte recoveries under various conditions, interference studies, applicability to expected variants of the analyte in anticipated matrices, comparison with previously available methods, calibration procedures, sources of standards, and developing a clear description of the actual procedure. In addition, specifications for equipment, reagents, and adsorbants must be established.

The report also gives the following advice with respect to significant figures: round standard deviations (and relative standard deviations) to two significant figures and round the mean to accommodate that parameter, that is, to the same number of places before or after the decimal point.

Until recently, few realized that permitting statisticians to analyze data any way they wished introduced as much disagreement in the interpretation of results as permitting chemists to conduct analyses any way they wished. A particular point requiring agreement was that collaborative assays must be analyzed by one-way analysis of variance, material by material. This is very important since the design of collaborative studies makes them appear to be excellent candidates for a two-way analysis of variance procedure. Without the one-way analysis of variance specification there might be disagreements among statisticians as to where to place an interaction term, if this term turns out to be significant.

Another essential point of agreement is how to handle outliers. It is a shocking testimony to laboratory performance that fully 25% of collaborative assays published by the AOAC during the past 75 years contain suspect data, and approximately 3% of the reported analytical data are gross outliers. Outliers are values that are far outside the bulk of the data. The IUPAC harmonization workshop standardized on a specific procedure for the evaluation and elimination of suspect data. First the results must be screened to remove invalid data. Invalid data are results from laboratories that did not follow instructions, reported malfunctioning instruments, did not achieve the expected separations, or observed phenomena not covered by the instructions. Invalid data must be distinguished from outliers found by statistical analysis. Invalid data can be traced to a specific cause; outliers are obtained when the cause cannot be determined and a computation shows that it is very improbable that the value belongs with the bulk of the data.

The data are first tested by the Cochran extreme variance test that discards results from laboratories that cannot check themselves as compared with the bulk of the laboratories. Then data are removed from laboratories that show extreme values by the single value Grubbs test. Since an extreme value may not trigger the Grubbs test when a second extreme value is present (a phenomenon known as "masking"), a "paired Grubbs test" is applied to check for the presence of two extreme values on the same side of the distribution and for one extreme value at each end of the distribution. All flagged outliers are removed if they exceed the critical value at the 1% probability (P) level. This is an extremely conservative outlier removal protocol and is expected to make a mistake (remove a value which is really part of a normal distribution) only 1 time in 100.

A program on the Food and Drug Administration computer is available to handle the calculations and to flag outliers automatically. However, for our use we do not permit the program to actually remove the outliers. Removing outliers requires a deliberate decision on the part of the examiner. We look at the statistical parameters before outlier removal. If these parameters are "acceptable" with no values removed, we do not remove values flagged at even a 1% probability value. This inaction is based on our observation that precision parameters are not meaningfully different unless they differ by a factor of approximately two for the typical collaborative study of eight laboratories. We call this protocol the "historical outlier removal treatment." On the other hand, if we wish the statistical parameters as developed by the harmonized procedure just outlined, we instruct the computer to calculate the precision parameters according to the "IUPAC-1987 Harmonized Procedure," which does remove all values flagged at the $P = 0.01$ level.

The impetus for the development of the harmonized protocol was the fact that many of the organizations participating in the workshop were also interested in supplying their approved methods of analysis to the Codex Alimentarius Commission (CAC) of the Food and Agricultural Organization of the United Nations. In fact, so many methods of analysis were submitted that it was almost impossible to make a choice among them. Consequently, a set of principles was developed which stated, among other things, that methods with reliability parameters that had been established through collaborative studies would be preferred over those without such a pedigree. Since considerable prestige was attached to having a method approved by the CAC, organizations began to perform collaborative studies. Each organization soon realized that selecting its own method and performing its own collaborative study was very inefficient, and that a single collaborative study, when there was general agreement on the design and interpretation, should be sufficient. Therefore, organizations began to look for ways to cooperate with each other. Organizationally, this led to the development of the working party of representatives of a number of organizations working under the auspices of the highly respected umbrella chemical organization—IUPAC. To date, only AOAC has approved the harmonized protocol.

The delay in other organizations is probably related to the time of annual meetings and the mechanics of approval within individual organizations.

Quality Control

The high level of outliers that appear in collaborative studies is very disturbing. Collaborative studies are conducted under such conditions that analysts know that their work will be scrutinized by other analysts as well as by their supervisors. Therefore they can be expected to perform their best work in this situation. We can only guess what the outlier rate would be under less stringent conditions, but certainly it would be no better.

The outlier problem has probably developed because in the course of a single generation of scientists, analytical chemistry has changed precipitously from a science of macroanalysis to one of trace analysis. In macrochemistry, gross errors were relatively easy to detect by the principle of consistency. A misplaced decimal point in concentration usually meant that the physical, chemical, biological, or sensory properties of a substance were changed sufficiently to arouse suspicions as to the accuracy of the analysis. A misplaced decimal point at the parts per million level can be discovered only by a completely independent analysis. Duplicate analyses in this situation are of little help because simultaneous operations usually result in repetition of the same mistakes. This is understandable when the cause of a mistake is an incorrect standard solution, a fluctuating line voltage, or an impurity in a reagent. Arithmetic errors are also likely to be perpetuated. Analysts must take very seriously the necessity for incorporating positive controls into their work to provide some assurance that no gross blunder is being made. Such controls can be certified reference materials, when available, but more frequently they must be specially formulated materials or simply a "house" standard that is analyzed repeatedly with each batch of "unknowns."

Methods of analysis must be written with quality control in mind to provide analysts with guideposts that indicate that the analysis is proceeding satisfactorily. Such a signal is easily incorporated into spectrophotometric methods by specifying an absorbance that should be obtained from a working standard solution. In chromatographic methods, an internal standard providing a signal of approximately the same intensity as a typical analyte concentration usually should be preferred over an external standard. In inductively coupled plasma and other types of emission spectroscopy, the ratios of intensities of peaks of various elements [*2*] are sufficiently constant to be useful as quality control indicators. Developers of methods must be educated to the necessity of inserting such guideposts into methods. Analytical chemists become so familiar with their technique that they think others are just as proficient. But even the most proficient of analysts occasionally produces outliers, and any assistance in avoiding potential damage before the report leaves the laboratory should be welcome.

Conclusions

International organizations are considering the use of a common protocol for the determination of the operating characteristics of chemical methods of analysis. These characteristics establish the expected limitations of each method, but the frequent occurrence of outliers even during the most careful analytical work suggests that more attention should be given to the incorporation of quality control guidelines into the operating directions. These guides are needed to indicate that the operations are proceeding satisfactorily or to warn the operator that an unobserved deviation in the system has occurred.

References

[*1*] Youden, W. J., *Journal of the Association of Official Analytical Chemists*, Vol. 50, 1967, pp. 1007–1013.

[*2*] Botto, R. L., *Spectrochimica Acta*, Vol. 39B, 1984, pp. 95–113.

Llewellyn R. Williams[1]

Harmonization of Quality Assurance—An Interagency Perspective

REFERENCE: Williams, L. R., "**Harmonization of Quality Assurance—An Interagency Perspective,**" *Aquatic Toxicology and Hazard Assessment: 12th Volume, ASTM STP 1027,* U. M. Cowgill and L. R. Williams, Eds., American Society for Testing and Materials, Philadelphia, 1989, pp. 11–18.

ABSTRACT: Quality assurance requirements and activities vary widely among federal agencies and among testing and measurement programs within agencies. Partly as a result of these differences, our knowledge of the quality of environmental data produced under these programs varies widely. Hence, the acceptability of test data (its "transportability") from one agency or group to another is greatly hindered. That acceptability or transportability of data can be greatly enhanced if the data quality objectives, quality assurance approaches, and data reporting requirements are more closely harmonized among the producing organizations. In recent years attempts to reach consensus agreements on measurement methodology have been given strong impetus by such standardization organizations as the American Society for Testing and Materials (ASTM) and the Association of Official Analytical Chemists (AOAC). However, the quality control (QC) and quality assurance (QA) practices associated with implementation of those "standardized" methods have been left up to the individual organizations performing testing and measurement. In an attempt to initiate efforts to harmonize QA/QC approaches and the associated reporting of the quality characteristics of measurement data, a workshop was held during February 1987 in Las Vegas, Nevada. Cosponsored by the U.S. Army Toxics and Hazardous Materials Agency, the U.S. Environmental Protection Agency, and the Department of Energy, the workshop brought together experts in many fields from the sponsors and other agencies. The workshop format was useful in developing a slate of recommended areas for potential harmonization and triggered a series of subsequent positive steps to achieve standardization in approaches to acquiring, characterizing, and reporting analytical data. An effort is underway to use the limited successes experienced in such ad hoc work groups as a stepping-stone to further interagency cooperation in the quality assurance arena. A draft management framework has been developed for an interagency steering committee. Under the aegis of the proposed committee, the activities of such ad hoc work groups as described above can be more formally recognized and coordinated and the results and recommendations of their efforts can be more effectively disseminated to meet mutual needs. Negotiations are currently underway to gain interest and support for this concept at the highest levels in various federal agencies that have environmental testing and measurement responsibilities.

KEY WORDS: quality assurance, quality control, environmental measurements, data quality objectives, method validation, method equivalency, reference materials, performance evaluation

Introduction

Background

Quality assurance (QA) procedures are used to determine the quality of data produced in environmental measurement and monitoring programs. In one sense, the term "quality assurance" is a misnomer—one cannot assure quality with the superimposition of procedures upon a

[1]Acting director, Quality Assurance and Methods Development Division, U.S. Environmental Protection Agency, Las Vegas, NV 89193-3478.

measurement program. What one can do is verify and document the quality of data that is being produced and use that knowledge to revise, if necessary, the experimental or program designs by incorporation of specific procedures, for example, proper replication of sample analyses, that will increase the likelihood that study objectives are met.

Many QA functions are oversight in nature and serve to monitor the quality of analytical data produced or the performance of the laboratories producing such data. Specific activities and procedures vary from study to study within organizations; they tend to vary even more widely between organizations with similar responsibilities, for example, regulatory agencies. If the goals of QA programs are generally the same, then similar QA functions should be operable to a greater or lesser extent in every program.

As there are many ways to accomplish similar data quality goals or objectives, agencies and organizations occasionally come into conflict because the approaches of one agency are either nonstandard or foreign to the other and are therefore not recognized. Such situations can become very uncomfortable when each agency or organization is content with and confident in their own approach, have built their QA/QC infrastructure about it, and have difficulty in understanding the necessity to change their modus operandi. To do business in another way is not only costly to the compromising agency, but may, in fact, be less satisfactory because of a lack of familiarity with alternative approaches and procedures. One possible answer lies in an attempt to harmonize the ways in which many key QA operations are performed by various organizations. Whose procedures should be adopted in such cases? Hopefully, the answer can be found by uncovering and adopting the most satisfactory and cost-effective procedures with which each organization can be comfortable and confident. This is perhaps best accomplished by opening broad communications among affected organizations to foster active QA/QC information exchange (it is hoped that the precedent of annual QA harmonization meetings is now established and will continue); expanding those communications with other agencies and organizations not currently participating; bringing consensus findings and recommendations to the attention of the highest levels of organizational management; and, effecting meaningful changes to facilitate better transport of quality environmental data among organizations with mutual interests. The observations and recommendations of the first of a series of proposed interagency QA work group meetings are included in this paper.

ASTM has, for decades, demonstrated the values of standardizing testing methods, materials, and procedures. Its continued success in so many arenas is proof that consensus standards, arrived at through the conscientious efforts of dedicated professionals, can achieve broad acceptance among regulators and regulated alike. While such professional societies as the American Society of Quality Control serve as effective vehicles for the exchange of ideas on new and more effective QA/QC procedures, they have not been very effective in promulgating standard procedures that receive such broad acceptance.

The ability to "transport" data and data bases among different organizations with similar needs and interests depends upon the documented quality of that data. Because of the different approaches to assessing and documenting the quality of data in various environmental data bases, there is frequently little confidence in the data produced by "others" and hence a very real reluctance to make use of otherwise potentially valuable information resources. The archival and shared values of many data bases are subsequently lost, and each new monitoring program bears the cost of reinventing all or a portion of a previously invented "wheel."

The equivalency of QA programs used in various testing and monitoring studies is rarely, if ever, compared. As a result, it is safer to retest than to rely on suspect data. That assumes, of course, that retesting is a possible alternative. Where familiar QA practices and procedures are used, our confidence in the interpretation of the resulting data is greatly enhanced. With increased confidence in the usefulness of data produced by other agencies and organizations comes new opportunities for comparative data analysis and effective use of the literally millions of environmental measurements made by various organizations.

Approach

This paper proposes a possible mechanism to facilitate communication among agencies and organizations to the mutually desirable end of sharing and adopting cost-effective and "smarter" ways of monitoring and correcting defects in our environmental measurement and testing programs. That mechanism is an interagency workgroup on quality assurance.

The first step in this effort was taken in February 1987 with the convening of an interagency "working panel on quality assurance in environmental measurements" in Las Vegas, Nevada. Cosponsored by the U.S. Army's Toxic and Hazardous Materials Agency (USATHAMA), the U.S. Environmental Protection Agency (EPA), and the Department of Energy (DOE), the meetings featured plenary lectures by QA experts, overview presentations of the QA programs of the USATHAMA and the EPA's Superfund Program, and working groups to address specific QA issues in detail.

The 60 attendees represented approximately 30 different agencies, organizations, or organizational units. The backgrounds and experiences of the participants were quite diverse, but the recognition of a need for consensus, standardized QA approaches appeared to be universal. The concept of quality assurance presented at the meetings was broadly defined by the meeting organizers. This breadth is reflected in the following list of subjects (and topic areas for the working groups):

1. Quality assurance management and data quality objectives.
2. On-site auditing, data review, and evaluation.
3. Performance evaluation and reference materials.
4. Method validation and equivalency.
5. Sample management, holding times, and chain of custody.
6. Statistics and chemometrics.
7. Documentation and data communication.

A similar slate of topics was addressed at the second interagency panel meetings held 25–26 May 1988 in Baltimore, Maryland. An emphasis on QA requirements for field sampling and field data collection was added for this meeting.

Panel Observations and Recommendations

Highlights and recommendations of the various working groups follow:

WORKGROUP 1—QA Management and Data Quality Objectives

In addressing the differences between the USATHAMA and EPA QA programs presented, it was recommended that the agencies meet at the highest practicable levels to establish the equivalency of various QA procedures used in the respective programs. The results of such "harmonization" should then be conveyed to each of the ten EPA regional offices to foster a more uniform pan-agency approach.

The key decisionmakers must be clearly identified. QA program similarities and differences must be clearly recognized and addressed. The "type" of data needed by the regulatory agency has to be defined. Three such types of data were delineated: compliance data, engineering design data, and discovery data. Subsequent meetings should be held to determine the precise nature of data requirements and the areas of program departure in addressing these requirements. Data requirements for laboratory analyses, field measurements, and sampling need to be addressed and considered in the context of overall data quality requirements and the additivity of variance.

WORKGROUP 2—On-site Auditing, Data Review, and Evaluation

Little commonality was found between the various agencies on how and why laboratory precertification takes place. Some groups require successful analysis of a performance evaluation sample prior to contract award, while others do not. Some programs have enabling legislation allowing them to confer actual legal certification to testing or analytical laboratories, but most do not. Some of the others simply recognize certain labs as competent and eligible to analyze samples or conduct tests.

For on-site laboratory evaluations, performance evaluation samples are used by some, but not others, for periodic assessment of performance. Where they are used, the data is (EPA) or is not (USDA) a major topic of on-site audit discussions.

Most participants agreed that certain elements should be reviewed during an on-site audit:

1. Personnel.
2. Instrumentation.
3. Standard operating procedures.
4. Facilities.
5. Test and QA data.
6. Specific problems.

Frequency of on-site audits varies from one every two to three years to more often than once a year.

Actions resulting from audits vary widely as well; some never use the audit as a discovery process that can result in halting sample analysis by the laboratory, while others have ordered a "stop work" on the basis of on-site findings.

The need to extend the on-site concept into the field sampling arena was recognized by many.

The relative value of sharing agencies' information on laboratory performance was discussed. No clear consensus was reached on the merits of routinely exchanging such information.

Although the time frame allowed for data review may range from days to months, the key elements are the same: identification of outlying data, evaluation of controls, identification of suspicious calibration, etc. The vast majority of critical data are at or near the detection limit.

The same data, reviewed the same way, will be applied differently by different end users. However, data involved in litigation and chain of custody invariably takes longer to review. The reviews can be facilitated by:

1. Computerized data scans.
2. Organized, standard data packages.
3. Access to lab personnel, sampling information, etc.
4. Exercise of a deadline, such as a holding time.

WORKGROUP 3—Performance Evaluation and Reference Materials

This workgroup decided to break into three subgroups because of the issues to be considered and the large number of participants in the group. The three subgroups were: (1) definitions; (2) performance evaluation materials; and (3) reference materials.

The definitions Subgroup established the following working definitions:

Reference materials—a general term for characterized substances used for the following three functions: (1) instrument calibration; (2) intralab QC; and (3) interlaboratory performance evaluation.

Calibration—process by which the relationship between the response of a measurement instrument and the concentration of selected analytes is established. Calibration is accomplished by using calibration standards which are well characterized as to purity, stability, and concentration, single or multicomponent, neat or in solution. Examples include standard reference materials from the National Bureau of Standards (NBS) and the standards from the U.S. EPA's repository.

Intralaboratory quality control—system of processes and procedures designed to provide an assessment of data quality within a laboratory. This information is developed, in part, by daily analysis of a laboratory check sample. In addition, the laboratory will periodically analyze externally supplied check materials of known composition, such as those supplied by EPA or NBS.

Performance evaluation—an assessment of the comparability of analytical results between laboratories. Performance evaluation materials (PEMs) are periodically supplied to the laboratory as unknowns by the sponsoring agency. PEMs may be derived from natural materials or may be synthetically prepared. The evaluation of laboratories using PEMs may be based on comparison with known values or by comparison of individual results against the average performance.

Performance Evaluation Materials Subgroup.

Need for solid organic performance evaluation samples:

a. Nonvolatile PEMs are relatively easy to prepare.
b. Preparation of volatiles in solid matrices is very difficult.
c. PE samples for explosives in soil are needed.
d. Contaminated natural matrices are preferable to spiked matrices.
e. Spikes into natural or synthetic matrices are sometimes needed.

Frequency and character of PE samples:

a. Case-specific PEs of a blind nature are planned for the Superfund's Contract Laboratory Program (CLP).
b. Introduction of blind samples is difficult because of the necessity of duplicate labeling of samples.
c. PE samples should be "as close to blind as possible."
d. At least one blind should be provided for each lot of samples.
e. Optimum frequency should be based upon studies of lab operations.
f. Control charts should be combined with PE samples for the best approach.

Selection of analytes, matrices, and concentrations:

a. Use levels consistent with the intended end use; multiple levels are recommended; high and low levels are probably optimum.
b. Include both easy and difficult-to-analyze compounds.
c. Use well-characterized analytes that have low intrinsic variability.
d. Select concentrations for the specific methods to be used.
e. Share of PE results among agencies (clear out any proprietary information first).
f. Include test of ability to differentiate between easily confused analytes (qualitative).
g. Select matrices to be representative of source of samples to be analyzed. A limited number of matrix types will allow this and minimize the ability to visually defeat the "blind" nature of the samples.
h. Develop catalog of sources of environmental PE samples. Most labs don't know where to go for such materials.

Evaluation of performance of labs from PE data:

a. End use of the data impacts selection and use of PE samples and data resulting from their analysis.
b. Make data quality objectives consistent with legal needs.

c. PE results are only one part of laboratory evaluation!
d. Much of the value of PE sample analyses is lost where the detection limit of the method approaches the regulatory action level.
e. High-level PE samples are occasionally subject to dilution errors.
f. Handle "high" outliers differently from "low" outliers.

Reference Materials.

a. Level of effort needed to "certify a material" should be defined.
b. For purity documentation of materials considered for standards, one identity and two purity assays should be performed, and only designated labs should be involved in that certification process.
c. Use traceability to NBS where practicable.
d. Identify and share reference materials among agencies to offset costs of certification.
e. Certificate of SRM analysis should include preparation and expiration dates along with analytical information.
f. Interagency study on feasibility of soil reference material should be investigated.

WORKGROUP 4—Method Validation and Equivalency

The method validation process starts with a definition of the method analytes and sample matrices. Next, a method is selected, optimized, and evaluated in a single-laboratory study to determine minimum detection limit (MDL), precision, and bias in a range of matrices of interest, and a test of "ruggedness" such as described by Youden and Steiner in the AOAC statistical manual. A complete method protocol is then written. Where it is possible, a collaborative study involving a minimum of six to eight laboratories should follow to determine interlaboratory variability.

Key issues identified included the selection of suitable test materials for use in the method validation program. While the composition of a totally synthetic sample matrix is known with the greatest confidence and can be expected to be highly homogeneous, it often does not serve as a true test of a method to be applied to environmental samples. The fortified natural matrix is a compromise; analytes of interest can be added at desired concentrations while the sample still poses the difficulties of an environmental matrix. The natural matrix is preferred, but the expense of locating and characterizing a specific matrix with the desired analytes at the proper concentrations may exhaust the resources available for the entire method validation effort. Obtaining volunteer laboratories to participate in interlaboratory studies is also difficult. Although laboratories may commit to serve in the belief that it will make them more competitive in the marketplace, there are often problems in getting the labs to deliver data in a timely manner.

Dynamic validation—The use of dynamic (in use) validation in lieu of a full collaborative study was discussed. It was recommended that dynamic validation be restricted to: those situations where at least 20 laboratories are available for participation; the method is based upon tested methodology for which there is a high degree of confidence that its performance will meet or exceed program requirements; and the method protocol contains a strong quality control component.

Method equivalency—No mechanism currently exists for demonstrating the equivalency of alternative methods to those in the Superfund CLP. The workgroup recommended that EPA staff meet with impacted agencies and adopt a policy allowing equivalent methods to be accepted within Superfund in certain limited circumstances. The existence of equivalency guidelines for other EPA programs was pointed out. Problems that might arise from a liberal equivalency policy include:

1. Development of "equivalent" methods that have different reporting requirements from those of Superfund.
2. Proliferation of different methods that makes data review difficult.

WORKGROUP 5—Sample Management, Holding Times, and Chain of Custody

Seven issues were identified by this workgroup:

Sample definition—Different agencies and organizations define "samples" in different ways. A uniform definition, while desirable, may not be feasible. Good communication and definition will enable each group to understand the meaning of the term as used by others.

Operation of sampling—Depending on data quality objectives, an enhanced effort could be made to address QA of field sampling and to unify procedures.

Field log—Production and use of a uniformly formatted field log would provide important information.

Chain-of-custody documentation—Differences exist in current practices. Increased use of automated analyses fosters unattended operations and therefore can present a problem in chain of custody that may need to be addressed.

Holding times—Specifications differ somewhat among agencies but are largely quite similar. More information on the scientific validity of holding times would be very useful.

Response to problems—Similar levels of data checking are performed by different agencies, but requirements for reanalysis when defects occur vary widely.

Information sharing—A centralized data base might be considered in which the summary information on current standing, date of most recent audit, date and results of most recent PE analysis, etc., could be maintained for each laboratory.

WORKGROUP 6—Statistics and Chemometrics

The following issues were addressed by this workgroup: detection limits; sampling; developing PE criteria; design/analysis comparability; and validation of reference materials.

Detection limits—The need for broad education on the meaning of detection limits was recognized. The detection limit represents a concentration where decisions about presence or absence of an analyte are usually made. The "limit of quantitation" or "reporting limit" is at some concentration above the detection limit. The detection limit is highly dependent on the individual characteristics of the apparatus, analyst, method, and, of course, the matrix of the individual sample. The methods used by the EPA and the USATHAMA yield detection limit values that are generally quite close, but can differ by as much as 50% in unusual cases. More research is needed to determine how limits should be set for multivariate methods. More information is also needed on the effects of matrix variation on detection limit variability.

Sampling—More investigation of the potential for use of composite sampling for cost-effective sampling designs is needed. Such investigation should include both compositing in the field and compositing of laboratory subsamples. The use of colocated samples for estimation of sampling variability should be considered. There is a clear need for improved laboratory subsampling protocols. More effective use of nested experimental designs should provide objective estimates, based on real samples, of: (1) field sampling variability; (2) laboratory subsampling and sample preparation variability; and (3) analytical variability.

PE criteria—Performance standards should be developed—that ultimately include various methods applied to different types of PE materials—and applied to many testing and analytical laboratories. One of the problems recognized was the handling of "outlier" data points without unnecessarily censoring the data sets and rejecting an excessively large percentage of data. The Biweight Robust Estimation Procedure [*1*], the basis for EPA performance evaluation limits for water analyses, was suggested as one alternative. Double-blind PE sample programs are ideal where feasible. The frequency of PE checks should be consistent with the needs of the program. The frequency should be selected based on the volume of data one is willing to discard in the worst-case analysis.

Design/analysis comparability—Laboratory and field personnel must communicate better. It is necessary to carefully formulate all of the protocols for sampling and analysis in advance to be sure that the statistical computations are compatible with the way in which the experiments were actually performed. Knowledge of types of uncontrolled variables in the procedures is also desirable to aid in the design process. It will likely require different experimental designs when optimizing methods for qualitative (presence/absence) vice quantitative analysis.

WORKGROUP 7—Documentation and Data Communication

The level of detail of documentation should vary with the purpose of the study. For characterization and general information gathering, the documentation should be "quick and dirty" with only what is required to make effective and rapid decisions. The key is a conscious decision as to what constitutes necessary and sufficient data for a given study objective.

Sampling and analytical contracts should require the level of documentation necessary for the purposes of that program, not assume that it will be provided.

A long-term objective should be for various agencies to reach some consensus on documentation (format specific) required of contract laboratories. Could the contractor laboratories who are frustrated by the similar yet different reporting requirements of the various agencies to which they respond be challenged to produce a consensus output format for review by the agencies involved? Can software systems be developed at contract labs to process certain QC elements common to all agencies, for example, calibration, or to provide standardized documentation?

Summary

The work group meeting produced as many or more questions than it answered. Each representative took much new information home on how other agencies and organizations were dealing with problems common to virtually all testing and analytical laboratories. However, there was an excitement generated with the realization that the answers to many of the common problems were not beyond reach, that the regulatory agencies could listen and positively respond to the problems implicit in contract laboratory operations, and that an ongoing dialogue was established to improve the manner in which we produce and review environmental data. Because of the high degree of interest expressed by the participating organizations in continuing the information exchange initiated with this meeting, a management framework has been developed for a standing QA oversight committee. The proposed committee structure is in the process of being circulated among federal agencies. It is hoped that the concept will be well received and supported at the highest levels of management, where the quest for data quality must begin. The coordination fostered through this mechanism should further enhance the standardization of approaches to generating and monitoring the quality of data from environmental programs throughout the country.

Reference

Kafadar, K., "A Biweight Approach to the One-Sample Problem," *Journal of American Statistical Association*, Vol. 77, No. 378, June 1982, pp. 416–424.

Review of Common Modes of Toxic Action

Ivan T. Show,[1] Marti B. Show,[1] and Llewellyn R. Williams[2]

Modeling the Contributions of Cross-Reactant Chemicals in Immunoassays

REFERENCE: Show, I. T., Show, M. B., and Williams, L. R., **"Modeling the Contributions of Cross-Reactant Chemicals in Immunoassays,"** *Aquatic Toxicology and Hazard Assessment: 12th Volume, ASTM STP 1027,* U. M. Cowgill and L. R. Williams, Eds., American Society for Testing and Materials, Philadelphia, 1989, pp. 21–33.

ABSTRACT: Immunoassay techniques often involve reagents that react with constituents of samples other than the target compound. The phenomenon is called cross-reactivity. This paper describes a study with three objectives aimed at overcoming some of the problems related to cross-reactivity: (1) to investigate the effects of cross-reactivity on the accuracy and precision of analytical measurements; (2) to develop mathematical techniques to "deconvolute" analytical measurements that contain cross-reactions; (3) to define the minimum analytical and statistical experimental design required to implement the mathematical techniques developed.

It was found that cross-reactivity affects method accuracy, but not necessarily precision; an additive bias occurs because what is measured is actually the sum of the reactions of the target analyte plus nontarget cross-reactions. The study was successful in developing a stochastic model by which the desired "deconvolution" could always be accomplished under certain experimental conditions. The model and estimation procedures are algorithmically simple and extendable to any number of cross-reactants. The minimum experimental design requires that (1) the sample mixture be qualitatively prescreened for the identity of potential cross-reactants, (2) associative rate constants be determined for the target analyte and cross-reactants, (3) the associative rate constants not be equal for the specific reagent used, and (4) the analytical runs produce a time series of measurements over the period during which binding occurs.

KEY WORDS: immunoassay, cross-reactivity, modeling, simulation

Recently, interest has grown in the use of immunochemical assay methodologies for environmental applications. These interests include the detection of organic as well as inorganic compounds in environmental samples. The appeal of immunochemical methodologies, particularly those utilizing monoclonal antibody reagents, is their potential for high levels of specificity. However, no such reagent is entirely specific to any given target-analyte; they will inevitably cross-react with other compounds. Since it is impossible to control all constituents of environmental samples, the cross-reactivity problem is especially serious and can give rise to misleading biases in immunochemical assay results. The purpose of this project was therefore to develop mathematical tools to analyze the results of immunoassays where cross-reactions between target and nontarget analytes occur. The objectives of the project were three-fold:

1. To investigate the implications of cross-reactivity in relation to the accuracy (bias) and precision (variability) of analytical results.
2. To develop mathematical and statistical techniques to "deconvolute" immunochemical results where cross-reactions are observed or expected.

[1]Independent consultant, 1604 Calle Plumerias, Encinitas, CA 92024.
[2]Office of Research and Development, U.S. EPA, Las Vegas, NV.

3. To define the minimal statistical experimental design that can be used to separate target-analyte from cross-reactant measurements.

Mathematical evaluation of the problem of cross-reactivity indicates three important facts. First, cross-reactivity affects method accuracy and not precision. An additive bias occurs because what is measured is actually the sum of the reactions of the target-analyte plus nontarget cross-reactants. The fundamental additive nature of this bias is obscured, however, by the nonlinear nature of the reaction kinetics, as will be shown later in this paper. Second, it is always possible to separate target-analyte and cross-reactant measurements under the proper experimental conditions given an adequate probabilistic model of the reaction kinetics. Third, the "deconvolution" technique can be used to obtain accurate target-analyte measurements in immunochemical assays other than the ELISA types for which it was originally developed.

Background

This section contains a brief description of the processes and assumptions used to develop the cross-reactivity deconvolution technique. Certain facts required for further extension of the technique are also presented.

The analytical methodology explored herein is immunoassay. In its precise definition, immunoassay refers to the use of antibodies (that is, biological molecules with a very high degree of structural specificity) as the analytical reagent. Therefore, the methodology is defined by the nature of the assay reagent and not the nature of the analyte. Immunoassay target analytes have traditionally been large, complex biological substances; however, recent work conducted under EPA sponsorship has demonstrated that the approach is applicable to smaller organic and inorganic compounds not necessarily of biological origin.

Antibodies represent only one of a group of biological compounds with a high capacity for molecular recognition. Other examples include hormone receptors, transport proteins, and transport enzymes. All of these highly specific macromolecules can be termed "binding compounds." Particular binding compounds can be used as assay reagents for different target-analytes in methods that are virtually identical in experimental design and underlying analytical principles. The deconvolution methods presented in this paper would, therefore, be equally applicable to any immunochemical technique that relies on a binding compound as a specific reagent.

All assays rely on the ability of a specific analytical reagent to bind or otherwise interact with the target-analyte. The problem of cross-reactivity arises from the fact that virtually no reagent is absolutely specific to any single analyte; that is, no antibody is so specific that only one compound will bind to it.

In this paper, the following definitions are used:

1. Cross-reactivity is the binding of more than one analyte to a specific antibody.
2. The analyte of interest is called the target-analyte.
3. All other analytes are called cross-reactants.
4. That portion of the measurement apparatus containing the antibody is called the assay system.
5. That portion of the measurement system containing the analyte is referred to as the sample.

In addition, the following variable definitions are used:

N_a = number of target-analyte molecules.
N_c = number of cross-reactant molecules.
N_b = number of antibody binding sites.
$N_T = N_a + N_c$ = total number of antibody binding molecules.

It is assumed that all extraction or purification steps specified in the immunochemical methods have occurred before the sample is presented to the assay system. Therefore, preliminary procedures, including the use of solvents and other preparatory reagents, play no role in this discussion. In addition, methods by which the analyte measurements are quantified (fluorescence, colorimetry, etc.) are not considered.

For the purpose of developing deconvolution techniques, an ideal assay system is assumed. In the ideal assay, cross-reactivity occurs only as a result of lack of specificity of the antibody reagent and is unaffected by the presence of ions or other substances that alter the binding kinetics. The ideal assay does not, however, assume a homogeneous population of antibodies in the assay system. Antibody heterogeneity is accepted as a source of variation and is accounted for in the statistical estimation techniques. Although the method developed in this paper is based on an ideal assay, it can be extended to nonideal conditions. Immunoassay techniques considered in this paper fall into two categories that differ in their experimental design; they are called Type I and Type II.

Type I assays require an excess of antibodies (N_b); that is, N_b in the assay system must exceed N_T in the sample. Analytical measurements are based on quantification of the number of binding sites occupied. Consider the case of a target-analyte and one cross-reactant. If the sample is left in contact with the assay system for a period of time long relative to the binding rate of the antibody, the relative number of N_a and N_c bound to the antibodies depends only on the relative proportions of N_a and N_c in the original sample

$$\lim_{t \to \infty} \frac{B_a(t)}{B_c(t)} = \frac{N_a}{N_c} \qquad (1)$$

where $B_a(t)$ is the number of bound analyte molecules and $B_c(t)$ is the number of bound cross-reactant molecules at time t.

If the assay is stopped before all N_T sample molecules have become bound, then the relative proportions of N_a and N_c bound will depend on the associative rate constants for the target-analyte (K_a) and the cross-reactant (K_c) as well as N_a and N_c in the original sample

$$\lim_{t \to \infty} \frac{K_a \cdot B_a(t)}{K_c \cdot B_c(t)} = \frac{K_a \cdot N_a}{K_c \cdot N_c} \qquad (2)$$

Type II assays require an excess of analyte. Here, N_T in the sample must exceed N_b in the assay system. Analytical measurements are based on the ratio of bound and unbound analyte molecules. Whether the assay is stopped before all N_b binding sites in the assay system are occupied or not, the relative proportion of N_a and N_c bound will depend on the relative proportions of N_a and N_c in the original sample and the relative magnitudes of K_a and K_c. Equation 2 expresses this relationship.

The problem with separating the binding reactions of target-analyte from cross-reactant is that both Type I and Type II assays always measure the total number of analyte molecules bound. Because both target-analyte and cross-reactant are bound and never the target analyte alone, immunoassay results are inevitably biased high if a cross-reactant is present.

Results and Discussion

The properties of Type I assays can be exploited to deconvolute target-analyte and cross-reactant measurements by utilizing an alternative statement of the behavior of Type I assays (see discussion in previous section). Because Type I assays are characterized by an excess of antibody in the assay system, the number of target-analyte plus cross-reactant molecules bound will increase with time until an asymptote is reached that represents $N_T = N_a + N_c$. Based on the following definitions:

1. $F_a(t)$ and $F_c(t)$ = the number of free target-analyte and cross-reactant molecules at time t, respectively;
2. $B_a(t)$ and $B_c(t)$ = the number of bound target-analyte and cross-reactant molecules at time t, respectively;

a probabilistic model is developed that describes the reaction kinetics of the assay-sample system. The amount of target-analyte is then obtained by solving the model for the following boundary conditions

$$F_a(0) = N_a \qquad F_c(0) = N_c \tag{3}$$

A precondition and minimum condition under which N_a and N_c can be separated is that K_a and K_c be known and K_a not equal K_c; more on this later.

The model is based on the set of probabilities

$$\begin{aligned} Pr(a) &= \text{probability of a target-analyte molecule binding} \\ Pr(c) &= \text{probability of a cross-reactant molecule binding} \end{aligned} \tag{4}$$

By the law of total probability, these may be expressed as

$$\begin{aligned} Pr(a) &= Pr(a \mid b) \cdot Pr(b) \\ Pr(c) &= Pr(c \mid b) \cdot Pr(b) \end{aligned} \tag{5}$$

where

$$\begin{aligned} Pr(a \mid b) &= \text{probability of a target-analyte binding given that a binding occurs, and} \\ Pr(c \mid b) &= \text{probability of a cross-reactant binding given that a binding occurs.} \end{aligned} \tag{6}$$

and

$$Pr(b) = \text{probability of any binding occurring} \tag{7}$$

Now, by the properties of the reaction kinetics, $Pr(a \mid b)$ is a function of the associative rate constant of the target-analyte binding on the antibody used in the assay system (K_a) and the amount of unbound target-analyte present at a given time t, $[F_a(t)]$. A corresponding argument holds for $Pr(c \mid b)$.

The probabilities of Eqs 6 are represented in the model as

$$\begin{aligned} Pr(a \mid b) &= \frac{K_a \cdot F_a(t)}{K_a \cdot F_a(t) + K_c \cdot F_c(t)} \\ Pr(c \mid b) &= \frac{K_c \cdot F_c(t)}{K_a \cdot F_a(t) + K_c \cdot F_c(t)} \end{aligned} \tag{8}$$

In addition

$$Pr(a \mid b) + Pr(c \mid b) = 1 \tag{9}$$

so that the probability of either a target-analyte or cross-reactant molecule binding is equal to one, a condition that must be true in a Type I assay because $N_b > N_T$.

$Pr(b)$, the probability that a binding occurs, is taken to be distributed exponentially as

$$Pr(b) = r(t) \cdot e^{-r(t) \cdot t} \tag{10}$$

Here, $r(t)$ must be related to the rate of reaction. In addition, $Pr(b)$ must be a monotonically nonincreasing function on the number of remaining binding sites and unbound analyte molecules.

The function selected for the current model is

$$r(t) = \frac{N_T - B_T(t)}{N_T + N_b - B_T(t)} \tag{11}$$

where

$$B_T(t) = B_a(t) + B_c(t). \tag{12}$$

Now, recalling that N_T is the total number of target-analyte plus cross-reactant molecules in the original sample and N_b is the total number of binding sites in the assay system, the right-hand side of Eq 11 is proportional to the number of unbound analyte molecules remaining, since

$$F_T(t) = N_T - B_T(t) \tag{13}$$

and

$$r(t) = \frac{F_T(t)}{N_T - F_T(t)} \tag{14}$$

$Pr(a \mid b)$ and $Pr(c \mid b)$ are analogous to the standard "ball-in-the-box" model where (1) there are N_b boxes that can receive at most one ball and (2) two types of balls with N_a of one type and N_c of the other. At any given instant in time (or if the binding process is taken to completion with excess analyte), of the $B_a(t) + B_c(t)$ binding sites occupied, Eq 9 together with

$$\frac{K_a \cdot F_a(t)}{K_c \cdot F_c(t)} Pr(c \mid b) = Pr(a \mid b) \tag{15}$$

yields

$$\frac{K_a \cdot F_a(t)}{K_c \cdot F_c(t)} [1 - Pr(a \mid b)] = Pr(a \mid b) \tag{16}$$

This implies that

$$Pr(a \mid b) = \frac{K_a \cdot F_a(t)}{K_a \cdot F_a(t) + K_c \cdot F_c(t)} \tag{17}$$

which is the model given as the first half of Eq 8. $Pr(c|b)$ in Eq 8 follows directly from Eq 9. Therefore, if K_a and K_c were expressed as bindings per time, one would expect $K_a \cdot Pr(a)$ target-analyte molecules and $K_c \cdot Pr(c)$ cross-reactant molecules to bind in a time interval $t - t_0$. This can be written as

$$\frac{dB_a(t)}{dt} = K_a \cdot Pr(a)$$
$$\frac{dB_c(t)}{dt} = K_c \cdot Pr(c) \tag{18}$$

Now, substituting from Eq 5

$$\frac{dB_a(t)}{dt} = K_a \cdot Pr(a|b) \cdot Pr(b)$$
$$\frac{dB_c(t)}{dt} = K_c \cdot Pr(c|b) \cdot Pr(b) \tag{19}$$

or

$$\frac{dB_a(t)}{dt} = K_a \cdot Pr(a|b) \cdot Pr(b)$$
$$\frac{dB_c(t)}{dt} = K_c \cdot (1 - P) \tag{20}$$

which can be rewritten as the following recursive relationship based on the finite difference analogues of Eqs 20

$$B_a(t+1) = B_a(t) + K_a \cdot Pr(a|b) \cdot Pr(b) \cdot \hat{t}$$
$$B_c(t+1) = B_c(t) + K_c \cdot [1 - Pr(a|b)] \cdot Pr(b) \cdot \hat{t} \tag{21}$$

where $Pr(a|b)$ is given in Eq 8 and $Pr(b)$ is given in Eq 11. Equations 21 have the initial boundary conditions

$$B_a(0) = 0$$
$$B_c(0) = 0$$
$$F_a(t) = N_a \tag{22}$$
$$F_c(0) = N_c = N_T - N_a$$

Given the proper experimental and immunochemical analytical procedures, Eqs 21 and the initial boundary conditions (Eqs 22) lead to a fully determined model with only one unknown. To make this last point clear, consider the following. By substituting

$$F_a(t) = F_a(0) - B_a(t) \tag{23}$$

and

$$F_c(t) = N_T - N_a - B_c(t) \tag{24}$$

and Eq 11 into Eqs 21, it can be seen that the only unknown variable is the initial boundary condition $F_a(t) = N_a$. This condition is, in fact, the value we are seeking in our analysis, namely, the amount of target-analyte (N_a) in the original sample.

Figures 1 through 4 give the results of assay simulations under a variety of model conditions. In each case, the boundary conditions of Eqs 22 as well as the two associative rate constants (K_a and K_c) are specified. In the figures, the curve labeled "Total Analyte" is $B_T(t)$, "Target Analyte" is $B_a(t)$, and "Cross-reactant" is $B_c(t)$. A careful comparison of these figures will show that the relative behaviors of $B_a(t)$ and $B_c(t)$ are rather complex functions of the ratios of K_a to K_c and N_a to N_c. $F_T(t)$ is well-behaved and relatively consistent from figure to figure. Subtle differences in $F_T(t)$ do occur, however, and are the sole basis for the separation of target-analyte and cross-reactant measurements.

If Figures 1 through 4 were actual analytical runs, one would know only $F_T(t)$ or "Total Analyte." One would also know (1) the total number of binding sites (N_b) and (2) the associative rate constants (K_a and K_c). The latter would be known from previously acquired experimental or QA data. In addition, one would calculate the total amount of analyte (N_T) in the original sample as

$$N_T = \lim_{t \to \infty} B_T(t) \tag{25}$$

The following segment of pseudocode demonstrates the process of simulating assay binding for some arbitrary value of N_a. A means of solving for N_a is presented later in this paper. The

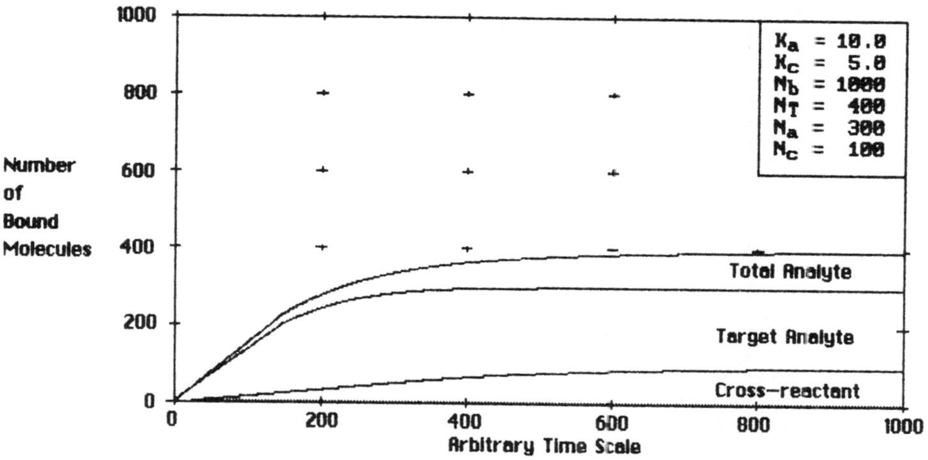

FIG. 1—*Example output from the numerical model describing the mutual behavior of the target-analyte and a single cross-reactant. The number of analyte molecules exceeds the number of cross-reactant molecules.*

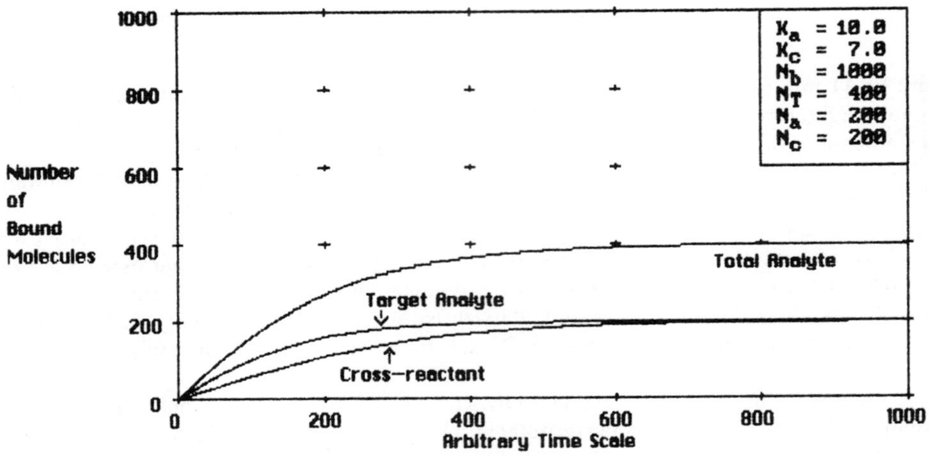

FIG. 2—*Example output from the numerical model describing the mutual behavior of the target analyte and a single cross-reactant. The number of analyte molecules equals the number of cross-reactant molecules.*

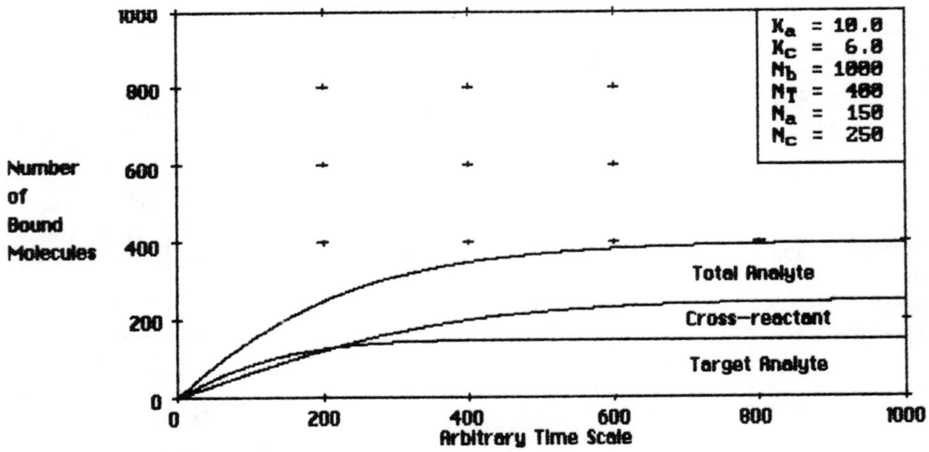

FIG. 3—*Example output from the numerical model describing the mutual behavior of the target analyte and a single cross-reactant. The number of cross-reactant molecules exceeds the number of target molecules.*

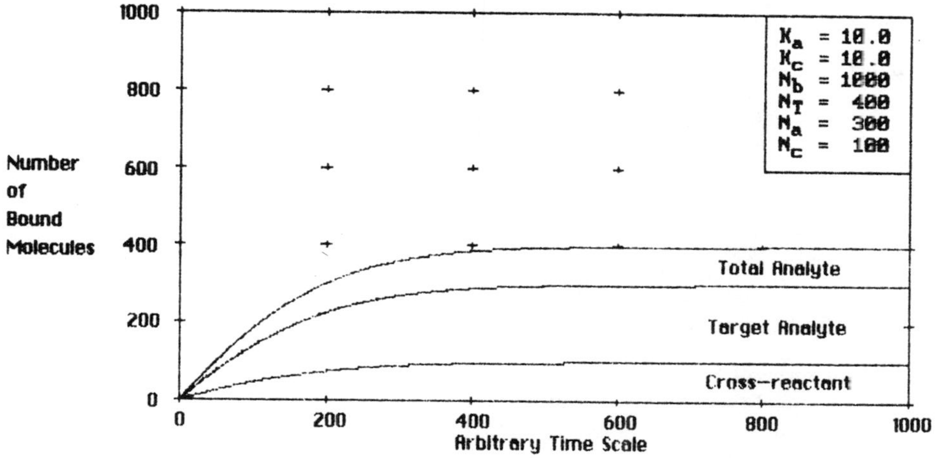

FIG. 4—*Example output from the numerical model describing the mutual behavior of the target analyte and a single cross-reactant. Reaction rate coefficients are equal.*

purpose of the following pseudocode segment is to show that N_a is, in fact, the only unknown in the model.

Set the following constants:

N_b = constant
N_T = constant
K_a = constant
K_c = constant
$\hat{t}$ = constant

Choose an arbitrary value for N_a:

$$N_c = N_T - N_a$$

$$B_a(t) = 0$$

$$B_c(t) = 0$$

Initialize: $t = 0$

Begin Iteration:

$$F_a(t) = N_a - B_a(t)$$

$$F_c(t) = N_T - N_a - B_c(t)$$

$$Pr(a|b) = K_a \cdot F_a(t)/[K_a \cdot F_a(t) + K_c \cdot F_c(t)]$$

$$r(t) = [N_T - B_a(t)]/[N_T + N_b - B_a(t)]$$

$$Pr(b) = r(t) \cdot \exp(-r(t) \cdot \hat{t})$$

$$B_a(t + 1) = B_a(t) + K_a \cdot Pr(a|b) \cdot Pr(b) \cdot \hat{}t$$

$$B_c(t + 1) = B_c(t) + K_c \cdot [1 - Pr(a|b)] \cdot Pr(b) \cdot \hat{}t$$

$$B_T(t + 1) = B_a(t) + B_c(t)$$

$$t = t + 1$$

End Iteration

Now we will consider the actual method for separating target-analyte and cross-reactant measurements. To begin the estimation process, the following must be available:

1. A time series of $B_T(t)$ as in Figs. 1 through 4.
2. A measure of the total number of binding sites (N_b).
3. The associative rate constants (K_a and K_c).
4. Calculated N_T (from Eq 25).

Choose a value for $F_a(0) = N_a$ and use this value in the pseudocode segment given above. At each iteration, calculate

$$Y_T(t) = \text{estimated } B_T(t) \tag{26}$$

Now, for "n" corresponding time points from the real analytical data and from the model simulated data, calculate

$$X^2 = \sum_{t=0}^{n} [Y_T(t) - B_T(t)]^2/B_T(t) \tag{27}$$

where $Y_T(t)$ is from the simulation model and $B_T(t)$ is from the same time points in the real analytical data.

Record X^2 for the initial choice of $F_a(0) = N_a$, then choose a new value of N_a and repeat the process, calculating a new value of X^2. Predicted N_a is that value of N_a at which the X^2 statistic is minimized. Figures 5 through 8 show the predicted solutions for N_a corresponding to the simulation of Figs. 1 through 4, respectively.

The predictions of Figs. 5 through 8 are accurate, as might be expected, since the same model was used for both simulation and prediction. However, this exercise is a very necessary preliminary to testing real data since it shows that the prediction method is quite accurate (given that the model is adequate) and that the overall prediction methodology produces unique solutions under the proper experimental and analytical conditions.

Figures 4 and 8 demonstrate a condition under which no solution is possible. This condition is where $K_a = K_c$. This points out one of the necessary experimental conditions for successful deconvolution of target and nontarget analytes, namely, that K_a may not be equal to K_c. A small amount of random variation was introduced into the simulations of Figs. 1 through 4. This variation is reflected in the width of the prediction bars in Figs. 5 through 7. Note again that Fig. 8 shows a no-solution situation.

It was found that the model algorithm (implemented on an IBM AT in MicroSoft FORTRAN 77) was fast enough that an exhaustive search for predicted N_a was possible. This was performed by setting the initial guess of N_a at zero and then incrementing N_a by approximately $0.01 \cdot N_T$ for each successive run.

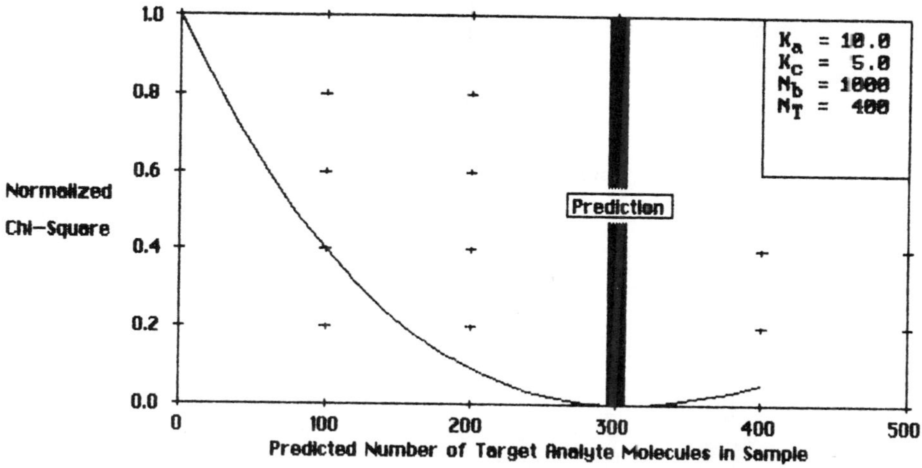

FIG. 5—*Predicted model solution based on the simulation of Fig. 1.*

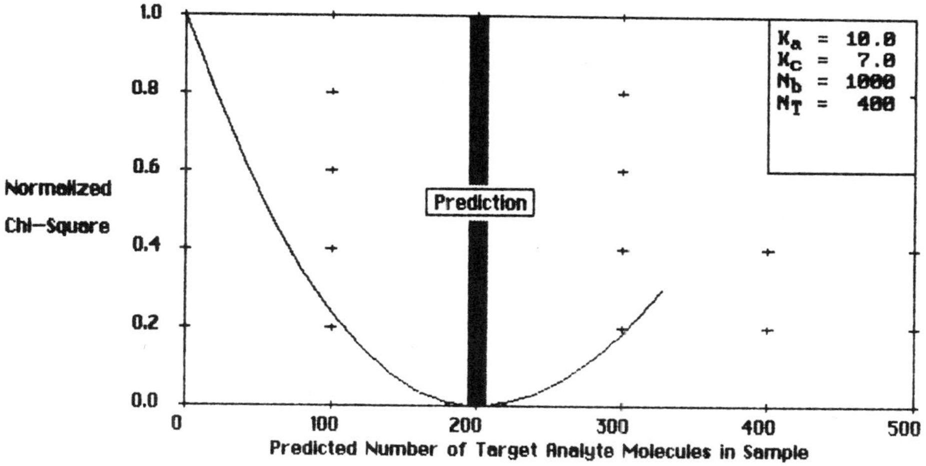

FIG. 6—*Predicted model solution based on the simulation of Fig. 2.*

Summary and Recommendations

The major conclusions to date are as follows:

1. The model is clearly a reflection of the reaction kinetics of immunoassay analytical systems and, at the same time, possesses properties necessary to yield an accurate estimate of the amount of target-analyte in the original sample.
2. The probabalistic model adequately estimates the variability present in all real analytical measurements.

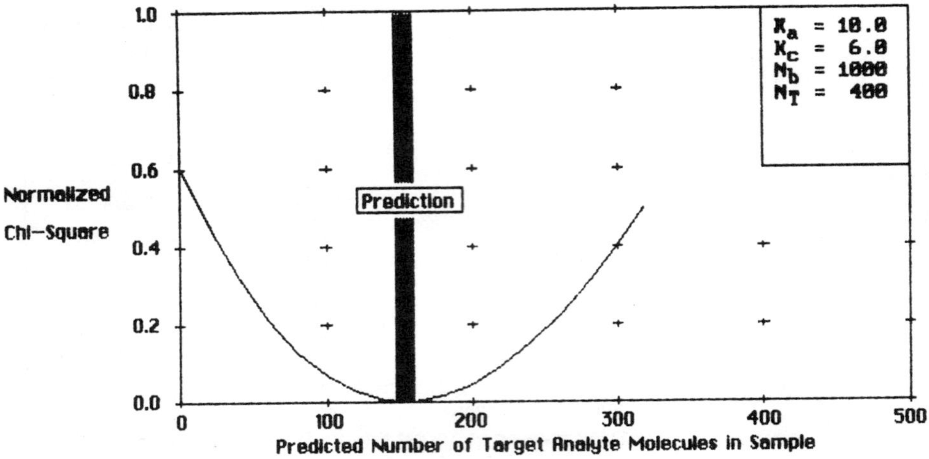

FIG. 7—*Predicted model solution based on the simulation of Fig. 3.*

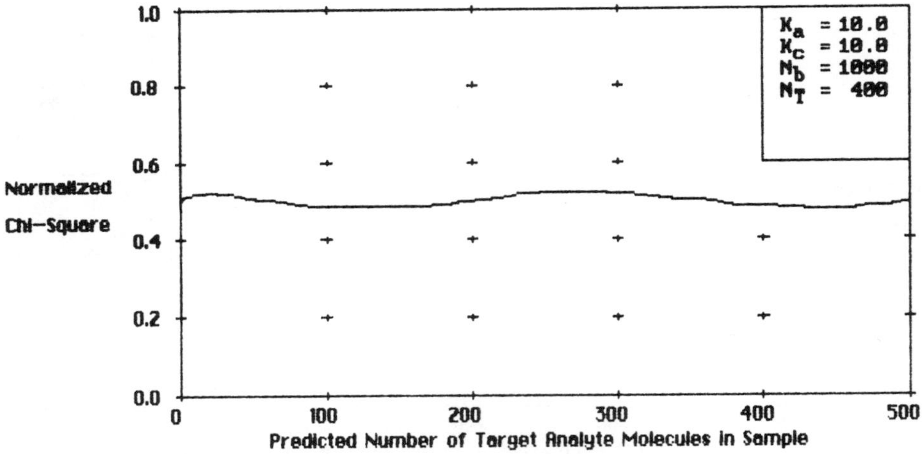

FIG. 8—*Prediction showing that no unique solution exists based on the simulation of Fig. 4.*

3. The model and estimation procedures are algorithmically simple and can be implemented on personal computer hardware.
4. The methods are readily extendable to more than one cross-reactant and by the application of multivariate methods to more than one target-analyte.

The minimum experimental design and conditions under which the methods in this paper can be implemented are as follows:

1. Preliminary experiments must be performed in which the associative rate constants are determined. This could be accomplished by presenting the assay system with a known quantity of analyte, measuring the reaction curve (as in Figs. 1 through 4), then solving Eqs 21 for K_a

with $Pr(a|b) = 1$. The solution methods presented above would be used here also. The entire process would be repeated for K_c.

2. A knowledge of K_a and K_c implies that a sample would have to be qualitatively screened to ascertain the presence and identity (or, at least, the chemical nature) of cross-reactants. An alternative would be an antibody reagent that has an associative rate constant that one is willing to assume is different from that of any cross-reactant that might be present in the sample. One would also have to assume an associative rate constant for the cross-reactant(s). An alternative form of the model would alleviate this problem to some extent by requiring that only the ratio of K_a to K_c be known.

3. Under any of the above conditions, K_a and K_c cannot be equal. The greater the difference between them, the greater the ability of the method to resolve target-analyte and cross-reactant measurements, especially in the presence of large analytical variability.

4. Analytical runs must produce a time series of measurements over the period during which binding occurs.

Recommendations for continuation of this project are as follows:

1. Test the model and estimation methods on the results of blind experiments in which quantities of target-analyte and cross-reactant known only to the referee are introduced into an assay system. The resultant time-series and values of K_a, K_c, and N_T would then be used to test the adequacy of the model.

2. Adjust the model and estimation procedures for any inadequacies discovered during the above tests.

3. Investigate the much more difficult problem of assay results that are presented only as end results after the analysis has run to conclusion.

Edmund H. Smith[1] *and Howard C. Bailey*[2]

The Application of Avoidance/Preference Testing in Aquatic Toxicology

REFERENCE: Smith, E. H. and Bailey, H. C., **"The Application of Avoidance/Preference Testing in Aquatic Toxicology,"** *Aquatic Toxicology and Hazard Assessment: 12th Volume, ASTM STP 1027,* U. M. Cowgill and L. R. Williams, Eds., American Society for Testing and Materials, Philadelphia, 1989, pp. 34–45.

ABSTRACT: Preference/avoidance behavioral testing was evaluated as a method to determine the effects of low levels of olfactory attractant, single toxicant, and mixed effluent wastes from domestic and industrial dischargers on selected organisms. Various dilutions of effluents and a reference toxicant (phenol) were tested using linear velocity, locational position, and other factors as a behavioral response of fish and invertebrates. A video-based computerized system was used to capture and analyze the behavioral data in real time. Test fish species used were: Chinook salmon, *Oncorhynchus tshawytscha;* striped bass, *Morone saxatilis;* steelhead, *Salmo gairdneri;* fathead minnow, *Pimephales promelas;* and the invertebrate *Daphnia magna.* Both attraction and avoidance responses were observed at different concentrations of the toxicant and effluents studied with responses detected as low as 500 : 1 for effluents. *Daphnia* was especially sensitive to phenol (0.5 mg/L), while salmonids responded to 10^{-4} olfactory stimulant (amino acid mix). A discussion of behavioral testing as a tool to assess potential impacts of toxicants, mixed effluents, and olfactory masking is presented.

KEY WORDS: aquatic toxicology, avoidance/preference, behavioral bioassay, olfactory stimulants, phenol

The importance of behavioral testing as a tool to determine animal responses to various stimuli for environmental hazard assessment has grown with the advent of reliable measures of behavior. Recent advances in video and computer technology have provided methods to obtain large amounts of data for reliable analysis of behavior [*1–3*].

Behavioral response to toxicity occurs when the introduction of a chemical, sewage waste stream, or other potentially stressful condition induces a behavioral change that exceeds the normal range of variability. The more widely studied behavioral responses (locational data, x and y coordinates) have been used in this study as well as new parameters developed to test different aspects of behavioral stress gradients.

Concern by regulatory agencies about the possible alteration of upstream and downstream fish migration because of discharge of waste materials into rivers and bays has increased in recent years. In addition, there is concern about more subtle, chronic effects upon resident populations. Preference/avoidance testing, when combined with other behavioral response parameters (rate of change of direction, linear acceleration), can provide comprehensive insight into the stress pathways induced by the toxicants.

It should be noted that behavioral response to toxicants may manifest itself in other ways than avoidance [*4*]. Organisms detect the presence of a chemical toxicant and respond to it by moving toward the chemical (preference or attraction), away from it (avoidance), or by ignoring it

[1]4090 Harrison Grade Rd., Sebastopol, CA 95472.
[2]Research associate, Department of Animal Science, University of California, Davis, CA 95616.

completely (although it may be toxic). There are no typical preference/avoidance reactions of aquatic organisms to chemicals because organisms detect and avoid some chemicals but not others [5]. The behavioral parameters can determine whether the organisms have sensed the toxicants and can measure different behavioral patterns which could effect their survival in the field (predator avoidance, prey capture, reproductive attraction responses).

Four fish species and an invertebrate were used in the studies. Examples of data from a series of studies are presented and discussed to form a background for evaluating various behavioral responses. The use of preference/avoidance testing as a method to detect behavioral responses to toxicants is evaluated and discussed. The importance of behavioral testing is evaluated, and the need to start preliminary efforts to standardize protocols is presented.

Material and Methods

Fish Test Systems

A more detailed description of the test facilities and chambers has been presented by Smith and Bailey [6]. (The flow-through chamber used in this investigation was modified from the system developed by Westlake and Lubinski [20].) The chamber is made of glass (91.4 cm long, 48.3 cm wide, 7.6 cm deep) and is divided into three chambers: control, test, and avoidance/preference (Fig. 1). Test fish are placed in all three chambers, which receive flow down both sides of the aquaria and through the avoidance/preference chamber. After the water has flowed through the two divisions, it enters the avoidance/preference area where the fish have three choices: avoid the test material, be attracted to it, or exhibit no deviation from control conditions. A steep gradient rather than a shallow gradient between test and control solutions was used for the avoidance/preference tests. The steep gradient provides the best opportunity for the fish to discriminate between opposing bodies of water [4].

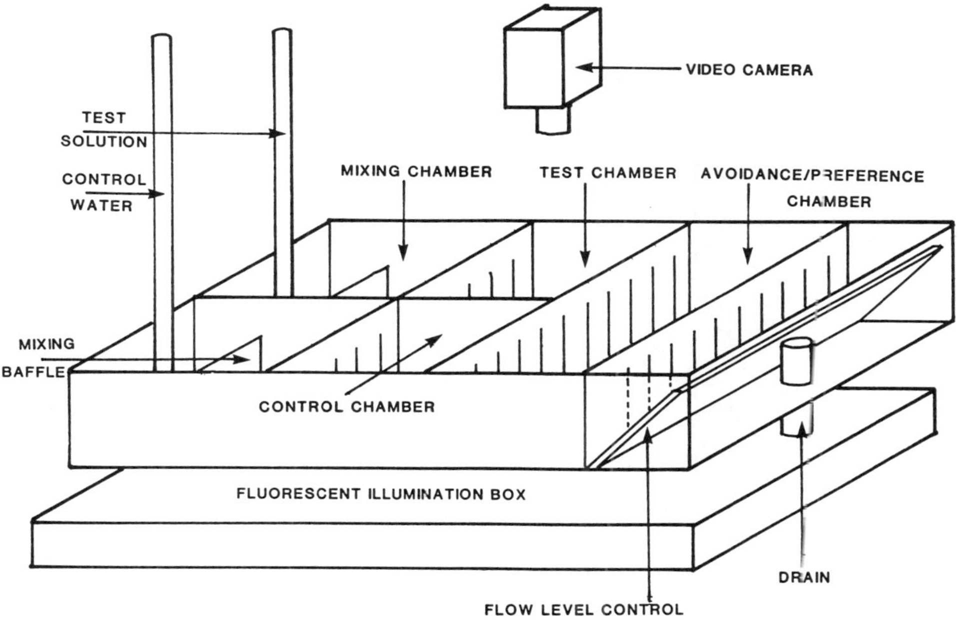

FIG. 1—*Diagram of the fish avoidance/preference tank.*

Based upon dye tests, it was determined that the zone of dilution on either side of the theoretical center division of the water mass was 3 to 4 cm wide in the avoidance/preference chamber. The best demarcation between water masses was obtained at 4 L/min. through each chamber for a total flow of 8 L/min.

Ten fish were placed in each of the divided areas for a total of 30 organisms per test. The chamber was illuminated with red-filtered, diffused fluorescent light from below throughout the test period. To determine the behavioral response of fathead minnows to phenol, a modified flow-through system was employed (the chamber was not divided for avoidance/preference). In this system only ten fish were placed in the video viewing area of the chamber.

Test System Used for Invertebrates

Control water was supplied from a gravity tank at a rate of 60 mL/min. into the test chamber. The algae *Selenastrum capricornutum* was continuously injected into the control water before entering the test chamber. The stock concentration and flow rate was adjusted so that the final concentration of approximately 500 cells/mL was achieved in the test chamber. The test toxicant was continuously mixed with a magnetic stirrer and was supplied to the test chamber incoming water supply from a metered pump.

The *Daphnia* test chamber was constructed from acrylic plastic and measured 12.7 cm wide by 17.8 cm long and 1.9 cm deep. The aquarium was sealed with a food grade neoprene gasket. Teflon fittings were threaded into the unit and attached to Nelgene tubing, allowing incurrent and excurrent flows to enter and leave the container (Fig. 2). A third Teflon fitting allowed the air to escape when the chamber was filled with water. The chamber was illuminated from behind by a infrared light source with a wave length of 820 nm.

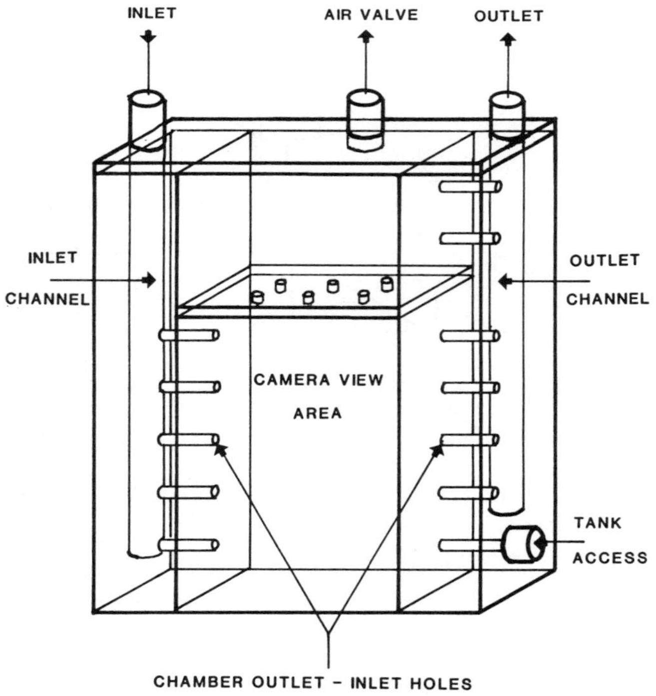

FIG. 2—*Diagram of the* Daphnia *test chamber.*

A changing frontal toxicant gradient was used in the *Daphnia* chamber. Dye tests were conducted to establish the mixing patterns within the chamber, time to completely fill the chamber with toxicant, and the pattern of the advancing toxicant gradient.

Ten adult *Daphnia magna* were used in each test. They were placed in the tank and acclimated before the tests were conducted.

Video-Computer System

The video-computer system used in these experiments was first described by Greaves [*8,9*] and later used by Miller et al. [*2*] to test the behavioral response of invertebrate larvae to copper. The system can tract the motion of multiple organisms (up to 50) in real time. Behavioral responses of the test organisms were calculated using a number of functions such as linear velocity, rate of change of direction, frequency of occupancy, and positional coordinates.

During the laboratory test period, the behavioral response was captured on video tape rather than directly on the hard disk of the computer. This step allowed the later visual review of behavior patterns and provided a chance to reanalyze the tapes in different ways. One hour before the test material was added, control behavior was recorded on video tape. These control data were compared statistically to a seven-day acclimation period in which behavioral information was collected continuously every 60 s. Based upon the behavioral parameters used in this study, it was found that between 12 and 24 h were necessary for the fish to develop a stable behavioral pattern with low variability. If the test control data were significantly different from the seven-day base line, the test was started over. The test data (toxicant added) were recorded for 1 h. After the toxicant was stopped, an additional half an hour of data was recorded.

Fifteen seconds of video data were captured every 60 s during the duration of the test. The computer software then produced the paths that each organism followed during each 15-s interval from the video images. For this series of tests, a mean of behavioral activity was established for the ten-test organism each 60 s. This mean and the variance around the mean were used for the calculation of behavioral response.

Water Quality and Temperature Control

Measurements of pH, temperature, dissolved oxygen, ammonia, alkalinity, hardness, and conductivity were taken daily from supply water and organism holding systems. The control water was identical to the holding water with respect to basic water quality parameters. The hardness and alkalinity of those waters were 120 mg $CaCO_3$ and 130 mg/L, respectively. Temperature was controlled in the recirculation system and during test periods at 12°C for salmonids, 17°C for striped bass, and 20°C for fathead minnows. *Daphnia* were maintained and tested at 22°C.

Chemical Analysis

Samples of the toxicant were removed from the discharge side of both the fish and *Daphnia* chambers and analyzed for phenol using the absorptiometric method [*10*]. The effluents from an oil refinery and domestic sewage plant were collected from the respective plants and transported to the laboratory by stainless steel tank truck. The effluents were held in large plastic tanks under controlled temperature until they were used in the flow-through systems. Two different stimulants were tested for the salmonid olfactory sensitivity experiments. These included homogenated krill (shrimp) and an amino acid mixture (Histidine 10^{-4} *M* and Threorine 10^{-4} *M* as a 50-50 mix). The krill was composed of frozen Euphausiids (San Francisco Bay Brand, Newark, California). The krill homogenate was processed by adding 1/3 water to 2/3 Euphausiids on a volume-to-volume basis. The mixture was reduced in a Waring blender and filtered through three layers of cheesecloth. The homogenate, after filtration, was considered

100% food stimulant concentrate. The amino acids were selected as a known attractant based upon the literature. The amino acids were obtained from Sigma Chemical.

Data Analysis

The parameters selected for analysis of behavior follows those suggested by Giattina et al. [*11*], as well as some additional measurements made possible with the video-computer system used for the testing. Linear velocity and the locational coordinates (x and y) are standard behavioral measurements. The frequency of occupancy was measured by counting the number of times the mean location (for the ten fish) occurred in the control or test areas. Comparisons were made between the pretest and test periods. The rate of change of direction is important as an index of turning regardless of the directions of the turns. The rate of change of direction of an organism moving in a straight line (regardless of speed) is equal to zero.

The significance of the linear velocity, locational coordinates, and rate of change of direction between controls and tests was established by using a two-sample t statistic, Mann-Whitney U test, mean difference test, and the Kolmogorv-Smirnov two-sample test. A Chi-square test was used to determine significance between control and test conditions in the frequency-of-occupancy test. The results of the control period were compared to the test conditions using the two-sample t statistic. While behavioral data are not normally distributed, the large sample size (40 to 110) used in these tests tends to bring the distribution curve toward normalcy. This assumption was verified by checking the results of the t statistics by using the nonparametric analysis (Mann-Whitney U test and Kolmogorv-Smirnov test). The tables presented in this paper reflect this approach and have been checked by using both parametric and nonparametric analyses.

Results

Preference/avoidance behavioral data were obtained from exposing juvenile striped bass *M. saxatilis* to oil refinery discharge waste at various dilutions (Table 1). The striped bass detected all dilutions of the effluent. The bass were attracted to the lower concentrations of the waste. At 10:1 the bass sensed the presence of the effluent by changing their vertical position, by increasing velocity, and by a significant increase in variability, but did not respond by changing their location in the preference/avoidance chamber. Velocity increased as the fish sampled the changing incoming water down one side of the chamber. The fish moved toward the incoming water (both control and test sources) as demonstrated by an attraction response in the vertical coordinate. The other four dilutions produced preference responses. The response to the reference toxicant (8.4 mg/L) showed a strong avoidance.

In the case of the domestic discharge steelhead, *S. gairdneri* showed a strong tendency to avoid the effluent (Table 1). At all dilutions except at 10:1, the frequency-of-occupancy parameter showed a significant movement away from the effluent.

The steelhead responded in a similar manner to the striped bass when exposed to 8.4 mg/L of phenol. The dramatic increase in velocity and magnitude of the response by the steelhead is depicted in Fig. 3, which shows the mean difference from the control for linear velocity.

The avoidance/preference technique was also applied to detect the olfactory response of Chinook salmon, *O. tshawytscha,* to chemical stimulants. The salmon showed significant attraction to the krill extract at the 1:200 and 1:300 dilutions (Table 2), while detecting the presence of the krill at 1:400. At this lower level the fish did not change position as shown by the frequency of occupancy results but did change their swimming patterns as indicated by the change in rate of direction. The salmon were unable to detect the amino acid mixture at 10^{-5} M but did respond to the 10^{-4} M concentration by moving to the side of the aquaria which contained the amino acids.

Tests were run to determine the level of detection of a single toxicant (phenol) by fathead minnows, *P. promelas.* These tests were not conducted in an avoidance/preference chamber

TABLE 1—*Behavioral response of striped bass and steelhead to oil refinery and domestic sewage discharges.*

Dilution	Vertical Position	Horizontal Position	Velocity	% Occupancy[a]
		STRIPED BASS—OIL REFINERY WASTE		
10:1	(+)	N	(+)	N
30:1	N	(+)	(+)	P
100:1	N	(+)	N	P
500:1	(−)	(+)	(+)	P
Phenol	(−)	(−)	(+)	A
		STEELHEAD—DOMESTIC WASTE		
100:1	(−)	(−)	(+)	A
3:1	(+)	(−)	(+)	A
5:1	(−)	(−)	(+)	A
10:1	(+)	N	(+)	N
20:1	(−)	(−)	(+)	A
100:1	(−)	N	(−)	A
Phenol	(−)	(−)	(−)	A

(+) = significant increase (velocity); significant attraction (position).
(−) = significant decrease (velocity); significant avoidance (position).
N = no significant effect.
[a] = Based upon Chi-square test $p < 0.05$.
A = avoidance; P = preference.

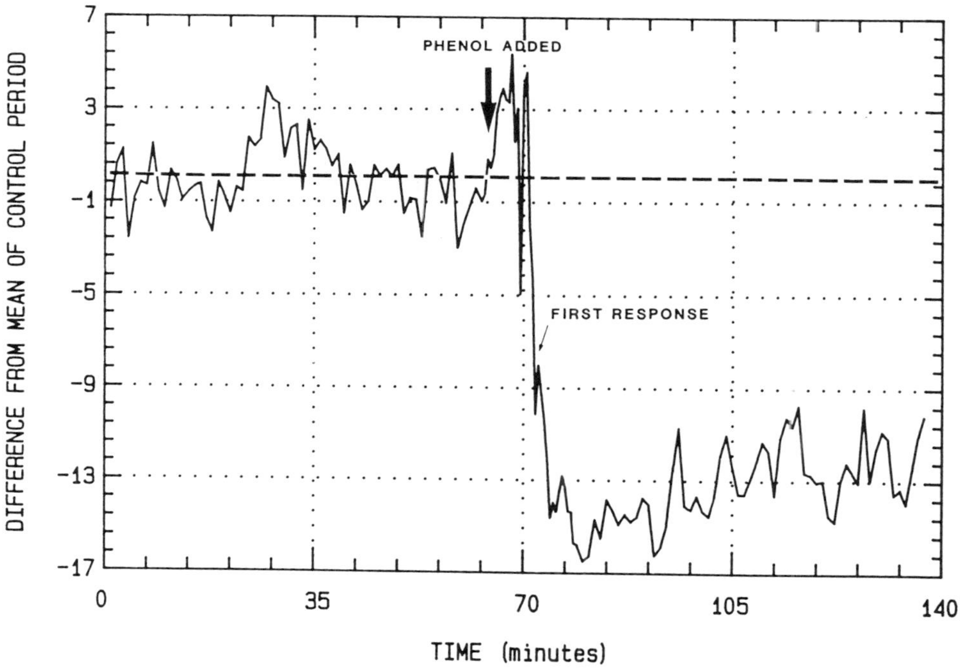

FIG. 3—*Plot of the mean difference from the control of steelhead exposed to phenol (8.5 mg/L). The elevated section of the plot represents control condition. The dramatic change in behavior occurred during the first contact with the toxicant, while the reduced activity is in response to continued contact with the toxicant.*

TABLE 2—*Olfactory response of Chinook salmon to chemical stimulants.*

Dilution	Vertical Position	Horizontal Position	Velocity	Rate CD	% Occupancy
1:200 K	(+)	(−)	(+)	(−)	P
1:300 K	(+)	(+)	(+)	(−)(1)	P
1:400 K	(+)	(−)	(−)	(+)(2)	N
10^{-5} *M* A	(−)	(−)	(−)	(−)	N
10^{-4} *M* A	(+)	(−)	(−)	(−)	P

Results expressed as: (+) = significant ($p < 0.05$), (−) = not significant, (1) = decrease in rate of change of direction, (2) = increase in the rate of change of direction.

Frequency of occupancy test: P = preference, N = no response.

A = amino acid mixture.

K = krill extract.

Rate CD = rate of change of direction.

but in a shallow aquarium (same dimensions as the avoidance chamber) without divisions. The toxicant was injected into the supply water; thus, the toxicant eventually filled the entire aquarium to the predetermined concentration. Unlike the avoidance/preference chamber, the fish had no region of the aquarium in which to avoid the toxicant.

The fish detected the presence of phenol at the 0.5 mg/L level moving away from the toxicant source as indicated by the change in vertical and horizontal coordinates. The rate of change of direction decreased significantly (Table 3).

At the higher concentrations, velocity was increased (except in Replicate 2, 4.5 mg/L), and hyperactivity was indicated by the increase in the rate of change of direction.

For vertical and horizontal coordinates at 4.45 mg/L phenol and the horizontal position at 2.25 mg/L phenol, the differences in the variance were significant (f = 2.75, 3.31, and 5.02 at $p = 0.05$). Thus, even though the means were not significant, the differences in variance indicate that the fish sensed the toxicant and increased their activity in response to the toxicant without significantly changing their overall locational coordinates.

The behavioral response of *Daphnia* to phenol and their recovery after purging the test chamber of toxicant were examined. The *Daphnia* responded within 1.8 min of exposure to 0.5 mg/L of phenol. The *Daphnia* showed avoidance of the moving gradient of toxicant as demonstrated by the significant change in the vertical and horizontal coordinates (Table 4) in all dilutions tested. Linear velocity also increased at every concentration.

TABLE 3—*Behavioral response of fathead minnows to phenol.*

Dilution, mg/L	Vertical Position	Horizontal Position	Rate CD	Velocity
0.5	N	(−)	(−)	N
0.5	(−)	N	(−)	N
2.25	(−)	(−)	N	(+)
2.25	(−)	N	(+)	(+)
4.45	(−)	(−)	(+)	(+)
4.45	N	N	(+)	(−)

(−) = significant decrease (velocity); significant avoidance (position); significant decrease in the rate of change of direction.

(+) = significant increase (velocity); significant attractant; significant increase in the rate of change of direction.

N = no significant effect.

TABLE 4—*Behavioral response of* Daphnia *to phenol.*

Dilution, mg/L	Vertical Position	Horizontal Position	Velocity
0.5	(−)	(−)	(+)
0.8	(−)	(−)	(+)
1.2	(−)	(−)	(+)
1.6	(−)	(−)	(+)
2.0	(−)	(−)	(+)

(−) = significant decrease (velocity); significant avoidance (position).
(+) = significant increase (velocity); significant attraction (position).

This same set of tests was used to determine the time it took for the organisms to return to near control levels. Figure 4 shows the linear velocity increase of *Daphnia* when exposed to phenol. The large dots represent the mean of ten *Daphnia,* while the bars indicate the variance around the mean. The response to the phenol can be clearly seen with a dramatic increase in swimming velocity at the onset of the toxicant gradient moving across the chamber. Variance also increases during the exposure time. During the postexposure period (purging of phenol from the chamber), swimming velocity returned toward the preexposure level. In the case of the horizontal coordinates, the *Daphnia* moved away from the source of the toxicant entering the left side of the aquaria. A return to a spatial distribution similar to the preexposure period can be seen during the purging of the phenol.

Discussion

The execution of any avoidance/preference response presupposes the presence of a directive factor [*4*]. Directive factors as defined by Fry [*13*] include all cues that enable an organism to determine its own location and the location of other organisms in the environment. In addition, avoidance/preference behavior depends upon the organism's ability to discriminate between two bodies of water.

A behavioral response may be the initial reaction of an organism to chemical exposure in the natural environment [*5*]. A chemical may be detected by an organism tested and responded to by moving toward it (attraction), away from it (avoidances), or by ignoring it completely (although it may be toxic). In the latter case, an organism may detect it (short-term response) and then not move out of the influence of the chemical, which could lead to physiological changes causing deleterious behavioral or morphological effects.

Behavioral testing protocols using undirected locomotor activity can be used to determine subtle effects of toxicants in a number of ways.

Avoidance/preference tests are designed to determine response of organisms to toxicants based primarily on changes in locational positions. These tests are often employed to determine if a discharge, toxicant, or stimulant would interfere with migration (anadromous fish) or effect the population dynamics of resident organisms.

Behavioral tests using other parameters such as linear velocity, rate of change of direction, and acceleration are often used to assess variation in behavior patterns that might alter survival success. Tests are often conducted to determine if chemicals entering the environment can interfere with the chemoreceptors' ability to detect alarm substances, food location, homing and migratory cues, and finding mates. These experiments can often be set up to test directed as well as unidirected locomotor activity.

Behavior data that are used in continuous biomonitoring systems for the early detection of

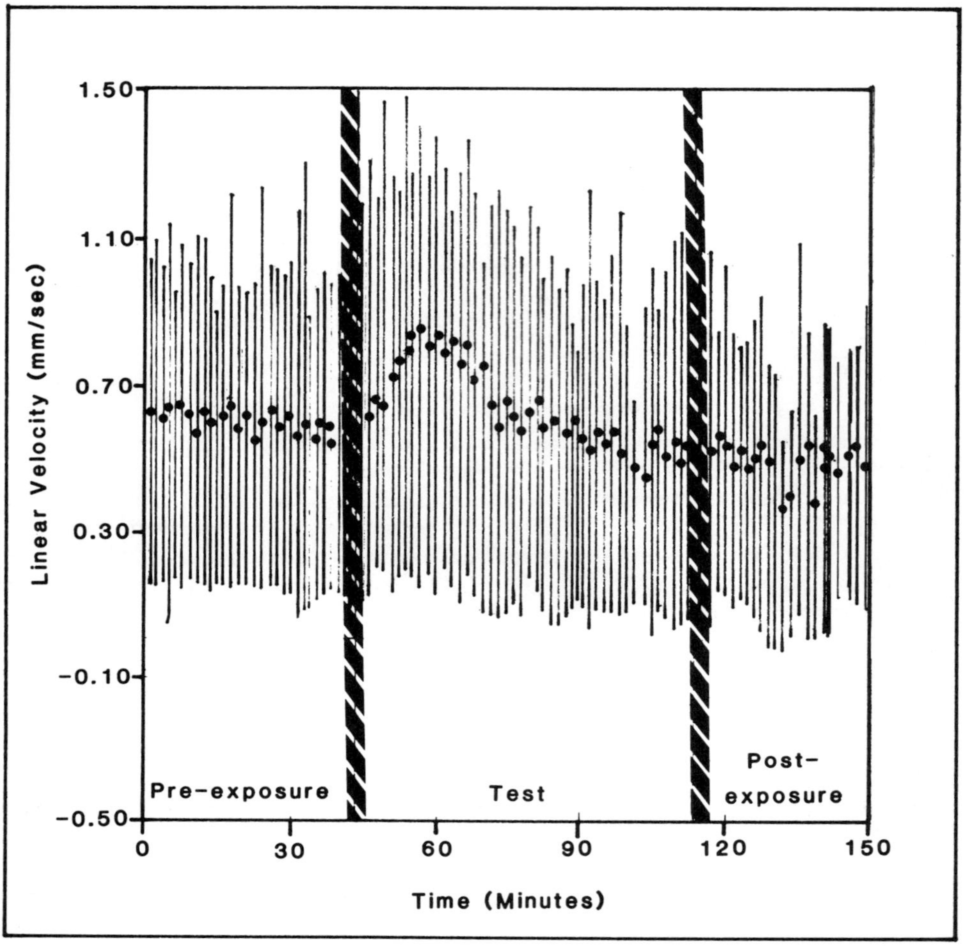

FIG. 4—*Plot of the mean and standard deviations for linear velocity of* Daphnia *exposed to phenol (1.2 mg/L). The dots represent the means of ten fish and the bars indicate the variance around the mean. The hatched lines divide the control, test, and posttest results.*

contaminants place emphasis on response time and the ability of the organisms to detect low levels of contaminates [*17*]. If the system is used for the protection of drinking water, most any sensitive organism can be used. If the system is designed to detect potential problems in discharge systems, a species which is endemic to the receiving water would be more important. Response parameters are similar to those outlined above.

The application and interpretation of the various behavioral measurements presented above provide one of the challenges of understanding locomotor activity. Detecting variances of organism responses, different from those established under control conditions, can be difficult when the organisms are exposed to stress conditions. Interpretation of results that at first seem highly random and variable can be accomplished with a good understanding of the behavioral patterns of the test organisms in the laboratory and, if possible, in their natural environment.

A change in the spatial patterns of swimming can be indicative of the presence of a pollutant [*12,14*]. These changes can be detected by a change in either the vertical or horizontal coordinates or both. A change in vertical position during the test will indicate that the organism has moved toward the source of the change in water quality, as seen in the striped bass avoidance/preference data. If the swimming pattern becomes more aberrant, with increased looping or head shaking, the rate of change of direction measurements will change significantly from the control.

The results of our studies on oil refinery wastes (from two separate refineries) indicate strong attraction of the test species; however, other behavioral parameters measured may indicate more subtle changes in behavior which potentially could effect survival of natural populations. Petersen et al. [*18*] found effects of petroleum refinery wastewater on the agonistic behavior of male sunfish. Changes in fin erections, chases, bites, and avoidance could have effects on the reproductive vitality of the population in the natural ecosystem. The lose of fins and tails in striped bass exposed to one of the oil refinery effluents during standard 21-day chronic flow-through testing was found after 10 days exposure. The same morphological alterations were found in striped bass caught in trawls around the outfall.

The decrease in rate of change of direction (klinokinesis) is seen in the response of the salmon to 1:300 krill dilution, indicating that the fish swam toward the source of the attractant in a more direct manner. This movement was done quickly, as indicated by the increase in linear velocity. The 1:400 krill dilution stimulated a move toward the source of the krill (change in vertical position) but with more of a searching motion (increase in rate of change of direction). The fish did not stay in the side of the chamber containing the attractant long enough to effect the frequency of occupancy measurement.

Smith and Lawrence [*15*] have also observed these combinations of behavioral response to alarm substances by fathead minnows. They observed three behavioral responses: freezing (motionless), slowing (reduced rate of movement), and dashing (short rapid increases in movement).

Daphnia responded by moving toward the top of the chamber and toward the side away from the source of the toxicant, as shown by the significant change in positional coordinates. Velocity also increased at all levels of dilution tested. In tests using barnacle nauplii exposed to copper, Giattina et al. observed increases in linear velocity and swimming patterns [*11*]. These authors felt that hyperactivity may affect swimming patterns but concluded that increased velocity was a more reliable indicator of low-level copper contamination. Their response data indicate that the threshold of sensitivity of *Daphnia* falls in the range of 0.067 to 0.0022 of the LC_{50}; Davis [*19*] found similar values using other invertebrates.

It should be noted that behavior is a continuum of activities of which we are measuring only discrete segments of the locomotor processes. These measurement, therefore, leave to our interpretation the recreation of the overall behavior of the organism. It is important to measure as many behavioral parameters as possible in order to better understand the response gradients obtained. One of the important aspects in behavioral testing is to establish a reliable control data base. In our application, all the comparisons of test data were made against controls; therefore, it is important to make sure the test organisms are acclimated to the test chamber before experiments are conducted. Avoidance/preference tests employing the Y tube protocol use a visual interpretation to determine when the fish have acclimated [*4*]. The standard time used in these tests seems to be about 15 min [*16*]. Our studies indicate that, in the system we used, the acclimation period is 12 to 24 h depending upon the species tested. Control behavior should be established each time new organisms are placed in the test system to avoid stressed behavioral patterns which can reduce the sensitivity of the tests or provide erroneous data.

The behavioral data collected (for this study) at any instant in time are not the data for each individual organism, but rather a mean for all organisms observed at that time. According to the central limit theory, the distribution of means from a population tends to be normal even when the underlying distribution of the population itself is not normal. Moreover, many parametric tests are robust against violations of initial assumptions. Many of the tests were analyzed by both parametric test and their nonparametric analogs. Results were the same for both types of analysis. As a consequence, parametric methods were used for the results given in this report. The use of the mean of the behavioral activity of ten organisms grouped into control and test data sets represents a robust approach to data analysis. Independent measurements of behavior were made by collecting 15 s of locomotor activity captured each 60 s. However, the combining of data taken uniformly over time looses much of the sensitivity of the data collected. We have started to apply various time series analyses to the data to better understand behavioral changes during control and test periods. The use of time series analysis would detect either the short-term dashing effect of the organisms when they detect a toxicant or the change of swimming patterns over time. The use of variance as an indicator of changing response patterns is also a valuable analysis. The activity of searching and rapidly returning to the same location is not always detected by the mean but does show up in the variance around the mean.

Most of the effects discussed in this paper and in many of the behavior papers reviewed [*5,12,18*] would not have been detected by standard, acute, or chronic testing. Behavioral tests should be considered as one of several types of stress indicators to be incorporated in an effluent monitoring program [*12*]. In fact, based upon a review of the literature, there has been an increase in behavioral studies related to pollution problems within the last two years.

Behavioral testing has become an important method for detecting stress response to toxicants. With new method development in rapid capture and analysis of data and increased interest in behavioral testing, the need to standardize methods becomes more important. Research work has begun to provide information on the response sensitivity of behavioral tests from which the most promising can be chosen for pollutant effects monitoring. Additional research should be done in comparing the behavioral tests with existing standard sublethal pollutant stress indicators (growth rates, reproductive success) to evaluate their utility for toxicant response assessment (speed of response, threshold, relative sensitivity, and ease of use) [*12*]. The significance of behavioral change in ecology needs to be better understood. This understanding is essential if behavioral testing is to be used to predict environmental effects. It is important to begin preliminary steps to standardize the identified behavioral tests which offer the most promise.

Conclusions

1. *Daphnia magna* responded to phenol at 0.5 mg/L within 1.8 min of sensing the toxicant. Changes in location data showed strong vertical and horizontal avoidance reaction.

2. Domestic effluent elicited statistically significant avoidance response by steelhead as low as 100:1 dilution.

3. The oil refinery effluent showed some strong attraction responses in striped bass as low as 500:1 dilution.

4. In salmonids, response to olfactory attractants was detected at levels of 10^{-4} *M* for amino acids and 1:400 dilution for krill extract.

5. Frequency of occupancy and rate of change of direction are important behavioral parameters in preference/avoidance testing.

6. Speed and x-y locational data are strong indicators of toxicant response by the test organisms.

7. Other authors using different methods of preference/avoidance testing indicated that a 15-min acclimation period was considered sufficient. Results from this study indicate that 12 to 24 h are necessary to acclimate the test organisms to the chambers before behavioral testing is started.

8. Evaluation of the variance can be as strong an indicator of behavioral change as the use of the mean. Time series analysis is likely to show a more sensitive response than the more robust analysis used in this study.

9. Interpretation of the laboratory results should be compared to field investigations. The design of the behavioral testing must be responsive to the field questions being addressed. The

selection of the behavioral parameters to measure is also dependent upon the behavioral strategies of the organisms in their natural ecosystem.

10. Behavioral response testing is becoming an important tool in assessing the potential impact of low-level toxicants upon organisms, and the need to assess present methods and to develop preliminary test standards is emerging.

Acknowledgments

We thank Bridgette Hejtmanek for her assistance in conducting these experiments and to Carol Lee for reviewing the paper. Special thanks to Dennis Logan for a critical review of the statistical analysis used in this paper.

References

[1] Cairns, J. and van der Schalie, W. H., *Water Research*, Vol. 14, 1980, pp. 1179–1196.
[2] Miller, D. C., Lang, W. H., Greaves, J. O. B., and Wilson, R. S. in *Aquatic Toxicology and Hazard Assessment (Fifth Conference), ASTM STP 766*, 1982, pp. 206–220.
[3] Pitt, W. W., *Journal Water Pollution Control Federation*, Vol. 56, No. 6, June 1984, pp. 548-550.
[4] Larrick, R., Dickson, K. L., Cherry, D. S., and Cairns, J., *Hydrobiologia*, Vol. 61, No. 3, 1978, pp. 257–265.
[5] Rand, G. M. in *Fundamentals of Aquatic Toxicology Methods and Applications*, 1st ed., G. M. Rand and S. R. Petrocelli, Eds., Hemisphere Publishing Corp., New York, 1985, Chap. 9, pp. 221-256.
[6] Smith, E. H. and Bailey, H. C. in *Automated Biomonitoring: Living Sensors as Environmental Monitors*, D. Gruber and J. Diamond, Eds., John Wiley & Sons, New York, 1988, Chap. 13, pp. 182-205.
[7] Lubinski, K. S., Dickson, K. L., and Cairns, J. in *Aquatic Toxicology (Third Conference), ASTM STP 707*, 1980, pp. 328–340.
[8] Greaves, J. O. B., *Proceedings of the IEEE*, Vol. 63, No. 10, October 1975, pp. 1415–1425.
[9] Greaves, J. O. B. and Wilson, R. S., "Development of an Interactive System to Study Sub-lethal Effects of Pollutants on the Behavior of Organisms," EPA Report 600/3-80-010, U.S. Environmental Protection Agency, Narragensett, RI, January 1980, pp. 1-35.
[10] Ochysnski, L., *Analyst*, Vol. 85, 1960, p. 278.
[11] Giattina, J. D., Gorton, R. R., and Steven, D. G., *Transactions of the American Fish Society*, Vol. 111, 1982, pp. 491–509.
[12] Miller, D. C., *Rapp. P. V. Cons. Int. Explor. Mer.*, Vol. 179, 1980, pp. 154–160.
[13] Fry, F. E. J., "Effects of the Environment on Animal Activity," Toronto University Studies: Biological Series, No. 55, 1947, pp. 1-62.
[14] Vernberg, W. R., DeCoursey, D. J., and Podgett, W. J., *Marine Biology*, Vol. 22, 1973, pp. 301-312.
[15] Smith, R. J. F. and Lawrence, B. J., *Environmental Toxicology and Chemistry*, Vol. 7, 1988, pp. 329–335.
[16] Folmar, L. C., *Bulletin of Environmental Toxicology*, Vol. 15, No. 5, 1976, pp. 509–514.
[17] Cairns, J. and Gruber, D., *Water Resources Bulletin*, Vol. 16, No. 2, April 1980, pp. 261-266.
[18] Petersen, J. C., Burks, S. L., and Miller, R. J., *Environmental Toxicology and Chemistry*, Vol. 5, 1986, pp. 463–471.
[19] Davis, J. C., *Journal of the Fish Research Board of Canada*, Vol. 33, 1976, pp. 2031-2035.
[20] Westlake, G. F. and Lubinski, K. S. in *Proceedings*, 1976 National Conference on Control of Hazardous Material Spills, Rockville, MD, Information Transfer, Inc. 1976, pp. 64–69.

John A. Bantle,[1] Douglas J. Fort,[1] and Douglas A. Dawson[2]

Bridging the Gap from Short-Term Teratogenesis Assays to Human Health Hazard Assessment by Understanding Common Modes of Teratogenic Action

REFERENCE: Bantle, J. A., Fort, D. J., and Dawson, D. A., **"Bridging the Gap from Short-Term Teratogenesis Assays to Human Health Hazard Assessment by Understanding Common Modes of Teratogenic Action,"** *Aquatic Toxicology and Hazard Assessment: 12th Volume, ASTM STP 1027,* U. M. Cowgill and L. R. Williams, Eds., American Society for Testing and Materials, Philadelphia, 1988, pp. 46-58.

ABSTRACT: Embryonic development depends on the timely expression of genetic information. As development proceeds, embryonic cells undergo processes such as cell division, interaction (induction), migration, differentiation, and selective death. Developmental toxicants affect one or more of these mechanisms. Whole embryo teratogenesis screening assays are best suited for hazard identification because all the developmental mechanisms are represented in the test. Cell culture and biochemical tests are better suited for determining which biochemical or cellular mechanisms are adversely affected by the toxicant. By combining data from a validated whole embryo teratogenesis test with data from cellular and biochemical tests in a battery format, a better indication of the potential human health hazard is afforded. We have tried this approach in the case of DNA synthesis inhibitors. These inhibitors are teratogenic in virtually all species tested regardless of where in the metabolic pathway they inhibit replication. DNA synthesis inhibitors, such as hydroxyurea, cytosine arabinoside, and 5-fluorouracil, were teratogenic in the Frog Embryo Teratogenesis Assay *Xenopus* (FETAX). These substances also inhibited DNA synthesis in a fluorometric assay using cleaving *Xenopus* eggs. However, they did not inhibit gene expression as measured by their inability to prevent the expression of the cloned herpes thymidine kinase gene microinjected into *Xenopus* Stage 6 oocytes. The combined data not only show that malformations are caused by these agents but pinpoint the developmental mechanism affected. This suggests that all species, including man, may be affected. A battery of simple tests could be developed which measures effects on the developmental mechanisms shared by all species. Risk assessment studies would still be necessary to consider dose levels and types of exposure needed to observe an effect in man.

KEY WORDS: *Xenopus,* teratogenesis assay, FETAX (Frog Embryo Teratogenesis Assay—*Xenopus*), DNA synthesis, mechanisms of teratogenicity

Numerous short-term teratogenesis screening assays have recently been devised in order to detect developmental toxicants present in the environment [*1-6*]. These assays range from cell culture systems [*7,8*] to primitive life forms [*3,4*] to whole embryo bioassays [*1,2,5*]. To date, the main argument for using these assays for human health hazard assessment has been validation with compounds of known mammalian (and human) teratogenic action. For tests lacking metabolic competence, the assay can be supplemented with an in vitro metabolic activation system to simulate mammalian metabolism. This procedure helps reduce the number of false positives

[1]Professor and graduate research assistant, respectively, Department of Zoology, Oklahoma State University, Stillwater, OK 74078.
[2]Postdoctorate fellow, Department of Animal Science, University of Tennessee, Knoxville, TN 37901.

and negatives and raises the predictive accuracy of the assay. As laudable as this approach is, the specificity of the test may still fall short. Consequently, additional tests having endpoints representing developmentally important processes must be designed to augment the primary screening assay. While the predictive accuracy of the test may never approach 100%, it will be good enough upon which to make decisions as to which sites warrant further testing as part of risk analysis.

Embryonic development depends on the timely expression of genetic information. The genetic program that guides embryonic development is likely to be evolutionarily conservative [*9*], especially the part that deals with early stages of embryonic development. As development proceeds, embryonic cells undergo cell division, interaction (induction), migration, differentiation, and selective death. Developmental toxicants affect one or more of these mechanisms. Teratogenesis screening assays utilizing whole embryos are likely to be superior to organ or cell culture systems because all cellular mechanisms function in the entire embryo. These latter systems are best utilized in detecting which cell mechanism has been affected. We feel the best approach to developmental toxicity hazard assessment is to use a validated whole embryo screening test that has an appropriate in vitro metabolic activation system. Coupled with this test would be additional tests that measure the effect of the toxicant on cell or molecular mechanisms essential to the development of all living organisms. DNA replication and expression tests are presented here that may provide valuable additional information when coupled with data from FETAX (Frog Embryo Teratogenesis Assay—*Xenopus*). Taken together, the tests may provide enough information to make hazard assessment possible.

Materials and Methods

Animal Care

The *Xenopus* adults were obtained from *Xenopus* I and maintained in glass aquaria and/or fiberglass raceways in dechlorinated tap water. This water was filtered through activated carbon and aerated for 48 h prior to use. It was periodically tested to ensure that the pH, dissolved oxygen content, hardness, and content of heavy metals and total organic carbon were at acceptable levels [*10*]. Adult frogs were fed beef liver and lung supplemented with liquid vitamins (Polyvisol).

Rat Liver Microsome Preparation

Five days prior to microsome preparation (metabolic activation system), an adult Sprague-Dawley (S/D) C/D strain male rat (200 to 280 g) was injected intraperitoneally with 500 mg/kg Aroclor 1254 in corn oil [*11*]. Microsomes were isolated as previously described [*12,13*]. P-450 activity was inferred by the measurement of formaldehyde generated from the N-demethylation of aminopyrine by the method of Nash [*14*]. Assay conditions were described by Lucier et al. [*15*]. Other components of the metabolic activation were present in concentrations as previously described [*13*].

FETAX Procedure

The conduct of FETAX has been previously described [*16*], and no changes were made in the setup of the assay, the scoring of mortality and malformation, and in the analysis of data.

Oocyte Expression Assay

The procedure for the Oocyte Expression Assay was used as previously described without modification [*17*].

Replication Assay

Embryos and Exposure Conditions—Embryos (40 to 50/group) which were between 2- to 64-cell stage were incubated in test solutions for 5.5 h at 23°C. Embryos were Stage 9 or 10 after incubation with the sample. For concentration/dose response data, all embryos can be processed at the end of 5.5 h.

The procedure for nuclei isolation and separation from the yolk platelets of the vegetal hemisphere of *Xenopus* embryos was adapted from Gaudette et al. [*18*]. The following solutions were used. Buffer A was composed of 0.5% citric acid w/v plus 1.5-mM spermine. Buffer B contained 100-mM KCl plus 1.5-mM spermine. Buffer C contained 50-mM Tris-HCl, pH 8.0, and 1.5-mM spermine, while Buffer D was composed of 10-mM Tris-HCl, pH 8.0, and 1.5-mM spermine; Buffer E contained 100-mM NaCl, 1-mM EDTA, and 10-mM Tris-HCl, pH 7.5.

Preparation of Nuclei and DNA Extraction—Embryos were homogenized in Buffer A (the pH was not adjusted) using 3 mL per 50 embryos. A 100-μL aliquot of homogenate was removed and frozen for a BioRad protein assay. The homogenate was centrifuged at 1000 $\times$ g avg, 10 min, 4°C. The pellet contained yolk, some lipids, and nuclei. The pellet was resuspended gently in an equal volume (3 mL) of Buffer B using a Pasteur pipet. The presence of nuclei in the suspension was confirmed by phase contrast microscopy. The suspension was again centrifuged at 1000 $\times$ g avg, 10 min, 4°C. Then, the nuclei were resuspended sequentially in Buffers C and D and pelleted by centrifugation as above. The pellet was finally resuspended in Buffer E and frozen at -20°C until extraction and DNA assay.

Samples were thawed by placing in a 37°C water bath, and 10 μL of a 10 mg/mL ribonuclease A solution was added. The samples were incubated for 1 h at 37°C. Then, 50 μL of a 20 mg/mL Proteinase K solution was added and incubated at 37°C for 3 h. The tubes were gently agitated every 30 min. Following incubation, 10 μL of 10% w/v sodium dodecyl sulfate (SDS) was added and the sample mixed gently. The sample was then incubated an additional 30 min at 37°C. Each sample was subjected to a phenol: chloroform extraction (1:1) for 15 min. The aqueous (top) phase was recovered and reextracted with an equal volume of chloroform:isoamyl alcohol (24:1). After phase separation by centrifugation, the aqueous phase was extracted with water-saturated ether. The phases were again separated by centrifugation, and the aqueous phase was placed in a 65°C water bath for 15 to 20 min to evaporate excess ether.

Fluorometric Determination of DNA—The total volume of each sample was measured, and the sample was placed on ice. Mithramycin C was selected as the fluorescent dye to measure the amount of DNA [*19*]. Mithramycin, an antibiotic, binds to double-stranded DNA and fluoresces in proportion to the quantity of DNA present. It does not bind to RNA. Mithramycin C was donated from Miles Pharmaceuticals.

For an expected DNA range of 0 to 10 μg/mL, the final concentration of Mithramycin C used was 5 μg/mL, while for a concentration of 10 to 30 μg/mL DNA, 10 μg/mL Mithramycin C was used. Twenty-five microlitres of the appropriate stock concentration of Mithramycin C was added to 975 μL of the DNA sample. Samples were then agitated for 10 s in a vortex mixer at high speeds. The fluorescence was read using a fluorescence spectrophotometer at an excitation wavelength of 440 nm and an emission wavelength of 540 nm. The DNA content of unknown samples may be calculated using linear regression analysis and a standard curve made up for each experiment. A BioRad protein analysis was performed using the aliquot removed from the initial homogenate [*20*]. The amount of DNA/mg protein/embryo was then calculated.

Results

FETAX Validation

Figure 1 shows the results from a typical FETAX validation experiment using N-ethyl N-nitrosourea (ENU) as the toxicant. Without the in vitro metabolic activation system, the 96-h LC_{50} was 0.29 mg/mL (95% confidence limits = 0.26 to 0.32), while the 96-h EC_{50} (malforma-

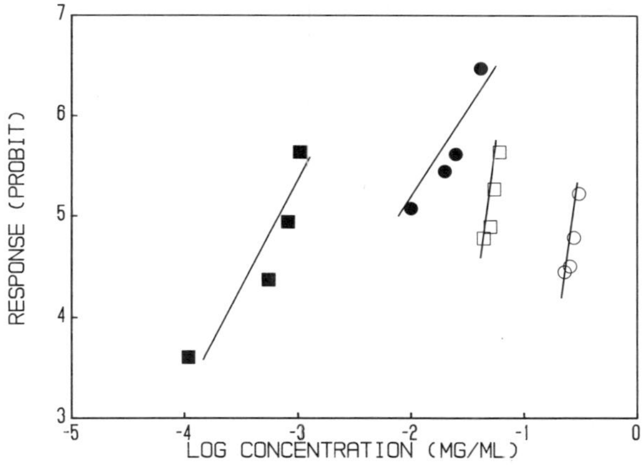

FIG. 1—*Effect of unactivated and activated N-ethyl N-nitrosourea on* Xenopus *embryo mortality and malformation. FETAX was performed using N-ethyl N-nitrosurea in the absence or presence of an in vitro metabolic activation system. Only data points used in calculating the 96-h LC_{50} and EC_{50} (malformation) are shown, although other concentrations were tested.* ○ *unactivated mortality,* □ *unactivated malformation,* ● *activated mortality,* ■ *activated malformation.*

tion) was 0.05 mg/mL (0.047 to 0.054). This yielded a teratogenic index (TI) of 5.8 (Table 1). A second trial using unactivated ENU (evaluated by a different operator) resulted in a 96-h LC_{50} of 0.25 mg/mL (0.24 to 0.26) and a 96-h EC_{50} (malformation) of 0.052 mg/mL (0.05 to 0.054) and a $TI = 4.9$. In the presence of the metabolic activation system, the 96-h LC_{50} was reduced to 0.0037 mg/mL (0.008 to 0.016) while the 96-h EC_{50} (malformation) was lowered to 0.00073 mg/mL (0.00049 to 0.0011). Bioactivation lowered the 96-h LC_{50} and EC_{50} (malformation) by a factor of approximately 73. However, the teratogenicity index remained approximately the same at 5.1. The data fit was good for all dose-response curves and the 95% confidence limits were narrow as typical for FETAX. Little genetic variation was observed from experiment to experiment. Table 1 shows the average *TI* values of our FETAX validation effort to date and how well the *TI* value serves as an indicator of teratogenic risk when compared to mammalian experiments. A value greater than 1.5 indicates the compound has a high teratogenic risk, as malformations occur at concentrations significantly lower than required for embryo lethality. For proteratogens such as cyclophosphamide [*13*], metabolic activation raised the *TI*, allowing the correct determination of the teratogenic risk of cyclophosphamide, while for ENU the *TI* was unchanged. The other compounds (Table 1) were not tested using metabolic activation as they were direct acting. Of the 19 compounds tested, only 2 did not yield responses similar to mammals.

Oocyte Expression Assay

The oocyte expression assay is typically used to detect inhibitors of gene expression [*17*]. However, specific inhibitors of DNA synthesis such as cytosine arabinoside caused an enhancement of expression (Fig. 2). Low concentrations of cytosine arabinoside nearly doubled the amount of herbes thymidine kinase expression in *Xenopus* Stage 6 oocytes microinjected with the pHSV-106 plasmid. However, cytotoxic concentrations (2 mg/mL) caused an inhibition of expression. The negative control was uninjected oocytes, while the positive control was pHSV-106 injected with no inhibitor. In a modification of the standard assay, the plasmid itself was treated with ENU for 2 h at 37°C (Fig. 3). Even though very high concentrations (near the 96-h LC_{50} and EC_{50} (malformation) of this direct-acting alkylating agent were used, it failed to affect the expression of the gene when compared to the control.

TABLE 1—*Validation of FETAX using compounds having known mammalian or human teratogenicity.*

Compound	CAS No.	Teratogen ($TI \geq 1.5$)	Non-teratogen ($TI < 1.5$)	Source
Actinomycin D[d]	50-76-0		0.9	Courchesne and Bantle [*21*]
6-aminonicotinamide[a]	329-89-5	560.0		Dawson, D. A. (unpublished)
Caffeine[a]	58-08-2	2.1		Dawson and Bantle [*16*]
Cycloheximide[c]	66-81-9		1.3	Courchesne and Bantle [*21*]
Cyclophosphamide unactivated[b]	50-18-0		1.3	Fort et al. [*13*] Bantle and Dawson [*12*]
Cyclophosphamide activated[a]		3.6		Fort et al. [*13*]
Cytosine arabinoside[c]	147-94-4	7.1		Courchesne and Bantle [*21*]
Diphenhydramine[b]	147-24-0	10.3		Dawson, D. A. (unpublished)
Ethanol[a]	64-17-5	1.6		Dawson and Bantle [*16*]
N-ethyl N-nitrosourea unactivated[a]	759-73-9	5.8		Fig. 1
N-ethyl N-nitrosourea activated[a]		5.2		Fig. 1
5-fluorouracil[a]	51-21-8	10.7		Dawson and Bantle [*16*]
Hydroxyurea[a]	127-07-1	4.2		Courchesne and Bantle [*21*]
Naphthalene[e]	91-20-3		1.1	Bantle and Courchesne [*22*]
Penicillin G[a]	69-57-8		<1.3	Bantle and Dawson [*16*]
Propylene glycol[e]	57-55-6		1.2	Bantle, J. A. (unpublished)
Puromycin[c]	66-81-9		1.0	Courchesne and Bantle [*21*]
Saccharin[a]	81-07-2		1.0	Dawson and Bantle [*16*]
Sodium cyclamate[a]	139-05-9		1.1	Dawson, D. A. (unpublished)

[a]Agrees with Smith et al. [*23*].
[b]Does not agree with Smith et al. [*23*].
[c]Does agree with literature. See reference for discussion.
[d]Does not agree with literature. See reference for discussion.
[e]Agrees with the Chernoff/Kavlock mouse assay [*24,25*].

Replication Assay

The replication assay provided a more direct way of detecting inhibitors of DNA synthesis than the oocyte expression assay. Early cleaving stages of *Xenopus* embryos were exposed to the toxicant for various time periods, and the amount of DNA was measured fluorometrically using mithramycin. Figure 4 shows a dose-response curve using the DNA synthesis inhibitor hydroxyurea. Increasing concentrations of hydroxyurea resulted in a 3.5-fold decline in DNA per embryo until cytotoxic concentrations were reached at concentrations greater than 1 mg/mL. Although the results were excellent, we felt that the 24-h exposure period was too long in order to determine a specific effect. Larger numbers of eggs at the 2 to 64 cell stage were exposed to higher doses of toxicant, and DNA content was measured after different time periods (Fig. 5). In this experiment, a concentration of hydroxyurea 300% greater than the 96-h LC_{50} [*21*] reduced DNA content by greater than 2.5 fold, but the protein synthesis inhibitor cycloheximide had only a slight effect at a concentration greater than the 96-h LC_{50}. The same results were observed when the experiment was repeated (data not shown). Similar results were obtained when cytosine arabinoside was used as the DNA synthesis inhibitor and puromycin was used as the protein synthesis inhibitor (Fig. 6). Both of these inhibitors were used at concentrations 300% greater than their 96-h LC_{50}. A second experiment yielded the same results.

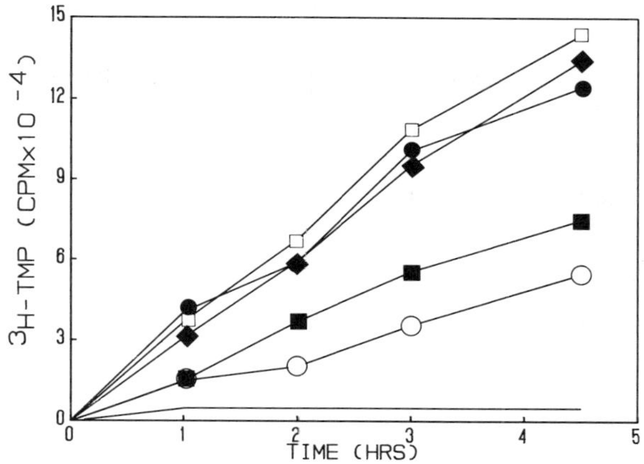

FIG. 2—*Enhancement of herpes thymidine kinase expression in* Xenopus *Stage 6 oocytes by cytosine arabinoside.* Xenopus *Stage 6 oocytes were exposed to cytosine arabinoside for 24 h. Thymidine kinase activity was then periodically assayed in the supernatant from oocyte homogenates by the conversion of* ^3H-*thymidine to* ^3H-*thymidine monophosphate. The positive control consisted of pHSV-106 injected oocytes while uninjected oocytes served as a negative control.* ■ *positive control,* —— *negative control, cytosine arabinoside:* ● *0.05 mg/mL,* ♦ *0.5 mg/mL,* □ *1.5 mg/mL,* ○ *2.0 mg/mL.*

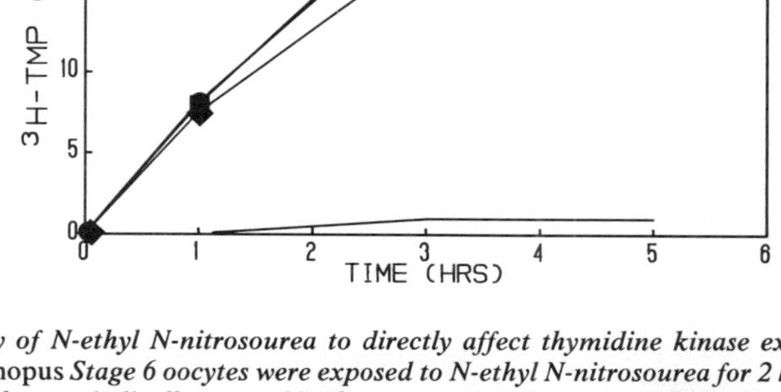

FIG. 3—*Inability of N-ethyl N-nitrosourea to directly affect thymidine kinase expression in* Xenopus *Stage 6 oocytes.* Xenopus *Stage 6 oocytes were exposed to N-ethyl N-nitrosourea for 2 h at 37°C. Thymidine kinase activity was then periodically assayed in the supernatant from oocyte homogenates by the conversion of* ^3H-*thymidine to* ^3H-*thymidine monophosphate. The positive control consisted of pHSV-106 injected oocytes while uninjected oocytes served as a negative control.* ● *positive control,* —— *negative control, N-ethyl N-nitrosourea:* ♦ *0.045 mg/mL,* ■ *0.5 mg/mL.*

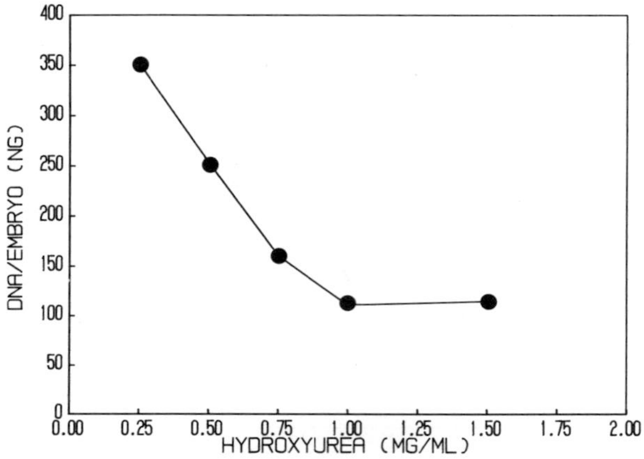

FIG. 4—*Dose-response curve showing the effect of the DNA synthesis inhibitor hydroxyurea on replication during* Xenopus *embryo development. Groups of* Xenopus *embryos (15 per group) starting at blastula were exposed to different concentrations of hydroxyurea at 24°C for 24 h. After exposure, embryos were removed and the DNA content selectively measured using the Mithramycin C technique.*

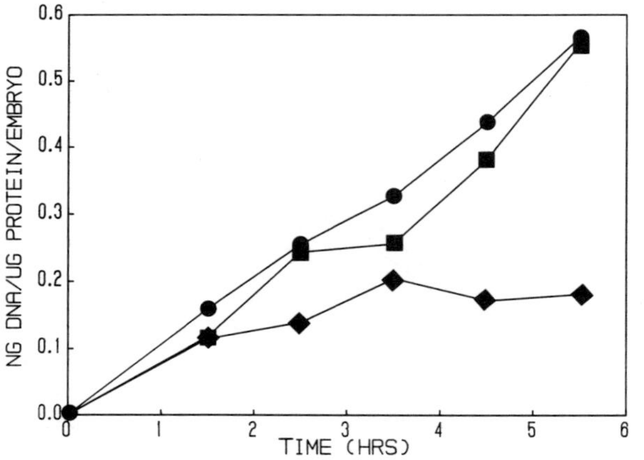

FIG. 5—*Effect of the DNA synthesis inhibitor hydroxyurea and the protein synthesis inhibitor cycloheximide on replication during early* Xenopus *embryo development. The concentrations selected were 300% of the FETAX 96-h LC_{50}. Groups of* Xenopus *embryos starting at the 2 to 64 cell stage were exposed to either cytosine arabinoside or puromycin at 24°C. At various time intervals, embryos were removed and the DNA content selectively measured using the Mithramycin C technique. The results of this experiment were normalized based on the amount of total protein in the homogenate. ● positive control, ♦ 5.46 mg/mL hydroxyurea, ■ 4.8 × 10^{-4} mg/mL cycloheximide.*

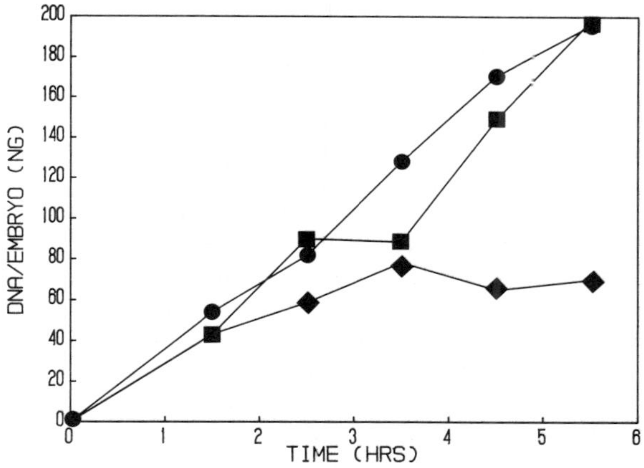

FIG. 6—*Effect of the DNA synthesis inhibitor cytosine arabinoside and the protein synthesis inhibitor puromycin on replication during early* Xenopus *embryo development. The concentrations selected were 300% of the FETAX 96-h LC_{50}. Groups of* Xenopus *embryos starting at the 2 to 64 cell stage were exposed to either cytosine arabinoside or puromycin at 24°C. At various time intervals, embryos were removed and the DNA content selectively measured using the Mithramycin C technique.* ● *positive control,* ♦ *16.8 mg/mL cytosine arabinoside,* ■ *0.26 mg/mL puromycin.*

Discussion

FETAX as an Teratogenesis Screening Assay

FETAX is an alternative test which offers some short- and long-term solutions to current problems associated with developmental hazard evaluation. Existing in vivo mammalian test systems are effective for the testing of drugs and cosmetics but are too lengthy for the screening of water samples and other complex mixtures. In addition, the number of chemicals in production requiring testing is increasing faster than is possible to test with in vivo test systems. In a study conducted for the National Institute of Health (NIH), the National Academy of Sciences recently concluded that the development and use of lower form models for toxicity testing should be emphasized [*26*].

Cost-effective, yet developmentally relevant, tests are needed for the detection of developmental toxicants in the environment as well as evaluation of the developmental toxicity of pure compounds and drugs. FETAX can detect the toxicity and teratogenicity of complex environmental mixtures where the individual action of each component may be known but the combined action cannot be predicted.

Xenopus can be considered a high connectivity model [*26*] due to the amount of biological information available on this species. There is a good chance that observations made on this species can be connected to data from other organisms, including man. *Xenopus* has been a standard lab animal for numerous developmental studies, and the *Xenopus* oocyte is a commonly used biomedical model. Consequently, much is known about the frog's morphology, physiology, and biochemistry. This cannot be stated for many of the alternative systems. *Xenopus* can be bred throughout the year providing numerous offspring. This facilitates statistical analysis. Since development is external, the speed of data collection is enhanced. The lack of a placental relationship is a disadvantage, but this problem is shared with other in vitro teratogenesis assays.

FETAX Test Performance and Developmental Relevance

Sabourin [*27*] has completed testing of 32 validation compounds using FETAX with 83% predictive accuracy. Our results (Table 1) suggest that the predictive accuracy of FETAX will be at least 85%. We have tested both known developmental toxicants and compounds that do not show developmental toxicity. However, most of the compounds tested to date are nonvolatile, water-soluble compounds that are direct acting.

Comparison of results within and between laboratories has been encouraging. The results reported here for ENU were very close even though different lots of ENU were used and the tests were performed by different individuals. Courchesne and Bantle [*21*] found a teratogenic index (96-h LC_{50}/96-h EC_{50} (malformation)) for hydroxyurea of 4.3, whereas Sabourin et al. recorded 4.5 for the same chemical [*28*]. In some cases, however, a lack of correspondence is apparent. Even though 5-fluorouracil tested strongly positive for teratogenesis in the laboratories of Bantle and Sabourin, the teratogenic indexes (*TI*) differed significantly. The reason for such discrepancies needs to be understood and may be due, in part, to variations in chemical purity and actual concentrations used as well as operator variability in scoring results. If the assay is to be used for purposes beyond screening, such as for ranking the teratogenicity of test chemicals, then the importance of minimizing variability in quantitative measurements emerges.

Additional evidence which supports the use of FETAX as a viable teratogen screen concerns (1) its performance compared with that of other potential test systems and (2) its developmental relevance. Sabourin et al. [*28*] found that FETAX was the assay of choice when compared to the planaria system [*3*] and the *Hydra* system [*4*]. These results can be attributed to the numerous endpoints of the frog assay as well as its higher phylogenetic position. We have also found that endpoints in the *Xenopus* assay may demonstrate developmental mimicry to those in mammalian systems. Dumont et al. [*29*] found that meclizine induced hydrocephalia in both frogs and mammals and that other teratogens produced similar abnormalities in both frogs and mammals. Courchesne and Bantle [*21*] reported that a number of genotoxic chemicals caused the same general types of malformations in both *Xenopus* and rodent embryos.

In Vitro Bioactivation

In humans, the maternal liver and placenta contain monooxygenases (cytochrome P-450 system) that detoxify some compounds while inadvertently activating procarcinogens and proteratogens to powerful carcinogens and teratogens. This mammalian activation must be duplicated in any in vitro screening test including FETAX.

We have developed an in vitro metabolic activation system for FETAX by using Aroclor 1254 induced rat liver microsomes cocultured with embryos. Initial attempts to use commercially available rat liver S-9 supernatant (Litton Bionetics) failed because of high S-9 toxicity. We then attempted to further purify microsomes from rat liver homogenate in order to reduce toxicity. We were successful in this approach and able to activate the proteratogen cyclophosphamide into its embryotoxic form as measured by FETAX [*12*]. Aroclor 1254 induction five days prior to microsome preparation increased the *TI* of cyclophosphamide from 1.3 to 3.6 [*13*]. We were also able to identify the cyclophosphamide metabolite thought to be responsible for the increase in teratogenicity and prove that residual Aroclor metabolite did not affect results [*13*]. The routine use of Aroclor induced rat liver microsomes should increase the predictive accuracy of FETAX. However, some problems may persist. We have recently tested isoniazid in FETAX and determined that it is teratogenic, although Smith et al. [*23*] list it as a nonteratogen. Isoniazid is normally metabolized by cytochrome P-450j, but Aroclor 1254 induction suppresses cytochrome P450j expression [*30*]. Therefore, while we would normally expect our in vitro metabolic activation system to detoxify isoniazid and give the same results as observed in the mammalian studies, the use of Aroclor 1254 induced microsomes may prove counterproductive in a few situations. Isoniazid is currently being tested with the metabolic activation system to see if this is true.

ENU was teratogenic in the absence of the metabolic activation system. This is in agreement with Smith et al. [*23*], who list ENU as direct acting. However, when we tested ENU in the presence of the metabolic activation system, there was a 73-fold increase in sensitivity although the *TI* remained essentially the same. ENU directly forms a alkyldiazonium ion by a hydroxyl ion-catalyzed decomposition even at neutral pH [*31*]. The in vitro effects are the same as in vivo [*31*]. Therefore we can postulate that the metabolic activation system caused the increase in sensitivity by enhancing the conversion of ENU to its ionic form. This may have provided a strong initial pulse of the ion producing increased effects.

The Oocyte Expression and the Replication Assay

By coupling a validated whole embryo bioassay with a battery of selected molecular or cellular tests, a better estimation of developmental hazard can be obtained than with whole embryo bioassays alone. Ideally, these suborganismal tests should indicate whether a chemical interferes with one of the cellular mechanisms important to the development of all animals. By coupling each test with the appropriate metabolic activation system, differences in metabolism can be accounted for. In order to test the feasibility of this approach we devised the Oocyte Expression Assay and the Replication Assay. The Oocyte Expression Assay is able to detect inhibitors of DNA, RNA, and protein synthesis. Specific inhibitors of DNA synthesis have proven to be teratogenic in virtually all species tested [*21,32*]. For example, replication inhibitors are teratogenic even though their molecular mechanism of action may be quite different. Inhibitors such as 5-fluorouracil and hydroxyurea inhibit replication by reducing the supply of deoxyribonucleotide triphosphates, while cyclophosphamide and cytosine arabinoside affect DNA structure such that DNA polymerase cannot replicate the DNA. As long as the net result is a significant reduction in DNA synthesis, abnormal development will occur. In this study, cytosine arabinoside caused an enhancement of thymidine kinase gene expression over most of its dose range. This increase indicated a specific replication inhibitor [*17*], whereas previous research indicated that indirect inhibitors of DNA synthesis such as protein synthesis inhibitors caused a decrease in thymidine kinase expression [*17*] and were not teratogenic [*21*].

Although ENU is an inhibitor of nucleic acid synthesis [*33*], concentrations near the 96-h LC_{50} and the 96-h EC_{50} (malformation) failed to alter thymidine kinase (TK) expression even though the plasmid bearing the TK gene was exposed directly to ENU prior to microinjection. In order for ENU to permanently affect the DNA, a number of rounds of cell division must occur [*31*]. The Oocyte Expression Assay is carried out in the absence of DNA synthesis. Because of this apparent deficiency, the replication assay was developed. In this assay, the amount of newly synthesized DNA is measured after several hours of development by a simple fluorometric technique. An early *Xenopus* blastula contains approximately 100 cells or 0.75 ng of DNA. Only 5 h later, the gastrula embryo has 2500 cells and contains 175 ng of DNA. This is a 250-fold increase in DNA such that 50 embryos will contain 8.75 μg of DNA. The fluorometric assay using Mithramycin C can easily detect as little as 0.5 μg of DNA per sample. If an inhibitor reduces the amount of DNA synthesis by 50% in 5 h, then the 4.4 μg of DNA still produced would be detected by this assay. The 4.4 ng of DNA difference between the 50% effect and the control provides enough signal to produce a dose-response effect. At these early stages of amphibian development, most development is independent of RNA synthesis and, therefore, the effect seen will be due to either a direct action on replication or the microtubular apparatus responsible for separation of chromosomes. Since the assay takes only 5 h to perform, doses need to be higher than those needed for the FETAX 96-h LC_{50} and the EC_{50} (malformation). We have found that 300% of the 96-h LC_{50} separates the action of specific DNA synthesis inhibitors such as cytosine arabinoside and hydroxyurea from protein synthesis inhibitors such as puromycin and cycloheximide (Figs. 5 and 6). The main advantage of coupling FETAX with these molecular tests is that the approach will prove cost-effective in that the same organism is used for all tests and there is no need to conduct expensive cell culture experiments.

Interdependence of the Assays

The ability to use a test battery to estimate possible human health hazards is shown by the ENU tests. FETAX initially suggested it was a direct-acting teratogen. The use of the metabolic activation system suggested that ENU activity could be enhanced by bioactivation, although different products were not made since the *TI* did not change. Most of the mortality took place during the first 48 h of the test, suggesting DNA synthesis was involved (data not shown). The Oocyte Expression Assay, modified to detect the direct effect of ENU on the DNA, failed to cause an effect, indicating that subsequent cell division may have been needed in order to affect replication [*31*]. Presently, ENU is being tested in the replication assay. A decrease in the amount of DNA synthesized is expected.

It will be more difficult to devise other suborganismal tests which indicate that an important cellular mechanism of development is altered by a toxicant. For example, cell interaction (induction) is a critically important mechanism of development. However, the nature of the signal molecule and the response of the target cell is likely to vary widely from one species to another. A test using cultured cells in the process of induction is likely to be required. Regardless of the test devised, dosing is going to be a problem since the test duration is likely to be different from FETAX or the Replication Assay. Basing the dosing on either the FETAX 96-h LC_{50} or EC_{50} (malformation) may be the best way to compensate for different assay conditions. Lastly, the same species of organism should be used whenever possible to keep costs down.

A complementary test battery employing a metabolic activation system will not only reduce the number of false positives and negatives but also serve to discover the mechanism of action for a particular toxicant. This could provide useful data for those studying structure-activity relationships.

Acknowledgments

This work was supported by a Reproductive Hazards in the Workplace grant number 15-30 from the March of Dimes Births Defects Foundation and The Oklahoma State University Center for Water Research. Douglas J. Fort and Douglas A. Dawson were recipients of Presidential Fellowships in Water Resources from Oklahoma State University.

References

[*1*] Greenberg, J., "Detection of Teratogens by Differentiating Embryonic Neural Crest Cells in Culture: Evaluation as a Screening System," *Teratogenesis, Carcinogenesis, and Mutagenesis,* Vol. 2, 1982, pp. 293-301.

[*2*] Schuler, R., Hardin, B. D., and Niemer, R., "*Drosophila* as a Tool for the Rapid Assessment of Chemicals for Teratogenicity," *Teratogenesis, Carcinogenesis, and Mutagenesis,* Vol. 2, 1982, pp. 293-301.

[*3*] Best, J. B. and Morita, M., "Planarians as a Model System for *in vitro* Teratogenesis Studies," *Teratogenesis, Carcinogenesis, and Mutagenesis,* Vol. 2, 1982, pp. 277-291.

[*4*] Johnson, E. M., Gorman, R. M., Gabel, B. E. G., and George, M. E., "The *Hydra Attenuata* System for Detection of Teratogenic Hazards," *Teratogenesis, Carcinogenesis, and Mutagenesis,* Vol. 2, 1982, pp. 263-276.

[*5*] Dumont, J., Schultz, T. W., Buchanan, M., and Kao, G. in *Short-Term Bioassays in the Analysis of Complex Mixtures III,* Waters, Sandhu, Lewtas, Claxton, Chernoff, and Nesnow, Eds., Plenum, New York, 1983, pp. 393-405.

[*6*] Sadler, T. W., Horton, W. E., and Warner, C. W., "Whole Embryo Culture: A Screening Technique for Teratogens," *Teratogenesis, Carcinogenesis, and Mutagenesis,* Vol. 2, 1982, pp. 243-253.

[7] Mummery, C. L., Van den Brink, C. E., Van der Saag, P. T., and De Laat, S. W., "A Short-Term Screening Test for Teratogens Using Differentiating Neuroblastoma Cells *in vitro*," *Teratology*, Vol. 29, 1984, pp. 271-280.
[8] Bournais-Vardiabasis, N. and Flores, J., "Drug Metabolizing Enzymes in *Drosophila melanogaster* Teratogenicity of Cyclophosphamide *in vitro*," *Teratogenesis, Carcinogenesis, and Mutagenesis*, Vol. 3, 1983, pp. 255-262.
[9] Cooke, J., "The Early Embryo and the Formation of Body Pattern," *American Scientist*, Vol. 76, 1988, pp. 35-41.
[10] "Methods for Acute Toxicity Tests With Fish, Macroinvertebrates and Amphibians," EPA-660/3-75-009, United States Environmental Protection Agency, 1975, pp. 1-69.
[11] Dent, J. G., Schnell, S., Graichen, M. E., Allen, P., Abernathy, D., and Couch, D. B., "Stability of Activating Systems for *in vitro* Mutagenesis Assays: Enzyme Activity and Activating Ability Following Long-Term Storage at −85°C," *Environmental Mutagenesis*, Vol. 3, 1981, pp. 167-179.
[12] Bantle, J. A. and Dawson, D. A. in *Aquatic Toxicology and Hazard Assessment: 10th Volume, STP 971*, W. J. Adams, G. A., Chapman, and W. G. Landis, Eds., 1988, pp. 316-326.
[13] Fort, D. J., Dawson, D. A., and Bantle, J. A., "Development of a Metabolic Activation System for the Frog Embryo Teratogenesis Assay: *Xenopus* (FETAX)," *Teratogenesis, Carcinogenesis, and Mutagenesis*, 1988, pp. 251-263.
[14] Nash, T., "The Colorimetric Estimation of Formaldehyde by Means of the Hantzsch Reaction," *Biochemistry*, Vol. 55, 1955, pp. 412-416.
[15] Lucier, G., McDaniel, O., Brubaker, P., and Klein, R., "Effects of Methylmercury Hydroxide on Rat Liver Microsomal Enzymes," *Chemico-Biological Interactions*, Vol. 4, 1971, pp. 265-280.
[16] Dawson, D. A. and Bantle, J. A., "Development of a Reconstituted Water Medium and Preliminary Validation of the Frog Embryo Teratogenesis Assay—*Xenopus* (FETAX)," *Journal of Applied Toxicology*, Vol. 7, 1987, pp. 237-244.
[17] Courchesne, C. L., Dawson, D. A., and Bantle, J. A., "Detection of Inhibitors of Protein and Nucleic Acid Synthesis Using Stage 6 Oocytes of *Xenopus laevis* Microinjected with the Herpes Thymidine Kinase Gene," *Chemico-Biological Interactions*, Vol. 60, 1986, pp. 13-30.
[18] Gaudette, M. F., Hiriyanna, K. T., Kaiserman, H., and Benbow, R. M., "Chromosomal DNA Replication in *Xenopus laevis* Embryos: Biochemical Analysis," *Journal of Cell Biology*, Vol. 101, 1985, p. 459a.
[19] Hukkelhoven, M. W. A. C., Vromans, E., Markslag, A. M. G., and Vermorken, A. J. M., "A Simple Fluorometric Microassay for DNA in Hair Follicles or Fractions of Hair Follicles," *Anticancer Research*, Vol. 1, 1981, pp. 341-344.
[20] Bradford, M. M., "A Rapid and Sensitive Method for the Quantitation of Microgram Quantities of Protein Utilizing the Principle of Protein-Dye Binding," *Analytical Biochemistry*, Vol. 72, 1976, pp. 248-254.
[21] Courchesne, C. L. and Bantle, J. A., "Analysis of the Activity of DNA, RNA, and Protein Synthesis Inhibitors on *Xenopus* Embryo Development," *Teratogenesis, Carcinogenesis, and Mutagenesis*, Vol. 5, 1985, pp. 177-193.
[22] Bantle, J. A. and Courchesne, C. L. in *Proceedings of the Second International Conference on Ground-Water Quality Research*, Oklahoma State University Press, Stillwater, OK, 1985, pp. 175-176.
[23] Smith, M. K., Kimmel, G. L., Kochhar, D. M., Shepard, T. H., Spielberg, S. P., and Wilson, J. G., "A Selection of Candidate Compounds for *in vitro* Teratogenesis Test Validation," *Teratogenesis, Carcinogenesis, and Mutagenesis*, Vol. 3, 1983, pp. 461-480.
[24] Kavlock, R. J., Short, R. D., and Chernoff, N., "Further Evaluation of an *in vivo* Teratology Screen," *Teratogenesis, Carcinogenesis, and Mutagenesis*, Vol. 7, 1987, pp. 7-16.
[25] Hardin, B. D., Schuler, R. L., Burg, J. R., Booth, G. M., Hazelden, K. P., MacKenzie, K. M., Piccirillo, V. J., and Smith, K. N., "Evaluations of 60 Chemicals in a Preliminary Developmental Toxicity Test," *Teratogenesis, Carcinogenesis, and Mutagenesis*, Vol. 7, 1987, pp. 29-48.
[26] "Models for Biomedical Research: A New Perspective," Committee on Models for Biomedical Research, Board on Basic Biology, Commission of Life Sciences, National Research Council, National Academy Press, Washington, DC, 1985, pp. 1-180.
[27] Sabourin, T. D. and Faulk, R. T., "Comparative Evaluation of a Short-Term Test for Developmental Effects Using Frog Embryos," Branbury Report 26: *Developmental Toxicology: Mechanisms and Risk*, Cold Spring Harbor Laboratory, 1987, pp. 203-223.
[28] Sabourin, T. D., Faulk, R. T., and Goss, L. B., "The Efficacy of Three Nonmammalian Test Systems in the Identification of Chemical Teratogens," *Journal of Applied Toxicology*, Vol. 5, 1985a, pp. 227-233.
[29] Dumont, J. N., Schultz, T. W., and Epler, R. G., "The Response of the FETAX Model to Mammalian Teratogens," *Teratology*, Vol. 27, 1983, p. 39a.

[*30*] Palakodety, R. B., Clejen, L. A., Krikun, G., Feierman, D. E., and Cederbaum, A. I., "Characterization and Identification of a Pyrazoleinducible Form of Cytochrome P-450," *Journal of Biological Chemistry*, Vol. 263, 1988, pp. 878–884.
[*31*] Singer, B. and Grunberger, D., *Molecular Biology of Mutagens and Carcinogens*, Plenum, New York, 1983, p. 70.
[*32*] Ritter, E. in *Handbook of Teratology*, J. G. Wilson and F. C. Fraser, Eds., Vol. 2, New York, Plenum Press, 1977, pp. 99–116.
[*33*] Lockhart, M. L., Deutsch, J. F., Yamaura, I., Cavalieri, L. F., and Rosenberg, B. H., "Termination of DNA Synthesis *in vitro* at Apurinic Sites But Not at Ethyl Adducts on the Template," *Chemico-Biological Interactions*, Vol. 42, 1982, pp. 85–95.

Steven P. Bradbury,[1] *Richard W. Carlson,*[1] *and Tala R. Henry*[2]

Polar Narcosis in Aquatic Organisms

REFERENCE: Bradbury, S. P., Carlson, R. W., and Henry, T. R., **"Polar Narcosis in Aquatic Organisms,"** *Aquatic Toxicology and Hazard Assessment: 12th Volume, ASTM STP 1027,* U. M. Cowgill and L. R. Williams, Eds., American Society for Testing and Materials, Philadelphia, 1989, pp. 59–73.

ABSTRACT: The majority of industrial organic chemicals lack identifiable structural characteristics that result in specific biological activity. These nonpolar-nonelectrolytes are acutely toxic to aquatic organisms via a nonspecific mode of action termed *narcosis.* The toxicity of industrial chemicals eliciting nonpolar narcosis can be reliably predicted by log *P* (baseline toxicity models). Using single chemical and joint toxic action models, several research groups have reported classes of polar compounds (for example, esters, phenols, and anilines) that elicit a narcosis-like syndrome; however, they are more acutely toxic than what is predicted using baseline toxicity models. An assessment of rainbow trout *(Salmo gairdneri)* in vivo respiratory-cardiovascular responses during intoxication by polar narcotic phenol and aniline derivatives established a toxicity syndrome unique to that elicited by nonpolar narcotics. This finding further suggests a mode of action unique to polar narcotics and supports the use of structure-activity relationships specific for these compounds. The proposition that there may be multiple mechanisms or sites of narcotic action is compatible with recent studies concerning the cellular and molecular mechanisms of anesthetic action.

KEY WORDS: aquatic toxicology, quantitative structure activity relationships, QSARs, narcosis, polar narcosis, mode of action, phenols, anilines, esters

The majority of industrial organic chemicals elicit their acute effects on aquatic species through the nonspecific mechanism of narcosis [*1–3*]. Narcosis can be defined as a reversible state of arrested activity of protoplasmic structures resulting from exposure to the appropriate xenobiotics. In the context of the intact organism, the terms *narcosis* and *general anesthesia,* or the pharmacologically induced and reversible stages of neural disruption [*4*], are used interchangeably. A diverse range of chemical structures have been reported to cause narcosis in mammals and aquatic organisms, including inert gases, aliphatic and aromatic hydrocarbons, chlorinated hydrocarbons, alcohols, ethers, ketones, aldehydes, weak acids and bases, and some aliphatic nitro compounds [*1–3,5–10*].

At least in the context of aquatic and environmental toxicology, recent studies have suggested that there are subclasses of narcotics that can be classified by either acute potency and/or the physiological and behavioral characteristics of the narcosis response. More specifically, narcosis induced by certain esters [*5*], phenols [*11,14*], and anilines [*11,14,15*] seem to be unique. These agents have been termed polar narcotics [*12,14*]; however, as will be discussed, there may be subclasses of compounds within the polar narcotics as well. The classification of narcotics into subclasses, based on potency and/or toxic responses, raises the question as to whether a single unitary theory of narcosis is relevant [*4*] or whether a theory of narcosis must include the possibility of multiple sites or mechanisms [*16*].

[1]Research toxicologist and research biologist, respectively, U.S. Environmental Protection Agency, Environmental Research Laboratory, Duluth, MN 55804.
[2]Biologist, American Scientific International, Inc., Duluth, MN 55804.

In this review the basis for establishing a unique group of narcotics, the polar narcotics, will be addressed by examining single chemical and joint toxic action studies and by assessing reports of whole-organism physiological responses to intoxication. These studies will be interpreted in light of current cellular and molecular hypotheses regarding narcosis.

Single Chemical and Joint Toxic Action Models

High quality quantitative structure-activity relationships (QSARs) and joint toxic action models can provide preliminary evidence for classifying xenobiotics by similar or common modes of action [*17,18*]. Several QSARs have been published that predict the acute lethality of nonpolar nonelectrolytes to fish [*1,2*] and aquatic invertebrates [*3*]. The lethality of these nonreactive compounds, which cover diverse classes of chemical structures (for example, alcohols, ketones, ethers, alkyl halides, certain substituted benzenes), can be accurately estimated with the *n*-octanol/water partition coefficient (log *P*). These models have been described as representing baseline toxicity estimates for chemicals acting through the nonspecific mechanism of narcosis [*1*].

Within the last several years, numerous studies have been published that indicated specific polar compounds, for example, esters [*5*], phenols [*12,13*], and anilines [*15*], were significantly more toxic that what was predicted by baseline-narcosis models. As a result, several specific structure-activity models were established for these compounds [*5,15*]. These QSARs use log *P* to estimate lethality, occasionally in conjunction with pKa or other electronic descriptors. Although in most cases signs of intoxication elicited by these polar compounds were not expressly reported in the above studies, the authors' general conclusions suggested a narcotic effect was expressed by the test organisms (signs of intoxication are described in the "Physiological Responses" section). Hence, QSARs for the appropriate esters, phenols, and anilines are generally described as models for polar narcosis, but considered distinct from baseline narcosis models that are appropriate for nonpolar compounds.

Subsequent to these empirical and statistical arguments for establishing separate narcosis models, Veith and Broderius [*14*] employed joint toxicity theory to provide more convincing evidence for the existence of different narcosis mechanisms. The underlying assumption for their approach is that toxicants acting through a similar mode of action should be concentration additive, which has been verified with compounds thought to act under baseline narcosis [*3,18,19*]. Using *n*-octanol and phenol, which together are less than strictly additive as reference baseline and polar narcotics, respectively, Veith and Broderius tested the joint action of a series of substituted phenols and primary anilines in the fathead minnow *(Pimephales promelas)*. Table 1 lists a representative set of phenols and anilines that were found to be strictly additive with phenol. Those compounds that are only additive with phenol generally have a log *P* less than 2.9 and are weakly acidic or basic with the presence of a strong hydrogen bond donating group. The findings suggest that hydrogen bonding at the site of action may be critical with these polar narcotics, while with nonpolar narcotics hydrophobic bonding probably dominates [*14*]. Phenols and anilines with log *P*s greater than 2.9 were found, in many instances, to be additive with both phenol and octanol (Table 1). Veith and Broderius [*14*] have proposed that for certain polar narcotic phenols and anilines with log *P*s greater than 2.9 the impact of hydrophobic bonding is equal to or greater than that of hydrogen bonding.

On the basis that the joint toxicity results provided supportive evidence of a unique narcosis mode of action, Veith and Broderius [*14*] proposed the following separate structure-activity relationship (Fig. 1) for narcosis elicited by phenols and primary anilines in the fathead minnow:

$$\text{Log LC}_{50} = -0.65\ (\pm 0.07)\ \text{Log}\ P - 2.29\ (\pm 0.22)$$

$$n = 39$$

$$r^2 = 0.90$$

These specific polar narcotics have been termed narcosis Type II compounds, while nonpolar chemicals acting under baseline narcosis have been termed narcosis Type I compounds [*14*]. As shown in Fig. 1, the narcosis Type I and narcosis Type II QSARs become essentially indistinguishable at a log *P* of about 3, indicating that either QSAR can reliably predict the lethality of more hydrophobic compounds. This finding is consistent with the joint toxicity results, where substituted phenols with a log *P* greater than 2.9 were additive with both phenol and octanol.

In the course of their study, Veith and Broderius [*14*] also reported several important classes of compounds that were not additive with phenol (or octanol). Generally, anilines and phenols that are acidic, with two or more nitro substituents or four or more ring substituted halogens, do not cause narcosis Type II intoxication (Table 2) and are in fact more toxic than what is estimated from the narcosis Type II QSAR. These compounds are likely to be uncouplers of oxidative phosphorylation and, therefore, should be modeled under a different QSAR. Recently other workers have also recognized that the acidity of phenols can be used to separate those compounds acting as polar narcotics or uncouplers [*13,20*]. Preliminary joint toxic action studies with monoesters suggest that these compounds are additive with octanol, even though they are more toxic than predicted by the narcosis Type I QSAR [*5*]. These seemingly contradictory results could be explained if the toxicity of monoesters were a function of in vivo ester hydrolysis [*21*]. Diesters are not additive with either phenol or octanol, even though they seem to elicit

TABLE 1—*Examples of phenol and aniline derivatives eliciting polar narcosis in fathead minnows, based on strictly additive joint toxicity with phenol.*[a]

Chemical	Log *P*
Phenol	1.46
2-chlorophenol	2.15
2-phenylphenol[b]	3.36
3-methoxyphenol	1.58
4-ethylphenol	2.58
4-methoxyphenol[c]	1.34
4-nitrophenol	1.91
4-phenoxyphenol[b]	3.75
4-propylphenol[b]	3.18
2,4-dimethylphenol	2.30
2,4-dichlorophenol[b]	2.92
1-naphthol	2.84
Aniline	0.90
2-chloroaniline[c]	2.65
4-chloroaniline[c]	1.83
4-ethylaniline[c]	1.96
4-nitroaniline[c]	1.31
3,4-dichloroaniline[c]	2.69
2-chloro-4-methylaniline[c]	2.58

[a]Data from Veith and Broderius [*14*].
[b]Chemicals also strictly additive with octanol.
[c]Personal communication from S. Broderius, U.S. Environmental Protection Agency, Environmental Research Laboratory, Duluth, MN 55804.

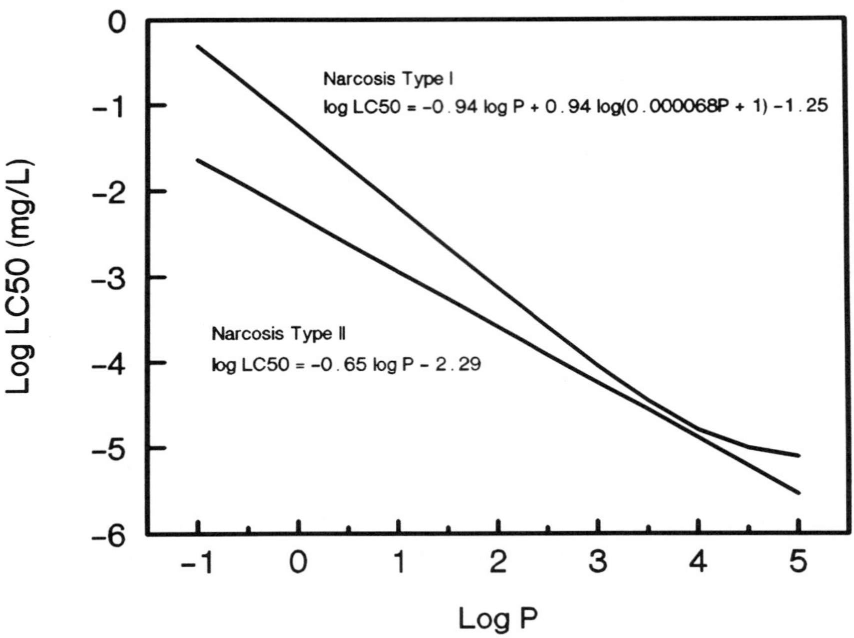

FIG. 1—*Relationship between nonpolar (Type I) and polar (Type II) narcotic fathead minnow 96-h* LC_{50}*s and log* P [1,14].

TABLE 2—*Polar phenol and aniline derivatives that do not elicit polar narcosis and show less than strictly additive toxicity with phenol and octanol.*[a]

Chemical	Log *P*
Pentachlorophenol	5.12
2,4-dinitro-1-naphthol	3.09
4-phenylazophenol	3.18
4,6-dinitro-*o*-cresol	2.56
2,4-dinitrophenol	1.54
1,4-dinitrobenzene	1.46
2,4-dinitroaniline	1.84
2,3,5,6-tetrachloroaniline	4.10
Pentachloropyridine	4.34

[a]Data from Veith and Broderius [*14*].

narcosis [*5*], indicating that there may be more than one mode of polar narcosis. Clearly, ester lethality should not be evaluated with either the narcosis Type I or narcosis Type II QSARs.

Physiological Responses

As described in a recent review [*22*], efforts in our laboratory to develop QSARs for use in predictive toxicology are centered on developing mechanism-based models. To properly implement structure activity models, separate QSARs must be formulated for chemicals with a common mode of action. Implementation of mechanism-specific toxicity models has required the development of qualitative structure-activity models that assign the proper mode of action and appropriate QSAR to a given chemical structure. Current efforts in developing the components of the model are centered on characterizing a toxic mode of action data base from which the structural requirements for specific mechanisms can be elucidated.

A major approach to defining common modes of action has been through the assessment of fish acute toxicity syndromes (FATS) [22]. FATS are distinct sets of in vivo rainbow trout *(Salmo gairdneri)* toxic responses that correspond to specific modes of action. By measuring a number of respiratory-cardiovascular variables (Table 3), FATS associated with nonpolar narcotics, oxidative phosphorylation uncouplers, repiratory membrane irritants, acetylcholinesterase (AChE) inhibitors, and the pyrethroid insecticide fenvalerate have been described [*23–25*]. Using the "FATS approach," we recently [*11*] assessed the physiological responses elicited by a series of phenol and aniline derivatives (Type II narcotics as defined by the joint action study of Veith and Broderius [*14*]).

The physiological response of rainbow trout to intoxication by phenol, 2,4-dimethylphenol, aniline, 2-chloroaniline, and 4-chloroaniline was assessed using the experimental approach described previously [*11,23–25*]. Spinally transected trout (0.6 to 1.0 kg) were exposed to lethal concentrations of the polar narcotics in specially designed respirometer-metabolism chambers, and a series of physiological variables were monitored until death (Fig. 2). Those variables used in the statistical analyses are listed in Table 3. Preceding toxicant exposure, measurements for all variables were made on each fish ($N = 4$ fish exposed per chemical and a separate control group of $N = 4$ fish) to provide predose values. Thus, each fish served as its own control. The mean percentage change of physiological parameters from their predose mean was determined for the 25 to 75% survival time interval for each fish. Measurements within this interval were used to minimize inclusion of responses due to initial stress and death itself.

Rainbow trout exposed to lethal aqueous concentrations of phenol (8.82 mg/L), 2,4-dimethylphenol (9.04 mg/L), aniline (68.8 mg/L), 2-chloroaniline (1.66 mg/L), and 4-chloroaniline (22.4 mg/L) died between 3.5 and 17.3 h. Visible signs of intoxication were similar for all five compounds and were elicited shortly after toxicant exposure. The most striking characteristic associated with intoxication was the development of tremors and clonic seizures, which were initiated with coughs. Because seizures were generally initiated with coughs, seizures and coughs were considered synonymous for data analysis. Between 1.5 and 10% survival time, the fish exhibited tremors that originated in the head and then moved posterially to the tail. Between 10 and 30% survival time, the tremors progressed to periods of clonic seizures, which culminated in episodes where the tail was raised out of the water. As the seizures became more intense the trout became unresponsive to outside stimuli. The intensity and severity of the sei-

TABLE 3—*Physiological variables monitored in rainbow trout to define fish acute toxicity syndromes.*

Variable	Units
Ventilation volume (V_G)	mL/min
Total oxygen consumption (VO_2)	mg/kg/h
Gill oxygen uptake efficiency (U_E)	%
Ventilation frequency (f_V)	no./min
Cough frequency (f_C)	no./min
Heart frequency (f_H)	no./min
Total blood oxygen (arterial) (TaO_2)	g/100 mL
Total blood carbon dioxide (arterial) ($TaCO_2$)	mmol/L
Blood pH (arterial) (pHa)	pH units
Hematocrit (Hct)	%
Hemoglobin (Hb)	g/100 mL

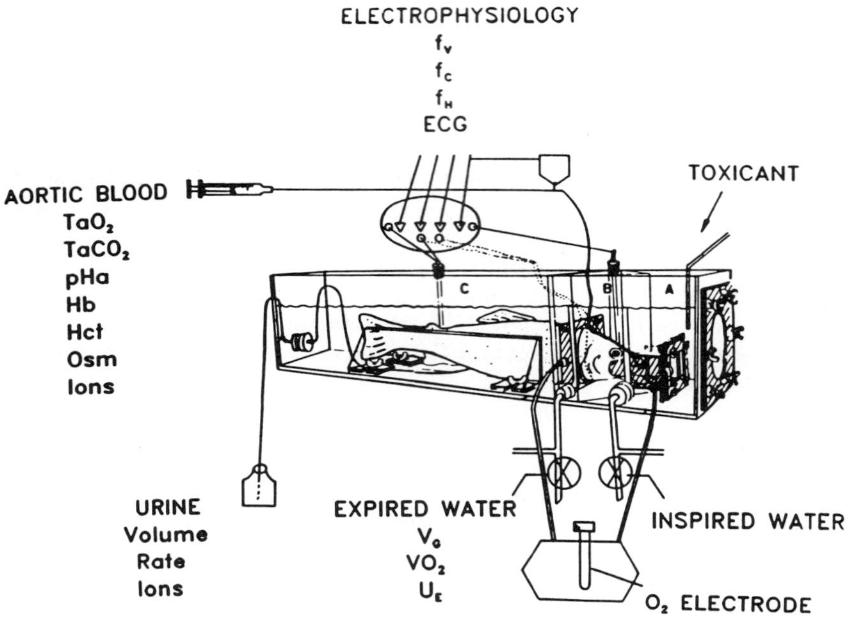

FIG. 2—*Experimental system for monitoring trout physiological responses to xenobiotics. See Table 3 for an explanation of abbreviations.*

zures increased through approximately 60% survival time after which they gradually subsided and the fish became mostly inactive. Interestingly, the seizures involved the entire fish even though the trout were spinally transected. The pyrethroid insecticide fenvalerate, a central nervous system seizure agent, elicited seizures in transected trout as well; however, the activity was confined to regions anterior to the transection site [*25*]. The response of these polar narcotics suggests that they were affecting the central nervous system, including the spinal cord posterior to the transection, or perhaps the peripheral nervous system. In mammals, the primary site of phenol stimulation in the central nervous system is thought to be the spinal cord. Tremors in mammals are seemingly due to reflex stimulation, while clonic convulsions result from increased sensitivity of the motor mechanism of anterior horn cells [*26*]. The visible signs of polar narcosis in the trout are similar to reports of phenol intoxication in a number of fish species [*27*], including the bluegill *(Lepomis macrochirus)* [*28*], the minnow *(Phoxinus phoxinus)* [*29*], the mullet *(Mugil auratus)* [*30*], the zebra fish *(Brachydanio rerio)* [*31*], and the fathead minnow (personal communication; R. Drummond, U.S. EPA, Environmental Research Laboratory-Duluth, Duluth, MN), and vertebrates in general [*32*].

Changes in rainbow trout physiological parameters (Figs. 3 and 4) were consistent with the increased muscular activity associated with seizures, as measured by f_C (mean percentage increases ranging from 385% in the phenol group to 2702% in the 2,4-dimethylphenol group). For phenol, 2,4-dimethylphenol, and 2-chloroaniline intoxicated fish, these changes were reflected by an increased oxygen demand to support the heightened activity as demonstrated by initial rises in VO_2 (mean increases of 15.7 to 73.3%) and V_G (mean increases of 49.5 to 126%). At approximately 20% survival time V_G and VO_2 began to decline, seemingly as a result of the increased frequency and severity of the seizures, which made respiration erratic and inefficient. Continued declines in V_G and VO_2 reflected an overall depression of respiratory activity with increasing exposure, consistent with a general narcotic effect [*23*]. The responses of the aniline

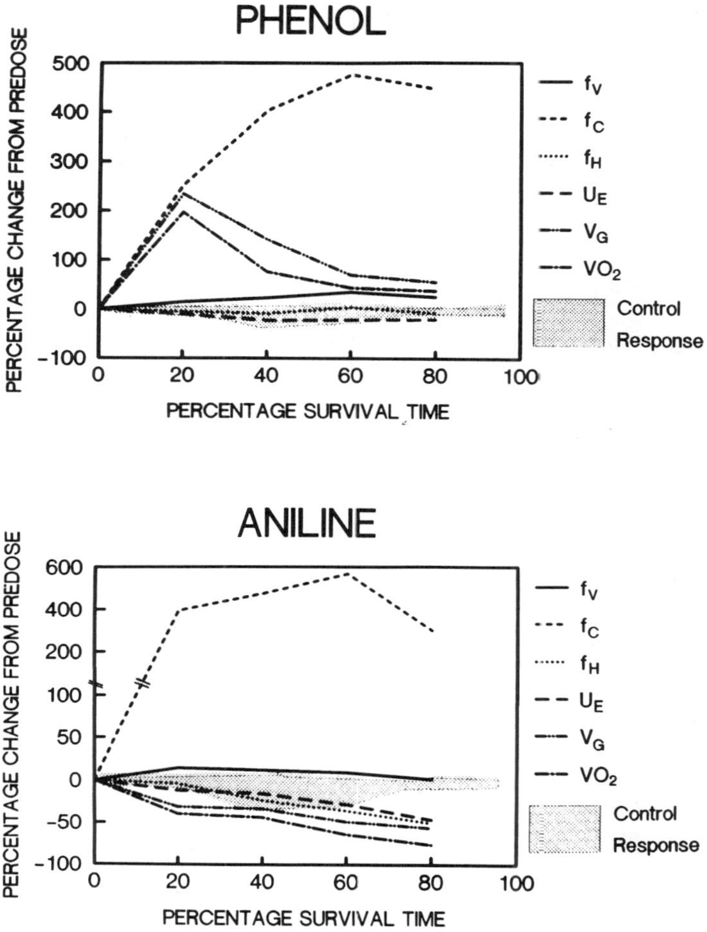

FIG. 3—*Representative respiratory-cardiovascular responses of rainbow trout exposed to acutely lethal aqueous concentrations of Type II polar narcotics. See Table 3 for an explanation of abbreviations. Each measurement represents the mean response of four trout. The stippled area represents the range of the mean response of four control fish (data from Ref* 11*).*

and 4-chloroaniline exposed fish were somewhat different in that these trout exhibited clonic seizures without a rise in V_G and VO_2 (mean declines of 40 to 50%). This response could be the result of a number of factors. Due to the steep dose response for these two compounds and the rapid time to death (ca. 4.5 h compared to ca. 12 h for the other three compounds), toxic responses were elicited much faster than what was observed with the other toxicants. A 15-min sampling interval was used for measuring V_G and VO_2, and it is possible that an initial rise in these parameters occurred prior to sampling. Alternatively, the time to onset and intensity of seizures in the aniline and 4-chloroaniline groups may have been sufficient to immediately disrupt normal ventilatory activity, whereas with the other compounds the disruption in ventilation required a longer period of time.

The major changes in blood-chemistry parameters (that is, decreasing TaO_2, $TaCO_2$, pHa, and increasing Hct) suggested a rapid shift toward anaerobic metabolism, which would be con-

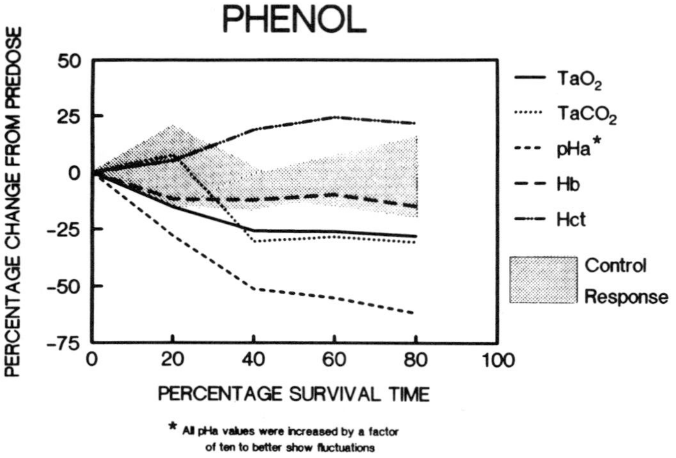

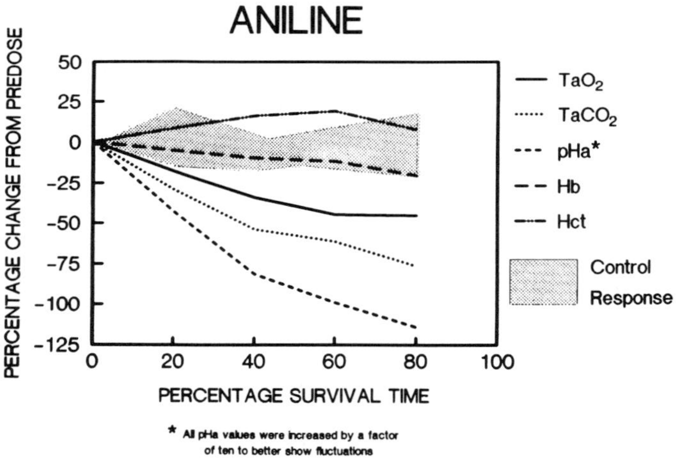

FIG. 4—*Representative blood-chemistry responses of rainbow trout exposed to acutely lethal aqueous concentrations of Type II polar narcotics. See Table 3 for an explanation of abbreviations. Each measurement represents the mean response of four trout. The stippled area represents the range of the mean response of four control fish (data from Ref* 11*).*

sistent with the initially heightened muscular activity [*33,34*] during seizures and the subsequent respiratory-cardiovascular depression [*35–37*]. The initial increases in V_G and VO_2 in the phenol, 2,4-dimethylphenol, and 2-chloroaniline exposed fish were seemingly not effective in supplying the exercising muscle with sufficient oxygen, as evidenced by declining TaO_2 (mean decline of 24.9 to 67.1%). In the case of the aniline and 4-chloroaniline exposed fish, initial increases in V_G and VO_2 were not observed, and the decreases in TaO_2 were two to three times larger than those seen in the other three toxicant groups. The development of a hypoxic state would initiate a shift to anaerobic metabolism [*38*] and a decrease in $TaCO_2$ [*25*]. Mean declines in $TaCO_2$ were between approximately 20 and 70%. Decreases in pHa in the intoxicated trout (4.6 to 10.1%) were likely a result of increasing lactic acid levels in the blood. Elevation of blood lactic acid concentrations and the associated depression of pHa [*38,39*] due to muscular activity [*33,34*] and narcosis (anesthesia) [*23,35–37*] has been extensively reported. Rising Hct (11.7 to 69.5%) without a rise in Hb, observed in all cases except for the 4-chloroaniline exposed trout, can be attributed to erythrocyte swelling induced by increasing blood lactic acid concentrations (decreasing pHa) and decreasing TaO_2 [*40,41*]. Swift [*42*] reported increasing Hct in rainbow trout exposed to phenol as well. Similar Hct and Hb trends were also described in *Notopterus notopterus* during acute phenol intoxication [*43*]. Increasing Hct in the 4-chloroaniline intoxicated fish was accompanied by increasing Hb, indicative of either hemoconcentration or an increasing number of erythrocytes. Plasma osmolality, which was also monitored in this study but not used in the multivariate statistical analyses, did not increase significantly in these fish. Thus, the increase in Hct in the 4-chloroaniline exposed trout was likely caused by cell swelling and a release of additional erythrocytes from hematopoietic tissues due to stress and/or hypoxia [*36,37,44*].

Using phenol as a model agent, the reversibility of the visible signs of intoxication and the primary respiratory-cardiovascular responses (V_G, VO_2, U_E, f_V, f_H, f_C) was also established. Trout could be brought to the point of respiratory collapse and then revived by artificial irrigation of the gills. Once revived, the respiratory-cardiovascular parameters and the visual appearance of the trout returned to predose status within 2 h. The reversible nature of the primary intoxication process fulfills an important criterion for defining the observed responses as a narcosis effect [*4*].

Based on these toxic responses, discriminant function analysis (DFA) [*22,24*] was used to assess whether the fish exposed to suspected polar narcotics (Type II narcotics) could be separated into a distinct FATS. The response sets obtained in this study were combined with FATS previously identified for respiratory uncouplers, nonpolar narcotics (Type I narcotics), AChE inhibitors, and respiratory irritants [*23,24*], giving a total of 52 trout exposed to 13 chemicals. The use of DFA resulted in 100% correct classification of five toxicity syndromes (Fig. 5), leading to the establishment of a narcosis Type I and a narcosis Type II FATS. Of the 11 responses measured and used in the DFA, f_C, pHa, V_G, VO_2 and f_H were the best discriminators for defining the five FATS. These variables discriminated 49 of the 52 fish (94%). However, three additional variables, f_V, TaO_2, and Hb, were required to achieve 100% discrimination. The best discriminating variable was f_C, which separated respiratory irritant and Type II narcotic exposed fish from the other three FATS groups because only the former chemicals elicited cough/seizure responses. pHa was the next best discriminating variable, separating Type II narcotic from irritant exposed fish. These two variables alone resulted in the discrimination of 90% of the Type II narcotic exposed fish and 65% of all 52 fish. The third best discriminating variable was V_G, which separated all the Type II narcotic and Type I narcotic exposed fish. The next five variables that were used in the DFA resulted in further separation of the original four FATS, which was previously described by McKim et al. [*24*].

The results of the Bradbury et al. [*11*] and McKim et al. [*23*] studies establish two distinct physiological definitions of narcosis in rainbow trout. These findings are similar to reports in the mammalian literature. Ljublina and Filov [*45*] have studied the narcotic effects of a large number of diverse industrial chemicals in mice and described two classes of narcotics distinguishable by their toxic effects and physicochemical properties. Polar agents were described as having a significant excitatory effect on the nervous system, followed by depression, while nonpolar agents caused only neuronal depression.

Reports of at least two types of narcosis action for industrial chemicals in fish and mammals are consistent with the multidirectional scheme of anesthesia (narcosis of the central nervous system) described by Winters [*46*]. Under this scheme, anesthetics induce an initial excitation and activated EEG pattern (Stage I). Some anesthetics then induce a depressant anesthesia (Stage III) which is characterized by a loss of responsiveness to stimuli, slow regular respiration, and central nervous system (CNS) depression as demonstrated by EEG patterns. Continued CNS depression in Stage III progresses to medullary paralysis (Stage IV), with further reduction

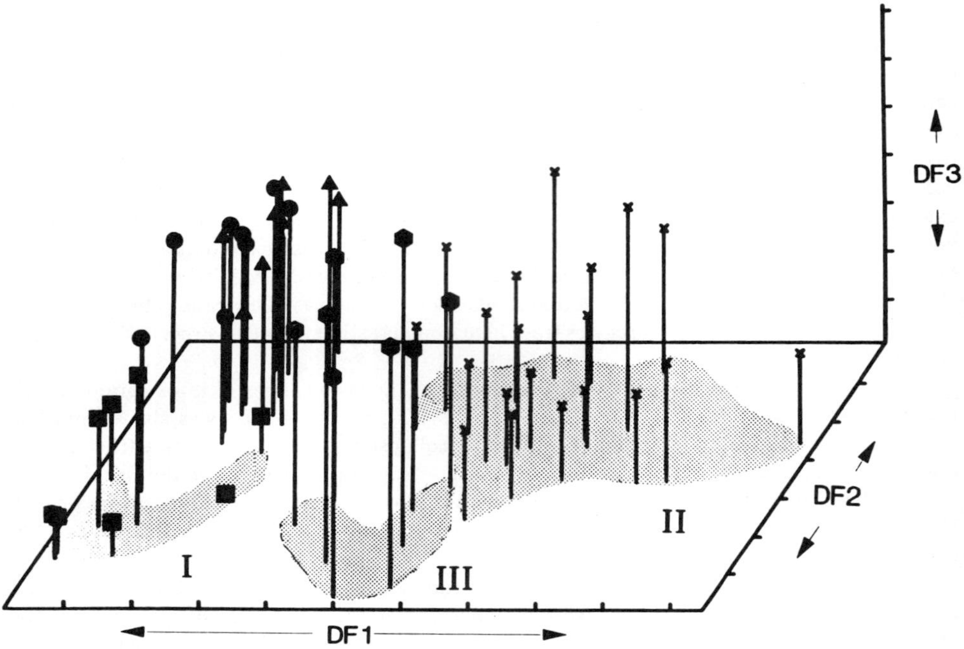

FIG. 5—*Three-dimensional plot of the first three discriminant functions that separate five fish acute toxicity syndromes—narcosis Type I syndrome (I), narcosis Type II syndrome (II), respiratory irritant syndrome (III), AChE inhibitor syndrome, and respiratory uncoupler syndrome. The AChE inhibitor syndrome and the respiratory uncoupler syndrome are correctly resolved as well* [24]; *however, because of the scaling required for the plot the separation of these syndromes is not readily discerned. Each point represents an individual fish exposed to one of the following 13 chemicals: MS-222 and 1-octanol* (■) [23], *pentachlorophenol and 2,4-dinitrophenol* (●) [23], *malathion and carbaryl* (▲) [24], *acrolein and benzaldehyde* (⬢) [24], *and phenol, 2,4-dimethylphenol, aniline, 2-chloroaniline, and 4-chloroaniline (X)* [11] *(modified from* [11]).

in cardiovascular-respiratory function leading to death. Some anesthetics elicit Stage II anesthesia, characterized by hallucinations and catalepsy, prior to Stages III and IV. During cataleptic anesthesia (Stage II-C) animals are unresponsive to stimuli, lose their righting reflex, and exhibit EEG patterns of a higher level of excitation than that noted in Stage I. Cataleptic anesthesia (Stage II-C) can then progress through Stages III and IV. Through Stage III, each of the steps is reversible if the chemical exposure is terminated. The narcosis Type I syndrome in rainbow trout [*23*] is seemingly consistent with depressant anesthesia (that is, Stages I, III, IV). Chemicals that elicit the narcosis Type II syndrome in trout [*11*] can be classified as cataleptic anesthetics. These compounds elicit Stage I excitability followed by increased muscular activity and incoordination that leads to cataleptic behavior and unresponsiveness to outside stimuli (Stage II). General depression (Stage III), medullary paralysis (Stage IV), and death follow.

Based on our underlying assumptions regarding FATS [*22*], the definition of two narcosis syndromes suggests at least two distinct modes of narcotic action. Based on his studies of anesthetic action, Winters [*46*] has also established that there is not a single anesthetic state and has proposed that the search for a single mechanism of anesthetic action is likely not to be successful.

Cellular and Molecular Mechanisms

As outlined in the previous section, the physiological response of fish to intoxication by Type II narcotics is different from that elicited by Type I narcotics. These findings, consistent with results observed in the mammalian literature, suggest that either there is more than one mechanism of narcosis or that the same molecular process occurs at different sites, resulting in different physiological responses.

Studies at the cellular and subcellular level support the likelihood of multiple mechanisms of anesthetic action. As an example, Haydon et al. [*47*] have recently reported differential effects of a diverse group of nonionic hydrophobic narcotics on the sodium channel from squid giant axons. Although all the narcotics tested suppressed sodium currents, hydrocarbons, alcohols, and ketones and carboxylic esters act in different ways to elicit kinetic and steady-state perturbations of voltage-gated currents. Hydrocarbons were found to cause hyperpolarizing shifts in steady-state inactivation and activation, while more polar compounds generally only cause a depolarization in steady-state activation. Haydon et al. [*47*] propose that the noted differences in voltage-gated conductances can be attributed to the polarity and structure (presence of double-bonded oxygen atoms) of the narcotics. The potential for a differential effect of phenol and aniline on the sodium channel is unknown at this time; however, these compounds are currently under study [*48*].

Although the cellular responses of phenol and aniline derivatives have not been deliberately investigated in conjunction with other narcotics, specific effects of convulsant phenolic compounds have been studied extensively and summarized by Kaila [*32*] in a recent review. Phenol and related compounds seemingly facilitate synaptic transmission through a prolongation of the presynaptic action potential. In turn, this effect on action potentials may be caused by an inhibitory action on voltage-dependent potassium conductance. The consistency of these findings with those of other narcotics awaits further study.

No intention is made here to draw a cause and effect relationship between the findings in cellular studies [*32,47*] and specific narcosis QSARs [*1,5,14*] and FATS [*11,23*] described previously. However, the findings by Haydon et al. [*47*] do underscore that by monitoring effects at both the subcellular and organismal level, similar distinctions in narcotic classes are being made and leading to proposals for the possibility of numerous sites or mechanisms of narcotic action.

The suggestions for multiple sites/mechanisms of narcotic action, raised by research on whole-organism and cellular responses, is compatible with current views on the molecular mechanisms of anesthesia, or narcosis of the central nervous system. Recent reviews are available in the literature that address the mechanistic aspects of anesthesia/narcosis [*49,50*] and the reader is referred to these sources for a detailed treatment of the subject.

The Meyer-Overton rule for narcosis (potency correlated with increasing chemical lipophilicity), developed at the turn of the century, is generally treated as the starting point for understanding the physical properties associated with narcosis. However, as presented by Miller [*50*], the concepts put forth by Ferguson [*51*], published in 1939, actually provide a better basis from which to investigate the physical properties of anesthetics and their molecular site(s) and mechanism(s) of action. Ferguson's rule [*51*] basically states that the potency of narcotics is correlated with thermodynamic activity. Thus, although quite varying concentrations of different narcotics may be required to cause a comparable effect, their thermodynamic activity at effect concentrations are essentially the same. For water soluble narcotics, Ferguson's rule can be presented as [*1*]:

$$A_{nar} \approx C_{nar}/C$$

where

C_{nar} = the aqueous concentration of the chemical that elicits narcosis,
C = the water solubility of the chemical, and
A_{nar} = the thermodynamic activity of the chemical at concentrations causing narcosis [*1*].

Essentially, Ferguson's rule implies that the potency of a narcotic is solely a function of a property of the chemical itself, for example, its water solubility or vapor pressure. Assuming a pharmacological/physicochemical basis for narcosis, potency should, of course, be viewed as a function of the strength of the chemicals' interaction at the appropriate receptor sites. However, if Ferguson's rule were followed by all narcotics, it would be meaningless to evaluate key physicochemical properties, as well as possible sites and mechanisms of action [*50*]. Thus, as pointed out by Miller [*50*], hypotheses regarding molecular mechanisms of narcosis must treat the potency of narcotics that are reliably predicted by Ferguson's rule and those which are not. Examples of anesthetics/narcotics that do not meet Ferguson's rule include perfluorohydrocarbons [*50,52*] and phenol and aniline [*51,52*]. In the former case, the perfluoro derivatives have a much lower potency than what would be predicted from their thermodynamic activity. Depending on the biological system being tested, phenol and aniline many times have thermodynamic activities significantly lower (that is, they are more potent) than narcotics that are consistent with Ferguson's rule. Mullins [*52*] has attributed the behavior of these compounds, in part, to their polarity. Interpretation of the response of perflurohydrocarbons, phenol, and aniline in light of Ferguson's rule suggests that these compounds may act differently at the molecular site(s) of action than nonpolar narcotics. The notion that the polarity of a narcotic may influence its molecular effects at a pharmacological site is, of course, compatible with conclusions drawn from physiologically based experiments at the whole organism and cellular/subcellular level, as described previously.

Consistent with the conclusions drawn above, researchers developing hypothesis for the molecular events of anesthesia/narcosis are increasingly recognizing that their models must accommodate the likelihood that there are multiple sites and mechanisms of action [*16*]. Current hypotheses regarding mechanisms of anesthesia can be divided into those that involve anesthetic-protein, anesthetic-lipid, and anesthetic-lipid/protein interactions [*50*].

Using firefly luciferase as a model protein, Franks and Lieb [*53–56*] have provided convincing arguments that the effects of narcotics are a function of absorption into hydrophobic pockets in key target protein(s) (not yet identified) and subsequent alteration of protein function. Conceivably, classes of target proteins, or protein sites, may be involved with varying hydrophobic and hydrogen bonding characteristics. Differential effects on target proteins, or proteins in different neuron classes, could provide an explanation for different narcosis/anesthesia effects at the cellular through organism level.

Anesthetic-lipid hypotheses also attribute anesthetic effects to perturbations in protein function, but as the result of changes in the proteins' lipid environment. Hypotheses supporting anesthetic-lipid interactions can be generally presented as [*50*]

$$E_{50} = C_{50} \cdot P_1 \cdot T_P$$

where

E_{50} = some functional change occurring when 50% of the organisms are anesthetized,
C_{50} = the corresponding anesthetic concentration in the membrane lipid,
P_1 = the perturbation per mole of anesthetic in the membrane bilayer, and
T_P = a function describing the change in protein structure and activity induced by the perturbation P_1.

It is becoming increasingly clear that membrane lipids should not be considered as homogeneous solvents, but as a nonrandom asymmetrical environment [*57*]. Thus, anesthetics with differing polarity may distribute quite differently in membranes, causing varying effects in membrane fluidity, volume, or phase transition and perhaps differing effects on target proteins [*57*]. Haydon et al. [*47*] have attributed the different effects of hydrocarbons and alcohols on sodium channel function (discussed previously in this section) to varying effects on membrane thickness and surface potential. These effects on membrane characteristics seemingly are a function of the narcotic's polarity.

The third class of mechanistic hypotheses, which are less well developed, propose that anesthetics interact at the lipid/protein interface in membranes [*50*]. Haydon et al. [*47*] have proposed that the narcotic action of carboxylic esters may be a result of hydrogen bonding of the xenobiotic with the target protein across the lipid/protein interface.

Certainly mechanistic research into anesthetic/narcotic mode of action is still far from providing a definitive resolution of the issue. However, the various hypotheses proposed do provide, either implicitly or explicitly, the basis for attributing anesthesia/narcosis to more than one mechanistic process.

Conclusions

A critical review of the aquatic toxicology literature and appropriate mammalian literature supports the characterization of a polar narcotic class of industrial organic chemicals. The results of single chemical and joint toxic action studies provide evidence that the potencies and responses elicited by certain esters and weakly acidic or basic phenols and anilines (the latter two groups of compounds being Type II narcotics) are unique to those associated with more nonpolar narcotics (Type I narcotics). These findings establish preliminary evidence for the existence of unique narcosis mechanisms and justify the establishment of specific QSARs. Interestingly, further use of joint toxic action studies has also established that ester narcosis [*5*] is seemingly distinct from the polar narcosis associated with phenol and aniline derivatives [*14*]. An assessment of the physiological responses of fish to intoxication by Type II narcotics provides additional support for the concept that more than one mode of narcotic action is likely. The narcosis Type II FATS is characterized by initial excitation followed by cataleptic behavior, cardiovascular-respiratory depression, medullary paralysis, and death [*11*]. The narcosis Type I FATS is characterized by excitability followed by cardiovascular-respiratory depression, medullary paralysis, and death [*23*]. The suggestions of multiple mechanisms/sites of narcotic action raised by these studies are compatible with recent cellular investigations and current views regarding molecular-based hypotheses of anesthetic action.

It is our belief that reliable QSARs in predictive toxicology must be mechanism specific. Research thus far sufficiently supports the notion that at least two modes of narcotic action are relevant in assessing acute toxicity in fish. In turn, these findings add validity to defining and implementing appropriate narcosis QSAR models. Additional efforts to define those chemicals that are associated with different narcosis syndromes could provide important information about the structural and electronic properties underlying the mechanisms of anesthetic action.

References

[*1*] Veith, G. D., Call, D. J., and Brooke, L. T., *Canadian Journal of Fisheries and Aquatic Sciences,* Vol. 40, 1983, pp. 743–748.
[*2*] Konemann, H., *Toxicology,* Vol. 19, 1981, pp. 209–221.
[*3*] Hermens, J., Canton, J. H., Janssen, P., and de Jong, R., *Aquatic Toxicology,* Vol. 5, 1984, pp. 143–154.
[*4*] Kaufman, R. D., *Anesthesiology,* Vol. 46, 1977, pp. 49–62.
[*5*] Veith, G. D., DeFoe, D., and Knuth, M., *Drug Metabolism Reviews,* Vol. 15, 1984, pp. 1295–1303.
[*6*] Roth, S. H., *Federal Proceedings,* Vol. 39, 1980, pp. 1595–1599.
[*7*] Hesser, C. M., Fragraeus, A., and Adolfson, J., *Undersea Biomedical Research,* Vol. 5, 1978, pp. 391–400.
[*8*] Crisp, D. J., Christie, A. O., and Ghobasky, A. F. A., *Comparative Biochemistry and Physiology,* Vol. 22, 1967, 629–645.

[9] Albert, A., *Selective Toxicity,* 6th ed., John Wiley and Sons, New York, 1979.
[10] Gero, A., "Intimate Study of Drug Action. III. Possible Mechanism of Drug Action," *Drills Pharmacology in Medicine,* J. R. DiPalma, Ed., McGraw Hill, New York, 1965, pp. 47-69.
[11] Bradbury, S. P., Henry, T. R., Niemi, J. G., Carlson, R. W., and Snarski, V., *Environmental Toxicology Chemistry,* Vol. 8, 1989, pp. 247-261.
[12] Schultz, T. W., Holcombe, G. W., and Phipps, G. L., *Ecotoxicology and Environmental Safety,* Vol. 12, 1986, pp. 146-153.
[13] Lipnick, R. L., Bickings, C. K., Johnson, D. E., and Eastmond, D. A., "Comparison of QSAR Predictions with Fish Toxicity Screening Data for 110 phenols," *Aquatic Toxicology and Hazard Assessment (Eighth Symposium), ASTM STP 891,* R. C. Bahner and D. J. Hansen, Eds., American Society for Testing and Materials, Philadelphia, 1985, pp. 153-176.
[14] Veith, G. D. and Broderius, S. J., "Structure-Toxicity Relationships for Industrial Chemicals Causing Type (II) Narcosis Syndrome," *QSAR in Environmental Toxicology—II,* K. L. E. Kaiser, Ed., Elsevier, Amsterdam, 1987, pp. 385-391.
[15] Hermens, J., Leeuwangh, P., and Musch, A., *Ecotoxicology and Environmental Safety,* Vol. 8, 1984, pp. 388-394.
[16] Miller, K. W., *Trends in Neuroscience,* February, 1986, pp. 49-51.
[17] Hermens, J. L. M., *Pesticide Science,* Vol. 17, 1986, pp. 287-296.
[18] Broderius, S. J. and Kahl, M., *Aquatic Toxicology,* Vol. 6, 1985, pp. 307-322.
[19] Konemann, H., *Toxicology,* Vol. 19, 1981, pp. 229-238.
[20] Schultz, T. W., *Ecotoxicology and Environmental Safety,* Vol. 14, 1987, pp. 178-183.
[21] Kamlet, M. J., Doherty, R. M., Taft, R. W., Abraham, M. H., Veith, G. D., and Abraham, D. J., *Environmental Science and Technology,* Vol. 21, 1987, pp. 149-155.
[22] McKim, J. M., Bradbury, S. P., and Niemi, G. J., *Environmental Health Perspectives,* Vol. 71, 1987, pp. 171-186.
[23] McKim, J. M., Schmieder, P. K., Carlson, R. W., and Hunt, E. P.; *Environmental Toxicology and Chemistry,* Vol. 6, 1987, pp. 295-312.
[24] McKim, J. M., Schmieder, P. K., Niemi, G. J., Carlson, R. W., and Henry, T. R., *Environmental Toxicology and Chemistry,* Vol. 6, 1987, pp. 313-328.
[25] Bradbury, S. P., McKim, J. M., and Coats, J. R., *Pesticide Biochemistry and Physiology,* Vol. 27, 1987, pp. 275-288.
[26] Deichmann, W. B. and Keplinger, M. L., "Phenols and Phenolic Compounds," *Patty's Industrial Hygiene and Toxicology,* G. D. Clayton and F. E. Clayton, Eds., John Wiley and Sons, New York, 1981, pp. 2567-2627.
[27] European Inland Fisheries Advisory Commission, *Water Research,* Vol. 7, 1973, pp. 929-941.
[28] Troma, F. B., *Notulae Naturae,* Vol. 269, 1955, pp. 1-10.
[29] Jones, J. R. E., *Journal of Experimental Biology,* Vol. 28, 1951, pp. 261-270.
[30] Krajnovic-Ozretic, M. and Ozretic, B., *Bulletin of Environmental Contamination and Toxicology,* Vol. 40, 1988, pp. 23-29.
[31] Razani, H., Nanba, K., and Murachi, S., *Bulletin of the Japanese Society of Scientific Fisheries,* Vol. 52, 1986, pp. 743-748.
[32] Kaila, K., *Comparative Biochemistry and Physiology,* Vol. 73C, 1982, pp. 231-241.
[33] Black, E. C., *Canadian Journal of Fisheries and Aquatic Sciences,* Vol. 12, 1955, pp. 917-929.
[34] Black, E. C., *Canadian Journal of Fisheries and Aquatic Sciences,* Vol. 14, 1957, pp. 117-134.
[35] Nieminen, M., Laitinen, M., and Pasenen, P., *Comparative Biochemistry and Physiology,* Vol. 73C, 1982, pp. 271-276.
[36] Kirk, W. L., *Transactions of the American Fisheries Society,* Vol. 3, 1974, pp. 553-600.
[37] Houston, A. H., Madden, J. A., Woods, R. J., and Miles, H. M., *Canadian Journal Fisheries and Aquatic Sciences,* Vol. 28, 1971, pp. 625-633.
[38] Hille, S., *Journal of Fish Biology,* Vol. 20, 1982, pp. 535-569.
[39] Sovio, A., Nyholm, K., and Huhti, M. *Journal of Fish Biology,* Vol. 10, 1977, pp. 91-101.
[40] Sovio, A. and Nikinmaa, M., "The Swelling of Erythrocytes in Relation to the Oxygen Affinity of the Blood of the Rainbow Trout, *Salmo gairdneri* Richardson," *Stress and Fish,* A. D. Pickering, Ed., Academic Press Inc., London, 1981, pp. 103-109.
[41] Sovio, A., Westman, K., and Nyholm, K., *Journal of Fish Biology,* Vol. 6, 1974, pp. 763-769.
[42] Swift, D. J., *Journal Fish Biology,* Vol. 19, 1981, pp. 45-61.
[43] Verma, S. R., Rani, S., and Dalela, R. C., *Journal of Environmental Science and Health,* Vol. B16, 1981, pp. 273-282.
[44] Perry, J. W. and Conway, M. W., *Comparative Biochemistry and Physiology,* Vol. 56C, 1977, pp. 123-126.
[45] Ljublina, E. I. and Filov, V. A., "Chemical Structure, Physical and Chemical Properties and Biological Activity," *Methods Used in the USSR for Establishing Biologically Safe Levels of Toxic Substances,* World Health Organization, Geneva, 1975, pp. 19-44.

[46] Winters, W. D., *Annual Review of Pharmacology and Toxicology*, Vol. 16, 1976, pp. 413–426.
[47] Haydon, D. A., Elliott, J. R., Hendry, B. M., and Urban, B. W., "The Action of Non-ionic Anesthetic Substances on Voltage-Gated Ion Conductances in Squid Gaint Axons, *Molecular and Cellular Mechanisms of Anesthesia*, S. H. Roth and K. W. Miller, Eds., Plenum Publishing Corp., New York, 1986, pp. 267–277.
[48] Elliott, J. R., "The Perturbation of Voltage-Gated Sodium and Potassium Channels by Anaesthetic Molecules," paper presented at the Royal Society of Chemistry Conference on *The Interaction of Biologically Active Molecules with Membranes*, Cambridge, UK, 8–10 Sept. 1987.
[49] *Molecular and Cellular Mechanisms of Anesthetics*, Roth, S. H. and Miller, K. W., Eds., Plenum Publishing Corp., New York, 1986.
[50] Miller, K. W., *International Review of Neurobiology*, Vol. 27, 1985, pp. 1–61.
[51] Ferguson, J., *Royal Society of London. Proceedings. Series B, Biological Sciences*, Vol. 127, 1939, pp. 387–404.
[52] Mullins, L. J., *Chemical Reviews*, Vol. 54, 1954, pp. 289–323.
[53] Franks, N. P. and Lieb, W. R., "Do Protein/Anesthetic Interactions Underlie the Mechanism of General Anesthesia?," *Molecular and Cellular Mechanisms of Anesthesia*, R. S. Roth and K. W. Miller, Eds., Plenum Publishing Corp., New York, 1986, pp. 319–329.
[54] Franks, N. P. and Lieb, W. R., *Archives of Toxicology*, Vol. Suppl. 9, 1986, pp. 27–37.
[55] Franks, N. P. and Lieb, W. R., *Nature*, Vol. 316, 1985, pp. 349–351.
[56] Franks, N. P. and Lieb, W. R., *Nature*, Vol. 310, 1984, pp. 599–601.
[57] Miller, K. W. and Roth, S. H., "Inside the 'Black Box'," *Molecular and Cellular Mechanisms of Anesthetics*, S. H. Roth and K. W. Miller, Eds., Plenum Publishing Corp., New York, 1986, pp. 261–266.

Wayne G. Landis,[1] Robert S. Anderson,[1] Nancy A. Chester,[1] H. Dupont Durst,[1] Mark V. Haley,[1] Dennis W. Johnson,[1] and Renee M. Tauber[1]

The Organophosphate Acid Anhydrases of the Protozoan, *Tetrahymena thermophila,* and the Clam, *Rangia cuneata*

REFERENCE: Landis, W. G., Anderson, R. S., Chester, N. A., Durst, H. D., Haley, M. V., Johnson, D. W., and Tauber, R. M., **"The Organophosphate Acid Anhydrases of the Protozoan, *Tetrahymena thermophila,* and the Clam, *Rangia cuneata,*"** *Aquatic Toxicology and Hazard Assessment: 12th Volume, ASTM STP 1027,* U. M. Cowgill and L. R. Williams, Eds., American Society for Testing and Materials, Philadelphia, 1989, pp. 74–81.

ABSTRACT: Organophosphate acid anhydrases (OPA anhydrases) are a diverse and widespread group of enzymes that are able to hydrolyze the potent organophosphate acetylcholinesterase inhibitors 0,0-diisopropylphosphofluoridate (DFP) and 0-1,2,2-trimethylpropylmethylphosphonofluoridate (soman). Five enzymatic activities have been identified in the ubiquitous freshwater protozoan, *Tetrahymena thermophila.* In some cases the activities share characteristics of both the squid-type and Mazur-type DFPases. A chromogenic substrate related to parathion, 4-nitrophenylethylphenyl(phenyl)phosphinate (NPEPP), is hydrolyzed by Tetrahymena DFPase-1 and DFPase-2 and by the Mazur-type DFPase from hog kidney. The close DFP analog N,N′-diisopropyldiamidofluoridate (Mipafox) is a potent inhibitor of the Tetrahymena-DFPases. OPA anhydrase activities have also been found in the clam, *Rangia cuneata.* Three groups of activity can be identified at molecular weights of 20 000 to 30 000 daltons (D), 45 000 to 50 000 D, and 70 000 to 100 000 D. The higher molecular weight activity resembles a Mazur-type DFPase in that it hydrolyzes soman faster than DFP, is Mn^{2+} stimulated, and hydrolyzes NPEPP. The lower molecular weight activity is in the range of the squid-type DFPase. An enzymatic activity that hydrolyzes Mipafox is apparent at a molecular weight of 130 000 D. A hypothesis is presented that the wide variety of OPA anhydrases is in part due to the need for the metabolism of a wide variety of naturally occurring organophosphates and halogenated organics.

KEY WORDS: OPA anhydrase, organophosphate degradation, *Tetrahymena thermophila, Rangia cuneata,* biodegradation

One of the critical needs in ecotoxicology is an improved understanding of the mode of action and fate of an xenobiotic. The ability of an organism to compensate for a toxic insult is in part due to its ability to transform the material into a less toxic form. A classic case of such an enzymatic system are the phylogenetically widespread glutathione *S*-transferases that have the ability to conjugate and subsequently detoxify a wide variety of substrates [*1–4*]. In addition to the understanding of toxic insults on a molecular level, enzymes are important in the metabolic pathways that mineralize xenobiotics. This report concentrates on the comparatively poorly understood group of enzymes, the DFPases, that have recently been renamed the organophos-

[1]Environmental Toxicology Branch, Toxicology Division, Chemical Research, Development and Engineering Center, Aberdeen Proving Grounds, MD 21010-5423.

phate acid anhydrases (OPA anhydrases) to reflect the broad range of potential substrates (First DFPase Workshop held at the Marine Biological Laboratory, Woods Hole, MA, June 1987). The OPA anhydrases have been known since the 1946 report of Mazur [*5*], and extensive research was conducted with sources from mammalian tissues and numerous strains of bacteria [*6*]. Hoskin continued this research [*7–9*], discovering a second major type of OPA anhydrase in squid (squid type), performing purifications of the squid and other activities, and demonstrating that squid-type OPA anhydrase bound to an agarose support was capable of breaking down several common organofluorophosphate acetylcholinesterase inhibitors. Hoskin's research has led to recent revival of the investigation into the nature and role of the OPA anhydrases.

Enzymes capable of hydrolyzing diisopropylfluorophosphate (DFP), related acetylcholinesterase inhibitors such as 0-1,2,2-trimethylpropylmethylphosphonofluoridate (soman), and 3,3-dimethylbutylmethylphosphonofluoridate (Dimebu) have been reported in the tissues of many animals [*9,10*]. (Structures of the acetylcholinesterase inhibitors discussed in this paper can be found in Fig. 1.) Previously, two categories of organofluorophosphate hydrolyzing OPA anhydrases were recognized [*9*]. Typically, Mazur type is stimulated by manganous ion, hydrolyzes soman faster than DFP, is usually found to be dimeric with a molecular weight of approximately 62 000 D [*11*], and is competitively or reversibly inhibited by N,N′-diisopropylphosphorodiamidofluoridate (Mipafox) [*12*]. Mipafox is a strong structural analog to DFP. Activities resembling Mazur type are found in hog kidney and *Escherichia coli.* Squid type [*9*]

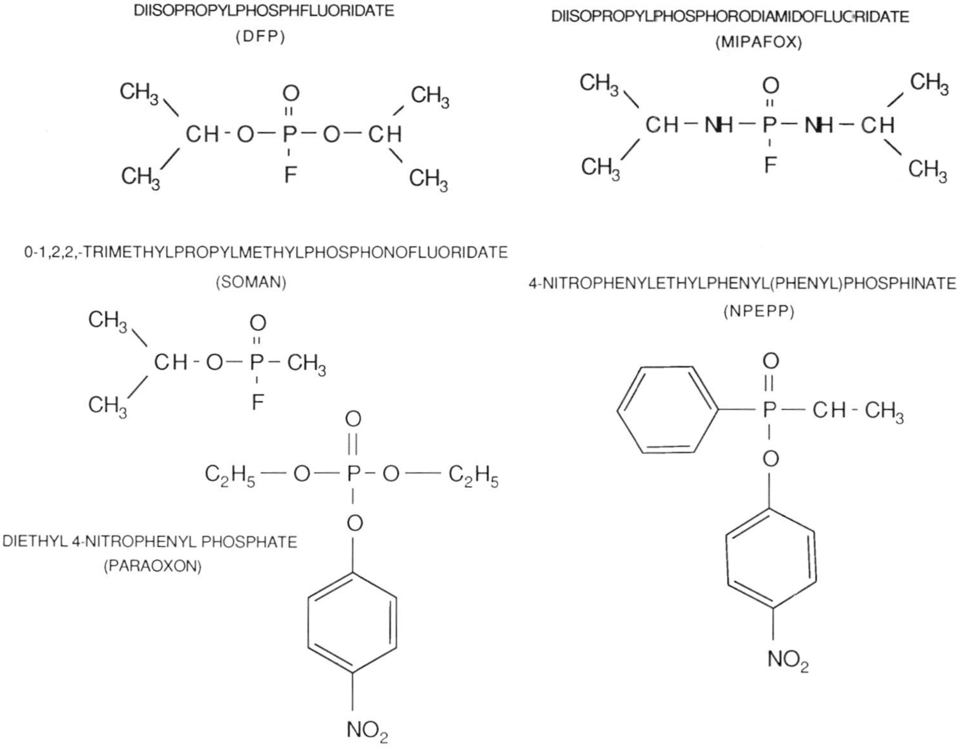

FIG. 1—*Structures of the acetylcholinesterase inhibitors and substrates for OPA anhydrases. Mipafox is a potent inhibitor of many of the OPA anhydrases. Paraoxon, although structurally similar, is not hydrolyzed by the Tetrahymena-OPA anhydrase.*

hydrolyzes DFP faster than soman, is stable, has a molecular weight of approximately 26 000 D, manganous ion has no effect or is slightly inhibitory, experiences no inhibition of DFP hydrolysis by Mipafox [*12*], and is present in the optic ganglia, giant nerve axon, hepatopancreas, and salivary gland of cephalopods [*13*].

In this report we will introduce and contrast the recently discovered enzymatic activities in the freshwater protozoan, *Tetrahymena thermophila* [*14*], and the estuarine clam, *Rangia cuneata* [*15*]. Several activities have been discovered in each organism, several of which do not fit the classical definitions of squid- or clam-type OPA anhydrases. The widespread nature and variety of the OPA anhydrases points to an important role in organismal metabolism. We hypothesize that the OPA anhydrases are important in the metabolism of naturally occurring organophosphates and in the detoxification of natural toxicants.

Methods and Materials

In this section we will briefly describe and review the methods used in the comparison of the OPA anhydrases of *Tetrahymena* and *R. cuneata*.

Initial culture and extraction: The culture, extraction, and purification of the Tetrahymena-OPA anhydrases has been reported previously [*14*] and is routinely produced by Bruce C. Harper (Dugway Proving Grounds, Dugway, Utah). Briefly, aerated 200-L cultures of *T. thermophila* stock BIV using a proteous peptone media were established. Cells were extracted using a continuous flow centrifuge, disrupted by freeze-thawing, and the protein concentrated by ammonium sulfate precipitation. A Sephadex G-100 column was used for desalting and additional purification. Lyophilyzed extract could be stored at 5°C with no loss of activity. In the studies described below, a solution of reconstituted lyophilized material of 15 to 20-fold purification and containing all five OPA anhydrases (partially purified extract) was used as the starting material.

R. cuneata were collected in sediment samples from the Chesapeake Bay near Aberdeen Proving Ground, Maryland and held in ambient water at 2°C for several hours prior to processing. Homogenates, 33% clam tissue by weight in Hank's balanced salt solution (HBSS) (Difco), were prepared from digestive gland or visceral mass pooled from 30 to 50 individuals and stored at 4°C. Prior to testing, the tissue was diluted with Hoskin's buffer (400 mM KCl, 50 mM NaCl, and 5 mM 1,3-bis [tris hydroxymethylmethylamino] propane (bis-tris-propane) in glass distilled water, pH 7.2) and centrifuged at 1500 rpm for 20 min, yielding supernatant to be used for the initial assays and for placement on the column.

OPA anhydrase assay: Hoskin's buffer was used in the assays, and all chemicals were reagent grade. Three substrate solutions were originally used to conduct the initial assays and to examine possible inhibition by Mipafox: 3.0×10^{-3} M DFP, 3.0×10^{-3} M DFP with 3.0×10^{-3} M Mipafox, and 3.0×10^{-3} M Mipafox. Two substrate solutions were used to examine the activity of the fractions: 3.0×10^{-3} M DFP and 3.0×10^{-3} M Mipafox. In each case the solution was added to a 30°C water-jacketed glass beaker equipped with a magnetic stirrer. Substrate hydrolysis was quantified using a fluoride electrode attached to an Orion 901 microprocessor ionanalyzer that recorded fluoride concentration at 1-min intervals. The meter was calibrated using appropriate NaF standards. After recording spontaneous hydrolysis, enzyme-mediated hydrolysis was measured following the addition of 100-μL crude extract. Determination of protein concentration of fractions was measured using the Biorad [*16*] or Whitaker and Granum ultraviolet absorption method [*17*]. Reaction rates were calculated using the Apple-compatible program, DFPASE2 (listing available upon request).

Column chromatography: The molecular sizing column was composed of Pharmacia Sephacryl S-300 superfine grade suspended in Hoskin-buffer solution maintained at 4°C. The column measured 2.5 cm in diameter by 75 cm. Flow rate was 0.43 to 0.46 mL/min. The column was calibrated using blue dextran (2 000 000 D), aldolase (158 000 D), and bovine serum albumin

(67 000 D). Fractions of 1 to 1.5 mL were collected. All fractions were stored at 5°C after collection, and assays to determine OPA anhydrase activity and Mipafox inhibition were completed within 72 h of collection. Samples were then stored at −70°C.

Results

The characteristics of the OPA anhydrase activity of the partially purified homogenates from the Tetrahymena and clam are presented in Table 1. The clam data reflect that of the clam digestive gland; other tissues have different characteristics but have not been as extensively examined. One of the striking aspects is the similarity of the activities at this stage. Both sets of activities would be classified as classical Mazur type since both hydrolyze NPEPP, hydrolyze soman faster than DFP, and are stimulated by the addition of manganous ion. The K_m, V_{max}, and the temperature and pH ranges are very close. The major difference is the lack of inhibition of Mipafox on the OPA anhydrase activity of the clam. Figure 2 quantifies this difference, comparing the inhibition of hog kidney OPA anhydrase (positive control) to the Tetrahymena and clam activities. The activities from the hog kidney and Tetrahymena are sharply reduced; however, the OPA anhydrase activity from the clam is not affected or even appears to be slightly higher than DFP alone.

The known characteristics of the activities that can be separated using a molecular sizing column appear in Table 2 with a comparison to typical squid-type and Mazur-type OPA anhydrases. The names of the various OPA anhydrases are prefaced by a *Tt* for *T. thermophila* and *Rc* for *R. cuneata* activities. Several activities are apparent in each organism examined. In *T. thermophila,* five activities have been identified, ranging in molecular weights from 96 000 to 67 000 D. DFP and soman hydrolysis occur with a ratio of approximately 1.0. NPEPP is hydrolyzed by *Tt* DFPase-1 and *Tt* DFPase-2. Soman hydrolysis by *Tt* DFPase-4 and *Tt* DFPase-5 is increased 8 to 30 fold upon the addition of Mn^{2+}. Mipafox is a reversible inhibitor of the Tetrahymena activities examined.

The digestive gland homogenate from *R. cuneata* contains several activities. The low molecular weight activities, *Rc* OPA-1 and *Rc* OPA-2, are not inhibited by Mipafox and are not stimulated by Mn^{2+}. *Rc* OPA-3 has a molecular weight in the range of the Tetrahymena OPA anhydrases, 73 000 to 82 000 D. *Rc* OPA-4 hydrolyzes the inhibitor Mipafox, is not stimulated by Mn^{2+}, and is relatively large, between 105 000 to 138 000 D.

TABLE 1—*Comparison of the OPA anhydrase activities of homogenates from* T. thermophila *and* R. cuneata.

Homogenates	*T. thermophila*	*R. cuneata*
Substrates hydrolyzed		
DFP	+	+
soman	+	+
NPEPP	+	+
Mipafox	+	+
Kinetics (DFP substrate)		
K_m (40°C, mM)	1.17	1.52
V_{max} (40°C, μmol/g protein/min)	43.70	43.24
temperature range	25 to 55	25 to 50
pH range	6 to 9	6 to 9
Soman/DFP ratio	2	15
Mn^{2+} stimulation	+	+
Mipafox inhibition	+	−

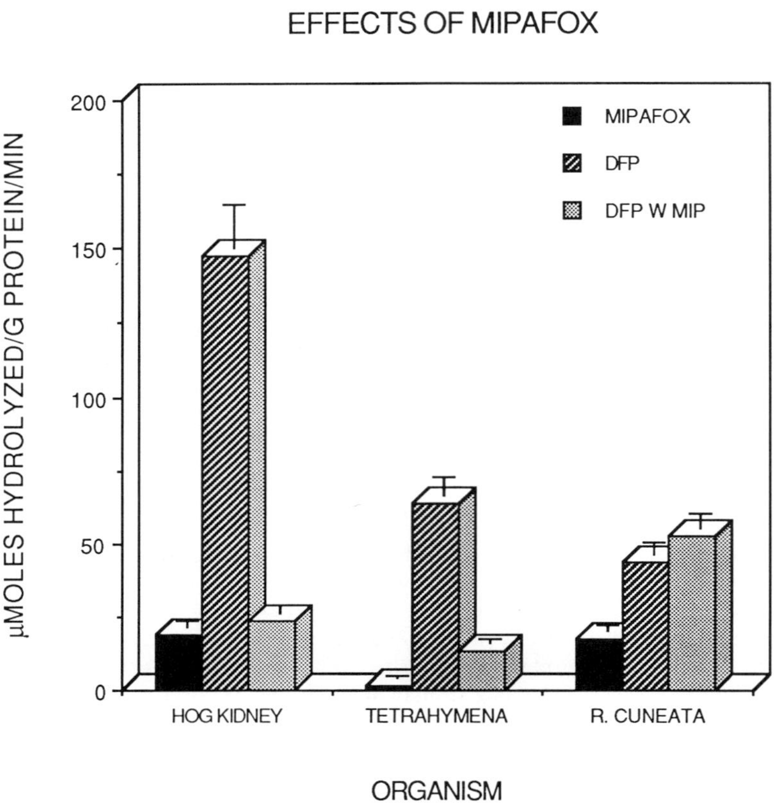

FIG. 2—*Comparison of hog kidney, Tetrahymena, and clam OPA anhydrase activities in their ability to hydrolyze DFP, Mipafox, and the combination DFP and Mipafox. As is readily apparent, the clam material does not appear to be inhibited by Mipafox.*

Discussion

Two dramatically different aquatic organisms, a protozoan and a clam, contain similar activities against compounds that are at most 40 years old and that are relatively rare in the environment. The production of organofluorophosphates began in World War 2, so the widespread phylogenetic distribution of these enzymes is more likely due to natural selection for the metabolism of natural substances. We would like to compare these activities to each other and to those found in a variety of other diverse organisms, propose a hypothesis for the evolution of these xenobiotic hydrolyzing enzymes for which there is no known substrate, and to comment on why these studies are crucial to the continued development of ecological toxicology and hazard assessment.

Diversity of the OPA Anhydrases

Although possessing very similar kinetics and characteristics in homogenate form, the OPA anhydrase activities of *T. thermophila* and *R. cuneata* are markedly different upon even a simple purification. The Tetrahymena OPA anhydrases fall within a narrow range, from 96 000 to 67 000 D. *Tt* DFPase-4 is the activity with the dramatic soman hydrolyzing activity upon the addition of 1 mM Mn^{2+}. *Tt* DFPase-5 also shows a substantial increase in soman hydrolysis

TABLE 2—*Comparison of the identified enzymatic activities of* T. thermophila *and* R. cuneata *with squid- and Mazur-type enzymes.*

		Substrate Hydrolysis						
Characteristic	mw[1]	DFP	Soman	NPEPP	Mipafox	Soman/DFP Ratio	Mn^{2+} Stimulation DFP or Soman	Mipafox Inhibition
ACTIVITY								
T. thermophila								
Tt DFPase-1	80 000	+	+	+	...	1.12	2.5–4.0	+
Tt DFPase-2	75 000	+	+	+	...	1.26	2.0	+
Tt DFPase-3	72 000	+	+	...	...	0.71	1.7–2.5	+
Tt DFPase-4	96 000	+	+	...	nt[2]	1.95	17–30	nt
Tt DFPase-5	67 000	+	+	...	nt	0.34	8	nt
R. cuneata								
Rc OPA-1	19–35 000	+	nt	...	...	nt	1	...
Rc OPA-2	38–43 000	+	nt	...	...	nt	1	...
Rc OPA-3[3]	73–82 000	+	nt	nt	+	nt	1	...
Rc OPA-4	82–138 000	...	nt	...	++	nt	nt	nt
Squid-type OPA anhydrase (*Loligo pealei*)	23–26 000	+	+		...	0.25	1	...
Mazur-type OPA anhydrase (hog kidney)	62–66 000	+	+		+	6.54	2	+

[1]Molecular weight as determined by column chromotography.
[2]Nt = not tested.
[3]Tentative designation.

upon the addition of Mn^{2+} but is of lower molecular weight and is poor at DFP hydrolysis. In contrast, *R. cuneata* has a low molecular weight activity *Rc* OPA-1 that is very similar in the characteristics examined to the classical squid-type OPA anhydrase. The clam also has a Mipafox hydrolyzing activity that hydrolyzes Mipafox faster than DFP.

The diversity of the OPA anhydrases in the clam and Tetrahymena attests to the variety of enzymes that exists. In addition to the molluscs and protozoa, numerous bacteria, fungi, and mammals contain enzymes that have activities that classify them as OPA anhydrases. The diversity of the OPA anhydrases is puzzling. Zech and Wigand [*18*] isolated two OPA anhydrase enzymes from *E. coli,* one of which hydrolyzed only DFP and the other paraoxon. Wild has isolated a paraoxon hydrolyzing OPA anhydrase that is capable of hydrolyzing DFP (James R. Wild, Texas A&M, personnel communication). The OPA anhydrases from *T. thermophila* and hog kidney do not hydrolyze paraoxon but do hydrolyze the related compound, NPEPP. In some instances, the OPA anhydrase hydrolyzes only soman and not DFP. The phylogeny of these diverse enzymes is unclear. Some of the diversity may be artifactual, the result of repeatable modification of the enzymes by proteases and handling during the extraction and purification of the activities. The consistency of the molecular weights, number of forms, and enzymatic characteristics bolster the conclusion that these results are not artifactual. Fundamental differences in substrate specificities and other characteristics support an explanation that real differences do exist. Perhaps we are looking at enzymes that are ancient and related phylogenetically, but have diverged significantly. The diversity of the OPA anhydrases makes the classification of squid type and Mazur type a very broad outline that describes two ancient lineages of OPA anhydrases, with perhaps a third being the Mipafox hydrolyzing activity discovered in *R. cuneata.* Antigenic relationships, sequencing, and the determination of the three-dimensional structure will help answer the question of the classification of the OPA anhydrases.

Natural Role of the OPA Anhydrases

An enzymatic activity that is so widespread must be important to the cellular metabolism or to the survival of the organism. However, the natural substrate and role(s) of the OPA anhydrases is unknown. Correlations to isethionate [*19*], pyruvate, and squid neurotoxin [*9,20*] exist, but no cause-effect relationship has been found to exist. However, many types of naturally occurring organophosphates have been identified from a variety of sources [*21*]. The amino acid analog, 2-aminoethylphosphonic acid (AEP), is synthesized by Tetrahymena [*22*]. Phosphonates are found free in cells, incorporated into glycerophosphonolipids, sphingophosphonolipids, and phosphonoproteins [*23–25*]. The linkage of AEP to phosphonolipids appears to be covalent, although this has not been conclusively demonstrated [*21,26*]. We [*10,14,27*] have previously suggested that the OPA anhydrases are parts of a metabolic system handling the various organophosphonates incorporated into the cellular matrix and encountered in food sources. That hypothesis must be expanded to include the role that some OPA anhydrases may also be important in dehalogenation of naturally occurring halogenated metabolites.

Not only are naturally synthesized organophosphates present in the biome, but halogenated organics also occur. Neidleman and Geigert [*28*] review the evidence that halometabolites are used as chemical defense in marine organisms. The organisms involved range from a green algae, *Avrainvillea longicalulis,* to the Nudibranch mollusc, *Diaulula sandiegensis.* One of the more interesting speculations of Neidleman and Geigert is the role that toxic halogenated compounds may play in prey-predator interactions. Perhaps the synthesis of active halogenated compounds is sufficiently damaging to a predator to reduce the efficiency of the predation or destroy the organism. Competitive relationships among microorganisms may also be mediated by the production of halogenated organics. Detection of the very low concentrations of these molecules appears to be the major stumbling block in further elucidating the role of halogenated organics in predator-prey and competitive relationships.

The discovery and elucidation of the variety of detoxification systems, their evolution, and role has a direct bearing on the prediction and amelioration of the impacts of xenobiotics on aquatic ecosystems. Apparently a wide variety of enzymatic systems exists spread over phylogenetic boundaries capable of degrading compounds that have only been recently produced by man. The evolutionary pressures that selected for many these pathways are unlikely to have been artificially synthesized compounds. Secondary metabolites, including those used to mediate competitive interactions, may have played an important role in the evolution of the detoxification machinery. Sufficient selective pressure may have been caused by toxic materials that exist in very small mean concentrations, too low to readily identify. The recent discovery of the numbers of halogenated organics is an example, and this is probably the tip of the iceberg. An understanding of the natural role and evolution of these systems will give ecotoxicologists the ability to predict the effects and fate of xenobiotics more accurately and from a mechanistic instead of an empirical viewpoint. Our increased capacity to understand the mechanisms and interactions of biodegradations should also enhance the effort currently being made to engineer biological systems to degrade *in situ* hazardous wastes.

References

[1] Stenersen, J. and Oien, N., *Comparative Biochemistry and Physiology*, Vol. 69, 1981, pp. 243–252.
[2] Awasthi, Y. C., Dao, D. D., and Saneto, R. P., *Biochemical Journal*, Vol. 191, 1980, pp. 1–10.
[3] Jakoby, W. B., *Advances in Enzymology*, Vol. 46, 1978, pp. 383–414.
[4] LeBlanc, G. A. and Cochrane, B. J., *Comparative Biochemistry and Physiology*, Vol. 82C, 1985, pp. 37–42.
[5] Mazur, A. J., *Biological Chemistry*, Vol. 164, 1946, pp. 271–289.
[6] Mounter, L. A. in *Hanbuch der Experimentellen Pharmakologie: Cholinesterases and Anticholinesterase Agents*, G. B. Kolle, Ed., Springer-Verlag, Berlin, 1963, pp. 486–504.
[7] Hoskin, F. C. G. and Trick, G. S., *Canadian Journal Biochemistry and Physiology*, Vol. 3, 1955, pp. 963–969.
[8] Hoskin, F. C. G. and Long, R., *Journal Archives Biochemistry and Physiology*, Vol. 150, 1972, pp. 548–555.
[9] Hoskin, F. C. G., Kirkish, M. A., and Steinman, K. E., *Fundamental and Applied Toxicology*, Vol. 4, 1984, pp. 5165–5172.
[10] Landis, W. G., Savage, R. E., Jr., and Hoskin, F. C. G., *Journal of Protozoology*, Vol. 32, 1985, pp. 517–519.
[11] Storkebaum, W. and Witzel, H., *Forschungsber. Landes Nordrhein-Westfalen*, Vol. 2523, 1975, pp. 1–23.
[12] Hoskin, F. C. G., *Biochemical Pharmacology*, Vol. 34, 1985, pp. 2069–2072.
[13] Hoskin, F. C. G. and Prusch, R. D., *Comparative Biochemistry and Physiology*, Vol. 75C, 1983, pp. 17–20.
[14] Landis, W. G., Durst, H. D., Savage, R. E., Jr., Haley, D. M., Haley, M. V., and Johnson, D. W., *Journal of Applied Toxicology*, Vol. 7, 1987, pp. 35–41.
[15] Anderson, R. S., Durst, H. D., and Landis, W. G., *Comparative Biochemistry and Physiology*, in press.
[16] Bradford, M., *Anales of Biochemistry*, Vol. 72, 1976, pp. 248–254.
[17] Suelter, C. H., *A Practical Guide to Enzymology*, John Wiley and Sons, New York, 1985, p. 30.
[18] Zech, R. and Wigand, K. D., *Experientia*, Vol. 31, 1975, pp. 157–158.
[19] Hoskin, F. C. G. and Long, R. J., *Journal of Neurochemistry*, Vol. 20, 1972, pp. 1317–1327.
[20] Hoskin, F. C. G. and Rouch, A. H., *Science*, Vol. 215, 1982, pp. 1255–1257.
[21] Hilderbrand, R. L. and Henderson, T. O. in *The Role of Phosphonates in Living Systems*, R. L. Hilderbrand, Ed., CRC Press, Boca Raton, Florida, 1983, pp. 5–30.
[22] Rosenburg, H., *Nature* (Lond), Vol. 203, 1964, pp. 299–300.
[23] Kitteridge, J. S. and Roberts, E., *Science*, Vol. 164, 1969, pp. 37–42.
[24] Rouser, G., Kritchevsky, G., Heller, D., and Lieber, E., *Journal, American Oil Chemical Society*, Vol. 40, 1963, pp. 425–545.
[25] Simon, G. and Rouser, G., *Lipids*, Vol. 2, 1967, pp. 55–59.
[26] Quin, L. D. and Shelburn, F. A., *Journal of Marine Research*, Vol. 27, 1969, pp. 73–84.
[27] Landis, W. G., Haley, M. V., and Johnson, D. W., *Journal of Protozoology*, Vol. 32, 1986, pp. 216–218.
[28] Neidleman, S. L. and Geigert, J., *Biohalogenation: Principals, Basic Roles and Applications*, Ellis Horwood Limited, West Sussex, England, 1986, pp. 128–130.

Gregory J. Thorpe[1] *and John D. Costlow*[2]

The Relation of the Acute (96-h) Uptake and Subcellular Distribution of Cadmium and Zinc to Cadmium Toxicity in Larvae of *Rhithropanopeus harrisii* and *Palaemonetes pugio*

REFERENCE: Thorpe, G. J. and Costlow, J. D., **"The Relation of the Acute (96-h) Uptake and Subcellular Distribution of Cadmium and Zinc to Cadmium Toxicity in Larvae of *Rhithropanopeus harrisii* and *Palaemonetes pugio*,"** *Aquatic Toxicology and Hazard Assessment: 12th Volume, ASTM STP 1027,* U. M. Cowgill and L. R. Williams, Eds., American Society for Testing and Materials, Philadelphia, 1989, pp. 82–94.

ABSTRACT: Newly hatched larvae of the mud crab *R. harrisii* and the grass shrimp *P. pugio* were exposed for 96 h to a range of cadmium free-ion concentrations using metal-buffered seawater. Survival, growth, and molting frequency were monitored. Results show that mud crab larvae are more sensitive to cadmium exposure than grass shrimp larvae. Mud crab larvae experienced decreased survival, reduced growth, and decreased molting frequency at cadmium concentrations in which grass shrimp had high survival and no reduction in growth or molting frequency. In a second series of experiments, newly hatched larvae of each species were exposed for 96 h to a matrix of three cadmium free-ion concentrations and a constant zinc concentration using cadmium-109 and zinc-65 as tracers. Crab larvae showed greater total uptake of cadmium as well as higher accumulation of cadmium in the cytosolic compartment compared to shrimp larvae at cadmium concentrations in which crab larvae suffered adverse toxic effects, as described above. Zinc uptake by both species decreased with increasing cadmium concentration at lower cadmium exposures; however, mean total zinc accumulation was greatest at the highest cadmium concentration. Crab larvae also accumulated more zinc than shrimp larvae at all cadmium concentrations, with a higher percentage of zinc accumulation occurring in the cytosolic compartment as well. These interspecific differences in the metabolism of zinc and cadmium may explain the greater sensitivity of crab larvae to cadmium compared to shrimp larvae.

KEY WORDS: cadmium, zinc, accumulation, subcellular distribution, larvae, mud crab, *Rhithropanopeus harrisii,* grass shrimp, *Palaemonetes pugio*

In recent years it has become evident that to predict the toxic effects of heavy metals on biological systems at the organismal level, it is essential to understand the physiological mechanisms controlling the metabolism of these metals at the subcellular level. Differences in metabolic strategies can affect the distribution and turnover of metals in the animal's tissues, thus affecting the potential toxicity of the metals in a dynamic biological system. The toxicity of the metal can be expressed as decreased survival or changes in growth rates or molting frequencies.

A considerable amount of information is available on the acute toxicity of cadmium to estuarine organisms [*1,2*]. Even though it has been demonstrated repeatedly that early life-history

[1]Environmental scientist, NC Division of Environmental Management, Water Quality Section, Raleigh, NC 27611.

[2]Professor of zoology and director, Duke University Marine Laboratory, Beaufort, NC 28516.

stages are generally more sensitive to environmental stressors than are mature stages [3-6], a vast majority of the acute toxicity studies have been done on adult animals. Of the 17 96-h LC_{50}s listed for marine crustaceans by McClurg [7], for example, only two are for immature stages. Furthermore, no studies have been done in which the sublethal effects of acute cadmium toxicity on early life-history stages of estuarine crustaceans are examined in relation to the metabolism of the metal. An investigation of this type would provide insight into the mechanism of toxicity of the metal. In an effort to bridge this gap, the acute toxicity assays and metal uptake studies described in this paper were conducted.

Recent studies of metal toxicity in aquatic organisms have consistently indicated that the bioavailability of metals is dependent upon their chemical form. The toxicity of metals has been shown to be related to the metal free-ion concentration and not to the total dissolved metal concentration [8-10]. The chemical form of a particular metal in water depends on a complex interaction of a variety of factors, including the chemical properties of the metal and the chemical composition of the water [11]. The metal free-ion concentration is determined by the total dissolved metal concentration and by the extent of metal-ligand complexation. The metal free-ion concentration in water can be controlled by the use of metal-ion buffers which "provide a controlled source of free metal ions in a manner similar to the regulation of hydrogen ion concentration by pH buffers" [12]. The relationship of pM, cadmium free-ion concentration, and total dissolved cadmium concentration is presented in Table 1.

Methods and Materials

Toxicity Assays

Experimental Animals—Ovigerous female *R. harrisii* were collected from the Neuse River near New Bern, North Carolina. Crabs were acclimated in a controlled-environment cabinet to experimental conditions of 20 0/00 seawater, 30°C, and a 14/10 light/dark cycle. The crabs were provided with filtered seawater at the same temperature and salinity daily. Crabs were maintained under these conditions until sufficient numbers of larvae were obtained for conducting an experiment from hatches occurring in a single night in a 4-h period from approximately 2000 to 2400 h. Ovigerous *P. pugio* were collected from Calico Creek, Morehead City, North Carolina, and were acclimated and maintained in the lab in the same manner as described for *R. harrisii.*

Experimental Protocol—Immediately following larval release, all adult females were gently removed from the culture bowls. The larvae were provided with clean seawater at the same temperature and salinity, fed newly hatched *Artemia salina* nauplii (Great Salt Lake, Utah), and returned to the culture cabinet.

Approximately 12 h posthatch, the larvae were pooled and randomly distributed to 8.0-cm-diameter glass culture bowls at a density of ten larvae per bowl. Each experimental series con-

TABLE 1—*The relationship of pM to cadmium free-ion concentration and total metal concentration.*

$p[CD]^{2+}$	μmol/L, Free Ion	$p[Cd]_{total}$	μmol/L, Total	μg/L[a], Free Ion
11.8	2×10^{-6}	7.5	0.03	0.0002
10.8	2×10^{-5}	6.5	0.3	0.002
9.8	2×10^{-4}	5.5	3.0	0.02
8.8	2×10^{-3}	4.5	30.0	0.2
7.8	2×10^{-2}	3.5	300.0	2.0

[a]Cadmium concentrations expressed as micrograms/liter (μg/L), or parts per billion, as the free ion.

sisted of five metal ion concentrations, five bowls per concentration, and each series was replicated five times for a total of 250 larvae per concentration.

A metal-buffered seawater system was used as the experimental media in this work. The metal-chelating ligand used was EDTA (ethylenediaminetetraacetic acid). Metal free-ion concentrations were computed from complex, computer-based metal ion-chelate equilibrium equations developed by W. G. Sunda et al. [*8*]. Metal free-ion concentrations are expressed as pM (or pCd, for cadmium), which is the negative logarithm of the free-ion concentration and is analogous to pH, as previously indicated. A pM of 11, for example, is equivalent to a metal free-ion concentration of 1×10^{-11} *M*, and as the metal free-ion increases the pM will decrease.

Known additions of chelate, essential trace elements, and cadmium (as cadmium chloride) were added to seawater of the same temperature, salinity, and pH as that in which the larvae had hatched (30°C; 20 0/00; pH 7.6). All experimental media contained 1×10^{-5} *M* EDTA, 4×10^{-5} *M* NaOH, 1×10^{-7} *M* Fe, 6×10^{-8} *M* Cu, 2.4×10^{-8} *M* Co, 3.4×10^{-8} *M* Mn, and 5.8×10^{-6} *M* Zn. The respective pM's of the metal free-ion concentrations for the essential trace elements are pCu = 13.8, pCo = 11.8, pMn = 9.2, and pZn = 9.6. Larvae were exposed to five concentrations of cadmium free-ion. The respective pM's of the cadmium free-ion concentrations to which the larvae were exposed are pCd = 11.8, 10.8, 9.8, 8.8, and 7.8. A cadmium-free (pCd = 0.0) metal-buffered seawater control series in which the free-ion concentrations of the essential trace elements copper, manganese, cobalt, zinc, and iron were held constant utilizing 10^{-5} *M* EDTA was run concurrently.

Larvae were transferred daily to clean culture bowls containing fresh experimental media and few newly hatched *Artemia*. All larvae were sampled at 96-h exposure. Survival and molting frequency data were recorded daily. Growth (expressed as dry weight per larva) was determined by dividing the dry weight of one replicate of larvae by the number of larvae in that replicate.

Metal Uptake Studies

Experimental Animals—Ovigerous female *R. harrisii* and *P. pugio* were collected and acclimated as previously described.

Experimental Protocol—Approximately 12 h posthatch, larvae were pooled and randomly distributed to 13-cm-diameter glass culture bowls containing 300-mL of experimental media at an approximate density of 100 larvae per bowl. Each experimental series consisted of three metal ion concentrations, three bowls per concentration, and a control series as described above. Radioisotopes of cadmium (^{109}Cd) and zinc (^{65}Zn) were incorporated into the metal-buffered experimental media at a specific activity of 1 microCurie per millilitre (μCi/mL). All experimental media were as described above.

Survival was estimated by subtracting the number of dead larvae removed during the experiment from the number of larvae initially distributed to the culture bowl. Molting frequency was monitored daily by stereomicroscopic examination of three subsamples of approximately ten larvae each from each culture bowl.

Larvae were sampled at 96-h exposure. Growth of larvae was not measured as dry weight per larva in these experiments. Samples of larvae were frozen at −80°C, but not lyophilized, in order to maintain the biochemical integrity of subcellular metal-chelating ligands. The total soluble protein content of each sample was determined by the Bradford [*13*] standardized dye-binding protein assay (Bio-Rad Protein Assay). Dry weight is directly proportional to total soluble protein content. All metal concentrations in larvae are expressed as nanomoles metal per gram soluble protein (nmol/g).

Samples of larvae which had been frozen at −80°C for at least 12 h were homogenized and fractionated by differential centrifugation into three subcellular fractions, according to the following protocol.

Each sample of larvae was homogenized for 1 min in 1 mL of TRIS buffer (0.05 *M*; pH 7.4). A 50-μL subsample of the homogenate was frozen for later gamma counting and protein determination. The remaining homogenate was centrifuged at 200 g for 10 min. The 200 g ("slow-spin" or "cellular debris") pellet was frozen at −80°C for later gamma counting. The supernate was transferred by pasteur pipet to a clean centrifuge tube and centrifuged at 100 000 g for 60 min. Following the second centrifugation, the 100 000 g supernate ("cytosol") and ("fast-spin" or "microsomal") pellet were transferred to separate clean test tubes and stored at −80°C.

Accumulation of cadmium and zinc radioisotopes was determined in homogenate subsamples and all subcellular fractions by gamma emission counting and was corrected for radioactive decay and counting efficiency. All radioisotope counts were converted to nanomoles metal per gram soluble protein based on the amount of protein present in the homogenate subsample.

Data were analyzed statistically by Multiple Comparison of All Means (MCHETV) tests, according to the methods of Sokal and Rolf [*14*]. All tests of statistical significance are at least to the 0.05 probability level.

Results

Toxicity Assays

Survival—*R. harrisii* larval survival decreased with increasing cadmium ion concentration, that is, as pCd decreases (Fig. 1). Larval survival at pCd 9.8, 8.8, and 7.8 was significantly different from larvae of the control series. Although larval survival at pCd 9.8 is significantly lower than that of control larvae, it does not differ significantly from survival at pCd 11.8 or

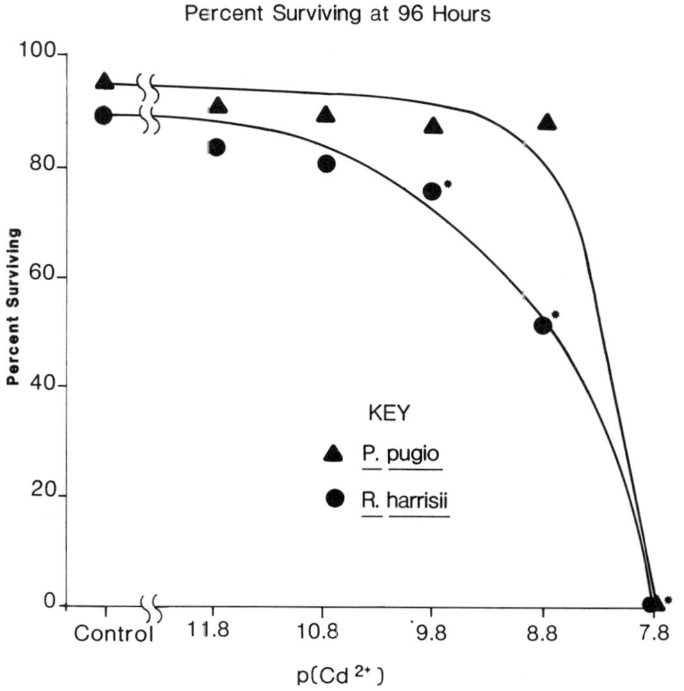

FIG. 1—*Relationship between 96-h survival and the negative logarithm of the free-ion concentration* (*p*[Cd^{2+}]) *for larvae of* R. harrisii *and* P. pugio. *Asterisks (*) indicate significant differences between treatment and control series.*

10.8. Survival at pCd 8.8 and 7.8 differs significantly from all other intraspecific groups tested (MCHETV; degrees of freedom (*df*) = 143; *Q* = 4.2).

In general, survival of *P. pugio* larvae decreased slightly as cadmium ion concentration increased (pCd decreased). The only statistically significant difference in survival of *P. pugio* larvae occurred at pCd 7.8 (MCHETV; *df* = 143; *Q* = 4.2). Percent survival at this concentration was 0.0% compared to 89.0% at pCd 8.8, delineating a critical range of cadmium ion exposure concentrations (2.0 E − 08 *M* to 2.0 E − 09 *M* cadmium free-ion). A similar critical range of exposure concentrations is shown by *R. harrisii* zoeae, though the initial decrease in *R. harrisii* survival occurs at a lower concentration and covers a wider range of concentrations (pCd 9.8 to 7.8) than for *P. pugio* larvae.

Survival of *R. harrisii* larvae differed significantly from that of *P. pugio* larvae at cadmium ion concentrations of pCd 9.8 and 8.8, two orders of magnitude lower than the lowest concentration at which significant mortality of grass shrimp larvae occurred (Fig. 1). No significant differences in larval survival occurred between the two species at pCd 10.8, 11.8, or controls (MCHETV; *df* = 143; *Q* = 4.2). No larvae of either species survived at pCd 7.8. The estimated 96-h LC_{50}s for *R. harrisii* and *P. pugio* are 0.2 μg/l and 1.1 μg/l cadmium, respectively.

Growth—*R. harrisii* larval growth, as measured by total dry weight at 96-h postexposure, generally decreased as cadmium ion concentration increased (Fig. 2). Analysis of nonsignificant subsets (MCHETV; *df* = 143; *Q* = 4.7) indicates that dry weight of *R. harrisii* larvae is significantly reduced at pCd 10.8, 9.8, and 8.8, compared to growth of control larvae. Dry weight of larvae exposed to pCd 9.8 is not significantly different from that at pCd 10.8, and pCd 10.8 results are not significantly different from those at pCd 11.8. Although dry weights of *P. pugio*

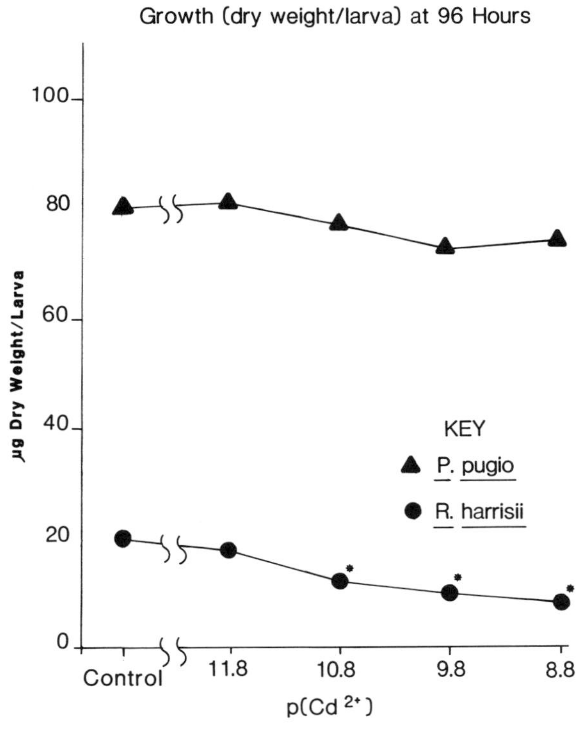

FIG. 2—*Relationship between growth at 96-h exposure and p[Cd^{2+}]. Asterisks (*) indicate significant differences between treatment and control series.*

larvae tended to be slightly reduced at higher cadmium ion concentrations, no statistically significant differences occurred. A comparison of interspecific differences in dry weight is not appropriate since *P. pugio* larvae are naturally larger than *R. harrisii* larvae.

Molting Frequency—Molting frequency of *R. harrisii* larvae was significantly reduced at pCd 9.8 and 8.8 (MCHETV; $df = 143$; $Q = 4.6$; see Fig. 3). First-stage zoeae comprised a substantially higher proportion of survivors at the higher cadmium concentrations. Many of the first-stage zoeae surviving in the higher cadmium concentrations 96-h postexposure showed limb deterioration and gross microbial growth on the carapace. These larvae were still active, and many were observed struggling to extract themselves from partially molted exuviae.

Although *P. pugio* larvae have a different molting frequency than *R. harrisii,* all *P. pugio* larvae surviving to 96-h molted to Stage III.

Metal Uptake Studies

General—Estimates of survival, growth, and molting frequency of larvae in this set of experiments were similar to those found for larvae from the toxicity assays described above (Table 2).

Cadmium Accumulation—Cadmium accumulation by mud crab larvae is shown in Fig. 4. Mean total cadmium accumulation increased 88-fold over a 100-fold increase in cadmium free ion in the medium, from pCd 11.8 to 9.8. Mean cadmium uptake in the cytosol > cellular debris pellet > microsomal pellet, at all exposures.

Cadmium accumulation in mud crab larvae increased at a higher rate in the cytosol than in the cellular debris pellet (Fig. 4). Cadmium uptake increased approximately 100-fold in the

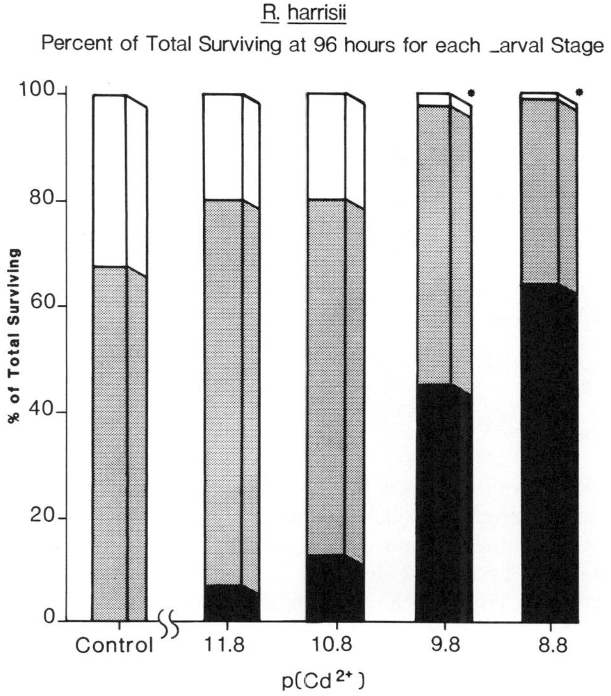

FIG. 3—*Relationship between p[Cd^{2+}] and the percent composition of each larval stage, as the percent of the total number of larvae surviving in each treatment at 96 h postexposure time, for larvae of* R. harrisii. *Zoeal stages are represented by shading of the bars as follows: Stage I, dark shading; Stage II, light shading; Stage III, white. Asterisks (*) indicate significant differences between treatment and control series.*

TABLE 2—*Estimates of survival, molting frequency, and growth at 96 h for larvae from metal-uptake experiments.*

Test Species	$p[Cd]^{2+}$	Percent Survival	Molting[a] Frequency	Growth, μg Dry Weight/Larva
R. harrisii	0.0	85	0,60,40	15
	11.8	85	10,70,20	15
	10.8	80	10,70,20	10
	9.8	75	30,60,10	10
P. pugio	0.0	85	0,0,100	80
	11.8	90	0,0,100	80
	10.8	80	0,0,100	80
	9.8	90	0,0,100	70

[a]Estimated molting frequency as percent of total number of surviving larvae at each larval stage, reported for larval Stages I, II, and III, respectively.

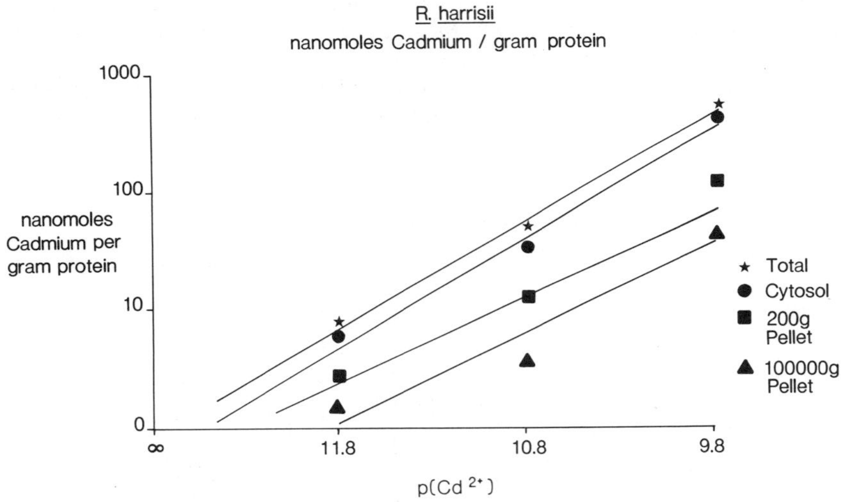

FIG. 4—*Cadmium accumulation by* R. harrisii *larvae at 96-h exposure. (See figure key and text for explanation.)*

cytosol over the exposure range of pCd 11.8 to 9.8 and 50-fold in the cellular debris pellet. Cadmium uptake in the microsomal pellet was similar to that in the cytosol, increasing approximately 115-fold over the cadmium exposure range of pCd 11.8 to 9.8.

Cadmium accumulation by *P. pugio* followed the same pattern as that seen for *R. harrisii.* Cadmium accumulation increased with increasing cadmium ion concentration. Cadmium uptake was greatest in the cytosolic compartment, followed by the cellular debris pellet and the microsomal pellet (Fig. 5).

The rate of total cadmium uptake was similar in the grass shrimp larvae to that in the mud crab larvae. Mean total uptake of cadmium increased approximately 85-fold in grass shrimp larvae from a cadmium exposure of pCd 11.8 to 9.8, compared to an 88-fold increase for mud crab larvae. Mud crab larvae accumulated about four times more total cadmium than grass shrimp larvae.

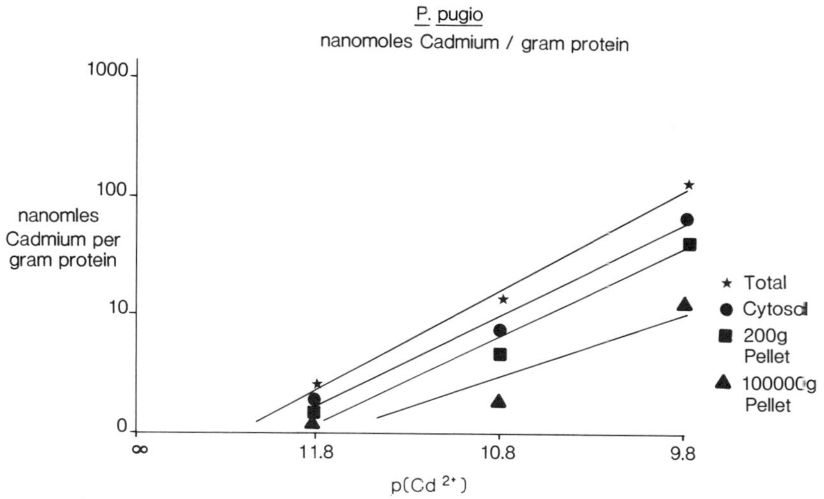

FIG. 5—*Cadmium accumulation by* P. pugio *larvae at 96-h exposure. (See figure key and text for explanation.)*

Mean cadmium accumulation rates in the subcellular compartments of grass shrimp larvae followed different relative patterns compared to those four mud crab larvae. Whereas cadmium uptake by mud crab larvae increased about 100-fold in the cytosol versus 50-fold in the cellular debris pellet, cadmium uptake in grass shrimp larvae cytosol increased 70-fold versus 130-fold for the debris pellet. In other words, mud crab cytosolic cadmium accumulation was about twice that in the debris pellet, while accumulation in the cytosol of grass shrimp larvae was about half that in the debris pellet.

The percent of total accumulated cadmium that was associated with the cytosol of mud crab larvae in this study increased from about 65% at pCd 11.8 and 10.8 to 75% at pCd 9.8. This increase was accompanied by a similar decrease in the percent total cadmium accumulated in the cellular debris pellet. Grass shrimp larvae showed an opposing trend with the percent cytosolic cadmium decreasing at the highest cadmium exposure concentration. Cadmium uptake into the cytosol of grass shrimp larvae comprised 60% of total cadmium uptake at pCd 11.8 and 10.8, and only about 50% at pCd 9.8. Although 50% of the cadmium accumulated by grass shrimp larvae at pCd 9.8 is present in the cytosol, mud crab larvae cytosol accumulated six times more cadmium at pCd 9.8 than did grass shrimp larvae cytosol. In addition, mud crab larvae had a higher percent of total accumulated cadmium in the cytosol than did grass shrimp larvae, ranging from 5% more cytosolic cadmium at pCd 11.8 and 10.8 to 25% more at pCd 9.8, due to relative shifts in the subcellular distribution of cadmium.

The average rate of cadmium accumulation in the grass shrimp larvae microsomal pellet was about 88-fold, compared to a 115-fold increase in cadmium uptake in the microsomal pellets of mud crab larvae. The latter uptake rate is similar to that for cadmium accumulation in the debris pellet of grass shrimp larvae.

Zinc Accumulation—Total zinc accumulation in mud crab larvae tended to decrease with increasing cadmium concentration from a pCd of 0.0 (control) to pCd 10.8, then increase at pCd 9.8 (Fig. 6). Zinc accumulation was highly variable, but no significant changes in total zinc uptake can be seen. Zinc accumulation in the cytosol and cellular debris pellet followed a similar pattern, whereas zinc uptake in the microsomal pellet increased with increasing cadmium concentration over the entire exposure range. Zinc accumulation in the subcellular compart-

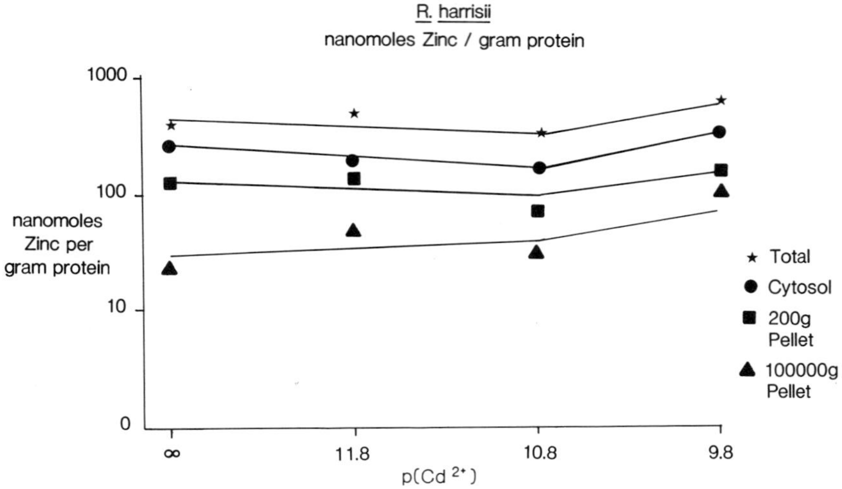

FIG. 6—*Zinc accumulation by* R. harrisii *larvae at 96-h exposure. (See figure key and text for explanation.)*

ments was greatest in the cytosol followed by the cellular debris pellet and the microsomal pellet, at all exposure concentrations.

Total zinc accumulation in the grass shrimp larvae showed a trend similar to that seen in mud crab larvae (see Fig. 7). Zinc uptake decreased slightly with increasing cadmium concentration from pCd 0.0 to 10.8, then increased at pCd 9.8. Zinc uptake rates were similar in the cytosol and cellular debris pellets of grass shrimp larvae at all cadmium concentrations. Zinc uptake was lowest in the microsomal pellets of grass shrimp larvae, at all cadmium concentrations, compared to the other subcellular compartments. Zinc accumulation in all the subcellular com-

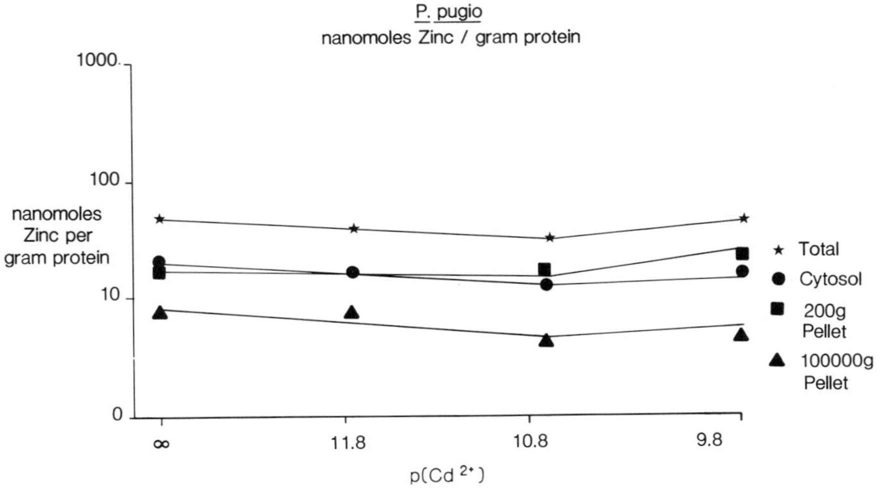

FIG. 7—*Zinc accumulation by* P. pugio *larvae at 96-h exposure. (See figure key and text for explanation.)*

partments in grass shrimp larvae followed the same general pattern as that shown for total zinc uptake.

Discussion

Acute Toxicity

Several previous investigators have determined the acute toxicity of cadmium to larvae of estuarine or marine organisms [*15–17*]. Numerous studies have demonstrated the importance of chemical speciation to biological availability and toxicity of metals [*8, 9, 18*]. These studies have shown that the metal free-ion concentration is the best measure of the most biologically available form of the metal to which the animals were exposed. Nonetheless, the majority of research investigations of metal toxicity to aquatic organisms continue to be based upon total metal concentration. For the sake of comparison, however, it can be assumed that the calculated cadmium free-ion concentrations (in parts per billion, or μg/L, cadmium as the free ion; see Table 1) utilized in this study are equivalent to the metal exposure concentrations reported in the literature by other investigators.

Table 3 lists LC_{50} values determined by other researchers for various crustaceans. No data relating acute toxicity of cadmium to larval shrimp survival, other than that reported in this study, have been found. These comparisons must be made with caution. In addition to differences in experimental temperatures and salinities, both of which are known to influence metal toxicity [*19*], the extent of metal chelation is unknown for the studies cited above, making estimates of biologically available metal concentrations difficult. These differences in experimental protocol could readily account for the variation observed in the magnitude of the acute toxicity responses given.

R. harrisii larvae showed a general trend of decreased growth at increased cadmium concentrations. Reduced growth is a commonly observed response to environmental stress and has been reported previously for larval crustaceans in response to heavy metal exposure [*20, 21*]. In their study of the response of *R. harrisii* larvae to free cupric ion exposure, Sanders et al. [*21*] observed reduced growth at high cupric ion concentrations, as well as a significant increase in larval growth relative to larvae of control series at cupric ion concentrations lower than those at which reduced growth occurred. Significant reductions in larval *R. harrisii* growth relative to control larvae occurred in larvae surviving to 96-h at a pCd (10.8) one order of magnitude lower than that concentration in which significant reduction in survival was observed. This response lends support to the generally accepted view that effects on growth are a more sensitive indicator of environmental stress than effects on survival. Growth may be a more sensitive indicator of

TABLE 3—*Acute cadmium toxicity values for larvae of marine and estuarine crustaceans reported in the literature.*

Test Species	Time, h	LC_{50}, μg/L	Reference Number
Paragrapsus quadridentatus	96	490	[*5*]
Homarus americanus	96	78	[*42*]
Cancer magister	96	247	[*43*]
Cancer irroratus	96	250	[*22*]
Tisbe holothuriae	48	645	[*44*]
Portunis pelagicus	48	380	[*45*]
P. sanguinalentus	48	250	[*45*]
Charybdis feriatus	48	250	[*45*]
Eurytemora affinis	96	150	[*12*]

stress than molting frequency as well, since significant reductions in the number of *R. harrisii* larvae molting to zoeal Stage III by 96-h exposure-time occurred only at pCd 9.8 and 8.8.

Other investigators have reported sublethal effects of heavy metals, including cadmium, on the growth, development, and survival of various crustaceans [*17,22–24*]. These "effects" are indicative of a hierarchy of responses to environmental stress which become measurable, or observable, when the organism(s) can no longer maintain the physiological integrity of the homeostatic feedback mechanisms controlling these processes (such as growth and molting frequency). The mechanisms controlling growth and molting in crustaceans are obviously closely integrated and regulated throughout the molt cycle, and a stressor affecting one process is likely to have an impact on the other.

Metal Uptake Studies

These uptake studies demonstrate that mud crab and grass shrimp larvae accumulate substantial quantities of cadmium and zinc during the early stages of larval development. Cadmium accumulation increased in direct proportion to external cadmium concentrations, indicating a lack of internal regulation of total cadmium uptake by either species. White and Rainbow [*25*] reported that adult shrimp *(Palaemon elegans)* did not regulate cadmium uptake over four orders of magnitude increase in cadmium exposure concentration. Other investigators have also concluded that cadmium uptake was not regulated by decapod crustaceans [*26–28*].

The four-fold difference in total cadmium accumulation observed between mud crab and grass shrimp larvae may be due to various factors, such as differences in integument permeability, molting frequency, or differences in the capacity to bind or store metals. Molting has been shown to be a mechanism of metal loss in crustaceans. White and Rainbow [*29*] found that adult shrimp *(P. elegans)* lost up to 18% of accumulated cadmium upon molting. Losses of up to 20 and 50% of accumulated cadmium have been reported for the euphausid *Meganyctiphanes norvegica* [*30*] and the shrimp *Lysmata seticaudata* [*31*], respectively. Grass shrimp larvae in this study all molted to the third larval stage just prior to sampling, whereas samples of cadmium-exposed mud crab larvae were comprised of from 10 to 20% Stage III, 60 to 70% Stage II, and 10 to 30% Stage I zoeae (see Table 2 and Fig. 3). Due to the lower relative number of Stage III larvae in the mud crab samples, less accumulated cadmium may have been lost just prior to sampling. In addition, the Stage I and Stage II larvae may have continued to accumulate more cadmium over longer intermolt periods, relative to Stage II grass shrimp larvae, contributing further to their net total cadmium accumulation.

The greater cytosolic cadmium accumulation shown by mud crab larvae could be due to greater induction of metallothionein during the 96-h exposure. Metallothionein is a low molecular weight, cysteine-rich metal binding protein completely lacking aromatic amino acids [*32*]. The synthesis of metallothionein has been shown to be induced by copper, zinc, and cadmium [*33,34*], as well as various other stressors [*35*]. Sanders et al. [*21*] studied the cytosolic distribution of copper in *R. harrisii* larvae exposed to a range of free copper ion activities. They found that most of the cytosolic copper (47 to 87%) was associated with metallothionein and that copper thionein increased at higher cupric ion activities.

Zinc Accumulation

Zinc accumulation by mud crab larvae was highly variable. The variability may be due to a combination of variable molting frequencies, individual genetic variability in zinc metabolic requirements, as well as interference with zinc metabolism by cadmium.

White and Rainbow [*36,37*] have observed high variation in zinc uptake among individual adult shrimp *(P. elegans)*. They could not determine any significant correlation between zinc uptake and size or sex of the organism, but speculated that the observed variation might be due to individual genetic variability. Variable losses of zinc at the time of ecdysis have been reported [*36–39*].

The interaction of zinc and cadmium has been studied previously. Ahsannulah et al. [*40*] observed that zinc uptake was enhanced by the presence of cadmium in the external medium. Most studies of cadmium/zinc interactions have been concerned with the effect of zinc exposure concentrations on cadmium toxicity, but not cadmium accumulation. In general, it has been shown that increased zinc concentrations reduced the toxicity of cadmium [*41*].

The accumulation of zinc by grass shrimp larvae decreased with increasing cadmium concentration up to pCd 10.8, then increased at pCd 9.8 (Fig. 7). This pattern of zinc uptake was observed for all subcellular compartments in grass shrimp larvae. This pattern may be indicative of interference with zinc metabolism by cadmium. At low concentrations, cadmium may be competing with zinc for metal uptake or binding sites, while at high cadmium concentrations metal regulation may begin to break down in general, allowing an increased influx of both the essential and nonessential metal. Mud crab larvae accumulated ten times or more zinc than grass shrimp larvae at all cadmium concentrations. Similar results were observed for all subcellular compartments. Mechanisms similar to those responsible for the higher cadmium accumulation observed in mud crab larvae are likely to be involved in the higher zinc accumulation by mud crab larvae as well.

Although cations such as cadmium and zinc cannot form complexes with one another, the retention of one is often increased by the presence of the other. One explanation is the induction of metallothionein by increased exposure to metals, as described for cadmium, and the subsequent binding of both metals by the newly synthesized metallothionein. The high variability in zinc uptake observed in mud crab larvae but not in grass shrimp larvae may be a further indication of mud crab larvae susceptibility to cadmium interference with zinc metabolism via this mechanism.

Summary

Evidence has been presented from this study which correlates interspecific differences in the toxicity of cadmium with the uptake and subcellular distribution of cadmium as well as effects on the metabolism of zinc. Crab larvae showed greater total uptake of cadmium, as well as higher accumulation of cadmium in the cytosolic compartment, compared to shrimp larvae, at cadmium concentrations in which crab larvae suffered adverse toxic effects, as described above. Zinc uptake by both species tended to decrease with increasing cadmium concentration at low cadmium exposures; however, the mean total zinc accumulation was greatest at the highest cadmium concentration. Crab larvae also accumulated more zinc than shrimp larvae at all cadmium concentrations, with a higher percentage of zinc accumulation occurring in the cytosolic compartment, as well. These interspecific differences in the metabolism of zinc and cadmium may explain the greater sensitivity of crab larvae to cadmium compared to shrimp larvae.

Acknowledgment

This research was supported by the Office of Health and Environmental Research, Ecological Division of the Department of Energy.

References

[*1*] Bryan, G. W. in *Marine Ecology*, Vol. 5, O. Kinne, Ed., Wiley, New York, 1984, pp. 1289–1431.
[*2*] "Ambient Water Quality Criteria for Cadmium—1984," U.S. Environmental Protection Agency, Washington, DC, 1985.
[*3*] Connor, P. M., *Marine Pollution Biology*, Vol. 3, No. 12, 1972, pp. 190–192.

[4] DeCoursey, P. J. and Vernberg, W. B., *Oikos,* Vol. 23, 1972, pp. 241-247.
[5] Ahsannulah, M. and Arnott, G. H., *Australian Journal of Marine and Freshwater Research,* Vol. 29, 1978, pp. 1-8.
[6] Castagna, A., Sinatra, F., Scalia, M., and Capodicasa, V., *Marine Biology,* Vol. 64, 1981, pp. 285-289.
[7] McClurg, T. P., *Water SA,* Vol. 10, 1984, pp. 40-45.
[8] Sunda, W. G., Engel, D. W., and Thuotte, R. M., *Environmental Science and Technology,* Vol. 12, No. 4, 1978, pp. 409-413.
[9] Sunda, W. G. and Guillard, R. R., *Journal of Marine Research,* Vol. 34, 1976, pp. 511-529.
[10] Zamuda, C. D. and Sunda, W. G., *Marine Biology,* Vol. 66, 1982, pp. 77-82.
[11] Stumm, W. and Morgan, J. J., *Aquatic Chemistry,* Wiley-Interscience, New York, 1970.
[12] Perrin, D. D. and Dempsey, B., *Buffers for pH and Metal Ion Control,* Chapman and Hall, London, 1974.
[13] Bradford, M., *Analytical Biochemistry,* Vol. 72, 1976, pp. 248-254.
[14] Sokal, R. R. and Rolf, F. J., *Biometry,* Freeman and Co., San Francisco, 1981.
[15] Reish, D. J., Martin, J. M., Piltz, F. M., and Word, J. Q., *Water Research,* Vol. 10, 1976, pp. 299-302.
[16] Watling, H. R., *Bulletin of Environmental Contamination and Toxicology,* Vol. 28, 1982, pp. 195-201.
[17] Sullivan, B. K., Buskey, E., Miller, D. C., and Ritacco, P. J., *Marine Biology,* Vol. 77, 1983, pp. 299-306.
[18] Borgman, U., *Canadian Journal of Fisheries and Aquatic Sciences,* Vol. 38, 1981, pp. 999-1002.
[19] McLusky, D., Bryant, V., and Campbell, R., *Oceanography and Marine Biology Annual Review,* Vol. 24, 1986, pp. 481-520.
[20] Brown, B. and Ahsanullah, M., *Marine Pollution Biology,* Vol. 2, 1971, pp. 182-187.
[21] Sanders, B. M., Jenkins, K. D., Sunda, W. G., and Costlow, J. D., *Science,* Vol. 222, 1983, pp. 53-55.
[22] Benijts-Claus, C. and Benijts, P. in *Sublethal Effects of Toxic Chemicals on Aquatic Animals,* J. H. Koeman and J. J. Strik, Eds., Elsevier, Amsterdam, 1975, pp. 43-52.
[23] Rosenberg, R. and Costlow, J. D., *Marine Biology,* Vol. 38, 1976, pp. 291-303.
[24] Johns, D. M. and Miller, D. C. in *Physiological Mechanisms of Marine Pollutant Toxicity,* W. B. Vernberg, A. Calabrese, F. P. Thurberg, and F. J. Vernberg, Eds., 1982, pp. 261-288.
[25] White, S. L. and Rainbow, P. S., *Marine Ecology—Progress Series,* Vol. 8, 1982, pp. 95-108.
[26] Vernberg, W. B., DeCoursey, P. J., and O'Hara, J. in *Pollution and Physiology of Marine Organisms,* F. J. Vernberg and W. B. Vernberg, Eds., Academic Press, New York, 1974, pp. 381-425.
[27] Wright, D. A., *Experimental Biology,* Vol. 67, 1977, pp. 137-146.
[28] Jennings, J. R. and Rainbow, P. S., *Marine Biology,* Vol. 50, 1979, pp. 131-139.
[29] White, S. L. and Rainbow, P. S., *Marine Ecology—Progress Series,* Vol. 32, 1986, pp. 17-25.
[30] Benayoun, G., Fowler, S. W., and Oregioni, B., *Marine Biology,* Vol. 27, 1974, pp. 205-212.
[31] Fowler, S. W., *Nature,* Vol. 269, 1977, pp. 51-53.
[32] *Metallothionein* (proceedings of a symposium), Kagi, J. H. R. and Norberg, M., Eds., Birkhauser Verlag, Boston, Stuttgart, 1979.
[33] Olafson, R. W., Kearns, A., and Sim, R. G., *Comparative Biochemistry and Physiology,* Vol. 62B, 1979, pp. 417-424.
[34] Overnell, J., *Comparative Biochemistry and Physiology,* Vol. 64C, 1982, pp. 547-553.
[35] Oh, S. W., Deagen, J. T., Whanger, P. D., and Weswig, P. S., *American Journal of Physiology,* Vol. 234, 1978, pp. E282-E285.
[36] White, S. L. and Rainbow, P. S., *Marine Ecology—Progress Series,* Vol. 16, 1984, pp. 135-147.
[37] White, S. L. and Rainbow, P. S., *Marine Ecology—Progress Series,* Vol. 19, 1984, pp. 153-166.
[38] Engel, D. H., *The Biological Bulletin,* Vol. 172, 1987, pp. 69-82.
[39] Al-Mohanna, S. Y. and Nott, J. A. in *Proceedings of the First Arabian Conference on Environment and Pollution,* Allen Press, Oxford, 1987, pp. 195-207.
[40] Ahsannulah, M. D., Negilski, D. S., and Mobley, M. C., *Marine Biology,* Vol. 64, 1981, pp. 311-316.
[41] Magos, L. and Webb, M., *Environmental Health Perspectives,* Vol. 25, 1978, pp. 151-154.
[42] Johnson, M. and Gentille, J., *Bulletin of Environmental Contamination and Toxicology,* Vol. 22, 1979, pp. 258-264.
[43] Martin, M., Osborn, K. E., Billig, P., and Gliekstein, N., *Marine Pollution Bulletin,* Vol. 12, 1981, pp. 305-308.
[44] Verriopoulas, G. and Moraitou-Apostolopoulou, M., *Marine Pollution Bulletin,* Vol. 13, 1982, pp. 123-125.
[45] Greenwood, J. G. and Fielder, D. R., *Comparative Biochemistry and Physiology,* Vol. 75C, 1983, pp. 141-144.

I. Cecil Felkner,[1] *Brenda Worthy,*[1] *Tom Christison,*[1] *Christine Chaisson,*[1] *Janis Kurtz,*[2] *and Philip J. Wyatt*[3]

Laser/Microbe Bioassay System[4]

REFERENCE: Felkner, C., Worthy, B., Christison, T., Chaisson, C. and Kurtz, J., **"Laser/Microbe Bioassay System,"** *Aquatic Toxicology and Hazard Assessment: 12th Volume, ASTM STP 1027,* U. M. Cowgill and L. R. Williams, Eds., American Society for Testing and Materials, Philadelphia, 1989, pp. 95–103.

ABSTRACT: The laser/microbe bioassay was developed to meet the needs of the Department of Defense and National Institutes of Environmental Health Sciences for a rapid detection system to detect low quantities of toxic contaminants in potable water (that is, the U.S. EPA's priority pollutants). The bioassay system consists of the Dawn B Laser Photometer, which makes 1200 measurements/s at 15 unique angles, and an isogenic set of *Bacillus subtilis* mutants. Toxicity to the bacteria is detected by differential light scattering (DLS) of the laser beam, which measures the response of the bacteria in terms of light intensity at various angles over 180°. The DLS pattern (liquid sample) shows the number of particles present, their size, shape, and the distribution of particles [that is, if a bacterial culture increases in numbers, there will be an overall increase or decrease in intensity (number of particles)]. If the bacteria increase or decrease in size, the intensity versus angles (DLS pattern) will shift. Therefore, shrinking, swelling, or failure of the bacteria to divide over time are the measured parameters of toxicity. Concentration of the toxicant is determined by the level of response by the bacteria in a given sample.

The 19 bacteria each differ by only one property, which causes them to be more sensitive than other set members based on the mechanism of toxicity. Therefore, a fingerprint or profile unique to each chemical is generated from the differential response of the member bacteria (for example, a profile unique for malathion, dibrom, lindane, 2,4,5, trichlorophenol, and chromate has been generated using this system). In addition, this system has been used to detect malathion extracted from oil base cards used in monitoring fields sprayed with malathion.

The time required for a complete assay is approximately 66 min (time required for untreated bacteria to divide twice in the absence of any toxicant). However, individual samples can be read and analyzed by the computerized system at a rate of one sample/min (that is, 5 to 10 concentrations of an aqueous sample are easily processed in just over an hour), generating data for: (1) identifying test material; (2) assessing concentration of test material; and (3) assessing whether test material is acutely or chronically toxic.

KEY WORDS: differential light scattering, laser/microbe monitor, environmental monitor, chemical fingerprint, laser biotechnology, *Bacillus subtilis,* microbial assay

The ability to efficiently and accurately monitor different environmental media (that is, water, soil, air, and food) for natural and man-made toxicants is one of the most crucial scientific requirements in the public health arena. The identification and quantitative measurement of such entities are necessary to assess risk to the public and are basic components of the regulatory constructs of society. The efficiency, dependability, and cost of conducting such monitor-

[1]Technical Assessment Systems, Inc., Washington, DC.
[2]U.S. Environmental Protection Agency, Washington, DC.
[3]Wyatt Technology Corp., Santa Barbara, CA.
[4]Originally published in the *Proceedings of the Fifth National Conference on Hazardous Wastes and Hazardous Materials,* 19–21 April 1988, Las Vegas, NV, pp. 81–84. Republished by permission of the Hazardous Materials Control Research Institute, Silver Spring, MD.

ing affects international and interstate trade, marketability of products, public health, and financial resources of commercial and governmental institutions.

The overall objective of the studies described in this document was the development of a cheaper, quicker, more sensitive, and yet specific analytical method for the detection of potentially toxic chemicals in various complex media; to develop the laser/microbial assay technology as a cost-effective and accurate method for detecting toxicants in a complex environmental media; and to ensure that a site cleanup demonstration is successful in proving that the relevant toxicants have been removed to a safe level and that their movement into the aqueous media has been controlled.

The laser/microbial system is a rapid, accurate, and qualitative bioassay system that can specifically identify chemicals, toxins, or radioactive substances which exhibit toxic effects on living organisms (toxicants). It uses an integrated set of mutant bacteria *(Bacillus subtilis)* to detect toxicity and differential light scattering (DLS) of a neon-helium beam (632.8 nm) to measure the response of the bacteria (in terms of light intensity at various angles over 180°). The response of the bacteria to a toxicant(s) being by DLS (approximately 400 measurements/s) is fed directly into a computer, which by virtue of mathematical software programs reduces the data to a form that is easily interpreted. The technique for making laser differential light scattering measurements on a suspension of bacteria and subsequent development of instrumentation is attributed to Wyatt et al. [*7–9*].

The DLS pattern produced from a liquid sample containing bacteria and the test article when analyzed by the system over a 66-min period tells the operator if the bacterial cells change in size, shape, or numbers (over time) and whether the cells receive a toxic insult from the test material. The general biological interpretations relative to DLS profile shifts, with time, on the y-axis (relative scattered intensity) and the x-axis (scattering angle in degrees) are summarized as follows:

1. Positive increase in both x and y coordinates indicates a nontoxic (normal) response, that is, normal growth.
2. Positive increase in y, but negative shift relative to x, indicates increase in bacterial numbers and a concurrent size increase. This indicates a minimum toxic response.
3. Decrease in y and negative shift relative to x indicates a toxic and lethal response due to cellular swelling.
4. Decrease in y and positive shift relative to x indicates a toxic and lethal response due to cellular shrinking.

The extent of changes in these parameters can be directly translated into concentration levels of toxicant. Display of responses from the complete set of 19 bacteria produces patterns (fingerprints) which are unique for specific toxicants. Once the fingerprint of a toxicant has been determined, however, the response of a smaller subset (isoset) consisting of three to five unique mutant bacteria is sufficient to identify a given chemical or other type of toxicant.

Since the system contains a program that automatically analyzes and displays (on command) the data, a complete result is available to the system operators at virtually the same time that these data are collected.

The bacteria used in the bioassay system [*1,2*] are of a single species, *Bacillus subtilis*, which has numerous advantages over other bacteria that traditionally are used in bioassays. First, *Bacillus subtilis* is the most widely studied and genetically mapped gram positive microorganism; it is second only to *Escherichia coli* among all organisms [*3*]. It is, therefore, thoroughly understood in terms of its morphology, biochemistry, genetics, and physiology. Its morphological features are of special interest because it can exist as a spore, a vegetative cell, or in a form that permits it to take up very large molecules (competence state). As a spore (dormant form), it is capable of surviving genetically unchanged while resisting heat, drying, or other adverse effects for an indefinite period (thousands of years). As a vegetative cell, it can divide rapidly to double itself in approximately 26 min (during logarithmic growth phase) and in this stage is much more sensitive to toxic material than are the gram-negative microorganisms (for example, *Salmonella typhimurium* and *Escherichia coli*). In the competent state, *Bacillus subtilis* readily takes up large molecules such as deoxyribonucleic acid (DNA); it also can integrate genes derived from members of the same species or even those derived from other life forms through a procedure referred to as recombinant DNA technology.

The microbial bioassay was developed by Felkner et al. [*2*] from a collection of several hundred *Bacillus subtilis* mutants; the selected strains have unique genetic defects which cause them to be very sensitive to the effects of specific chemical classes, toxins, and radiation. To make the system have great uniformity, the mutant genes from these strains were introduced (using recombination technology) into identical clones of the wild type (normal) *Bacillus subtilis* strain. Thus, a battery of 19 strains, differing only by a single character (immediate set of genes), was produced. The 19-member strains are, therefore, isogenic (that is, identical except for single-defined genes). Hence, the response differences between the strains can be attributed to the function of a single gene or an identified combination of functions; these functions are well-defined by numerous published articles in which the molecular basis for deficiencies in repair of genetic damage, synthetic processes, recombination, and morphology are established.

The unique function deficiencies in the 19 members of the isogenic set of bacteria ensure that each will respond quite differently to a given chemical, chemical class, or other toxicant and with great but specific sensitivity. This is because each mutant bacteria's response (or nonresponse) to a given toxicant is tied directly to the mechanism through which the toxicant imparts damage.

Collectively, the 19-member isogenic set of bacteria can be used to create a response pattern or fingerprint that is unique for any assay chemical, provided that the chemical is toxic. This fingerprint is expressed through the differential response of the member bacteria which, because of a single genetic difference, are more sensitive to the chemical based on the mechanism by which it imparts toxicity. In a liquid suspension and by using conventional bacteriological methods (for example, microscopy, spectrography, and colony or cell counts), it has been documented that chemical toxicants (for example, 4-nitroquinoline-1-oxide) [*4*] can block cell division, induce cellular swelling, and cause cell rupture as a consequence of binding to and damaging the deoxyribonucleic acid (DNA) of *Bacillus subtilis*. These parameters as well as the molecular interactions of many chemical classes have been documented for the tester bacteria used in this battery [*2,5,6*]. Furthermore, the plate and liquid methods using this bacterial battery meet the U.S. EPA Office of Pesticide Program's Guidelines as an acceptable way of monitoring genotoxic response (that is, DNA damaging effects, and some of the strains included in this battery are also acceptable for assessing mutagenic response). The assays have been used as a part of the data required for registration of pesticide products. Therefore, the biological test method has been validated as an assay for determining potential adverse human health effects, and, when integrated into the laser detector, even greater sensitivity, specificity, and speed are achieved.

Results

Figure 1 shows the light scattering profiles for three different bacteria, that is, two *Bacillus subtilis* mutants which differ in size, although both are rod-shaped, and *Staphylococcus aureus* strain Georgio, which is spherically shaped. Notice that the profiles of the two *B. subtilis* strains have the same shape; however, the profile for the smaller bacillus (WT168) is shifted toward the larger angles, that is, to the right on the x-axis. On the other hand, the *S. aureus* profile is entirely different. Therefore, the differences in size and shape of microorganisms can be readily determined using the DLS system. Also to be noted is the height on the y axis, which from this reading shows that more *B. subtilis* fh2006 cells than WT168 are present in the solution.

Figure 2 is an actual assay using a sensitive mutant of *B. subtilis* (strain 10 or fh2006-7) to assay for 4-nitroquinoline-1-oxide at three different concentrations, 0.15, 1.5 and 4.5 μg/ml,

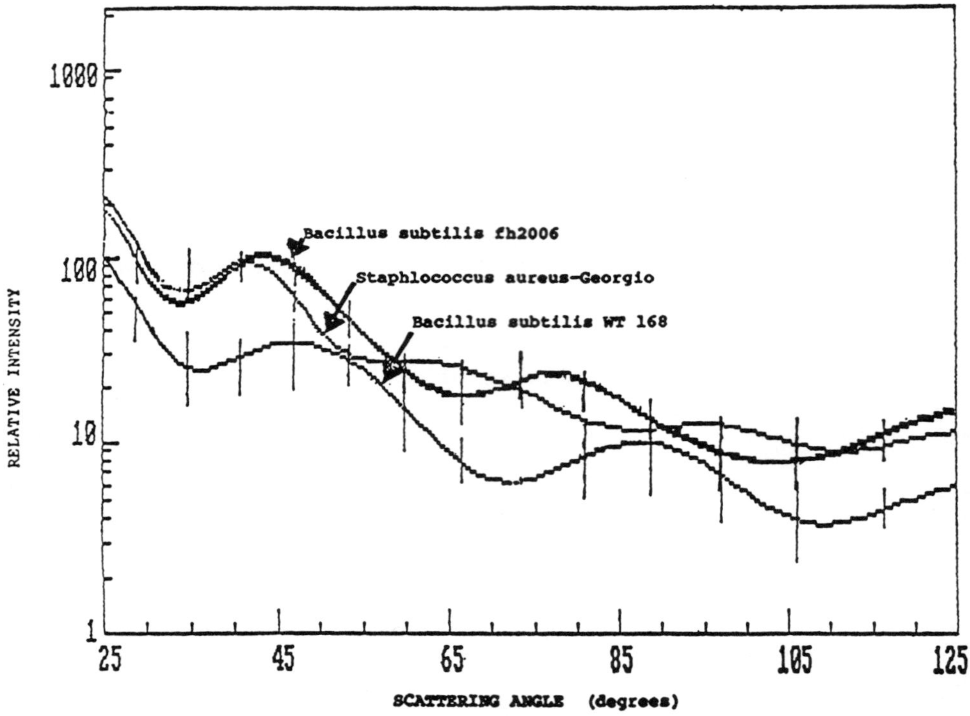

FIG. 1—*Comparative differential light scattering profiles of two* Bacillus subtilis *strains and a* Staphylococcus aureus *strain.*

respectively. The parameters of interest are N/N_0 (relative number of cells at time 0 versus 66 min), TAU (min, which represents the generation time or average time one cell takes to divide into two), and TAU/TAUC (the ratio of generation time at a given concentration of test material versus the untreated control). These results show that the number of cells after 66 min of growth (N/N_0) decreases as the 4-nitroquinoline-1-oxide (4NQO) concentration is increased and that the generation time becomes longer as the concentration of 4NQO is increased. Therefore, there is a dose-dependent inhibition of growth that is readily seen by the increase in the isogenic series.

TAU/TAUC with increasing concentrations of test material. Since this effect is much greater in this mutant strain than in the wild type (normal) bacteria or any other mutant in the isogenic set, the test material is mutagenic (and in the case of 4NQO, it is also carcinogenic). The important point to be made is that the assay can quantify the amount of test material and that the test material has a much greater toxicity for this mutant than any other in the isogenic set. Thus, specificity is demonstrated. However, the following figures (Fig. 3 and Fig. 4) show the specificity of response using two different pesticides to challenge one of the mutants in the isogenic series.

Figure 3 shows the treatment of *B. subtilis* mutant strain 12 with 1.5 mg/mL of the pesticide carbaryl. Notice that the N/N_0, TAU and TAUC for the treated sample versus the control remains unchanged. This result shows that the test material had no effect on this mutant strain. Figure 4 shows the treatment of *B. subtilis* mutant strain 12 with 1.5 mg/mL of malathion. Note, in the treatment that the N/N_0 is reduced, TAU is more than double, and TAU/TAUC has increased to 2.14 in the treatment sample. Thus, malathion is toxic to this mutant bacterial

```
SAMPLE0 1 03041987 SET= 1 STRAIN 10                    CONTROL
0 MINUTES
SAMPLE0 1 SET= 40 STRAIN 10                  CONTROL                66 MINUTES
SAMPLE0 2 03041987 SET= 2 10/4NQO                 .15UG/ML
0 MINUTES
SAMPLE0 2 SET= 41 10/4NQO                    .15UG/ML               66 MINUTES
SAMPLE0 3 03041987 SET= 3 10/4NQO                 1.5UG/ML
0 MINUTES
SAMPLE0 3 SET= 42 10/4NQO                    1.5UG/ML               66 MINUTES
SAMPLE0 4 03041987 SET= 4 10/4NQO                 4.5UG/ML
0 MINUTES
SAMPLE0 4 SET= 43 10/4NQO                    4.5UG/ML               66 MINUTES
```

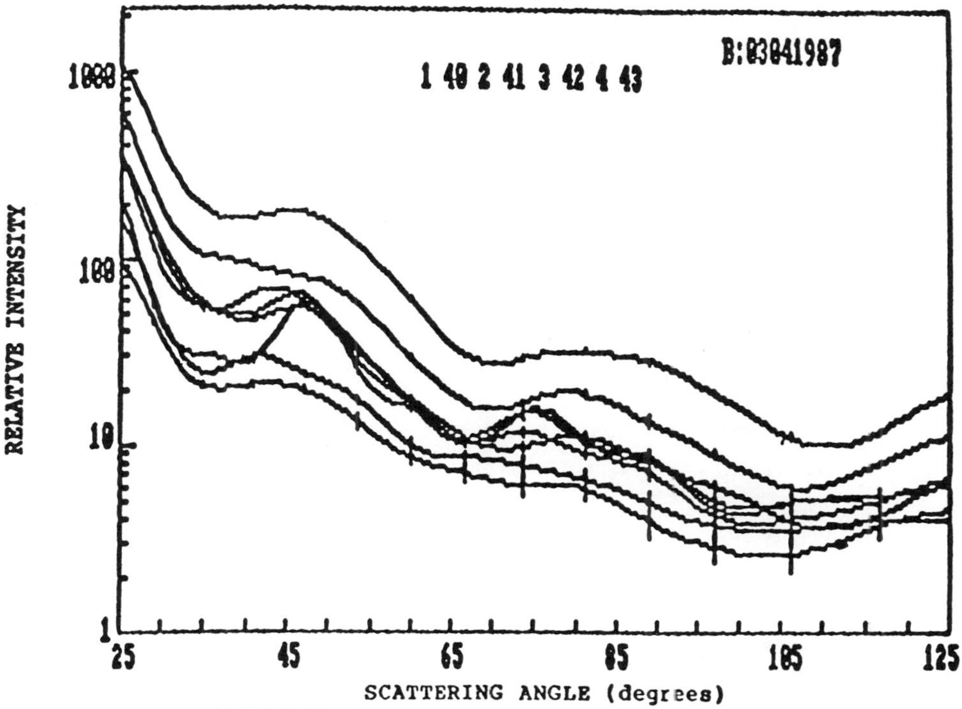

SCORING FOR FILES 03041987 control SAMPLE0 1
(Doubling time =0.69311TAU)

SN0	SET0s	TIME	N/NZERO	TAU(min)	DLEN1	DLEN2	DCHE0	TAU/TAUC	DEL(L1)	DEL(L2)	COMMENTS
1	1 40	66	5.2	40.06	5.11	3.09	0.15	1.00	0.000	0.000	STRAIN 10
2	2 41	66	2.1	90.94	0.31	0.80	0.15	2.27	3.200	-2.290	10/4NQO
3	3 42	66	1.6	134.15	-5.53	-0.28	0.04	3.35	-10.639	-3.365	10/4NQO
4	4 43	66	1.2	463.71	-9.19	-0.35	0.18	11.58	-14.302	-3.437	10/4NQO

FIG. 2—*Response of* Bacillus subtilis *strain fh2006-7 to variable concentrations of 4-nitroquinoline-1-oxide: measurement of dose response by laser differential light scattering.*

SAMPLE# 1 03031987 SET= 13 0 MINUTES
12/0 CONTROL 0 MINUTES
SAMPLE# 1 SET= 43 12/0 CONTROL 66 MINUTES
SAMPLE# 2 03031987 SET= 16 1.5UG/ML M
0 MINUTES 12/3
SAMPLE# 2 SET= 46 12/3 1.5UG/ML CARB. 66 MINUTES

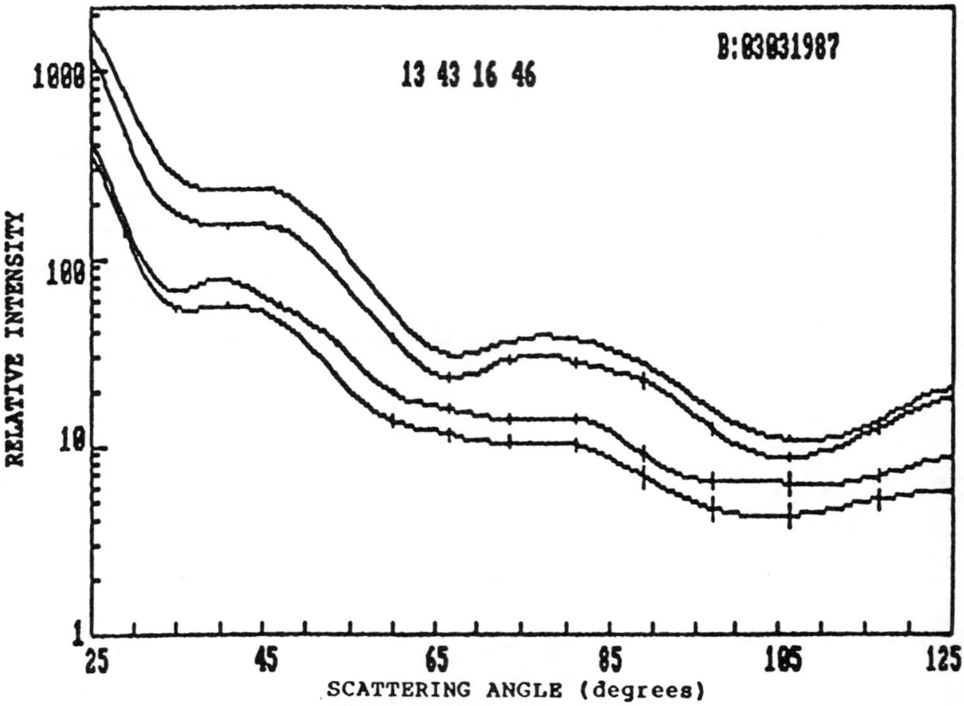

SCORING FOR FILES 03031987 control SAMPLE# 1
(Doubling time =0.6931*TAU)

SM#	SET#s	TIME	N/NZERO	TAU(min)	DLEN1	DLEN2	DCHEG	TAU/TAUC	DEL(L1)	DEL(L2)	COMMENTS
1	13 43	66	2.8	63.48	5.68	2.72	0.05	1.00	0.000	0.000	0 MINUTES
2	16 46	66	2.8	63.17	5.17	2.29	0.14	1.00	-0.518	-0.439	1.5UG/ML M

FIG. 3—*Response of* Bacillus subtilis *Strain 12 to 1.5 μg/mL carbaryl measured by laser differential light scattering.*

strain and by comparing the results from Figs. 3 and 4, it is apparent that different chemicals can be expected to show specificity with regard to their toxic effects on the bacterial assay strains. (Other mutant strains, not sensitive to malathion, are sensitive to carbaryl.)

Results such as these, using a battery of *B. subtilis* strains, provide unique fingerprints of a chemical by sensitive and specific monitoring of the cytotoxic and genotoxic effects on the strains. As each strain responds uniquely to the chemical in a dose-dependent manner, the responses are recorded and deciphered with relatively simple algorithms, assisted by portable or telephonically linked computers. Thus, the identity and concentration of chemical in a medium can be ascertained.

SAMPLE# 1 03031987 SET= 13 0 MINUTES
12/0 CONTROL 0 MINUTES
SAMPLE# 1 SET= 43 12/0 CONTROL 66 MINUTES
SAMPLE# 2 03031987 SET= 15 0
12/2 1.5 UG/ML 0 MINUTES
SAMPLE# 2 SET= 45 12/2 1.5UG/ML MALT. 66 MINUTES

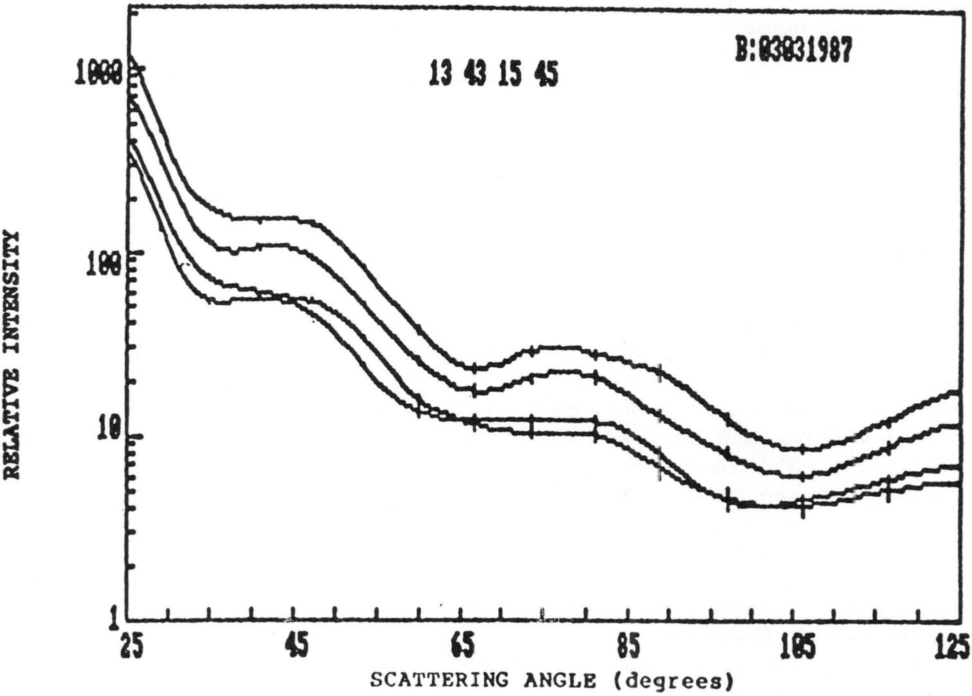

(Doubling time =0.693*TAU)

SM#	SET#s	TIME	N/NZERO	TAU(min)	DLEN1	DLEN2	DCHEB	TAU/TAUC	DEL(L1)	DEL(L2)	COMMENTS
1	13 43	66	2.8	63.48	5.68	2.72	0.05	1.00	0.000	0.000	0 MINUTES
2	15 45	66	1.6	135.83	3.42	0.78	0.01	2.14	-2.268	-1.947	0

FIG. 4—*Response of* Bacillus subtilis *Strain 12 to 1.5 μg/mL malathion measured by laser differential light scattering.*

Conclusions

Existing methods for identification and quantification of toxicants in our environmental media are dependent on several major factors:

1. Traditional chemical analytical methods require the utilization of rather sophisticated and expensive analytical tools. Laboratories, fully stocked with delicate and sophisticated instrumentation, must be utilized for detection of chemicals from the media. Prior to their identification and quantification, highly purified solvents and gases, as well as highly trained professionals, must be available for extraction and preparation of samples for analysis.

2. Well-trained technicians and analytical chemists must operate this equipment. Their constant vigilance over the standardization of the equipment for comparison with results from other laboratories is necessary. Frequently, no nationalized standardization exists; hence, results from different laboratories may provide conflicting answers.

3. Obviously, these analytical laboratories are geographically fixed; this dictates the necessity of transporting samples from food, water, air, or soil to the laboratory site for analysis. Many compounds which are volatile may bind to the specimen container or degrade under conditions of transportation; this step alone causes scientific difficulty and increases the expense of successful monitoring efforts.

4. Existing traditional analytical chemical techniques often require many hours or days from the onset of the analytical procedure to final determination. The cost for any given analysis may range from $10 to several thousand dollars. Techniques which employ methods for determination of groups of chemicals may be relatively insensitive to any one chemical, thus being inadequate for quantification of the toxicant. Adequate screening to see if toxicants are in the sample does not exist because the methods are relatively specific for given chemicals or chemical classes.

These requirements and attendant resource expenditures suggest a series of serious limitations on the efficiency and availability of truly adequate monitoring of the environment for toxicants. Such limitations have been recognized and discussed in forums such as scientific society meetings, regulatory agency budget and priority sessions, consumer advocate publications, congressional debates, and corporate boardrooms. Presently, Congress is considering legislation to require companies to develop cheaper, quicker, and more sensitive analytical methods for detection of their chemicals in the media in which the chemicals may exist. Consumer advocates are stressing the need for regulatory bodies to more thoroughly enforce existing regulations of pesticides and other contaminants in food and water and calling for more expanded monitoring. Therefore, corporations attempting to cleanse the land of toxic sites are frustrated and financially penalized by the expensive and time-consuming nature of existing analytical techniques which burden their efforts.

Until more efficient and dependable methods can be found to monitor the media and enforce existing regulations, the chance of human exposure to harmful levels of toxicants is optimized. This situation presents a crucial need for research and resources for alternative monitoring methods as traditional analytical chemistry may not be the vehicle by which these problems can be solved. It is extremely unlikely that new state-of-the-art analytical methods will reduce the need for sophisticated laboratory equipment or highly trained technicians and chemists. It is also unlikely that new methods can be developed which are quicker and cheaper than existing methods.

Meanwhile, highly beneficial chemicals may be restricted from use, not because of the health risk which is posed but because of the difficulties in identifying and quantifying their residues in our food and water. While regulatory agencies struggle to credibly monitor the media and enforce the protective laws, they must decide whether to commit a high proportion of their operating funds toward monitoring or monitoring with attendant credibility loss and public health risk. Industry can be crippled by the cost and skill demands necessary to comply with these regulations, recover land and water resources, and develop new products.

It is obvious that a new approach is necessary—one that will offer an inexpensive and efficient method to determine that some toxicant exists in the media (screening), to identify that toxicant (specificity), and to quantify the concentration in the medium (sensitivity). Ideally, such methodology should be portable. The technology should be capable of being transported into the field where samples can be taken and quickly analyzed. The new methodology should be accompanied by easy and practical methods for standardization and efficient comparison to equivalent methodologies at other sites. Methodology should be inexpensive and widely available to all groups wishing to monitor toxicants in these media. The level of sensitivity of this technology should be at least equivalent to existing traditional analytical chemistry methods and should work on all media in which the toxicant could occur, with limited necessity for extraction procedures. These objectives can be met by the laser/microbial bioassay system, which can be further developed for on-site toxic waste cleanup monitoring.

References

[1] Felkner, I. C., "Development of a Transformation Tester System for Screening of Environmental Mutagens/Carcinogens in *Bacillus subtilis*," *Modern Trends in Bacterial Transformation and Transfection*, A. Portoler, R. Lopez, and M. Espinoza, Eds., Elsevier/North Holland, New York, 1977, pp. 293-298.

[2] Felkner, I. C., Laumbach, A. D., and Harter, M. L., "Development of a *Bacillus subtilis* System to Screen Carcinogens/Mutagens; DNA, Damaging and Mutation Assays," *Microbial Testers: Probing Chemical Carcinogenesis*, I. C. Felkner, Ed., Marcel-Dekker, New York/NH, 1981, pp. 89-120.

[3] Henner, D. J. and Hoch, J. A., "The *Bacillus subtilis* Chromosome," *Microbiology Review*, Vol. 44, 1980, pp. 57-82.

[4] Laumbach, A. D. and Felkner, I. C., "Formations of a 4-Nitroquinoline-1-Oxide Complex in Normal and Repair Deficient Strains of *Bacillus subtilis*," *Mutation Research*, Vol. 15, pp. 233-245 and Vol. 16, pp. 444, 1972.

[5] Song, P. S., Ou, D. N., Tapley, K. J., Jr., "Photoactivation of Furocoumaryl Carcinogens/Mutagens and their Interactions with Nucleic Acids," *Microbial Testers: Probing Chemical Carcinogenesis*, I. C. Felkner, Ed., Marcel-Dekker, New York/NY, 1981, pp. 35, 66.

[6] Streips, U. N., Laumbach, A. D., and Yasbin, R. E., "Bacterial Mutation Monitors for Active Metabolites of Chemical Carcinogens: *Bacillus subtilis* Assays for Mutation and DNA Repair," *Microbial Testers: Probing Chemical Carcinogenesis*, I. C. Felkner, Ed., Marcel Dekker, New York, 1981, pp. 131, 143.

[7] Wyatt, P. J., "Differential Light Scattering: a Physical Method for Identifying Living Bacterial Cells," *Applied Optics*, Vol. 7, 1968, p. 1879.

[8] Wyatt, P. J., Pittillo, R. F., Rice, L. S., Woolley, C., and Mellett, L. B., "Laser Differential Light-Scattering Bioassay for Methotrexate (NSC740)," *Cancer Treatment Reports*, Vol. 60, 1976, p. 225.

[9] Wyatt, P. J., "Quantitative Quality Control of Food and Beverages by Laser Light Scattering Techniques," *Analysis of Foods and Beverages, Modern Techniques*, G. Charalambous, Ed., Academic Press, New York, NY, 1984.

T. Wayne Schultz[1]

Nonpolar Narcosis: A Review of the Mechanism of Action for Baseline Aquatic Toxicity[2]

REFERENCE: Schultz, T. W., **"Nonpolar Narcosis: A Review of the Mechanism of Action for Baseline Aquatic Toxicity,"** *Aquatic Toxicology and Hazard Assessment: 12th Volume, ASTM STP 1027,* U. M. Cowgill and L. R. Williams, Eds., American Society for Testing and Materials, Philadelphia, 1989, pp. 104–109.

ABSTRACT: Today it is generally accepted that a separate quantitative structure activity relationship (QSAR) can be developed for each mechanism of toxic action. This idea has rekindled an interest in modes of action, especially in aquatic systems. One of the first toxic mechanisms to be identified was simple or nonpolar narcosis, a common mechanism observed with most industrial organic chemicals. It is reversible and physical in nature. Within limits, it is independent of molecular structure, being associated with most nonionic, nonpolar, nonreactive chemicals that are either not metabolized or metabolized very slowly. For these reasons it is considered baseline toxicity. Nonpolar narcosis shows equal activity resulting from equal concentration at the site of action. Since this phenomenon is related to hydrophobicity, narcotic potency correlates well with lipid partitioning. The specific site of action is unknown at this time. However, two likely sites are the bimolecular layer of cell membranes and hydrophobic regions of membrane-bound proteins. A threshold volume is believed to be important. Nonpolar narcosis can be defined by a series of specific in vivo behavioral and/or respiratory-cardiovascular responses with fish and is modeled by QSAR in several test systems.

KEY WORDS: aquatic toxicity, mode of action, nonpolar narcosis, quantitative structure-activity relationship, fish acute toxicity syndrome

Studies of the mechanisms of toxic action, especially in aquatic systems, have enjoyed a rebirth in recent years. This has stemmed from a number of factors including the application of the principles of quantitative structure activity relationship (QSAR) to the field of environmental toxicology. This, in turn, has led to a renewed appreciation of the far-reaching studies of such toxicological pioneers as Overton, Meyer, and Baum. The major reason for all of this has been the recent general realization that a separate QSAR can be developed for each mechanism of toxic action, and, therefore, there is a need for a better understanding of mechanisms of toxic action.

An examination of the early literature concerning narcosis[3] reveals a number of key papers, the majority of which have been put in new perspective since the QSAR studies of Könemann [*1*] and Veith et al. [*2*]. Könemann [*3*], in an earlier synopsis of his three-part series on the relative toxicity of selected organic chemicals in the seven-day guppy *(Poecilia reticulata)* test [*1,4,5*],

[1]Associate professor, College of Veterinary Medicine, The University of Tennessee, Knoxville, TN 37901-1071.

[2]This work was supported by a University of Tennessee, College of Veterinary Medicine, Career Development Award to the author.

[3]The term narcosis as used in this review refers to simple or nonpolar narcosis. This is the same as Type I narcosis. It is not to be confused with polar or Type II narcosis.

felt that the QSAR, $\log LC_{50}^{-1} = 0.871 (\log K_{ow}) - 4.87$ ($n = 50$, $r^2 = 0.976$, $s = 0.237$),[4] "probably predicts a minimum toxicity" and this "toxic action is nonspecific and may be related to membrane perturbation." Veith et al. soon followed with their QSAR work on the toxic effects of "narcotic industrial chemicals" using the four-day fathead minnows *(Pimephales promelas)* bioassay [*2*]. They presented a similar log K_{ow}-dependent correlation and at the same time reintroduced the term ***narcosis*** to the aquatic toxicology literature. The term ***narcosis*** as it relates to environmental toxicology has a long and significant past. It was used as early as the 1860s by Richardson [*9*] in discussing the physiological effects of exposure to a series of fatty alcohols, and, of course, is found in the title of Overton's classic treatise "Studien über die Narkose" (Studies of Narcosis) [*10*]. In fact it was the use of this term, ***narcosis,*** that caused some delay in the publication of Veith's QSAR work (Veith, personal communication).

The works of Könemann [*1,3*] and Veith et al. [*2*], their citing of the studies of Meyer [*11*] and Overton [*10,12*], and the reviews of Ferguson [*13*] and Mullins [*14*], have caused the modern aquatic toxicologist to relearn the past. The work on narcosis performed by Overton, Meyer, and Baum was originally published in German [*10,11,12,15–17*]. However, thanks in part to the efforts of Lipnick [*18*], much of this original literature has been recently reviewed. This early work was followed by a number of other studies including those of McGowan [*19–22*]. More recent reviews by Albert [*23*], Crisp et al. [*24*], Seeman [*25*], Kaufman [*26*], and Roth [*27*] all point to narcosis as a basic mechanism of toxicity. In fact, today it is safe to say that nonpolar narcosis is the most commonly recognized mode of toxic action.

The Theory of Narcosis

The original narcosis theory was published simultaneously by Meyer [*11*], Overton [*12*], and Meyer's coworker, Baum, [*17*]. It has been readdressed a number of times, most notably by Ferguson [*13*] and Mullins [*14*]. The narcosis theory contains several points including:

1. Narcotic and anesthetic agents are synonomous.[5]
2. Narcosis is rapidly and completely reversible.
3. Narcosis does not depend on chemical class.
4. Narcotics produce no chemical change upon the biophase (that is, cellular membranes).
5. Instead they produce a physical change owing to the migration of the agent into the cellular membranes.
6. Narcosis is, therefore, correlated with lipophilicity.[6]
7. Its relative effect depends primarily on the quantity of agent absorbed.
8. It also possesses a qualitative component and, therefore, is related to the volume occupied at the site of action.

This theory of narcosis has its roots in the work originally done in the 1890s and reported in Overton's monograph [*10*]. He observed the responses of cells, plants, and animals, especially tadpoles of *Rana temporia,* to exposure from a variety of organic chemicals. Most of these

[4]While both Könemann [*1*] and Veith et al. [*2*] used log *P* to denote the 1-octanol/water partition coefficient, a term coined by Leo et al. [*6*], this author will use the term log K_{ow} in the present paper. Berthelot and Jungfleisch [*7*] defined partition coefficient as the ratio of concentrations at equilibrium of a solute distributed between two immiscible phases. Partitioning and QSAR have recently been a topic of review [*8*].

[5]Some authors note a distinction between an anesthetic which affects all cell types and a narcotic which affects only nerve cells. Since the results of the work of this laboratory with protozoans parallels those from whole animal studies, this author believes, like Overton [*10*], that this distinction is false.

[6]The use of the term lipophilicity has its roots in the early Meyer/Overton work [*10–12,15–17*] which noted that the narcotic effect of derivatives of various classes of organic chemicals increased with their olive oil/water partition coefficient. In light of the current debate over the specific site of action of narcotics (see section entitled "Site of Action"), this author prefers to use the term hydrophobicity.

observations have been validated a number of times over the last decade by this author as well as others. Such observations include:

1. The rapid and complete recovery of normal physiological activity by an experimental organism upon removal from contact with an agent.
2. Chemicals with widely varying molecular structure elicit the same physiological response (see section on in vivo signs).
3. The correlation between an increase in narcotic activity and an increase in length of alkyl substitution among a homologous series of organic chemicals.
4. However, beyond a certain length, activity is no longer observed (that is, methyl, ethyl, propyl, butyl, "futile").
5. A similar concomitant increase in activity as one compares derivatives formed by moving down the 7A column of elements (that is, the halogens) from fluorine through iodine.
6. An inverse relationship between activity and isomer branching (for example, *n*-butyl more active than [*tert.*]-butyl).
7. An absence of activity associated with chemicals that have little or no hydrophobicity (that is, polyhydroxylated chemicals).

For narcotic chemicals, relative toxicity is a constant fraction of water solubility [*2*]. Therefore, narcosis follows the "Ferguson principle" [*13*]. Stated in other terms for chemicals causing narcosis, an equal potency results when an equilibrium exists between the biophase (for example, a cell, a plant, aquatic or air breathing animal) and the environment (gas or liquid) and the agents have equal concentrations at the site of action (that is, same activities) rather than when the agents have equal environmental concentrations. Therefore, the rate limiting step in narcosis is the ability of the agent to reach the site of action. This point is discussed below.

Mullins [*14*] in his discussion of narcosis introduced a correction for molecular size. He rationalized that agents with larger molar volumes are more toxic because at equal concentrations they occupy a greater volume fraction. Moreover, narcosis occurs when a threshold volume fraction (about 1%) of the agent in the membrane is reached. Miller et al. [*28*] refined this concept and noted that narcosis occurs when the target site volume is "swollen" beyond a critical volume (approximately 0.5%) by the presence of an agent. Seeman [*25*] provided credence to this idea by showing that correcting the membrane molar concentration of agents by their molecular volume dramatically improves estimations.

Site of Action

While the particular site of action and, therefore, the actual mechanism of narcosis is still unresolved, recent reviews [*29–31*] have shed much light on the issue. The traditional lipid solubility theories of narcosis [*28,30*] follow the "critical volume" doctrine. Stated most simply, this doctrine says that narcosis is the result of an accumulation of an agent in the lipid bilayer in cell membranes. The resulting altered property (that is, membrane fluidity, membrane thickness, membrane surface tension, etc.) results in a change in membrane function. On the other hand, the "protein binding" doctrine states that narcosis is a result of the agent binding to an intrinsic or extrinsic hydrophobic region of membrane proteins. This binding results in an altered structure and function. Recent support for this doctrine is based on work with purified luciferase [*32*]. These doctrines are not necessarily in conflict since relative toxicity correlates with log K_{ow} for several reversible mechanisms of toxic action, as shown by work in this laboratory [*33*] as well as by others [*34,35*]. These observations support the idea of multiple sites of action and multiple modes of action [*36*].

Rate Limiting Step

Nonpolar narcosis is the most common mode of toxic action [*2*], and it is associated with a wide variety of chemical groups which have in common the facts that they are nonionic and nonreactive. Classic industrial narcotics include: (a) nonaromatic alcohols; (b) ketones; (c) ethers; (d) alkyl halides; and (e) alkyl and/or halogen substituted aromatic hydrocarbons. This reversible physiological effect is for the most part independent of molecular structure. Whether it follows the critical volume or protein pocket doctrine, the rate limiting step for narcosis is the ability of the agent to reach the site of action. This is a partitioning phenomenon and, especially in aquatic-based systems using eukaryotes, is best quantified by log K_{ow} [*6,8*]. When this parameter is applied to Overton's original tadpole data [*10*], it shows a better correlation ($r^2 = 0.913$) than other parameters [*37*]. This relationship, of course, also is exemplified by the earlier noted studies of Könemann [*1*] and Veith et al. [*2*].[7]

In Vivo Signs of Narcosis

Narcosis has been linked to a specific series of fish behavioral stages [*39*]. These include: (a) loss of reaction to external stimuli; (b) loss of equilibrium; (c) decline in respiratory rate; and (d) subsequent death. Drummond et al. [*40*] have taken the whole fish response one step further and established a standardized set of behavioral/morphological signs for adult fathead minnows and formulated "mode-of-action groups." The set of signs for narcosis is summarized as follows: "hypoactive, underreactive to outside stimuli, very dark (blackish) in color, respiration rapid and shallow."

More recently McKim et al. have expanded these efforts in an effort to define narcosis by a specific physiological response set [*41,42*]. They used spinally transected rainbow trout *(Salmo gairdneri)* as an in vivo model with which to monitor respiratory-cardiovascular responses to acute exposure to model chemicals and developed "fish acute toxicity syndromes" (FATS). Tricaine methanesulfonate (MS-222) and *n*-octanol were the model narcotics used to generate the "narcosis syndrome." The most evident toxic response to these narcotics was a dramatic slowing of all respiratory-cardiovascular parameters. The ventilation volume and total oxygen consumption decreased 40 to 50% from preexposure levels, whereas oxygen utilization increased 20 to 30%. At the same time, ventilation rate remained within the control range. In addition, a 40 to 50% reduction in heart rate was observed. As respiration declined arterial blood oxygen level and arterial pH showed a steady decrease to a level more than 50% below preexposure values at the time just prior to death. The arterial blood carbon dioxide level dropped steadily to about 20% below preexposure level just prior to death. In response to this hypoxia, hematocrit steadily increased to a level about 40% above preexposure level, whereas hemoglobin increased just slightly. This general set of responses defines the FATS for narcosis and separates it from other modes of toxic action [*41*].

Summary

In summary, we see that nonpolar narcosis, the most common mechanism of toxic action for industrial chemicals, is a reversible physical phenomenon which is independent of molecular structure. It represents baseline toxicity. Narcosis follows the "Ferguson principle" of equal potency resulting from equal concentration at the site of action and, therefore, is correlated with hydrophobic/hydrophilic partition coefficients. Whether the specific site of action is the bimolecular lipid layer of cell membranes or hydrophobic pockets associated with membrane

[7]QSAR have been generated from toxicity data on a number of congeneric series using a variety of collinear parameters (for example, number of carbon atoms, molecular weight, boiling point, various partition coefficients, various solubilities, molecular volumes, molecular connectivity indices, etc.) as independent variables. While a tabulation of this literature is well beyond the scope of this review, it is nevertheless interesting to note that the idea that toxicity is related to physicochemical parameters has been with us for a long time. Richardson [*9*] related the narcotic effect of alcohols to molecular weight. Richet [*38*] showed that narcotic potency of ethers, alcohols, and ketones was inversely related to water solubility. Overton [*10*] and Meyer [*11*] both, of course, showed that narcosis is directly related to oil/water partitioning.

proteins is currently unresolved. In any case, a critical or threshold toxicant molar volume appears to be important. Lastly, we see that, based on whole fish behavior and physiological-biochemical parameters, an acute toxicity syndrome can be formulated that defines narcosis.

References

[1] Könemann, H., *Toxicology*, Vol. 19, 1981, pp. 209-221.
[2] Veith, G. D., Call, D. J., and Brooke, L. T., *Canadian Journal of Fish Aquatic Sciences*, Vol. 40, 1983, pp. 743-748.
[3] Könemann, H., *Ecotoxicology and Environmental Safety*, Vol. 4, 1980, pp. 415-421.
[4] Könemann, H. and Musch, A., *Toxicology*, Vol. 19, 1981, pp. 223-228.
[5] Könemann, H., *Toxicology*, Vol. 19, 1981, pp. 229-238.
[6] Hansch, C. and Leo, A., *Substituent Constants for Correlation Analysis in Chemistry and Biology*, John Wiley & Sons, New York, 1979, p. 339.
[7] Berthelot, M. and Jungfleisch, E., *Annals of Chemistry and Physics*, Vol. 26, 1872, pp. 396-407.
[8] Dearden, J. C., *Environmental Health Perspectives*, Vol. 61, 1985, pp. 203-228.
[9] Richardson, B. J., *Medical Times and Gazzette*, 1869, pp. 703-706.
[10] Overton, E., *Studien über die Narkose, zugleich ein Beitrag zur allgemeinen Pharmakologie*, Gustav Fischer, Jena, 1901, p. 195.
[11] Meyer, H., *Archiv fuer Experimentelle Pathologie und Pharmakologie (Naunyn-Schmiedebergis)*, Vol. 42, 1899, pp. 109-118.
[12] Overton, E., *Vierteljahrsschrift der Naturforschenden Gesellschaft Zuerich*, Vol. 44, 1899, pp. 88-135.
[13] Ferguson, J., *Proceedings Royal Society London Series B*, Vol. 127, 1939, pp. 629-645.
[14] Mullins, L. J., *Chemical Reviews*, Vol. 54, 1954, pp. 289-323.
[15] Overton, E., *Vierteljahrsschrift der Naturforschenden Gesellschaft Zuerich*, Vol. 40, 1895, pp. 159-201.
[16] Overton, E., *Vierteljahrsschrift der Naturforschenden Gesellschaft Zuerich*, Vol. 41, 1896, pp. 386-406.
[17] Baum, F., *Archiv fuer Experimentelle Pathologie und Pharmakologie (Naunyn-Schiedebergis)*, Vol. 42, 1899, pp. 119-137.
[18] Lipnick, R. L., *Trends in Pharmacological Sciences*, Vol. 7, 1986, pp. 161-164.
[19] McGowan, J. C., *Journal of Applied Chemistry*, Vol. 1, (supplement 2), 1951, pp. 120-126.
[20] McGowan, J. C., *Journal of Applied Chemistry*, Vol. 2, 1952, pp. 323-328.
[21] McGowan, J. C., *Journal of Applied Chemistry*, Vol. 2, 1952, pp. 651-658.
[22] McGowan, J. C., *Journal of Applied Chemistry*, Vol. 4, 1954, pp. 41-47.
[23] Albert, A., *Selective Toxicity*, 3rd ed., John Wiley & Sons, New York, 1965, p. 380.
[24] Crisp, D. J., Christie, A. O., and Ghobasky, A. F. A., *Comparative Biology and Physiology*, Vol. 22, 1967, pp. 629-645.
[25] Seeman, P., *Pharmacological Reviews*, Vol. 24, 1972, pp. 583-655.
[26] Kaufman, R. D., *Anesthesiology*, Vol. 46, pp. 49-62.
[27] Roth, S. H., *Federation Proceedings, Federation of American Societies for Experimental Biology*, Vol. 39, 1980, pp. 1595-1599.
[28] Miller, K. W., Paton, W. B. M., Smith, R. A., and Smith, E. B., *Molecular Pharmacology*, Vol. 9, 1983, pp. 131-143.
[29] Miller, K. W., *International Review of Neurobiology*, Vol. 27, 1985, pp. 1-61.
[30] Franks, N. P. and Lieb, W. R., *Trends in Pharmacological Sciences*, Vol. 8, 1987, pp. 169-174.
[31] *Molecular and Cellular Mechanisms of Anesthesia*, S. H. Roth and K. W. Miller, Eds., Plenum Publishing Corp., New York, 1986.
[32] Franks, N. P. and Lieb, W. R., *Nature*, Vol. 316, 1985, pp. 349-351.
[33] Schultz, T. W., Holcombe, G. W., and Phipps, G. L., *Ecotoxicology and Environmental Safety*, Vol. 12, 1986, pp. 146-153.
[34] Veith, G. D., DeFoe, D., and Knuth, M., *Drug Metabolism Reviews*, Vol. 15, 1984-85, pp. 1295-1303.
[35] Veith, G. D. and Broderius, S. J., in *QSAR in Environmental Toxicology-II*, K. L. E. Kaiser, Ed., D. Reidel Publishing Co., Dordrecht, Holland, 1987, pp. 385-392.
[36] Miller, K. W., *Trends in Neurosciences*, Vol. 10, 1986, pp. 49-51.
[37] Leo, A., Hansch, C., and Church, C., *Journal of Medicinal Chemistry*, Vol. 12, 1969, pp. 766-771.
[38] Richet, C., *Comptes Rendus des Seances de l'Academie des Sciences Series III Sciences de la Vie*, Vol. 9, 1893, pp. 775-776.
[39] McFarland, W. N., "A Study of the Effects of Anesthetics on the Behavior and Physiology of Fishes," Institute of Marine Sciences, University of Texas, Port Aransas, TX, 1959, pp. 23-55.

[*40*] Drummond, R. A., Russom, C. L., Geiger, D. L., and DeFoe, D. L. in *Aquatic Toxicology and Environmental Fate, ASTM STP 921*, T. Poston and R. Purdy, Eds., American Society for Testing and Materials, Philadelphia, 1986, pp. 415–435.
[*41*] McKim, J. M., Bradbury, S. P., and Niemi, G. J., *Environmental Health Perspectives*, Vol. 71, 1987, pp. 171–186.
[*42*] McKim, J. M., Schmieder, P. K., Carlson, R. W., Hunt, E. P., and Niemi, G. J., *Environmental Toxicology and Chemistry*, Vol. 6, 1987, pp. 295–312.

Field Techniques

James F. Hobson,[1] *John W. Sherman,*[2] *and Mark A. Palmieri*[3]

Evolution of the Farm Pond Study

REFERENCE: Hobson, J. F., Sherman, J. W., and Palmieri, M. A., **"Evolution of the Farm Pond Study,"** *Aquatic Toxicology and Hazard Assessment: 12th Volume, ASTM STP 1027,* U. M. Cowgill and L. R. Williams, Eds., American Society for Testing and Materials, Philadelphia, 1989, pp. 113–119.

ABSTRACT: During the past decade, field studies to assess the potential impact of agricultural chemicals on aquatic ecosystems have increased in number and sophistication. As the questions addressed in these studies have become more complex, the traditional "farm pond" study has evolved into an elaborate fate and effects study. This paper will examine the increasing complexity of aquatic field studies over the past ten years using as examples studies conducted to support the registration of three pyrethroid insecticides: permethrin, cypermethrin, and bifenthrin. A permethrin study was conducted in a single season (1976) using a minimal regime of residue and biological samplings. The current bifenthrin study has involved extensive biological residue sampling with in situ populations of vertebrate and invertebrate species, on-site bioassays, and community metabolism over a three-year period (1985–1987). The future trends for aquatic field studies will also be discussed, including replicate treatment ponds, management of fish populations, and the use of experimental enclosures.

KEY WORDS: aquatic field testing, farm pond studies, pyrethroids, pesticides, mesocosm, littoral corrals, experimental enclosures

The past decade has seen a rapid evolution in the sophistication of field studies designed to examine the impact of agricultural chemicals on aquatic ecosystems. From a regulatory standpoint, the Federal Insecticide, Fungicide and Rodenticide Act (FIFRA), Pesticide Assessment Guidelines, Subdivision E [*1*], state that, under specified conditions, "an actual or simulated field study" may be required for registration of a pesticide. Until the advent of the mesocosm [*2*] as the "simulated field study," the examination of a farm pond or ponds surrounded by agricultural land was the study design of choice to fulfill this requirement. The advantage of such studies over laboratory studies or a simulated aquatic ecosystem was that they included the environmental fate processes (for example, biological, chemical, and physical) that may serve to mitigate the toxic potential of a pesticide in an aquatic ecosystem, even when these processes are not precisely understood. An actual use study was expected to accurately depict the hazard of a pesticide to a representative aquatic ecosystem [*2*].

One class of pesticides upon which the requirement for aquatic field studies has been imposed is the synthetic pyrethroids. As these compounds are extremely toxic to aquatic organisms under laboratory conditions, they have triggered ecosystem studies as part of the tiered testing scheme under FIFRA guidelines. A series of studies were conducted by FMC Corp. as part of the registration processes of three synthetic pyrethroids—permethrin, cypermethrin, and bifenthrin—during the period 1976 to 1987.

In this presentation, FMC's pyrethroid field studies are used to review the recent evolution of farm pond study protocols, with emphasis on the increasing complexity of both the field methodologies as employed and the questions being addressed concerning the fate and effects of pesti-

[1] Rhône-Poulenc, Research Triangle Park, NC 27709.
[2] The Academy of Natural Sciences of Philadelphia, Philadelphia, PA 19103.
[3] FMC Corp., Princeton, NJ 08543.

cides in aquatic ecosystems. The future use of farm ponds for ecological testing will be discussed.

Pyrethroid Field Studies

Four farm pond studies involving synthetic pyrethroids developed by FMC Corp. have been conducted over the past decade (Table 1). The compounds involved were permethrin (Studies I and II) [*3,4*], cypermethrin (Study III) [*5*], and bifenthrin (Study IV) [*6*].

During the period 1976 to 1987 the design of farm pond studies increased in complexity as a result of progressive negotiations between the Ecological Effects Branch (EEB) of the Environmental Protection Agency (EPA) and FMC and other manufacturers of agricultural chemicals. The EEB required increasingly more complex data in order to conduct more detailed risk analyses of pesticides in aquatic ecosystems. The objectives for each study reflect the increase in scope over this time period.

Study Objectives

The objectives of a 1976 (Study I) farm pond study were general in nature:

"... to provide data on the environmental impact of the insecticide permethrin when applied under field use conditions."

In this study, observations were designed to detect gross impact of a pesticide to an aquatic ecosystem by monitoring a limited number of biological and chemical parameters following a typical ground application.

In 1986 (Study IV) the expanded study objectives were well defined and explicit:

"... to provide empirical information regarding:

1. the distribution, including drift and runoff of [bifenthrin] residues resulting from a typical aerial application of this compound on a representative field crop;
2. the accumulation of [befenthrin] residues in the aquatic environment;
3. the movement of these soil-borne residues especially into an adjacent aquatic environment;
4. the presence of the residues in the aquatic environment;
5. the effects of these residues on indigenous aquatic populations; and
6. the longer-term stability and viability of exposed populations."

The protocol for Study IV included a detailed set of chemical and biological observations designed to address each of these specific objectives.

TABLE 1—*General study design of aquatic field studies: 1976 to 1987.*

Parameters	Study I	Study II	Study III	Study IV
Treatment year	1976	1979	1987[a]	1986
Compound	Permethrin	Permethrin	Cypermethrin	Bifenthrin
Method of application	Tractor	Aerial	Aerial	Aerial
Study period, weeks	19	52	>104	>104
Area treated, m^2	20 235	171 998	203 350	182 115
Pond area, m^2	12 141	28 329	28 329	13 355
Ratio, field : pond	1.7 : 1	6 : 1	8 : 1	15 : 1
Reference pond	Y	Y	—	Y
Applications	17	15	10	10
Interval of application, days	5	5	7	7

[a]The protocol for this 1987 study was designed several years earlier, but the implementation was delayed. Thus this study represents an older and less complex design as compared to the 1986 bifenthrin study. The letter Y indicates that a parameter was monitored in the respective study and a minus sign (—) indicates that the given parameter was not monitored.

General Study Design

The general study designs for four aquatic field projects conducted from 1976 to 1987 are compared in Table 1. The compounds tested are all synthetic pyrethroids. The method of application was generally aerial, which was considered to represent the worst case scenario. Application frequency and interval were according to label instructions at proposed maximum labeled rates or higher.

The study periods ranged from 19 weeks to over 2 years. These increases in the time frames in Studies III and IV were due to the extensive pretreatment and posttreatment periods. This includes monitoring (primarily of biological parameters) before and after application during the treatment year and extensive monitoring during the active biological season the years before and after treatment. Comparison of this additional data was designed to enhance the ability to differentiate between natural variability and chemical-induced perturbations in aquatic communities.

The size of treated fields ranged from 5 to 50 acres and increased in size from Study I to Study IV. Although the pond size remained relatively unchanged (that is, three to seven acres), the increased size of the treated field provided a more typical field-to-pond ratio as recommended by the EPA.

Reference ponds were utilized in three of the four studies being compared. However, in Studies I and II the reference ponds were sampled only to provide a cursory comparison. In Study IV the reference pond was monitored with extensive biological sampling for two years, concurrent with the treatment and posttreatment phases of the study. Reference ponds were incorporated into the study design to provide additional baseline information regarding natural temporal variation in biological communities.

Physicochemical Parameters

Selected physicochemical parameters were monitored in all four studies including: air and water temperature, dissolved oxygen, pH, conductivity, alkalinity, turbidity, and pond depth.

Chemical Residues

Residue analyses were performed on a range of biotic and abiotic samples (see Table 2). The list of matrices sampled increased from Study I to Study IV. The residue programs in Studies I and II involved relatively few samples and matrices. The protocols were expanded in Studies III and IV to better understand the fate of the compound. The sampling program for residue analysis became more elaborate. The number and frequency of samplings as well as the number of replicate samples per collection date increased dramatically in order to more precisely define exposure to each biological community within the pond ecosystem.

The water, sediment, and soil samples for Study I were collected at ten stations in the pond and field. These samples were composited for each sample type for each collection date, thus providing only a single residue value for each of eleven collection dates.

In Study IV the water, runoff water, runoff soil, sediment, and field soil were collected at five to eight sites depending upon the matrix, and only the field soil samples were composited.

Samples for the measurement of residues in biota increased both in the number of matrices sampled and in the number of samples per collection date over the period 1976 to 1987. In Study I only three matrices, fish, crayfish, and mussels, were collected for analysis. One pound of each matrix was collected and analyzed for each sampling date.

In Study IV biologicals for residue analysis involved multiple species collected in several sampling zones. Fish were collected at three sites with samples of two species represented for nearly all sampling dates over a 15-month period from June 1986 to September 1987. Other species of

TABLE 2—*Residue samples collected at regular intervals during aquatic field studies.*

Matrices	Study I	Study II	Study III	Study IV
		Abiotic		
Pond water	Y	Y	Y	Y
Pond sediment	Y	Y	Y	Y
Field soil	—	Y	Y	Y
Runoff water	Y	—	Y	Y
Runoff soil	Y	—	—	Y
Drift cards	—	Y	Y	Y
		Biotic		
Fish	Y	Y	Y	Y
Mussels	Y	—	—	Y
Crayfish	Y	—	—	Y
Cotton plants	—	—	—	Y

NOTE: The letter Y indicates that the given parameter was monitored in the respective study, and the minus sign (—) indicates that the given parameter was not monitored as part of the respective study.

fish (for example, gizzard shad, redear sunfish, etc.) were sampled frequently enough to provide a profile of residues in each respective species over time.

In addition to the collection of feral populations of fish, caged catfish were held at three stations throughout the study and sampled for residue analyses. Populations of mussels also were established in three sampling zones and subsampled for residue analysis. These introduced populations were incorporated into the study design to provide consistent biological residue data over the entire 15-month sampling period.

The residue analyses for Study IV included multiple sampling sites to examine the spatial distribution of biological residues within the pond system and a variety of species from different trophic levels. Among the collected fish species different ecological niches and life history habits were represented.

This extensive sampling program was designed to provide insight into the fate of the compound being studied. It should be noted that the extent of this type of data and the variety of species and trophic levels sampled was critical to the interpretation of this study.

The more recent pyrethroids are more potent and require less compound to be applied for effective use. As a result, the concentration of environmental residues are also reduced.

The analytical techniques required to monitor pesticide residues in these studies have required increasingly lower limits of sensitivity. Study I had a detection limit for permethrin in the range of 10 parts per trillion (ppt) and Study IV had a detection limit of less than 1 ppt bifenthrin in water. This decrease in the detection limit requires a significant increase in cost and effort required for sampling and analysis.

Biological Parameters

Monitoring of Natural and Caged Populations—The characterization of the biological components of the farm pond ecosystem was examined prior to application, during the application period, and following application through various sampling and monitoring techniques (see Table 3).

In Studies I and II, the macroinvertebrates were sampled using benthic grabs and kick nets. Zooplankton samples were collected using nets. In Study II, aquatic macrophytes were also sampled. In these early studies samples were composited into one sample per collection date.

TABLE 3—*Biological parameters monitored during aquatic field studies.*

Parameter	Study I	Study II	Study III	Study IV
Fish				
In situ survival	—	—	Y	Y
Tag/recapture	—	—	Y	Y
Larval traps	—	—	Y	Y
In situ reproduction	—	—	Y	Y
Invertebrates				
Benthic grabs	Y	Y	—	Y
Kick nets	Y	Y	Y	Y
Periphyton	—	—	Y	Y
Emergence traps	—	—	Y	Y
Zooplankton	Y	Y	Y	Y
Phytoplankton	—	—	Y	Y
In situ survival	—	—	—	Y
Biomonitoring				
7-day fish larval-growth	—	—	—	Y
21-day *Daphnia* chronic	—	—	—	Y
Ecosystem Metabolism				
$^{14}CO_2$ incorporation	—	—	—	Y
O_2 evolution	—	—	—	Y

NOTE: The letter Y indicates that the given parameter was monitored in the respective study, and the minus sign (—) indicates that the given parameter was not monitored as part of the respective study.

In Study IV, eight sampling zones were established and an array of biological communities were monitored for at least three of these zones. The location of sampling zones was designed to elucidate any potential gradients in pesticide impact within the ecosystem. This data could then be correlated with the spatial distribution of pesticide residues in the pond. Phytoplankton, zooplankton, and substrate colonization samples were collected in each of the eight zones. The other parameters were monitored at three or more sites. These included emergence of aquatic insects, survival of feral fish populations, and in situ survival of three vertebrate and four invertebrate species established as caged populations. In situ populations of aquatic organisms were included in the study design to provide acute toxicity data. In Study IV many of these populations were maintained throughout the 15-month observation period and were sampled for residue analyses (see "Chemical Residue" section). Reproduction was monitored using caged populations of fathead minnows *(Pimephales promelas)* and mosquito fish *(Gambusia affinis)* as well as with larval traps and larval tows to monitor reproduction/recruitment in feral fish populations.

An assessment of the survival and reproduction of fish populations is a critical part of any aquatic field study. Two approaches to the study of fish populations have emerged in the more recent farm pond studies (Studies III and IV). In Study IV the study site selected had well-established feral populations of fish representing a diversity of species with a variety of life history habits. Therefore, this site was thought to represent an optimal model for the study of the impact of a pesticide on fish populations.

A different approach to the study of fish populations was taken in Study III. A pond was chosen which was younger and where fish species were limited and populations were controlled by fisheries management practices. By controlling the number of species present and the population size, the results of fish data might be more interpretable.

Biomonitoring—No toxicological assessments of the ponds were conducted in the early studies (that is, Studies I and II) except for gross observations for fish mortality and the monitoring of caged populations of crayfish.

In Study IV an on-site bioassay trailer was used to supplement the gross mortality observations of natural and caged populations. Water samples were drawn at regular intervals from three sampling zones and used in flow-through bioassays. The tests conducted were the 21-day *Daphnia* life cycle and the 7-day larval growth with fathead minnows. These tests were designed to provide subtle toxicological data relevant to the impact of the pesticide and to document any subsequent recovery. These data are not clouded by natural variation observed in indigenous populations.

Metabolism Studies—These were not conducted in Studies I or II. In Study III chlorophyll-a to biomass ratio was monitored in phytoplankton samples. In the case of Study IV a more detailed approach was used. The primary production of phytoplankton was assessed utilizing ^{14}C bicarbonate incubations, and periphyton populations were monitored using biomass to chlorophyll-a ratios. The purpose of these elements of the study was two-fold. First, the direct effects of the pesticide on the primary producers of the ecosystem were examined. Equally important, however, was the ability to assess seasonal and spatial variability in the primary production of the system. This information is necessary for the interpretation of observed spatial and temporal fluctuations in the primary and secondary consumers (that is, invertebrates and fish). Finally, whole pond oxygen studies were conducted for the purpose of monitoring both photosynthesis and respiration on a pond-wide basis.

In summary, it is obvious that the number of parameters as well as the depth of examination for each parameter monitored in aquatic field studies has increased dramatically over the period 1976 to 1987. Furthermore, this increase in the complexity of the study design is a reflection of the need to better understand the potential impacts of pesticides on aquatic ecosystems.

The Future of Pond Studies

The future designs for this type of aquatic field study will likely be larger and even more complex. In fact, studies currently being conducted are utilizing, minimally, two treatments and two reference ponds. In these studies with replicate ponds every attempt to choose highly similar ponds will be made. Furthermore, attempts to manipulate various aspects of the replicate pond systems will likely be made.

For example, it is proposed that fish populations be removed using rotenone two years prior to the treatment phase. All ponds would then be restocked with balanced fish populations using the same stocking rates. This is an attempt to begin a replicated pond study with similar fish populations.

If the farm pond study design is to continue to be used in the ecological risk assessment for the purposes of registering pesticides, then more replication will likely be required by the EPA and other regulatory agencies. This requirement of replication is to ostensibly provide more reliability to data that is to be used in risk analysis.

The alternative "simulated field study," or fixed-pond mesocosm, described by Touart [*2*] is designed to provide an acceptable level of replication of ecosystem effects studies; however, these synthetic ecosystems may or may not represent mature or established ecosystems. Furthermore, this test system provides only effects data that are divorced from many of the real world fate processes which may serve to mitigate the potential adverse impact of a pesticide in an aquatic ecosystem.

In 1987, FMC and the Academy of Natural Sciences of Philadelphia developed a protocol which incorporated a two-treatment and two-control pond design in combination with a littoral corral (mesocosm) study [*7*] run concurrently in a fifth pond. The latter is an alternative mesocosm design which has been developed at the USEPA Environmental Research Laboratory—Duluth. Such a design provides one pond with a number of replicate (that is, 12 to 16) experimental enclosures which partition the preexisting pond ecosystem including the littoral zone. These littoral corrals can be treated with a range of replicate exposure concentrations and monitored using the same methods applied to any mesocosm.

The advantage of the combined farm pond/littoral corral design is that it provides replication of the full farm pond system, which incorporates both fate and effects under typical field use conditions. In addition, the littoral corral can provide dose-response data for ecosystem effects in a similar farm pond system under essentially the same environmental conditions. We believe that this test design would combine the best aspects of the farm pond test system (that is, fate and effects) with the replication and statistical power of a mesocosm (effects) in a unified approach. The results of such a study could provide a data base superior to the results of either the traditional farm pond study or the fixed-pond mesocosm alone.

Conclusions

During the past decade, the methods applied to farm pond aquatic field studies have become increasingly more complex, addressing more completely questions of pesticide effect on both the structure and function of aquatic ecosystems. Protocols of the future field tests will require greater replicability through increase in the number of ponds employed, or more realistically, through combinations of whole-pond studies and mesocosms such as littoral corrals.

Acknowledgments

The staffs of various organizations contributed to the studies reviewed including: The Academy of Natural Sciences of Philadelphia, USEPA Environmental Research Laboratories—Duluth, FMC Corp., Hydroqual, Inc., I.C.I. Americas, Springborn Life Sciences, Union Carbide Environmental Services, Wildlife International, and others.

References

[*1*] *Pesticide Assessment Guidelines, Subdivision E, Hazard Evaluation: Wildlife and Aquatic Organisms,* USEPA, Office of Pesticide and Toxic Substances, Washington, DC 20460, 1982, p. 86.

[*2*] Touart, L. W., 3 Sept. 1987 draft, *Aquatic Mesocosm Tests to Support Pesticide Registration,* USEPA, Office of Pesticide and Toxic Substances, Washington, DC 20460, p. 35.

[*3*] "The Application of Two Permethrin Formulations on a Cotton Field Adjacent to an Aquatic Ecosystem," unpublished report, FMC Corp., Princeton, NJ, 1977, p. 36.

[*4*] "Potential Aquatic Ecological Effects of Aerial Application of Two Permethrin Formulations on Cotton," FMC Corp., Princeton, NJ, unpublished report, 1980, p. 49.

[*5*] "A Study of the Aquatic Ecosystem of a Farm Pond in the Year of Cypermethrin Application on a Cotton Crop Surrounding the Pond," unpublished report, FMC Corp., Princeton, NJ, 1988.

[*6*] "Aquatic and Ecological Effects of FMC 54800," unpublished protocol, FMC Corp., Princeton, NJ, 13 Feb. 1986, p. 47.

[*7*] Siefert, R. E., Lazano, S. J., Brazner, J. C., and Knuth, M. L. in *Using Mesocosms for Assessing the Aquatic Ecological Risk of Pesticides: Theory and Practice,* Miscellaneous Publications, Entomological Society of America, Lanham, MD, 1988, in press.

James R. Clark[1]

Field Studies in Estuarine Ecosystems: A Review of Approaches for Assessing Contaminant Effects

REFERENCE: Clark, J. R., **"Field Studies in Estuarine Ecosystems: A Review of Approaches for Assessing Contaminant Effects,"** *Aquatic Toxicology and Hazard Assessment: 12th Volume, ASTM STP 1027,* U. M. Cowgill and L. R. Williams, Eds., American Society for Testing and Materials, Philadelphia, 1989, pp. 120–133.

ABSTRACT: A sampling strategy designed around contaminant source (agricultural runoff, direct discharge) and fate (solubles, particulates, sediments) and the hydrodynamics of the system studied is required to characterize the exposure of estuarine biota to contaminants. Field data obtained on contaminant effects should be applicable to risk assessment in order to verify approaches to predicting contaminant fate and effects in estuarine systems. Only through systematic evaluations of field and laboratory exposure-response relationships will we be able to quantify the limits of applicability of laboratory data used for ecological risk assessment. Survival of caged test animals at field test sites provides data for direct comparison with laboratory toxicity test results. Coupling survival and other effects data from caged animal studies with assessments of stocks and dynamics of populations of the same or a related species at the field site may allow extrapolation from simple laboratory and field test results (acute or chronic) to more complex and ecologically significant endpoints. This paper presents examples of various approaches to contaminant problems in estuaries and discusses their applications to risk assessment procedures.

KEY WORDS: estuary, field validation, ecotoxicology, pesticides, risk assessment, aquatic toxicology, contaminants, pollutants

In the past, much of our understanding of contaminant effects in aquatic ecosystems has been based on laboratory exposure-response studies. Currently, there is considerable interest in field validation of toxicological data and ecotoxicological predictions by conducting systematic exposure-response studies and ecotoxicological investigations in complex natural systems [*1–3*]. Studies of contaminant effects in estuarine ecosystems address the same exposure-response relationships germane to freshwater systems. However, the hydrological dynamics, physical scale, and biological complexity of estuarine systems demand specific aspects of field work often not found in freshwater studies [*4*]. Regular and predictable environmental changes (tidal movements, salinity gradients, temperature fluxes, etc.) and mixing processes (currents, wind, tidal flows, etc.) must be evaluated carefully when designing field studies in estuarine ecosystems. The size of an estuary can range from limited zones of freshwater input into an ocean, such as the Newport River estuary along the Pacific coast, to systems the size of Chesapeake Bay or Puget Sound. Studies to assess contaminant effects must be designed to complement the scale of the problem and system to be examined [*5*]. Large estuaries can be studied through analyses of smaller components that are more manageable. However, the significance of problems studied must be evaluated in terms of the dynamics of entire ecosystems [*6*]. Biological

[1] Research aquatic biologist, U.S. Environmental Protection Agency, Environmental Research Laboratory, Gulf Breeze, FL 32561.

complexity characteristic of estuaries complicates field studies because specific biota are present in limited zones of the estuary and may change habitats with seasons or life stages. Finally, commercial harvests, shipping activities, and development of waterfront property are likely to have ecological effects within an estuary as contaminant problems are being studied [7].

Need For Correspondence of Field and Laboratory Data

The characterization of anthropogenic impacts on estuarine systems has been ongoing since declines in harvest of food items were linked to human activities. However, the scientific approaches to hazard assessment, risk assessment, and risk management are relatively new areas of interest [8–10]. Currently, only limited information is used in problem evaluation or decision making. In order to improve our approaches to risk assessment and risk management, the kinds of data used in risk assessment must be applicable to field investigations, and field data that provide for direct evaluation of predictions from laboratory data must be obtained. I will provide an overview of approaches that could be used to characterize contaminant effects in estuarine systems and will focus on applications of these approaches relevant to risk assessment procedures. Specific examples of approaches which have been useful for field validation research at the U.S. EPA Environmental Research Laboratory at Gulf Breeze (ERL/GB) will be used to demonstrate the applicability of these data.

Approaches To Characterization of Exposure and Response

The key to assessment of contaminant effects in estuaries is the ability to document cause and effect relationships. Ecological changes are explained through analyses of single-species responses that are reflected in population and community dynamics [11]. Relating biotic responses to quantified contaminant concentrations in the environment is the basis for risk assessment and ecotoxicological research [8,9,12,13]. The scope of such research includes acute or chronic effects of contaminants on single species, populations, or communities (Table 1). The advantages and disadvantages of various approaches must be evaluated in light of specific study objectives and resources available. No single approach will be satisfactory, necessary, or cost-effective for all studies. The examples listed in Table 1 are but a few of the factors that must be considered.

Exposure

The transport and fate of contaminants in estuaries have been the subject of numerous studies and publications [3,14, and others]. Contaminant fate must be tied to biotic exposure characterizations in order to make predictive assessments of ecological response to contaminant input [15,16]. The persistence, transformation, and sorption dynamics of contaminants in water and sediment is a major factor controlling plant and animal exposures. Chemical sorption to particulate or dissolved organic matter and inorganic matter can be a considerable sink for estuarine systems and sources of exposure for biota. Chemical fate and transport will not be dealt with in this manuscript. However, this area is recognized as a scientific approach to characterizing bioavailability and exposure, essential aspects of risk assessment and exposure-response relationships.

Acute Effects

As a result of spills, intentional or accidental discharges, runoff events, or other direct or indirect inputs, contaminants entering estuaries can cause acute, lethal effects with local or widespread impact. Generically referred to as "fish kills" (to include mass mortalities of

TABLE 1—*Advantages and disadvantages of various approaches used to investigate contaminant effects in estuaries.*

Approach	Advantages	Disadvantages
Chemical surveys	(1) Field effects focused on sample collection (2) Allows integration of water, sediment, and biota residues (3) Bioavailability and exposure can be used to predict effects	(1) Biotic response to exposure is not quantified (2) Residue data are biased towards persistent chemicals (3) Exposure-response relationships are deduced or hypothetical
Fish kill investigations	(1) Effects obvious and indisputable (2) Important species recognized (3) Acute, lethal effects are assumed	(1) Total numbers of dead animals may be underestimated (2) Impact on plankton and microbiota not characterized (3) Exposure-response relationships are not quantifiable
Caged animal studies	(1) Animals tested are of known toxicological sensitivity and physiological health (2) Field exposure-response data are comparable to laboratory toxicity test data (3) Acute or chronic, lethal or sublethal effects can be examined	(1) Results may not be the same as ecological response (2) Problems with cage fouling, loss through vandalism, storms, or other events (3) Accurate field exposures difficult to quantify (4) Efforts must be coordinated with contaminant release
Population sampling	(1) Long-term effects can be estimated from population dynamics approach (2) Data are relevant to ecotoxicological interpretation (3) Biomonitoring for growth or gross pathological effects can be incorporated	(1) Sampling must be done over considerable time frame (2) Data interpretation more difficult than caged animal or fish kill studies (3) Resident test population sometimes is difficult to delineate due to species mobility (4) Exposure regime is difficult to quantify
Community studies	(1) Effects on multiple species can be assessed (2) Data are relevant to ecotoxicological interpretation (3) Biomonitoring approaches can be incorporated	(1) Sample collection and analyses require considerable effort (2) Sampling must be done over considerable time frame (3) Exposure regime is difficult to quantify
Ecosystem	(1) Provides ultimate interpretation of significance of effects	(1) Integrative methods for data analyses are not fully understood (2) Some estuaries are too large for intensive studies (3) Cost and scale of ecosystem studies are prohibitive

shrimp, crabs, or other invertebrates), these incidences are one of the most noticeable effects of contaminants that enter estuarine systems. Although these events provide obvious evidence of effects, causes of fish kills often are speculative. A full characterization of exposure-response relationships responsible for fish kills seldom is possible because (1) previous contaminant exposures are not known for animals counted among the dead and (2) assessment of contaminant concentrations responsible for the fish kill begins after dead animals are observed.

Studies using caged or transplanted sessile animals can provide quantitative exposure-response assessments. Animals of a known toxicological sensitivity can be tested to obtain acute exposure-response data in a manner comparable to laboratory toxicity tests. However, these studies are useful only when contaminant releases into estuaries can be anticipated, are regularly scheduled, or for bioassaying effects of in-place contaminants. Test animals also can be evaluated for sublethal effects indicative of toxicant exposure or physiological stress.

Chronic Effects

Caged-animal studies can be used to assess contaminant effects over extended intervals to develop chronic exposure-response data analogous to laboratory early life-stage tests with fishes following ASTM toxicity testing procedures. Growth or survival of larval stages of invertebrates or fishes can be quantitated with appropriately designed cages. If a reproductive cycle cannot be completed in cages, a combination of field exposures and subsequent observation in the laboratory could be used to quantitate effects of contaminants on reproductive success of estuarine biota. For example, gravid females could be placed in cages for field exposures, and effects on egg development or hatching success could be characterized in the laboratory. Tests with sessile or confined animals can be adapted for a variety of studies [*17*].

Because of (1) the need for constant attention and (2) potential for predator influences (human and nonhuman) and loss during storms, long-term, chronic exposures using caged animals may not be feasible for all estuarine environments. Abbreviated chronic test procedures [*18*] could be adapted to short-term, caged-animal studies to reduce the risk of interfering events. Alternatively, discrete or composite water samples may be collected at field sites and returned to the laboratory for toxicity tests. Caging animals in the field allows integration of all environmental variables that covary with contaminant exposures, while such laboratory tests capture only step increments of these covarying conditions. Use of an on-site toxicity testing laboratory that pumps estuarine water into a laboratory test system captures some aspects of in situ assessments with fewer risks of losing test animals. In order to obtain chronic exposure-response data in the field that parallel laboratory chronic or early life-stage tests, a commitment of resources for mobile laboratories or field tests is warranted.

Marked Animals

When a precise exposure history is not necessary for quantifying exposure-response relationships, capture-mark-recapture or transplant of sessile organisms can be used to obtain sample data on chronic effects of contaminants in estuarine systems. Initial size, weight, or condition measurements can be compared with measurements from individuals or test populations upon subsequent recapture or sampling [*19,20*]. Endpoints such as growth or external pathology frequency are especially applicable to this approach. Data from marked individuals can be obtained as part of larger population sampling efforts described below. Information on movement, avoidance, or rates of emigration or immigration obtained from marked individuals can assist in assessments of single species, population, and community changes.

Clinical Markers

Measures of stress or exposure to contaminants can include biochemical, physiological, and pathological parameters known to respond to organic or inorganic pollutants [*2,17,21*]. A variety of measurements are possible for different types of contaminants such as metals, PAHs, organochlorines, etc. [*22,23*]. Clinical markers currently are used as supplemental information or in screening efforts rather than ecological effects endpoints, although field studies are underway to correlate laboratory and field effects [*20*]. Field surveys of indigenous species and evaluations of clinical measures of stress or contaminant exposure can assist in site selection or narrowing the scope of suspected problems. Assessments of changes in healthy animals placed at test sites would allow comparison of laboratory and field exposure-response measurements for acute or chronic exposures [*20*].

Population Studies

One of the most meaningful aspects of field studies is the ability to quantify changes in indigenous populations. The significance of changes will answer questions that occur when laboratory data are used to predict field effects (for example, "So what if 50% of the animals die, only 10% live to be adults anyway."). Laboratory-based predictions of exposure-response can be evaluated in terms of changes in the abundance and age structure of species present at field study sites [*24–26*]. Transient effects such as localized fish kills may be masked by recolonization from unaffected areas. However, the ability to demonstrate long-term effects on the ability of a population to sustain itself or sustain biomass sufficient to support predator populations has significant ecological and regulatory implications [*27*].

Various techniques are available for measuring the abundance and dynamics of populations of selected estuarine species. Perhaps the most studied species are those of commercial or sport significance, species with restricted distribution, or those species important to estuarine food webs. Collection efforts for population studies can be combined with surveys on the incidences of gross pathologies among a population to assist in quantifying the health of organisms in an area [*21*].

Changes in abundance of populations following catastrophic events such as fish kills can be monitored to demonstrate the resiliency and recovery of estuarine populations. If populations are affected by contaminant inputs in localized habitats, field studies can quantitate recovery rates for these systems and evaluate the importance of refuge areas for sustaining estuarine populations. The choice of methods to quantify changes in population structure or dynamics is left to individual investigators because each habitat has site-specific aspects that may provide for advantages of one approach over another.

Community Studies

Analyses of estuarine communities must take into account the physical characteristics of habitats studied. The complexity of estuarine communities can range from very intricate collections of considerable numbers of species and individuals (common to habitats with high salinity and considerable-to-moderate physical energy input) to simple communities of few species and individuals (typical of habitats of low or widely ranging salinity and high-energy input) [*28*]. Characterizations of communities with diversity indices or other indices of community complexity may show reference or untreated areas with values typical of severely polluted habitats because the everchanging physical conditions common to some estuarine habitats may limit the numbers and kinds of species present [*29*]. Emphasis on species composition, biomass distributions, or functional guilds may provide different perspectives of contaminant effect. A variety of methods should be considered when characterizing contaminant effects over the range of physical conditions found in estuaries.

Studies of benthic communities that colonize "hard bottoms" are commonly reported in the literature because these communities are easy to sample through time. They are structurally complex with a variety of functions represented. Although infaunal communities of "soft bottoms" play an important role in estuarine food webs, the diversity of life forms and abundance of various individuals can be more limited relative to hard-bottom communities [*30*]. Studies of infaunal communities are extremely important for estuarine systems because soft bottom habitats often are contaminant sinks. Changes in the structure or biomass of communities in these areas may be indicators of contaminant effects [*30*].

Studies of planktonic communities also can be important when evaluating contaminant effects in estuaries. Because of tidal movements, direct effects on planktonic communities must be studied while following a contaminated water mass as it is moved through an estuary. Short-term exposures of planktonic organisms may be more intense than that of benthic communities because contaminant concentrations diminish through time only by mixing and dispersion, while exposures for benthic communities diminish as a result of tidal flow out of an area. Problems with spatial and temporal heterogeneity in the distribution of planktonic communities are significant, leading to intensive sampling regimes in order to obtain sample data sufficient for sensitive statistical analyses.

Ecosystem Responses

Field data on the response of estuarine systems to contaminant exposures can be interpreted through ecosystem structure-and-function analyses [*6,11,31*]. Such an approach would provide the ultimate assessment of contaminant effect in a risk assessment process, allowing an evaluation of the significance of changes seen in single species, populations, or communities. However, estuaries tend to be large dynamic systems and pose a considerable challenge for obtaining data bases appropriate for assessing system-level responses [*4*]. Costs and time demands have prohibited field assessment of system-level, risk-assessment procedures. System-level tests have been limited to modeling efforts or microcosm approaches which would then need field validation before final acceptance for risk-assessment decisions [*32*].

Experimental Design Problems

Estuaries are semienclosed systems with freshwater input at the headwaters and saltwater input at the mouth. Thus, studies of contaminant input and effects cannot "capture a piece of the system" and contain it for study as has been proposed for limnocorrals in larger lakes [*33*], limnetic enclosures, or freshwater ponds [*34*]. Attempts to partition an estuarine system into component habitats or follow effects of contaminated water masses through time cannot be completed with ecological reality if exchange or interaction with boundary areas is ignored. Thus, many estuarine systems are studied on a large scale with little resolution to achieve specific exposure-response relationships [*20*]. Rather, correlative trends through time are established between measures of contaminant dispersion and biotic components [*23*].

Field studies in estuaries contaminated by toxic substances often are limited by the absence of appropriate control or reference sites that are not influenced by contaminant sources. Our level of understanding of contaminant effects requires extensive baseline data for sufficient characterization of "healthy" systems [*11,20*]. At some level of resolution, each estuarine system is unique, and therefore an appropriate reference or control cannot be found. It is necessary that the intensity of study into contaminant effects must not go beyond the degree of difference between sites selected for comparison, otherwise site-to-site variability in distribution or condition of biota may mask contaminant effects. Fortunately, many estuaries have dendritic basins that allow selection of areas within the system for comparative studies of exposed and relatively unexposed habitats through time. Unfortunately, water exchange within an estuary often is extensive enough so that all areas within the basin are influenced to some degree by the same water masses. That degree of exchange provides an advantage because it tends to lead to greater biotic homogeneity at sites within a system as opposed to the variability induced by selecting reference or control sites in another system. However, mixing of water masses within an estuary complicates our assumptions about treated and control sites as we cannot rule out the fact that contaminants may be transported to the reference site within the system. Further, estuaries are sites of multiple stresses that are constantly affecting water quality and biotic processes. Thus, selecting a reference site that is not influenced by the contaminant of interest must be evaluated in light of other anthropogenic and ecological inputs into the estuary.

When studying contaminant input from short-term chemical applications or agricultural runoff events, synoptic sampling of preexposure, exposure, and postexposure conditions provide a means to compare each site to itself through time. When event and sampling intervals encompass hours or days, comparison of data can be considered valid [*35*]. However, when weeks or months are involved, the dynamic physical, chemical, and biological processes characteristic of estuaries may influence results and valid statistical comparisons [*35*], especially if the study is conducted during times of the year when storms are frequent or movements of biota through a system can affect the results.

ERL/GB Program Approaches

Characterization Of Field Exposure Regimes

Accurate quantification of the magnitude and duration of contaminant exposures are key aspects of field studies conducted by ERL/GB. Fundamental toxicological relationships established through laboratory tests should be predictive of field effects if field exposures are similar to laboratory test conditions. The dynamic mixing and transport processes characteristic of estuaries provide considerable challenge to field investigators whose analytical resources are limited. Our field validation studies have centered around pesticide spray applications or agricultural runoff events that have specific timing or advance notice associated with them, thus providing a reference point for when to begin sampling and when concentrations of contaminants probably have begun to diminish. Sampling procedures designed to characterize discontinuous discharges or unanticipated spills would require a considerably different approach.

When spray applications of salt marsh mosquito control insecticides were studied [*36*], we were able to minimize the number of prespray samples because we had scheduled the pesticide applications for specific days or hours. We sampled frequently after insecticide application to characterize the initial deposition and mixing processes, and then increased the time between sample collection as the rate of change of concentrations decreased.

Sampling in tidal creeks during pesticide runoff was intensive at the beginning of rainfall in order to quantitate the initial input of pesticides into the system [*37*]. Sampling was curtailed after 4 to 6 h when we perceived that subsequent dilution and rainwater runoff was dominating exposure conditions, and pesticide input diminished. What was not taken into account was the movement of contaminated water masses in and out of the creek with the tides. At low tide during runoff, the salinity of water samples taken from our creek-side stations was 10 to 15‰ with increases to 33‰ at high tide, representing tidal basin water with little input from runoff (Fig. 1). At low tide 6 h later, salinity was again low and could have had high pesticide concentrations retained from the initial runoff event or from newly contributed sources held in the field drainage ditches during high tide. It became obvious that we would have to follow initial water masses as well as new sources of runoff during subsequent tidal periods. The temporal changes at any one sampling site and the spatial ranges within an estuary at one point in time must be fully characterized if exposure regimes are to be related to effects on estuarine biota.

Movement of pesticide-contaminated runoff into a salinity-stratified estuary also can cause problems when trying to characterize animal exposures. When monitoring alachlor runoff from treated corn fields into the South River estuary near Beaufort, North Carolina, Sandstrom et al. [*38*] showed concentrations of alachlor ≤ 15 μg/L in low salinity (0 to 2‰) surface water and below detection limits (≤ 0.01 μg/L) in deeper waters of 5 to 8‰ salinity. Exposure of biota to alachlor during this and similar events was highly dependent upon where in the estuarine system destratification occurred. Animals held in floating or bottom cages would have considerably different exposures as would planktonic or benthic communities. Full characterization of contaminant dynamics in a complex system will require extensive analytical effort.

Exposure-Response Relationships

We have employed caged mysids *(Mysidopsis bahia),* shrimp *(Palaemonetes pugio* and *Penaeus duorarum),* and fish *(Cyprinodon variegatus)* to assay the field toxicity of mosquito control sprays of fenthion [*36*]. Effects of pesticide runoff from agricultural systems near tidal creeks have been studied by using caged shrimp *(Palaemonetes pugio)* and fish *(Fundulus heteroclitus)* [*37,39*]. These tests allowed direct comparison of laboratory and field exposure-response data

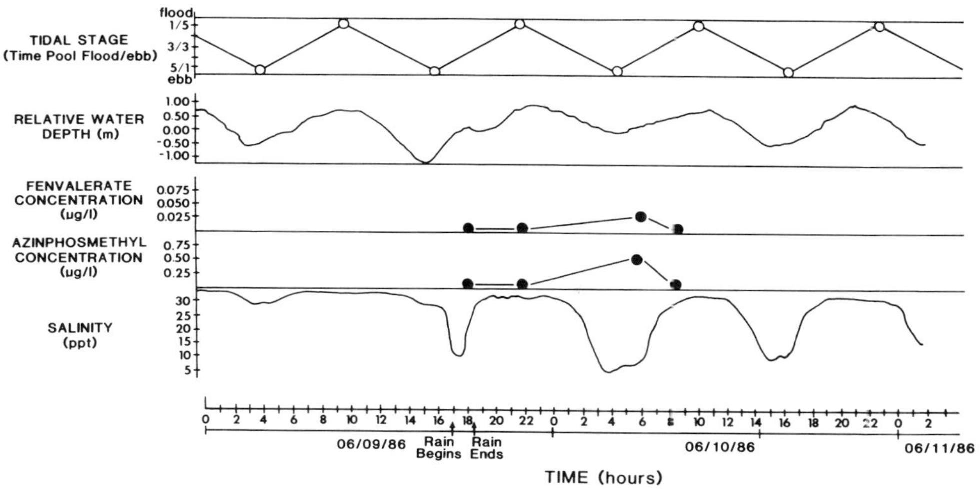

FIG. 1—*Relationship between tidal stages, relative water depth, measured concentrations of pesticides, and salinity in a tidal creek receiving agricultural runoff during storm events. Greater concentrations of pesticides in low-salinity water demonstrates input from farm areas. Figure adapted from Scott et al.* [37].

and an evaluation of our ability to predict acute lethal effects in the field from standard laboratory tests.

We found that standard acute toxicity tests provided a reasonable basis for predicting acute lethal effects of fenthion on caged mysids, grass shrimp, and pink shrimp in habitats where fenthion exposure concentrations changed slowly over time [*36,40*] (Table 2). For habitats where fenthion concentrations diminished rapidly as a result of mixing and tidal exchange, acute LC_{50}s overestimated the toxicity of fenthion to caged animals in the field. However, laboratory pulse-exposure tests, designed to mimic rapidly changing field exposure regimes, provided an accurate estimate of fenthion toxicity to caged animals. A similar approach of matching laboratory and field exposures was successful in evaluating the combined toxicity of fenvalerate and endosulfan or azinphosmethyl to shrimp and fish in tidal creeks that received runoff from tomato fields in South Carolina [*37*]. Fish kills and caged shrimp and fish mortality could not be explained through predictions based on LC_{50} data alone. Laboratory pulse-exposure tests provided a better basis for predicting field mortalities when exposures were modeled to represent rapidly changing insecticide concentrations.

By using several species of differing toxicological sensitivity to the contaminant(s) of interest at the field study site, caged-animal studies can assist in documenting the potential magnitude of effect associated with a contaminant release or contribution of individual contaminants responsible for effects during multiple contaminant exposures. The 24-h LC_{50}s for fenthion in tests with mysids, grass shrimp, and sheepshead minnows were 0.85, 9.5, and 2000 μg/L, respectively [*41*]. By placing these animals in cages at salt marsh habitats that received fenthion applications, we were able to show that lethal effects occurred only during some tests and always with the most sensitive species [*36*]. No unexplained mortality occurred among the less-sensitive animals, and most mortality could be explained by exposure to fenthion alone.

Scott et al. [*37*] used the relative differences in sensitivity of grass shrimp and mummichogs (*Fundulus heteroclitus*) to fenvalerate and their similar sensitivity to endosulfan to test for the predominant chemical effect of runoff containing both insecticides in tidal creeks (see Table 2). Runoff events when both caged fish and grass shrimp died implicated both endosulfan and fenvalerate as causative agents, whereas events when only grass shrimp died implicated fen-

TABLE 2—*Comparison of caged animal exposure-response studies in the field with laboratory acute toxicity data. Maximum field exposure concentrations are characterized as exceeding (>) or less than (<) laboratory-derived effect concentrations from standard LC_{50} tests and pulse exposure-response studies. Fenthion data from Clark et al.* [36]; *endosulfan, fenvalerate, and azinphosmethyl data from Scott et al.* [37].

Test Chemical and Species	Field Exposure Relative to Lab Test Concentration		Caged Animal Mortality Following Field Exposure	Comments
	96-h LC_{50}	MTPE[a]		
Fenthion				
Mysid	>	<	6%	Nonlethal pulse in field
Mysid	>	>	94%	Lethal pulse in field
Mysid	>	<	87%	96-h exposure to slowly changing concentrations
Mysid	<	<	10%	Nonlethal exposure for 72 h
Pink Shrimp	>	<	0%	Nonlethal pulse in field
Pink Shrimp	>	>	100%	Lethal pulse in field
Pink Shrimp	>	<	100%	96-h exposure to slowly changing concentrations
Pink Shrimp	<	<	10%	Nonlethal exposure for 72 h
Endosulfan and Fenvalerate				
Grass Shrimp	< endosulfan > fenvalerate	< >	52%	Mortality attributed to fenvalerate
Grass Shrimp	> endosulfan > fenvalerate	< >	100%	Mortality attributed to fenvalerate
Mumichog	> endosulfan < fenvalerate	no pulse data	Fish kill in tidal creek	Mortality attributed to endosulfan
Azinphosmethyl and Fenvalerate				
Grass Shrimp	> azinphosmethyl > fenvalerate	< >	90%	Mortality attributed to fenvalerate

[a]Maximum tolerated pulse exposure, maximum concentration in pulse exposure study that resulted in 10% mortality.

valerate. These assumptions were borne out when pesticide residues in water samples were compiled after caged animal mortality had been quantified. Laboratory toxicity tests showed no synergistic toxicities between endosulfan and fenvalerate [*42*]. There were no events when fish died from insecticide exposure and grass shrimp survived, results consistent with laboratory exposure-response data.

Factors Affecting Interpretation Of Field Exposure-Response Data

Field studies with caged animals allow quantification of exposure-response relationships but may not portray ecological effects accurately. In the laboratory, test animals are not allowed the opportunity to escape from toxicant exposure and animals held in cages in the field are denied similar means of minimizing exposure to contaminants. In the field, aquatic animals have been shown to avoid habitats influenced by contaminants, thus minimizing their exposure [*13,23*]. We cannot confidently extrapolate to ecological effects from responses of caged animals in the field.

Cages should be designed with mesh sizes sufficiently small to prevent animal escape but with mesh openings sufficiently large to allow adequate exchange of site water through the cage. Fouling on the cage surface can minimize exchange of test media or oxygen and export of animal wastes. Fouling also may provide a sorptive surface that reduces contaminant concentrations within cages. We compared concentrations of malathion inside and outside of mysid and fish cages at various intervals after a mosquito-control spray at an estuarine salt marsh (Table 3). At 1 h after the malathion application, concentrations inside fresh fish cages were similar to water concentrations outside the cage. Concentrations were considerably less in fish cages that had been in site water for 24 h before the spray. By 4 h postspray, malathion concentrations inside of fresh and 24-h incubated cages were similar to those measured outside of the cages. For mysid cages which had smaller mesh compared to fish cages, malathion concentrations in the 24-h incubated cages were consistently lower than those in the fresh cages. As with fish cages, concentrations inside and outside the cages were more comparable by 4 h postspray. It appears that contaminant concentrations inside cages constructed of small-mesh screen or those with biofilm development on the screen surface may not reach the immediate high concentrations as site water following pesticide applications. By 4 h postspray, sufficient mixing and diffusion had occurred to bring concentrations inside the cages to values comparable to those of site water. Our concern for sorption of malathion onto the biofilm on the cage surface was minimal for this organophosphate insecticide. However, the specific chemical traits of other

TABLE 3—*Concentrations of malathion measured in salt marsh waters following mosquito control application compared to concentrations measured inside of mysid and fish cages that had been placed at the site 1 h or 24 h before malathion application. Triplicate water samples were collected outside the cages and inside cages at 1 h and 4 h postspray. Values are μg/L.*

Sample time/ Replicate	Outside Cage	Mysid Cages, 450-μm mesh,		Fish Cages, 600-μm mesh,	
		1 h	24 h	1 h	24 h
1 h postspray					
a	5.8	7.5	1.7	3.1	2.3
b	8.3	6.7	2.4	3.3	1.9
c	4.1	3.1	2.1	3.2	3.4
1-h mean	6.1	5.8	2.1	3.2	2.5
4 h postspray					
a	0.53	0.92	1.0	0.91	0.60
b	1.2	0.48	0.81	0.56	0.47
c	0.49	0.52	0.46	1.1	0.51
4-h mean	0.74	0.64	0.76	0.86	0.53

contaminants of interest in field studies will dictate how stringently sorption and biodegradation by biofilm on the cages must be monitored.

When testing caged animals in systems where dissolved oxygen is limiting or minimal in ambient waters, exchange of water in cages is critical for the survival of test animals. The 450-μm mesh nylon screen used to hold 5- to 7-day-old mysids [*43*] has required frequent brushing to reduce fouling on the mesh surface and to allow sufficient exchange of test media. Without brushing, fouling became so extensive that it reduced exchange to the point that air could not enter through the mesh sides as cages were lifted from the water and cages would not sink upon replacement into water because air inside the cage could not escape to allow sinking. In highly productive creeks of South Carolina during June and July, daily brushing was sufficient for four consecutive days of field testing. Fresh cages had to be used after four days because the carryover biomass on the cage surface was not removed by brushing only. A thorough soap and water cleaning and bleaching was necessary before the cage could be used again. Fouling on cages for fish (600-μm mesh), grass shrimp (600-μm mesh), and penaeid shrimp (1000-μm mesh) was much less of a problem because of the larger mesh size used.

Population Studies

Scott et al. [*37*] have used block nets placed along 50-m stretches of tidal creeks at low tide to census fish and grass shrimp populations. Tidal ranges of up to 2 m allow easy collection and quantitation of these species when concentrated in the creek at low tide. By compiling changes in species abundance on a monthly basis in a reference and treated tidal creek, Scott et al. were able to correlate decreases in abundance of grass shrimp and mummichogs with incidences of fish kills or mortality of caged test animals. Samples collected for population census in months following toxic pesticide runoff events showed decreased numbers and biomass of grass shrimp and mummichogs. Recovery to densities similar to that of the reference site occurred over one to three months after the agricultural season had ended. This approach begins to address the "so what" question posed when risk assessment predictions show mortality of 20 to 80% among test species. Further comparisons of laboratory toxicity data, caged animal survival data, and population monitoring are planned. Evaluations of recovery potential through immigration and reproductive cycles also are planned.

Community Studies

Tagatz and Plaia [*44*] compared effects of fenthion on laboratory-developed benthic communities to effects on benthic communities in a salt marsh treated with ultra-low-volume applications of fenthion as a mosquito adulticide. No effects from the mosquito control sprays were seen on benthic communities studied at the field site. Laboratory studies showed that short-term exposures at concentrations two orders of magnitude greater than field concentrations were necessary to elicit effects on the numbers and kinds of individuals or species colonizing sandy substrates. In other laboratory studies, Tagatz [*45*] showed significant changes in benthic communities can result from exposure to contaminant concentrations comparable to acute LC_{50}s for the most sensitive species in standard laboratory toxicity testing programs.

Benthic invertebrates of soft-bottom communities have been studied by Livingston et al. [*46*] and Diaz et al. [*47*] as a means to evaluate effects of sediment-associated contaminants in estuarine systems. Because of the spatial and temporal heterogeneity associated with the distribution of these communities, it was necessary to group individuals into functional guilds in order to better describe statistically significant trends in the data. Contaminants in sediments and settling particles with associated contaminants were shown to have direct toxic effects on benthic invertebrates when analyzed in this manner. This approach, tested in the field and in laboratory microcosm test systems, provided a reliable means to investigate effects on benthic communities at two study sites and interrelate the mechanism and magnitude of effect.

Conclusions

Because of the dynamic nature of estuaries, a single approach to assessing contaminant effects may not provide the exposure-response data necessary to characterize fully the mechanism of effect or ecological significance of an effect. Rather, a variety of integrated approaches to single-species, population, and community studies will provide a means to validate risk-assessment procedures and predictions, evaluate applicability of laboratory-based toxicity data, and demonstrate the mechanism and/or significance of a measured change at one level of assessment relative to another. While there are few examples of comprehensive and integrative risk assessment approaches used in estuarine field studies in the past, ongoing efforts at ERL/GB will enable us to provide more extensive exposure-response assessments in the future.

Acknowledgments

I would like to thank F. L. Mayer, P. R. Parrish, D. Flemer, and D. Middaugh for their thoughtful review comments and input; S. Foss and Y. Ford for preparing the figure, and M. Stubbs for typing the manuscript. This paper is contribution number 638 of the U.S. EPA, Environmental Research Laboratory, Gulf Breeze, FL.

References

[*1*] Peakall, D. B. and Tucker, R. K. in *Methods for Estimating Risk of Chemical Injury: Human and Non-human Biota and Ecosystems,* V. B. Vouk, G. C. Butler, D. G. Hoel, and D. B. Peakall, Eds., Wiley, New York, 1984, pp. 611-636.

[*2*] Bayne, B. L., Brown, P. A., Burns, K., Dixon, D. R., Ivanovici, A., Livingstone, D. R., Lowe, D. M., Moore, M. N., Stebbing, A. R. D., and Widdows, J., *The Effects of Stress and Pollution on Marine Animals,* Praeger Publications, New York, 1985.

[*3*] *Concepts in Marine Pollution Measurements,* H. H. White, Ed., Maryland Sea Grant College, College Park, MD, 1984.

[*4*] O'Connor, J. S. and Flemer, D. A. in *New Approaches to Monitoring Aquatic Ecosystems, ASTM STP 940,* T. P. Boyle, Ed., ASTM, Philadelphia, PA, 1987, pp. 70-90.

[*5*] Livingston, R. J. in *Concepts in Marine Pollution Measurements,* H. H. White, Ed., Maryland Sea Grant College, College Park, MD, 1984, pp. 681-692.

[*6*] Levin, S. A., Kimball, K. D., McDowell, W. H., and Kimball, S. F., *Environmental Management,* Vol. 8, 1984, pp. 375-442.

[*7*] U.S. Environmental Protection Agency, "Chesapeake Bay: A Profile of Environmental Change," U.S. EPA, Region 3, Philadelphia, PA, 1983.

[*8*] Cairns, J., Dickson, K. L., and Maki, A. W., *Estimating the Hazard of Chemical Substances to Aquatic Life, ASTM STP 657,* Philadelphia, PA, 1978.

[*9*] Dickson, K. L., Maki, A. W., and Cairns, J., Jr., *Analyzing the Hazard Evaluation Process,* American Fisheries Society, Washington, DC, 1979.

[*10*] Urban, D. J. and Cook, N. J., "Hazard Evaluation Division Standard Evaluation Procedure, Ecological Risk Assessment," EPA-540/9-85-001, Office of Pesticide Programs, Washington, DC, 1986.

[*11*] National Research Council, *Testing For Effects of Chemicals on Ecosystems,* National Academy Press, Washington, DC, 1981.

[*12*] Duke, T. W. in *Concepts in Marine Pollution Measurements,* H. H. White, Ed., Maryland Sea Grant College, College Park, MD, 1984, pp. 1-6.

[*13*] Geckler, J. R., Horning, W. B., Neiheisel, T. M., Pickering, Q. H., and Robinson, E. L., "Validity of Laboratory Tests for Predicting Copper Toxicity in Streams," EPA-600/3-76-116, National Technical Information Service, Springfield, VA, 1976.

[*14*] *Estuarine Transport Processes,* B. Kjerfve, Ed., University South Carolina Press, Columbia, SC, 1978.

[*15*] Phelps, D. K., Katz, C. H., Scott, K. J., and Reynolds, B. H. in *New Approaches to Monitoring Aquatic Ecosystems, ASTM STP 940,* T. P. Boyle, Ed., ASTM, Philadelphia, PA, 1987, pp. 107-124.

[16] Pilson, M. E. Q. in *Concepts in Marine Pollution Measurements,* H. H. White, Ed., Maryland Sea Grant College, College Park, MD, 1984, pp. 575-588.
[17] Royal Society of London, *The Assessment of Sublethal Effects of Pollutants in the Sea,* Royal Society London, 1979.
[18] Horning, W. B. and Weber, C. I., "Short-Term Methods for Estimating Chronic Toxicity of Effluents in Receiving Waters to Freshwater Organisms," EPA 600/4-85/014, National Technical Information Service, Springfield, VA, 1985.
[19] Bayne, B. L., Moore, M. N., Widdows, J., Livingstone, D. R., and Salked, P. in *The Assessment of Sublethal Effects of Pollutants in the Sea,* Royal Society of London, 1979, pp. 165-183.
[20] *New Approaches to Monitoring Aquatic Ecosystems, ASTM STP 940,* T. P. Boyle, Ed., Philadelphia, PA, 1987.
[21] Hinton, D. E. and Couch, J. A. in *Concepts in Marine Pollution Measurements,* H. H. White, Ed., Maryland Sea Grant College, College Park, MD, 1984, pp. 7-32.
[22] National Research Council Canada, "The Role of Biochemical Indicators in the Assessment of Ecosystem Health—Their Development and Validation," NRCC 24371, Publications, NRCC/CNRC, Ottawa, Canada, 1985.
[23] Phillips, D. J. H., *Quantitative Aquatic Biological Indicators,* Applied Science Publishers, London, 1980.
[24] Barnthouse, L. W., Suter, G. W., II, Rosen, A. A., and Beauchamp, J. J., *Environmental Toxicology and Chemistry,* Vol. 6, 1987, pp. 811-824.
[25] Daniels, R. E. and Allan, J. D., *Canadian Journal Fisheries Aquatic Sciences,* Vol. 38, 1981, pp. 484-494.
[26] Goodyear, C. P., *Transactions American Fisheries Society,* Vol. 114, 1985, pp. 107-113.
[27] Hines, A. H., Haddon, P. J., Miklas, J. J., Wiechert, L. A., and Haddon, A. M. in *New Approaches to Monitoring Aquatic Ecosystems, ASTM STP 940,* T. P. Boyle, Ed., ASTM, Philadelphia, PA, 1987, pp. 140-164.
[28] Odum, E. P., *Fundamentals of Ecology,* Saunders, Philadelphia, PA, 1971.
[29] Gray, J. S. in *The Assessment of Sublethal Effects of Pollutants in the Sea,* Royal Society London, 1979, pp. 147-159.
[30] Boesch, D. F., *Marine Biology,* Vol. 21, 1973, pp. 226-244.
[31] Sheehan, P. J., Miller, D. R., Butler, G. C., and Bourdeau, P., Eds., *Effect of Pollutants at the Ecosystem Level,* Wiley, New York, 1984.
[32] Huckabee, J. W. in *Methods for Estimating Risk of Chemical Injury: Human and Non-human Biota and Ecosystems,* V. B. Vouk, G. C. Butler, D. G. Hoel, and D. B. Peakall, Eds., Wiley, New York, 1985, pp. 637-674.
[33] Solomon, K. R., Smith, K. E., Guest, G., Yoo, J., and Kaushik, N. K., "The Use of Limnocorrals in Studying the Effects of Pesticides in the Aquatic Ecosystem," Canadian Technical Reports Fisheries Aquatic Sciences, No. 975, Ottawa, Canada, 1980, pp. 1-9.
[34] Touart, L. W., "Hazard Evaluation Division: Technical Guidance Document, Aquatic Mesocosm Tests to Support Pesticide Registrations," PB88-172465, NTIS, Springfield, VA, 1988.
[35] Greene, R. H. in *Concepts in Marine Pollution Measurements,* H. H. White, Ed., Maryland Sea Grant College, College Park, MD, 1984, pp. 647-655.
[36] Clark, J. R., Borthwick, P. W., Goodman, L. R., Patrick, J. M., Jr., Lores, E. M., and Moore, J. C., *Environmental Toxicology and Chemistry,* Vol. 6, 1987, pp. 151-160.
[37] Scott, G. I., Moore, D. W., Fulton, M. H., and Hampton, T. W., "Agricultural Insecticide Runoff Effects on Estuaring Organisms: Correlating Field and Laboratory Toxicity Testing with Ecotoxicological Biomonitoring," unpublished annual progress report to U.S. EPA ERL/GB, cooperative research agreement No. CR-813138-01, 1988.
[38] Sandstrom, M. W., Eisenreich, S. J., and Kirby-Smith, W. W., Abstract, Society Environmental Toxicology Chemistry, 8th annual meeting, Pensacola, FL, 1987, paper No. 111.
[39] Baughman, D. S., Moore, D. W., and Scott, G. I., *Environmental Toxicology and Chemistry,* in press.
[40] Clark, J. R., Goodman, L. R., Borthwick, P. W., Patrick, J. M., Jr., Moore, J. C., and Lores, E. M. in *Aquatic Toxicology and Environmental Fate: Ninth Volume, ASTM STP 921,* T. M. Poston and R. Purdy, Eds., ASTM, Philadelphia, PA, 1986, pp. 161-176.
[41] Borthwick, P. W., Clark, J. R., Montgomery, R. M., Patrick, J. M., Jr., and Lores, E. M. in *Aquatic Toxicology and Hazard Assessment: Eighth Symposium, ASTM STP 891,* R. C. Bahner and D. J. Hansen, Eds., ASTM, Philadelphia, PA, 1985, pp. 177-189.
[42] Scott, G. I., Baughman, D. S., Trim, A. H., and Dee, J. C. in *Pollution Physiology of Estuarine Organisms,* W. B. Vernberg, A. Calabrese, F. P. Thurberg, and F. J. Vernberg, Eds., University South Carolina Press, Columbia, SC, 1987, pp. 251-273.
[43] Goodman, L. R. and Cripe, G. M., *Journal American Mosquito Control Association,* Vol. 3, 1987, pp. 109-110.

[44] Tagatz, M. E. and Plaia, G. R., *Journal Florida Anti-Mosquito Association*, Vol. 56, 1985, pp. 76–81.
[45] Tagatz, M. E. in *Community Toxicity Testing, ASTM STP 920*, J. C. Cairns, Jr., Ed., ASTM, Philadelphia, PA, 1986, pp. 18–29.
[46] Livingston, R. J., Diaz, R. J., and White, D. C., "Field Validation of Laboratory-Derived Multispecies Aquatic Test Systems," EPA 600/4-85/039, National Technical Information System, Springfield, VA, 1985.
[47] Diaz, R. J., Luckenback, M., Thorton, S., Roberts, M. H., Jr., Livingston, R. J., Koenig, C. C., Ray, G. L., and Wolfe, L. E., "Field Validation of Multi-Species Laboratory Test Systems for Estuarine Benthic Communities," EPA 600/3-87/016, National Technical Information Service, Springfield, VA, 1987.

J. Reese Voshell, Jr.,[1] *Raymond J. Layton,*[1] *and Stephen W. Hiner*[1]

Field Techniques for Determining the Effects of Toxic Substances on Benthic Macroinvertebrates in Rocky-Bottomed Streams

REFERENCE: Voshell, Jr., J. R., Layton, R. J., and Hiner, S. W., **"Field Techniques for Determining the Effects of Toxic Substances on Benthic Macroinvertebrates in Rocky-Bottomed Streams,"** *Aquatic Toxicology and Hazard Assessment: 12th Volume, ASTM STP 1027,* U. M. Cowgill and L. R. Williams, Eds., American Society for Testing and Materials, Philadelphia, 1989, pp. 134–155.

ABSTRACT: The heterogeneity of shallow lotic environments presents formidable obstacles for comparative evaluations that are intended to quantify the effects of toxic substances on benthic macroinvertebrates. A review of the literature indicates that contemporary hazard assessments are conducted by the same techniques that have been used since the early 1970s. Improved design techniques that are available, but are not being used, include: narrow habitat stratification, small sampling units, and selected study populations. A multitude of innovative sampling methods have been developed for these design techniques. The importance of using fine mesh nets and sieves has been recognized, but even finer mesh sizes may be necessary if the entire benthic community must be quantified. A better design for drift studies is to take numerous replicate samples at one time shortly after dark, rather than taking fewer samples at periodic intervals throughout a diel cycle. Secondary production measurements offer advantages of greater precision and ecological meaning. These advances have been developed largely in response to needs for studying benthic macroinvertebrates in basic aquatic ecology. The same needs exist in hazard assessment, but the advances in techniques have not been put into use in this area of applied aquatic ecology.

KEY WORDS: benthos, macroinvertebrates, lotic, streams, methods, design, sampling, toxic effects

In order to manage the environmental risks from toxic substances, decision makers depend upon scientists to assess hazard reliably. Thorough environmental hazard assessment must include a combination of laboratory and field studies, and frequently field studies need to be conducted in streams. Benthic macroinvertebrates have long been studied under field conditions to determine the effects of many types of pollution. Unfortunately, it is not an easy task to conduct reliable field studies of benthic macroinvertebrates in flowing waters [*1,2*]. However, the need to be able to conduct reliable field studies as part of hazard evaluation is greater than ever. Such studies of toxic substances are necessitated by accidents, effluent permits, and registration of new substances. In addition, field validation of laboratory results is presently an area of great concern in hazard assessment [*3,4*].

The purpose of this paper is to discuss techniques for conducting field studies in lotic ecosystems to determine the effects of toxic substances on benthic macroinvertebrates. We will

[1]Department of Entomology, Virginia Polytechnic Institute and State University, Blacksburg, VA 24061.

emphasize sampling design and methods. Our objectives are to: summarize the techniques that have been used, describe improved techniques that are available but are not being put to use, and identify unresolved issues. Our approach will be a combination of literature review and presentation of original results.

Rationale and Scope

Before proceeding further, it would seem prudent to elaborate on why environmental scientists attempt the difficult task of studying benthic macroinvertebrates in lotic ecosystems, and why these studies are often conducted in rocky habitats. There are several reasons why benthic macroinvertebrates are among the biota most commonly investigated in field studies designed to assess the hazard of toxic substances.

1. *Occurrence.* Benthic macroinvertebrates are abundant in almost all freshwater environments.
2. *Collection.* Sedentary habits and comparatively large size, in combination with abundance, make them relatively easy to collect.
3. *Identification.* Most can now be identified to genus with comprehensive taxonomic works that are readily available (for example, Refs *5,6*). Species-level identifications are not always possible, but considerable progress is being made, usually by revisions of individual families and genera.
4. *Life histories.* Most benthic macroinvertebrates have life cycles that last from several months to a year or more, and they are largely confined to the water during this time. Although insects emerge from the water for their adult stage, the immature aquatic stages comprise a far greater proportion of the total life span. Thus, the benthic macroinvertebrates that exist in a given area are indicative of the environmental conditions that have occurred for at least several months previously [*7,8*].
5. *Diversity.* Taken as a group, benthic macroinvertebrates have diverse habitat and food preferences, but many individual taxa have narrowly defined niches. Therefore, it is likely that some taxa will demonstrate a change in response to any perturbation that might take place [*8,9*].
6. *Functions.* Activities of benthic macroinvertebrates affect major ecological processes of freshwater ecosystems. These significant activities include grazing on primary producers, decomposing organic matter, preying on smaller invertebrates, temporarily storing materials (spiralling), and providing food for fish [*8–11*].

Field studies for assessing hazard are often conducted in lotic ecosystems simply because the overall extent of streams and rivers makes them the surface waters most likely to receive toxic substances. In particular, small to medium-sized streams of Orders 1 through 6 [*12*] are very frequently investigated because, on average, they account for almost 97% of the length and almost 100% of the area of watersheds [*1*]. Flowing waters of this size are the most likely to be impacted by toxic substances from agriculture, forestry, light industry, urbanization, and land transportation accidents, among other perturbations.

As the title indicates, the scope of this paper is restricted to *rocky-bottomed* streams. This is merely for the sake of practicality. Most stream studies emphasize shallow sections that can be reached by wading (riffles, runs, glides). The channel bottom in these wadable segments consists of mineral substrate ranging in size from bedrock and boulders to pebbles and gravels (categories according to Hynes) [*1*].

Traditional Approaches

We have arbitrarily chosen the early 1970s as a point of reference for describing what we call the "traditional" approaches for conducting field studies to assess hazard in streams. Environmental legislation caused a considerable increase in the number of field studies during this period. The manual prepared by the USEPA [*13*] was generally accepted as the standard reference for how to go about studying macroinvertebrates in relation to water quality and, hence, will be the focus of this section. In addition, Cummins [*14*] and Hynes [*1*] had written excellent reviews of techniques for studying benthic macroinvertebrates from the perspective of basic aquatic ecology.

Design

The USEPA [*13*] identified three basic types of biological field studies: reconnaissance surveys, synoptic surveys, and comparative evaluations. Reconnaissance surveys are conducted only once, with the primary intent being to obtain information necessary for planning more comprehensive studies. Synoptic surveys are more detailed than reconnaissance surveys and are conducted as an end in themselves. They are intended to determine the kinds and abundances of all organisms in all recognizable habitats in the environment being studied. Synoptic surveys are particularly useful for long-term surveillance programs. Comparative evaluations attempt to determine the effects of specific perturbations by studying the biota with and without the perturbation and with different levels of the perturbation. Comparisons may be made in time (pretreatment versus posttreatment) or space (upstream versus downstream, adjacent streams). Comparative evaluations are the type of biological field study most commonly conducted in hazard assessment, and they will be the emphasis of this paper.

Both Cummins [*14*] and USEPA [*13*] identified three major sampling designs in biological field studies: systematic, simple random, and stratified random. Systematic sampling is performed when the entire fauna needs to be sampled quantitatively. It requires equal sampling effort in all available habitats. In streams this is usually accomplished by establishing several transects across the channel and taking samples at periodic intervals along each one. Systematic sampling is usually employed for reconnaissance and synoptic surveys. Comparative evaluations require random sampling so that statistical procedures can be used to analyze differences in the biota according to times or places. In simple random sampling each available unit in the entire study area has an equal chance of being selected. This is usually accomplished by superimposing a grid on the study area and selecting the desired number of units with a random number table. Simple random sampling is effective for determining the factors that affect the distribution of benthic macroinvertebrates within a study area, but is not efficient for detecting differences between times or places that are caused by perturbation. Benthic macroinvertebrate communities are very closely related to physical factors such as substrate, current, and depth. These physical factors produce so much variability among the fauna that the effects of a perturbation cannot be distinguished between times or places by simple random sampling. Stratified random sampling reduces this problem by only collecting samples in areas with particular physical characteristics (strata) which contain a less variable fauna.

The USEPA [*13*] endorsed stratified random sampling for comparative evaluations and recommended considering substrate type, current velocity, and depth when choosing strata. However, the EPA manual suggested only broad stratification, such as riffles versus pools, and the methods described for sampling riffle areas did not stratify the substrate any further than the entire assemblage of mineral particles.

The EPA manual recommended that a minimum of three replicate samples be taken in each stratum. Both Cummins [*14*] and Hynes [*1*] mentioned a study by Needham and Usinger [*15*] indicating that 73 samples would be required to be 95% confident of total numbers of organisms. Chutter [*16*] reanalyzed Needham and Usinger's data and found that the required number of samples for 95% confidence could be as high as 448. In the early 1970s these pessimistic results were largely ignored, and most studies were conducted with three replicate samples as recommended in the EPA manual.

Methods

For quantitative sampling of the natural substrate, the EPA manual recommended two basic types of devices: coring and sieving. Coring devices completely enclose a specific area of bottom. For stream riffles, corers must be stovepipe-type devices, 17 to 30 cm in diameter. The EPA manual mentioned only the Wilding sampler as a specific example of a coring device. Sieving devices delimit a specific area of bottom, but are open on at least two sides so that the current washes the organisms, which are dislodged by hand, into a catch net. The EPA manual mentioned only the Surber square-foot sampler as a sieving device. The mesh size of the catch net was not mentioned, but it can be assumed that 595 μm was recommended because that was the size used in the definition of macroinvertebrates (those retained on a U.S. Standard No. 30 sieve). The EPA manual recommended two artificial substrates as quantitative devices: multiple-plate samplers made of hardboard and rocks contained in metal baskets. Sampling drift by means of nets anchored on the bottom was also described as a quantitative technique, but was recommended primarily for synoptic surveys. Qualitative techniques suggested in the EPA manual included hand-held screens, dip nets, rakes, and hand picking from rocks and debris.

In addition to the quantitative techniques mentioned by the USEPA [*13*], Hynes [*1*] reviewed how hand nets and kick nets could be made sufficiently quantitative for numerical assessment of the differences between areas. Techniques for quantifying these devices include general collecting for a specified length of time or disturbing an area of stream bottom in front of the net in a standard way. A kicking technique was evaluated favorably by Frost et al. [*17*]. These kick net procedures were used commonly for pollution studies in streams during the early 1970s.

Contemporary Approaches

We decided arbitrarily that papers published during the past five years (1982 to 1987) would indicate what we consider to be "contemporary" approaches for assessing hazard in streams. In order to obtain applicable papers published from 1982 to 1986, we searched the annual bibliographies published by the North American Benthological Society and the Water Pollution Control Federation. The former society emphasizes the basic aspects of aquatic ecology, and the latter emphasizes applied aspects. Papers published in 1987 were obtained by searching Current Contents (Institute for Scientific Information, Inc.). No attempt was made to search for technical reports and other types of "gray" literature. The criteria for including papers in our review were: (1) the goal was to determine the damage caused by a toxic substance; (2) the scope included studies of benthic macroinvertebrates; (3) the study area was a natural stream with rocky bottom. A total of 45 papers were selected for review here; these were certainly not all of the papers published on the subject from 1982 to 1987, but the major journals are included and the papers are representative of the studies being conducted. The results of our literature search are presented in Table 1. Note that the number of reports for some categories may exceed the total number of papers reviewed (45) because some papers included more than one study or more than one method was employed.

Design

Almost all field assessments of hazard in streams are being done by means of comparative evaluation; synoptic surveys account for only 4% of reported studies. The control or reference that is used most commonly (almost ⅔) for comparison is an upstream site on the same stream; comparisons in time (pretreatment versus posttreatment) are used slightly more frequently than comparisons with separate unimpacted streams.

Two thirds of all studies are of sufficient duration for sampling to be repeated over time in order to follow trends, but most studies (38%) are for less than a year (usually only several

TABLE 1—*Profile of contemporary techniques for assessing the hazard of toxic substances to benthic macroinvertebrates in rocky-bottomed streams. See text for explanation of literature search.*

Characteristics	Number of Reports	Percent of Reports
DESIGN		
Type of study	48	
Comparative evaluation	46	96
Synoptic survey	2	4
Control or reference for comparative evaluation	46	
Same stream (up versus down)	29	63
Separate stream	7	15
Time (pre- versus posttreatment)	10	22
Duration of study	45	
Repeated: ≥2 years	5	11
Repeated: ≥1 year, <2 years	8	18
Repeated: <1 year	17	38
One-time event	15	33
Frequency of sampling for repeated studies	30	
1 year	2	7
>1 month, <1 year	12	40
1 month	6	20
≥1 week, <1 month	6	20
≥1 day, <1 week	4	13
Stratification of sampling	26	
General riffle	24	92
More specific (stones 1; leaves 1)	2	8
Number of samples	46	
2	5	11
3	12	26
4	13	28
5	9	20
6	2	4
8	1	2
10	2	4
30	2	4
METHODS		
Mesh size to retain organisms (net or sieve, μm)	44	
≥601	8	18
501-600	5	11
401-500	10	23
301-400	12	27
201-300	5	11
101-200	1	2
≤100	3	7
Sampling devices		
Benthic	52	
Natural substrates	40	
Corer, pipe, box (fully enclosed)	3	6
Surber	17	33
Enclosure with net (Hess 5; portable invertebrate box sampler 3; T-sampler 2)	10	19
Airlift	1	2
Kick net	7	13
Hand picking specific substrate	2	4

TABLE 1—*Continued.*

Characteristics	Number of Reports	Percent of Reports
METHODS		
Artificial substrates	12	
Standardized	8	
Multiple-Plate	2	4
Other (plastic strips 2; plastic plates 1; artificial turf 1; polypropylene rope 1; concrete blocks 1)	6	12
Representative (containers of rocks)	4	8
Drift (anchored nets)	18	

months). Only 18% of studies last long enough for seasonal events to be followed for a complete year, and only 11% of studies last long enough to follow trends for two or more years. One third of all studies are very short term and include only a single observation of the effect of the toxic substance. In the studies that are repeated to observe trends, samples are not collected very frequently. The intervals between sampling are usually longer than one month. Studies with sampling intervals of one year are conducted for the purpose of long-term monitoring, and those with sampling intervals less than one week are drift studies.

Sampling stratification is seldom mentioned, or else not clearly explained, but in the 26 papers where this aspect of design was understandable, only two studies conducted stratified sampling that was restricted any further than general riffle areas. Although benthic macroinvertebrate communities are known to be highly variable in riffle areas, in 3/4 of all studies only three to five samples are taken, with four being the most frequent number.

Methods

One half of all reports on the mesh sizes of nets and sieves being used to retain organisms are between 301 and 500 μm. Mesh sizes of 301 to 400 μm are used slightly more frequently than mesh sizes of 401 to 500 μm (27 and 23%, respectively). It should be noted that in 1/4 of all reports of contemporary studies, the mesh size is still the same as that recommended by the USEPA [*13*] (approximately 595 μm or larger). Our review revealed that mesh sizes of 200 μm or less are rarely used.

Devices that sample natural substrate are used much more frequently than artificial substrates to study bottom dwelling organisms (40 versus 12 reports, respectively). Clearly, the Surber sampler remains the most commonly used device to sample natural substrate (33% of reports). Enclosures with nets are the second most often used devices (19%), with three different types being reported in the literature. Kick net techniques are used occasionally (13%). Devices that fully enclose an area of natural substrate without having a catch net (corers, pipes, boxes) and hand picking from specific substrates are rarely used.

Multiple-plate samplers are not used frequently; however, a variety of other artificial substrate samplers that provide standardized surfaces for the attachment of organisms are used occasionally (12% of all reports for bottom-dwelling organisms). They are followed in usage by rock baskets (8%), which are considered to be artificial substrates that are representative of natural substrate.

Drifting organisms are frequently sampled as part of studies of the effects of toxic substances in streams (18 reports). Nets anchored in the current are consistently used for this purpose. Drift sampling is usually undertaken in conjunction with bottom sampling.

Availability of Improved Techniques

It is apparent from the preceding discussion and Table 1 that the field techniques currently being used to assess the hazard of toxic substances in streams are little different from those employed in the early 1970s. It is our contention that innovations in design and methods have been put forward in basic aquatic ecology, but they have not been applied towards improving hazard assessment in streams. To support our claim, we will review some of the suggestions that have been published and report some of our findings from the New River.

Design

Benthic—As pointed out earlier, stream ecologists have known that many samples are required for reasonable precision ever since the often-cited paper by Needham and Usinger [*15*]. They conducted an intense study in Prosser Creek, California, involving 100 Surber square-foot samples (0.09 m^2) taken from a single riffle on one occasion, and reported that 73 samples would be required to be 95% confident of total numbers of organisms. When Chutter [*16*] corrected some errors made by Needham and Usinger, he reported that instead of 73 Surber samples, 448 samples would be necessary to obtain a standard error equal to 5% of the mean at the 95% confidence level for total numbers of organisms. Such results have changed what is now considered to be "reasonable precision." Elliott [*18*] indicates that for benthic macroinvertebrates a standard error equal to 20% of the mean (equivalent to confidence limits of ±40% of the mean) is reasonable. However, analyses of several data sets (Table 2) reveals that even this more liberal definition of reasonable precision does not always reduce the required samples to a practical number.

TABLE 2—*Precision of density measurements for selected individual taxa and total organisms obtained by using devices measuring 0.09 to 0.10 m^2 in general riffle habitats. See text for details of studies (CL = confidence limits).*

Study	No. of Samples	Taxa	Mean	Standard Deviation	*n* for 95% CL ± 40%	95% CL for *n* = 4
1	50	*Baetis*	2.18	1.89	19	±87
		Ephemerella	11.78	7.54	10	±64
		Hydropsyche	1.12	1.39	39	±125
		Glossosoma	13.06	18.93	53	±146
		Simuliidae	2.52	6.40	163	±255
		Chironomidae	4.58	4.57	25	±100
		Total	72.54	34.32	6	±47
2	10	*Baetis glaucus*	22.80	11.36	5	±45
		Baetis harrisoni	19.20	20.55	37	±121
		Cheumatopsyche thomasetti	114.50	157.74	61	±156
		Macronema capense	101.90	76.08	18	±85
		Simulium spp.	27.60	15.73	10	±64
		Chironominae/Orthocladiinae	77.70	34.89	6	±51
		Total	1632.90	701.96	6	±48
3	10	*Isonychia*	3.47	3.60	30	±109
		Stenonema	7.94	11.10	16	±81
		Ceratopsyche	5.06	11.30	6	±51
		Cheumatopsyche	17.87	28.20	13	±72
		Optioservus	5.03	12.80	5	±44
		Chironomidae	1.95	2.70	17	±82
		Total	33.03	77.90	6	±48

The required number of samples (n) reported in Table 2 was calculated from the general formula in Elliott [*18*]:

$$n = t^2 s^2 D^{-2} x^{-2}$$

where

t = Student's t value for 95% probability,
s = standard deviation,
D = relative error in terms of confidence limits of the mean (0.40), and
x = mean.

Study Number 1 is derived from Needham and Usinger [*15*]. When Chutter [*16*] reexamined their study he found that 50 of the 100 Surber samples had been collected in water too deep for correct use of the device, and that the samples collected by Operator H had significantly higher numbers of organisms ($P < 0.01$) because he included stones lying outside the frame. In Table 2 we have included only the results from samples less than 30.48 cm deep, and the values from samples collected by Operator H have been reduced by 0.6363 as suggested by Chutter [*16*]. Study Number 2 is derived from Chutter and Noble [*19*], who collected ten Surber samples from a riffle on one occasion in the Vaal River, South Africa. Study Number 3 is unpublished data that we obtained from ten samples with a 0.10-m^2 portable invertebrate box sampler (PIBS) on one occasion in an unimpacted riffle of the Piney River in Virginia.

In each study reported in Table 2 the number of samples required for the total numbers of organisms to have 95% confidence limits of ±40% of the mean (equivalent to standard error of 20% of the mean) is six. Although six samples are not often collected in contemporary studies (Table 1), this would not be an unreasonable number of Surber or PIBS samples; however, determining changes in total numbers without determining changes in the numbers of individual taxa comprising the community would not provide sufficient explanation for assessing hazard. When representative individual taxa are considered in Table 2, in almost every case the number of Surber or PIBS samples required for the same level of precision is not reasonable, averaging about 30 samples and ranging as high as 163 samples. Based upon their analysis of a large set of samples taken with a 0.09 m^2 Hess sampler, Schwenneker and Hellenthal [*20*] reported even higher numbers of samples being required to detect 100% differences in populations of individual taxa.

The right column of Table 2 reports the 95% confidence limits of the mean obtained by substituting $n = 4$ (the usual number of samples in contemporary studies, Table 1) in the formula and solving for D. This indicates that changes in total numbers less than about 50% and changes in numbers of individual taxa less than about 100% would not be detectable. Changes of such great magnitude would only occur from extreme acute toxicity and would probably not have required the use of statistics in the first place. Schwenneker and Hellenthal [*20*] also concluded that "three or four samples generally are incapable of detecting less than catastrophic environmental changes."

Obviously, greater precision must be obtained if stream-dwelling benthic macroinvertebrates are to be useful for hazard assessment. Resh [*21*] demonstrated conclusively that habitat stratification reduces the number of samples necessary for a specified level of precision. He reported a population study of the caddisfly *Ceraclea ancylus* in which the samples were not stratified and stratified according to general riffle area, riffle area with specified substrate size (phi = 5) [phi is a standard unit of measurement for substrate size and is defined as the negative log to the base of the particle size diameter in millimeters], and individual stones of specified size (phi = 5). The number of samples required for a specified level of precision was distinctly higher for the unstratified design, intermediate for general riffle area and riffle area with specific substrate size, and much lower for individual stones. The concept of using individual stones as sampling units has been supported by other studies, which have also included methods for collecting stones with their constituent fauna and measuring the surface area of the stones [*1,22–26*].

Taking small samples is also a design technique that can be used to improve precision, particularly when used in conjunction with habitat stratification. Cummins [*27*] successfully used a 0.016-m^2 grid to sample larvae of *Glossosoma nigrior,* which graze algae on the surfaces of rocks. Elliott [*18*] lists several advantages of small sampling units over larger ones: (1) more samples can be collected and analyzed for the same effort; (2) with more degrees of freedom, the statistical error is reduced; and (3) a wider range of the specified habitat is covered, making the results more representative. The idea of taking greater numbers of smaller samples to increase precision without increasing the effort required for sample processing has been supported by two statistical analyses conducted on large data bases obtained from the literature [*28,29*]. Both of these studies reported that small sampling units were particularly effective for populations with high densities. Some precautions must be taken to assure that sampling units are not too small. Coffman et al. [*30*] found that 0.09-m^2 samples were adequate for estimating the standing stock of most benthic macroinvertebrates but were too small for estimating the standing stock of larger organisms such as crayfish. Elliott [*18*] cautions that with small sampling units the sampling error at the edge of the unit is proportionally greater.

Another narrow habitat stratum that can be sampled in small units to reduce variability is aquatic macrophytes. Hynes [*1*] suggested that moss can be made quantitative by either scraping a definite area or by plucking tufts at random and then relating volume of plant material to area of bottom by a previously established mathematical relationship. He mentioned that the inhabitants of moss are usually not lost during sampling because of their habits of hiding and clinging. Gregg and Rose [*31,32*] have examined the relationships between macroinvertebrates and macrophytes and have provided methods for measuring the surface area of the plants available for colonization. The accuracy, precision, and cost efficiency of sampling techniques for macrophyte standing stock and epiphytic invertebrate populations in the littoral zones of lentic environments were analyzed by Downing and Anderson [*33*] and Downing and Cyr [*34*], respectively. Quadrat samples of small area were recommended both for the plants and the invertebrates residing on them.

A very promising approach to reducing the number of samples required for reasonable precision within reasonable costs is to choose certain populations for study rather than the entire community of benthic macroinvertebrates [*20,21,27*]. Some species occur at such low densities with exceptionally clumped distributions that, realistically, they could never be studied with any reasonable degree of precision for making decisions. Unless these troublesome species have some particular importance, such as rare or endangered status or a unique ecological function, they probably should not be given much attention in hazard assessment. Failure to focus studies on certain species likely will result in not being able to draw conclusions or obtaining conclusions at higher costs and effort than necessary. The choice of study populations with useful sampling characteristics should not be misconstrued as studying only "easy" species [*21*]. With some preliminary sampling it should be possible to choose one or more study organisms that permit a larger series of samples to be collected and analyzed with less effort than fewer samples of the entire community, without sacrificing any information that could be used for making decisions. Narrowing the choice of study populations will often be achieved conjointly with narrow habitat stratification. Desirable attributes of study populations include: high density, distinct habitat preference, habitat that can be sampled reliably, predictable phenology, and ecological significance. Examples of ecological significance might be shredders providing fine detritus for collectors, grazers regulating primary producer populations, large invertebrate predators regulating prey populations, and any organisms serving as food for fish. The concept of focusing on a few study populations is not new, but different terminology has been used, such as "indicator," "target," and "keystone" species [*35,36*]. What is proposed here is closely related to the concept of keystone species. Keystone species not only indicate community characteristics, but also play a role in controlling community structure out of proportion to their numbers [*36*].

Drift—Although drift measurements may be very useful for assessing hazard in streams and are presently being used for that purpose (Table 1), little attention has been given to appropriate design for comparative evaluations. However, Allan and Russek [*37*] made a major contribution towards filling this void in our knowledge. These authors analyzed four years of data from Cement Creek, Colorado and concluded that drift sampling requires fewer replicates than benthic sampling for comparable precision. For an abundant mayfly species, 6 to 7 replicates were required to obtain 95% confidence limits of ±50% of the mean. Allan and Russek [*37*] also concluded that it is not necessary to quantify the periodicity of drift by taking frequent samples over a 24-h period if the goal is to make a comparative evaluation of times or places. By regressing drift density from the first night collection against drift density of seven samples collected at 3-h intervals over the remainder of the night and day, they demonstrated that 60 to 90% of the variation was accounted for by the single sample just after dark. Therefore, the effort normally put into frequent sampling over 24 h should be invested instead in replicated sampling just after dark, when drift normally is greatest. This design greatly enhances the usefulness of drift sampling for hazard assessment.

Methods

Natural Substrates—The Surber sampler remains the most commonly used device for general riffle areas, even though at least three other devices with catch nets (Hess, T, PIBS) are superior. All four devices are illustrated in Merritt et al. [*38*]. The Surber sampler is not enclosed, hence organisms can easily escape from the frame without being washed into the net [*21*]. The open construction also allows current to bring organisms and debris into the sample from upstream [*39*]. In addition, the frame does not fit closely to irregular stream bottom, providing another mechanism for escape. The only advantages of the Surber sampler are small size for easy transportation and minimum depth required for sampling in streams with low flows. The former advantage is merely one of personal convenience, which should take much lower priority than scientific reliability. Being able to take samples in very shallow water with the Surber may not be scientifically advantageous because such low flow conditions would make it more likely for the animals to escape by swimming out of the open sampler. We suggest that only those devices which completely enclose the sample area, either with solid walls or nets of appropriate size mesh, should be used for quantitative sampling of benthic macroinvertebrates.

The Hess and T-samplers are round devices with mesh-covered incurrent openings and nets attached to excurrent openings. They can be rotated and pushed down into substrate consisting of pebbles and small cobbles to effectively enclose the sample. Hess [*40*] first described the 0.1-m^2 cylindrical metal sampler that has come to bear his name, but it has undergone several modifications [*39,41*]. The T-sampler [*42*] is similar to the Hess device except the former is much smaller, being constructed from a "T" coupling for plastic pipe. Resh et al. [*43*] effectively used a 0.0075-m^2 T-sampler to study the impacts of geothermal energy development on a caddisfly inhabiting gravel substrate.

We prefer the PIBS for sampling general areas of riffles, because it works better on very irregular large cobble substrate than either the Hess or T, while also being very effective on small cobbles and pebbles. The foam-lined bottom establishes a thorough seal by taking the shape of the bottom rather than having to be pushed down into the substrate. The square shape of the PIBS and the flange at the bottom make it easy to attach to a backpack frame for transporting into remote areas.

Although the Wilding or stovepipe sampler (illustrated in Ref *38*) has not been widely used for studying the effects of toxic substances in streams, it should collect accurate samples because the solid walls guarantee that no organisms will escape after the device is pushed into the substrate. The probable reasons for it not being used are difficulty with removing the organisms and voluminous, time-consuming samples that contain all of the mineral substrate and coarse detritus. These difficulties have been overcome by modifying the Wilding device into a "vacuum benthos sampler" [*44*]. They added a battery-operated pump and hose that are used to transfer the organisms into a mesh collecting bag (as fine as 243 μm) while the substrate is stirred. The vacuum benthos sampler collects most taxa as effectively as a modified Hess sampler, while offering several advantages: (1) independent of flow; (2) mesh size can be quickly changed; (3) small volume samples; and (4) little damage to organisms. A major limitation of this device is that it can only be used in gravel beds which can be penetrated by stovepipe. Thus, the range of habitat that can be sampled effectively is approximately the same as the T-sampler. The Hess and especially the PIBS can sample a wider range of benthic habitats that include much coarser substrate.

Special devices have been developed for narrowly stratified sampling. Doeg and Lake [*23*] described a wire mesh box with a catch net for quantitatively sampling the macroinvertebrates of individual large stones in streams. The device is lowered over a randomly selected rock, enclosing it on all sides. The rock is placed on a wire frame in the mouth of the catch net; then the wire frame and catch net are removed and the fauna on the rock brushed and washed into the net. The authors compared the device to the usual method of sampling individual stones, lifting them in front of a kick net, and found that significant numbers of mobile organisms escape the kick net by swimming around the sides. Wrona et al. [*25*] developed a complete protocol for using individual stones as sample units and extrapolating the results to an entire study area to calculate population density and error terms.

Hudson and Oliver [*45*] described a device specifically for sampling outcrops. They devised a "portable suction sampler" that simultaneously collects macroinvertebrates and periphyton from 0.0023 m^2 of rock surface. The sampling area is delimited by a 20 to 25 cm section of 5.4 cm ID PVC pipe with foam rubber attached to form a better seal. Macroinvertebrates and algae are removed from the rock surface with a toothbrush and transported to a 1-L collection bottle by means of a hand vacuum pump. The authors listed several advantages over other substrate samplers: (1) samples rocks effectively; (2) samples both benthos and periphyton; (3) quantitative; (4) small area permits quick processing of samples; (5) avoids problem of exposure time that must be considered for artificial substrates; and (6) portable, inexpensive, and easy to use. They also pointed out several disadvantages: (1) limited to stable, smooth rocks with no macrophyte growth; (2) may be avoided by larger macroinvertebrates; and (3) use is limited by depth.

Investigators have continued to evaluate kick net methods for assessing hazard in streams (for example, Refs *46,47*), but little new information has been contributed beyond what was known in the early 1970s. Although these methods are sometimes regarded as quantitative, it is very difficult to relate the catch to a measurable area of stream bottom. We suggest that the only appropriate uses of kick net methods are for determining lists of taxa and perhaps for relative community composition; therefore, kick net methods should be used only for reconnaissance and synoptic surveys, not comparative evaluations.

Artificial Substrates—Artificial substrates have attracted the attention of aquatic scientists studying benthic macroinvertebrates for over 50 years. Flannagan and Rosenberg [*48*] compiled a complete and annotated bibliography of more than 120 types of artificial substrates. A very thorough review of their use was provided by Rosenberg and Resh [*49*]. The latter authors divided all artificial substrate samplers into two major categories: (1) representative artificial substrates (RAS) that closely resemble the natural substrate where they are placed (for example, a tray filled with rocks placed in a stream riffle); and (2) standardized artificial substrates (SAS) that differ from the natural substrate of the habitat where they are placed (for example, hardboard multiple plate samplers). Rosenberg and Resh [*49*] evaluated, and corrected when necessary, the many advantages and disadvantages of using artificial substrates that had been proposed in the literature. Several of the advantages and disadvantages are germane to our discussion of sampling techniques for rocky-bottomed streams.

The major advantage that is usually cited for using artificial substrates is a higher level of precision because of reduced sampling variability. Rosenberg and Resh [*49*] reanalyzed data from previous studies and also concluded that both RAS and SAS generally require fewer sample replicates than direct sampling devices to achieve a given sampling precision. However, Morin [*29*] contradicted that conclusion based upon analyses of a large data set established from published literature. He found that when all artificial substrates are pooled, the variability is almost identical to that observed on natural substrates. Morin [*29*] reported that some SAS are significantly more variable than average, but certain RAS (rock-filled trays and baskets) are significantly less variable.

A specific criticism of hardboard multiple plate samplers is that they have been discovered to be an attractive food source for some xylophagic (wood eating) chironomid larvae [*50*]. In nature, these species occur in leaves, stems, and petioles of vascular hydrophytes and decomposing wood substrates. Species of the genus *Stenochironomus,* and possibly seven other genera, occur in exceptionally high numbers on hardboard samplers as a result. If chironomids are not identified to the genus level (and they rarely are), an increase in density at the family level could be attributed incorrectly to a treatment effect. The problem becomes greater if the samplers are immediately returned to the water after cleaning them, and if the samplers are reused over several years, because decomposition of the wood by fungi is enhanced and the samplers offer ever-increasing food quality. Ferrington and Christiansen [*50*] also found that the supposedly higher precision of these SAS was at least partially due to presence of the organisms which were being attracted by the availability of food. When the authors removed the xylophagic chironomids from analyses, the within site sample variability almost always increased.

Another advantage of artificial substrates listed by Rosenberg and Resh [*49*] is for collecting data from habitats difficult to sample by other methods. While this is an important issue for many large river and lentic environments, it is not of any consequence for sampling in most small- to medium-sized, rocky-bottomed streams. As indicated in the introduction of this paper, most sampling in these environments occurs in shallow riffle areas, which are wadable.

Rosenberg and Resh [*49*] concluded that two disadvantages of artificial substrates were correct as claimed: (1) relatively long exposure times are required for artificial substrate samples; and (2) colonization is subject to seasonal variation. As a result of Number 1, artificial substrate sampling requires two visits: one to place the devices and one to retrieve them. This usually presents another problem because the times required for colonization are not known. In regard to Number 2, artificial substrates sampled during periods of cold temperature are usually colonized by fewer individuals and species than ones sampled during periods of warm temperature. Sampling natural substrates, where populations can always be assumed to be in equilibrium with existing conditions, would alleviate both of the problems associated with colonization of artificial substrates.

For the reasons given above, we conclude that there is little reason to use artificial substrates to determine the effects of toxic substances in shallow rocky-bottomed streams. A notable exception is rock-filled baskets and trays. Clements et al. [*51,52*] have recently described some innovative modifications and demonstrated several advantages of using small plastic trays (10 by 10 by 6 cm) filled with pebble and small cobble (2 to 6 cm diameter). These small RAS were considered to be effective for determining impacts because of: short colonization time, low variability, high density, and little processing time. Only six samples were necessary to detect a 15% reduction in density and a 12% reduction in number of taxa at impacted sites. This approach seems to obtain high precision, similar to narrowly stratifying sampling to single stones, and to have an additional advantage of collecting a more diverse community (average of 18 taxa with average dbar of 3.28 at unimpacted sites). A particularly interesting aspect of this approach is the ability to conduct manipulative experiments by moving colonized trays to laboratory streams for toxicity tests [*52*] or moving them to other field sites (Clements, Miami University, personal communication).

Drift—There have been no major improvements in methods for sampling drifting macroinvertebrates, and apparently none are needed. Sampling the water is an easier task than sampling the bottom of a stream. A typical drift sampler is illustrated in Merritt et al. [*38*]. The opening of the device usually measures from 0.05 to 0.10 m^2 and may be either rectangular or square. The net varies in length from 1 to 2 m and should be widened behind the opening before tapering to minimize backwash. Mesh sizes of 300 to 400 μm are usually used. The bottom of the opening should be placed slightly above the substrate (approximately 6 cm), and the top should be above the water surface [*37*].

Mesh Sizes of Nets and Sieves—While there has been a shift towards using finer mesh nets and sieves since the 1970s (Table 1), many studies of the effects of toxic substances may still be employing mesh that is too coarse for accurately quantifying benthic macroinvertebrates. The importance of using fine mesh has been known for over 30 years. Jonasson [*53*] investigated the efficiency of different mesh sizes for retaining lake benthos and found that, in comparison to 600-μm mesh, 510-μm mesh increased the total number of organisms by 114 to 118%, 260 μm by 351 to 394%, and 200 μm by 432 to 641%. The organisms most affected were oligochaetes and chironomids. The EPA manual [*13*] recommended using only a 595-μm sieve to retain all macroinvertebrates for water quality studies. However, investigators studying life history and production have come to realize that much of the benthic community will be missed with mesh of that size. Benke et al. [*54*] used a 100-μm mesh sieve for lotic fauna living in snag and sand habitats and a 250-μm mesh sieve for the fauna in mud habitat. They reported that if they had used a sieve as large as 600 μm, they would have overlooked virtually the entire sand community and a major portion of the mud and snag communities. Their results applied to numbers, biomass, and production and were mostly due to small chironomids. In a study of lotic chironomids, Storey and Pinder [*55*] found that 125-μm mesh would retain 94% of preserved specimens but only 61% of live specimens. Live chironomids will pass through any mesh larger than the width of their heads, whereas preserved specimens become curled and will be retained on coarser mesh. They found that 70-μm mesh was necessary to retain live chironomids. In addition to significantly underestimating (or perhaps entirely missing) the overall quantity of benthic macroinvertebrates, community structure and life history can be misinterpreted [*53*]. Although sampling precision may not be affected by mesh size, the accuracy with which samples reflect the actual community may be greatly affected. For example, the appearance and disappearance of taxa during the course of a study could be interpreted as treatment effects, when in reality these events might merely be the result of mesh size. We conclude that, if the objectives of a study require that the entire benthic community (all taxa and ages) be quantified, then the mesh size can be no larger than 100 μm.

In spite of the importance of mesh size as documented above, there are practical limitations on fineness. For example, Benke et al. [*54*] found that they could not wash mud samples from river backwater habitats through a 100-μm mesh sieve because of the great amount of fine organic matter, and had to increase the mesh size to 250-μm for those samples. We have found that 350-μm mesh is about as fine as can be used for the catch nets of samplers such as the Surber or PIBS when they must be used in very fast current. Any finer mesh makes it difficult to hold the sampler in place, and backwash prevents organisms from being carried into the catch net. In moderate current it is feasible to use such samplers with catch nets having mesh as fine as 250 μm. Whenever possible it would certainly be desirable to use nets and sieves with mesh sizes of approximately 100 μm. It can be seen from Table 1 that this is seldom done in hazard assessment, but it could be if "nontraditional" methods were used, such as the ones we have described for taking very small, narrowly stratified samples of specified study populations. Such an approach would assure both precision and accuracy. However, we conclude that in most hazard assessments the practical requirements for using mesh sizes larger than 100 μm will justifiably supercede precision and accuracy. In these cases: (1) as fine a mesh as possible should be used; (2) the mesh size should be clearly stated in the published results; and (3) the results should be interpreted cautiously, with the recognition that certain taxa and ages were probably missed in sampling.

Case History: New River Studies

The New River is a good example for the topic of this paper because decision makers are seeking an answer to an important question about the introduction of a toxic substance: What will be the overall environmental effect of consistently applying a pesticide to control black fly larvae? We have conducted field studies in the 16-km section of the New River below Hinton, West Virginia since 1982, in an attempt to answer this question. Of primary concern is a valuable sport fishery. Forage fish and early stages of sport fish feed extensively on black flies and the other aquatic insects in the New River [*56*]. Although the bacterial pesticide being used, *Bacillus thuringiensis* var. *israelensis* (Bti), is reasonably specific for black flies, there is some evidence for toxicity to nontarget taxa, especially chironomids [*57–59*]. Indirect effects on the benthic macroinvertebrate community through detritus processing and trophic dynamics have never been investigated [*60*].

In addition to having a toxic substance applied to it, the New River is a good example for the topic of this paper because the physical and biological characteristics present a "worst case" scenario for difficulty of study. The New River is a sixth order, medium-sized stream, about 300 m wide. The depth and current make it barely wadable, even during the low flows of summer and autumn. The bottom is rocky throughout most of its course, but much of the substrate is large, irregular boulders and outcrops, with cobble/pebble riffles interspersed. The habitat is made more heterogeneous by the abundance of both a submerged and an emergent macrophyte (*Podostemum ceratophyllum* and *Justicia americana,* respectively). The heterogeneity of the habitat results in highly variable distribution of benthic macroinvertebrates. The community is characterized by low diversity and high productivity, with a few taxa being highly successful because of an upstream reservoir that releases a rich supply of seston [*60,61*].

Design—We have used many of these innovative approaches in the sampling design of our New River studies: numerous (20), small (0.01-m^2) samples taken in one species of macrophyte *(Podostemum ceratophyllum)* growing attached to rock outcrops in current. Only five taxa are consistently collected in this habitat, and three of these achieve exceptionally high densities (Table 3). Some of the advantages of the design are demonstrated in Table 3, which summarizes data from one occasion (29 Aug. 1983) at one study site (immediately below Bluestone Dam). For total numbers of organisms only four samples are necessary for $\pm 40\%$ confidence limits of the mean, and only five to nine samples are necessary for four of the five major taxa. Because the samples are only 0.01 m^2 in area, approximately 40 could be processed with the same effort as four samples taken with larger samplers such as the Surber or PIBS. We have found it

TABLE 3—*Precision of density measurements for selected taxa and total organisms obtained by using a device measuring 0.01 m^2 in macrophyte-covered rock outcrops in the New River, West Virginia. See text for further details. (CL = confidence limits).*

Taxa	Mean	Standard Deviation	n for 95% CL ± 40%	95% CL for n = 20
Baetis intercalaris	15.50	8.89	9	±27
Heterocloeon curiosum	1.20	1.91	69	±75
Cheumatopsyche campyla	412.55	174.96	5	±20
Simulium jenningsi	39.75	21.27	8	±25
Chironominae/Orthocladiinae	665.70	325.84	7	±23
Total	1134.80	451.59	4	±19

feasible to process 20 samples on each occasion. Representative confidence limits from one date are shown in the right column of Table 3. Comparing Table 3 with Table 2 shows that we obtain confidence limits that are approximately 2X better for total number of organisms and approximately 2 to 6X better for most individual taxa, with less effort than would be required for four "traditional" samples.

All of the populations except *Heterocloeon curiosum* (Ephemeroptera: Baetidae) demonstrate the desirable attribute of high density. Because of its relatively low density and high variance, *H. curiosum* should not be chosen as a study population. The high degree of sampling precision for the other four taxa is probably the result of two other desirable attributes of study populations: distinct habitat preference and habitat that can be sampled reliably. These taxa probably exhibit a strong preference for the habitat provided by *Podostemum* because: (1) *Simulium jenningsi* (Diptera: Simuliidae) larvae attach to the apical portions of the branches to position themselves directly in the current for filtering; (2) *Cheumatopsyche campyla* (Trichoptera: Hydropsychidae) larvae use the sturdy basal portions of the branches to construct their capture nets in a zone of reduced current; (3) *Baetis intercalaris* (Ephemeroptera: Baetidae) nymphs cling and climb about on the basal branches to avoid excessive current and predators; and (4) Chironominae/Orthocladiinae (Diptera: Chironomidae) larvae inhabit the sediments that accumulate among the roots. This habitat can be sampled reliably because the *Podostemum* can be scraped off and the surface of the outcrops brushed clean without any variability that might be introduced by the occurrence of a hyporheic fauna. The outcrops can usually be waded and sampled, while the cobble/pebble riffles are often too deep.

In addition to providing precise data, the narrowly stratified sampling design for our New River studies also has an ecological rationale that is appropriate for the overall purpose of the study: to determine the ecosystem-level effects of consistently removing black fly larvae. Our surveys of the New River have shown that the *Podostemum*/outcrop habitat contributes 91% of the total density of benthic macroinvertebrates and 82% of the total standing stock in the 16-km section of the New River where there are environmental concerns [*62*]. All dominant taxa belong to the collector-filterer and collector-gatherer feeding functional groups [*6*], probably because of the seston released from the Bluestone Lake impoundment just upstream [*60*]. It is possible that the organisms living within the close physical confines of *Podostemum* mats could interact by transfers of materials through detritus pathways [*60*]. Hess [*56*] reported that in two New River sportfish, smallmouth and spotted bass, mayflies comprise 44% of the diet, black flies 23%, net-spinning caddisflies 16%, and chironomids 13%. For forage fish (shiners), he reported that the diets were 40% black flies, 35% chironomids, 4% net-spinning caddisflies, and 2% mayflies. Therefore, our ecological rationale for narrowly stratifying sampling to four study populations inhabiting *Podostemum*-covered outcrops is that this design includes: the greatest abundance of the target black fly, the greatest abundance of chironomids and other nontarget benthic macroinvertebrates, the greatest abundance of the dominant feeding functional groups, taxa among which ecological interactions would be likely, and taxa that comprise the majority of the diets of insectivorous fish.

Secondary Production

As part of our New River studies, we are analyzing the usefulness of secondary production measurements (rate of biomass elaboration by consumer populations) for assessing the hazard of toxic substances. We suggest that functional parameters, such as production, can be measured more precisely and will yield more meaningful ecological information. To date we have analyzed the precision of production measurements made during a baseline period prior to Bti applications.

According to our literature review (Table 1), almost ½ of all studies include sampling over a period of at least several months. The problem of high variability and low precision becomes even greater over time because of life history factors [*20*]; thus, the optimistic results presented in Table 3 do not necessarily hold true for studies of longer duration. Figure 1 compares the precision of production measurements for a six-month period in the New River (June to November 1983), density measurements for the same period, and the density measurements for one date (29 Aug. 1983). Precision was determined by using the equations from Elliott [*18*] in combination with the Krueger and Martin [*63*] method for estimating production. It takes about 20X fewer samples to achieve the same precision for production measurements of the dominant taxon (Chironominae/Orthocladiinae) as compared to density measurements over a six-month period. Even a comparatively rare taxon, such as *Baetis intercalaris,* required about 5X fewer samples to estimate production as compared to density for a specified level of precision. From the alternative view obtained by calculating D with $n = 20$ using all of the samples obtained over the six-month period, our sampling efforts obtained 95% confidence limits of ±20% of the mean for production measurements of chironomids as compared to ±80% of the mean for density measurements (Fig. 1). For the other three of our selected study taxa in the New River (*B. intercalaris, C. campyla, S. jenningsi),* the production measurements had 95% confidence

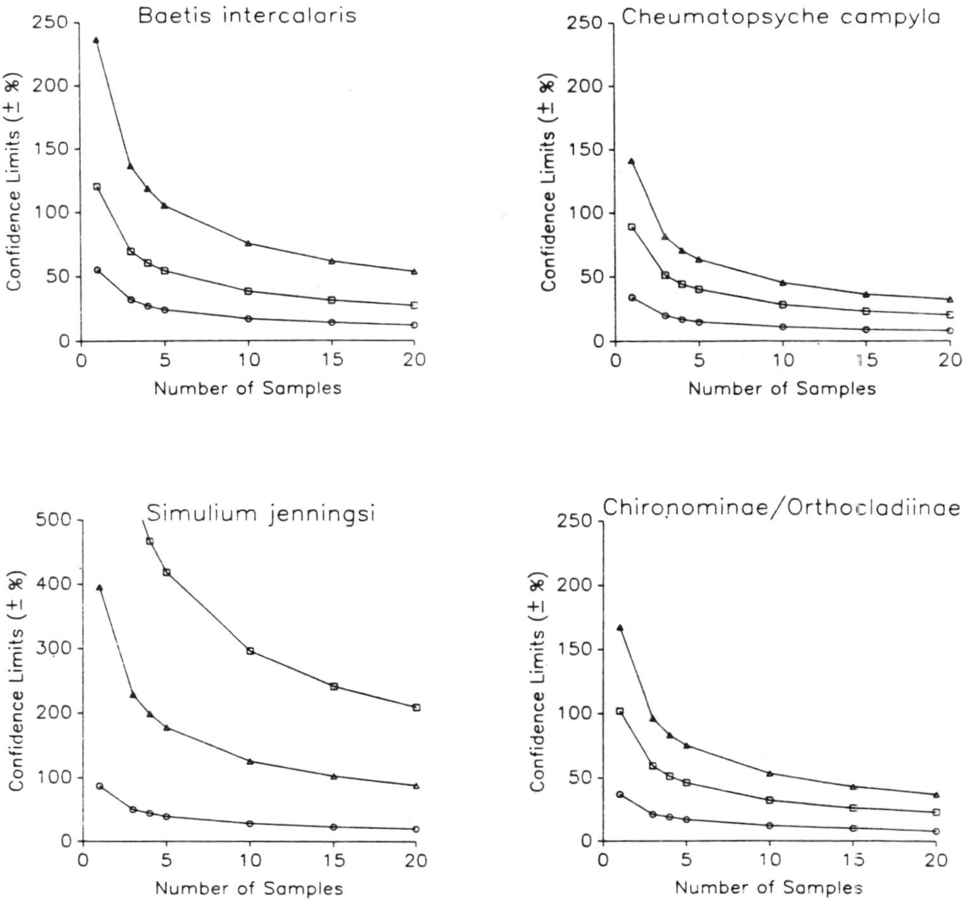

FIG. 1—*Relationship of precision (± confidence limits) and number of samples for different measurements of selected study populations. Production (Apr. to Nov. 1983) = ○; density (Apr. to Nov. 1983) = △; density (29 Aug. 1983) = □.*

limits within ±40% of the mean, which is a reasonable level of precision for biological studies [*18*]. The 95% confidence limits for density measurements, however, averaged only ±120% of the mean.

Production has received much attention recently because it combines into a single value both individual growth and population survivorship [*64*]. Density measurements only provide information on survivorship without providing any information as to the fitness of a population or its function in community energetics. Some tests have been conducted on the response of production to perturbation with the general response being depressed production [*64*], but other responses have included a shift in productive species [*65*] and an increase in production apparently due to reduced competition [*66*]. With the availability of new methods, production measurements are now becoming practical, accurate, and statistically testable [*63,64,67,68*] and should be given serious consideration for studying the impacts of toxicants in streams.

Methods—In the New River we had to construct a unique sampler that would enable us to collect very small samples (0.01 m^2) from *Podostemum* on rock outcrops in fast-moving water up to 75 cm deep. In addition, it was necessary to retain early instars for life history and production analyses. The Rock Outcrop Community Sampler (ROCS) (Fig. 2) is a square box 10 cm on each side and is 80 cm high. When placed over *Podostemum* on an outcrop, a 6.5-cm-wide strip of foam around the bottom edge assures a tight seal. Because of the narrow shape of the ROCS, long-handled tools are used to rub and scrape the substrate. A lightweight diaphragm pump that is designed to handle slurries is used to transport the sample contents into a 100-μm mesh nylon net. After the ROCS has been pumped dry, approximately 2 L of water is poured into the sampler and pumped out twice to make sure that all macroinvertebrates and vegetation are collected. The pump, net, and a ring clamp to hold sample containers are attached to the top of the ROCS for expedient collection of samples.

We have described the ROCS and several other original samplers to make the point that it is often preferable to construct unique devices rather than using ones that are readily available. This is especially true for narrowly stratified sampling. Cummins [*14*] made the observation that the number of different samplers is nearly proportional to the number of investigations. We suggest that this is desirable as long as certain criteria are met: (1) accuracy for the goals of the study; (2) reasonable precision documented; and (3) results reported in units that can be compared with other studies.

Unresolved Issues

Even though many improved techniques have been suggested, there are still problems with using benthic macroinvertebrates to assess the hazard of toxic substances in streams.

Costs

Analysis of benthic samples is expensive because of the labor and time required [*10*]; however, progress is being made at lowering costs. Many of the design techniques suggested in this paper for improving precision will also lower costs (small size samples, narrow habitat stratification, selected study populations). Resh and Price [*69*] have proposed that sequential sampling can significantly reduce the number of samples required to reach a decision. Rather than depending on a fixed number of samples, parameter values measured from individual samples are cumulatively compared with thresholds that are based on specified degrees of precision. Sheldon [*70*] has described a method for relating costs to sample size and precision for use in research planning. Techniques for improving the efficiency of various aspects of sample analysis continue to be reported (for example, Refs *71–73*). An exciting development in life history and production studies is interfacing electronic measuring and weighing instruments with microcomputers for automated analysis and data base management [*67,74*].

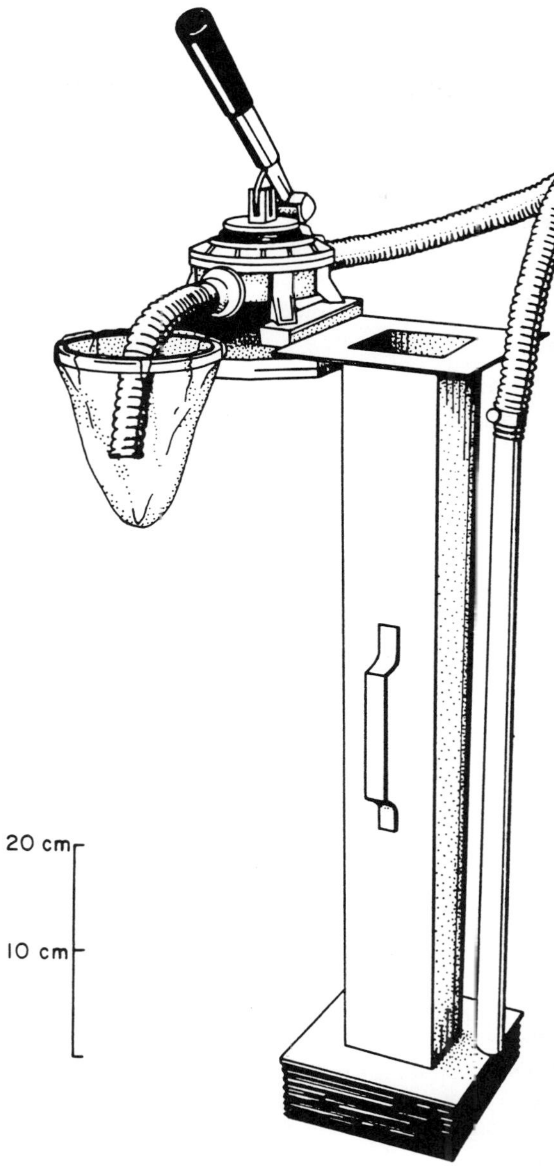

FIG. 2—*Rock Outcrop Community Sampler (ROCS). See text for explanation of construction and operation.*

Structural Versus Functional Parameters

Ecosystems have both structural (for example, density, number of species, diversity, biomass) and functional (for example, growth, production, nutrient dynamics, energy flow) characteristics. It has been proposed that assessments of the hazard posed by toxic substances in aquatic systems should include ecosystem-level effects [75], but our literature review revealed that functional analyses are seldom conducted as compared to structural analyses (11 versus 102 reports, respectively). The few reports of functional analyses consist of placing the fauna into groups with ecological similarity (feeding, behavior), with the grouping being performed by means of literature information on the constituent taxa. We found no reports of actually measuring and comparing functional parameters that result from the metabolic activities of the organisms.

Advances in our abilities to study functional aspects of ecosystems, such as the secondary production of benthic macroinvertebrates, now allow us to pursue these potentially more useful measures of ecosystem-level effects with only slightly greater effort than required for structural measures. There is some indication that functional analyses will provide information that is more meaningful for making decisions [*76, 77*]; however, there is considerable disagreement on this subject [*10, 78*], and more research is clearly needed.

Statistical Problems

There are several statistical problems facing scientists who attempt to conduct comparative evaluations of the effects of toxic substances in streams. Inferential statistical tests are based on three general assumptions: normal distribution of data, equality of variances, and independent replicates. Benthic macroinvertebrates tend to be clumped in their distribution and transformations must frequently be used to satisfy the first two assumptions. In our review of the literature, however, we found that few investigators mentioned transforming their data or verifying that these assumptions were met.

It has also been documented that the assumption of independent replicates is frequently violated in studies of benthic communities [*79*]. Samples obtained from within the same sample unit (that is, stream riffle) are not truly independent and should not be used in inferential statistics. According to Hurlbert [*79*], tests using inferential statistics to compare samples that are separated spatially (upstream versus downstream) or temporally (pretreatment versus post-treatment), but are obtained from the same experimental system (the treated stream), are *invalid.* Allan [*80*] is only slightly less pessimistic, suggesting that "statistical analysis applied to such experimental data must be viewed cautiously, at best." This problem, called *pseudoreplication* [*79*], applies to measurements of both structure and function and appears to be insurmountable. While artificial streams and experimental ponds can be used to provide true replicates for manipulative experiments, it appears unlikely that we would be able to truly replicate natural streams affected by toxic substances.

Not everyone is in complete agreement with Hurlbert. Stewart-Oaten et al. [*81*] dispute some of his objections about the use of inferential statistics for environmental impact studies and propose a design that they claim to be appropriate for assessing the impact of a discharge into a river at a single point. Their "BACI" design requires taking samples, replicated in time, *B*efore and *A*fter the impact begins at both the *C*ontrol and *I*mpact sites. Sampling times are used as true replicates, but with the important qualification that sites are sampled simultaneously and each sampling time is represented in the analysis by only one number, the difference between the impact and control samples for that time. Stewart-Oaten et al. [*81*] are generally complimentary to Hurlbert [*79*] regarding his efforts to explain statistics to biologists; however, they cite numerous references to make the point that other disciplines, particularly economics and sociology, have also had to cope with unavoidable imperfect experiments, and they make a noteworthy suggestion that a "more ecumenical spirit is called for."

Conclusions

Although some problems remain unresolved, there have been significant advances in field techniques for studying benthic macroinvertebrates in streams. The needs for greater accuracy and precision at reasonable costs have been recognized, and innovative designs and methods have been developed. These advances have been developed largely in response to needs for studying benthic macroinvertebrates in basic aquatic ecology. The same needs exist in hazard assessment, but the advances in techniques have not been put into use in this area of applied aquatic ecology. Contemporary hazard assessment in streams is little different from what it was at least 15 years ago.

We can only speculate on possible reasons for this situation. Perhaps basic and applied aquatic ecologists simply do not communicate as a result of not reading the same journals or not attending the same meetings. The decisions that are made as a result of hazard assessments may have serious economic impacts, so there might be a tendency for applied aquatic ecologists not to do anything that is not in a "cookbook." Also, it could be that persons employed in the practice of hazard assessment came out of graduate school in the early 1970s and have not continued their education.

Regardless of the reasons for the present situation, it is important to be able to assess the hazard of toxic substances reliably in streams, and benthic macroinvertebrates are well-suited for this purpose. Therefore, it is essential that applied aquatic ecologists break with tradition and either develop their own techniques for studying benthic macroinvertebrates or seek out and employ the techniques that have proven to be worthwhile in basic aquatic ecology.

Acknowledgments

Our research in the New River has been sponsored by several agencies: U.S. Army Corps of Engineers, Huntington District; U.S. National Park Service, New River Gorge National River; and West Virginia Department of Natural Resources, Wildlife Resources Division. Van D. Christman kindly assisted with the preparation of the manuscript.

References

[1] Hynes, H. B. N., *The Ecology of Running Waters*, Liverpool University Press, England, 1970.
[2] Wurtz, C. B., *Nautilus*, Vol. 73, 1960, pp. 131–135.
[3] Boyle, T. P. in *Validation and Predictability of Laboratory Methods for Assessing the Fate and Effects of Contaminants in Aquatic Ecosystems, ASTM STP 865*, T. P. Boyle, Ed., American Society for Testing and Materials, Philadelphia, 1985, Introduction, p. 1.
[4] Cairns, J., Jr. in *Community Toxicity Testing, ASTM STP 920*, J. Cairns, Jr., Ed., American Society for Testing and Materials, Philadelphia, 1986, Overview, pp. 1-5.
[5] Pennak, R. W., *Fresh-Water Invertebrates of the United States*, 2nd ed., Wiley, New York, 1978.
[6] *An Introduction to the Aquatic Insects of North America*, 2nd ed., Merritt, R. W. and Cummins, K. W., Eds., Kendall/Hunt, Dubuque, 1984.
[7] Gaufin, A. R. in *Biological Methods for the Assessment of Water Quality, ASTM STP 528*, J. Cairns, Jr. and K. L. Dickson, Eds., American Society for Testing and Materials, Philadelphia, 1973, pp. 96–116.
[8] Hynes, H. B. N. in *The Ecology of Aquatic Insects*, V. H. Resh and D. M. Rosenberg, Eds , Praeger, New York, 1984, Chap. 19, pp. 578-588.
[9] Berkman, H. E., Rabeni, C. F., and Boyle, T. P., *Environmental Management*, Vol. 10, No. 3, 1986, pp. 413-419.
[10] Rosenberg, D. M., Danks, H. V., and Lehmkuhl, D. M., *Environmental Management*, Vol. 10, No. 6, 1986, pp. 773-783.
[11] Coulson, R. N. and Crossley, D. A., Jr., *Bulletin of the Entomological Society of America*, Vol. 33, 1987, No. 2, pp. 64-68.
[12] Strahler, A. N., *Transactions of the American Geophysical Union*, Vol. 38, 1957, pp. 913–920.
[13] "Biological Field and Laboratory Methods for Measuring the Quality of Surface Waters and Effluents," Weber, C. I., Ed., USEPA, Cincinnati, 1973.
[14] Cummins, K. W., *The American Midland Naturalist*, Vol. 67, No. 2, 1962, pp. 477-504.
[15] Needham , P. R. and Usinger, R. L., *Hilgardia*, Vol. 24, No. 14, 1956, pp. 383-409.
[16] Chutter, F. M., *Limnology and Oceanography*, Vol. 17, 1972, pp. 139-141.
[17] Frost, S., Huni, A., and Kershaw, W. E., *Canadian Journal of Zoology*, Vol. 49, 1971, pp. 167-173.
[18] Elliot, J. M., *Some Methods for the Statistical Analysis of Samples of Benthic Macroinvertebrates*, 2nd ed., Scientific Publication No. 25, Freshwater Biological Association, Cumbria, England, 1977.
[19] Chutter, F. M. and Noble, R. G., *Archiv Fur Hydrobiologie*, Vol. 62, No. 1, 1966, pp. 95-103.

[20] Schwenneker, B. W. and Hellenthal, R. A., *Environmental Entomology*, Vol. 13, No. 3, 1984, pp. 741-750.
[21] Resh, V. H., *Journal of the Fisheries Research Board of Canada*, Vol. 36, No. 3, 1979, pp. 290-311.
[22] Shelly, T. E., *Ecological Entomology*, Vol. 4, 1979, pp. 95-100.
[23] Doeg, T. and Lake, P. S., *Hydrobiologia*, Vol. 80, 1981, pp. 3-6.
[24] McElhone, M. J. and Davies, R. W., *Canadian Journal of Zoology*, Vol. 61, No. 10, 1983, pp. 2300-2304.
[25] Wrona, F. J., Calow, P., Ford, I., Baird, D. J., and Maltby, L., *Canadian Journal of Fisheries and Aquatic Sciences*, Vol. 43, No. 10, 1986, pp. 2025-2035.
[26] Clements, W. H., *Journal of Freshwater Ecology*, Vol. 4, No. 1, 1987, pp. 83-91.
[27] Cummins, K. W. in *River Ecology*, B. Whitton, Ed., Blackwell Science, England, 1975, Chap. 8, pp. 170-198.
[28] Downing, J. A., *Journal of the Research Board of Canada*, Vol. 36, No. 12, 1979, pp. 1454-1463.
[29] Morin, A., *Canadian Journal of Fisheries and Aquatic Sciences*, Vol. 42, 1985, pp. 1530-1534.
[30] Coffman, W., Cummins, K. W., and Wuycheck, J. C., *Archiv Fur Hydrobiologie*, Vol. 68, 1971, pp. 232-276.
[31] Gregg, W. W. and Rose, F. L., *Aquatic Botany*, Vol. 14, 1982, pp. 309-324.
[32] Gregg, W. W. and Rose, F. L., *Hydrobiologia*, Vol. 128, 1985, pp. 45-56.
[33] Downing, J. A. and Anderson, M. R., *Canadian Journal of Fisheries and Aquatic Sciences*, Vol. 42, 1985, pp. 1860-1869.
[34] Downing, J. A. and Cyr, H., *Canadian Journal of Fisheries and Aquatic Sciences*, Vol. 42, No. 10, 1985, pp. 1570-1579.
[35] Cairns, J., Jr., *Water Resources Bulletin*, Vol. 10, 1974, pp. 338-347.
[36] Herricks, E. E. and Cairns, J., Jr., *Water Research*, Vol. 16, 1982, pp. 141-153.
[37] Allan, J. D. and Russek, E., *Canadian Journal of Fisheries and Aquatic Sciences*, Vol. 42, No. 2, 1985, pp. 210-215.
[38] Merritt, R. W., Cummins, K. W., and Resh, V. H. in *An Introduction to the Aquatic Insects of North America*, 2nd ed., R. W. Merritt and K. W. Cummins, Eds., Kendall/Hunt, Dubuque, 1984, Chap. 3, pp. 11-26.
[39] Canton, S. P. and Chadwick, J. W., *Progressive Fish-Culturist*, Vol. 46, 1984, pp. 57-59.
[40] Hess, A. D., *Limnological Society of America Special Publication*, Vol. 6, 1941, pp. 1-5.
[41] Waters, T. F. and Knapp, R. J., *Transactions of the American Fisheries Society*, Vol. 90, 1961, pp. 225-226.
[42] Mackie, G. L. and Bailey, R. C., *Journal of Freshwater Ecology*, Vol. 1, 1981, pp. 61-70.
[43] Resh, V. H., Lamberti, G. A., McElravy, E. P., Wood, J. R., and Feminella, J. W., "Quantitative Methods for Evaluating the Effects of Geothermal Energy Development on Stream Benthic Communities at the Geyers, California," University of California, Contribution No. 190, California Water Resources Center, Davis, 1984.
[44] Brown, A. V., Schram, M. D., and Brussock, P. P., *Hydrobiologia*, Vol. 153, 1987, pp. 241-247.
[45] Hudson, P. L. and Oliver, J. L., *Progressive Fish-Culturist*, Vol. 45, No. 2, 1983, pp. 123-124.
[46] Pollard, J. E., *Journal of Freshwater Ecology*, Vol. 1, No. 2, 1981, pp. 215-224.
[47] Mackey, A. P., Cooling, D. A., and Berrie, A. D., *Journal of Applied Ecology*, Vol. 21, 1984, pp. 515-534.
[48] Flannagan, J. F. and Rosenberg, D. M. in *Artificial Substrates*, J. Cairns, Jr., Ed., Ann Arbor Science, Ann Arbor, 1982, Chap. 7, pp. 237-266.
[49] Rosenberg, D. M. and Resh, V. H. in *Artificial Substrates*, J. Cairns, Jr., Ed., Ann Arbor Science, Ann Arbor, 1982, Chap. 6, pp. 175-236.
[50] Ferrington, L. C., Jr. and Christiansen, C., *Journal of the Kansas Entomological Society*, Vol. 58, No. 4, 1985, pp. 724-726.
[51] Clements, W. H., Cherry, D. S., and Cairns, J., Jr., *Hydrobiologia*, in press.
[52] Clements, W. H., Cherry, D. S., and Cairns, J., Jr., *Environmental Toxicology and Chemistry*, Vol. 7, No. 9, 1988, pp. 715-722.
[53] Jonasson, P. M., *Oikos*, Vol. 6, 1955, pp. 183-207.
[54] Benke, A. C., Van Arsdall, T. C., Jr., Gillespie, D. M., and Parrish, F. K., *Ecological Monographs*, Vol. 54, No. 1, 1984, pp. 25-63.
[55] Storey, A. W. and Pinder, L. C. V., *Hydrobiologia*, Vol. 124, 1985, pp. 193-197.
[56] Hess, L. in *Proceedings of the New River Symposium*, W. E. Cox and M. Kegley, Eds., National Park Service, Oak Hill, WV, 1983, pp. 15-21.
[57] Ali, A., Baggs, R. D., and Stewart, J. P., *Journal of Economic Entomology*, Vol. 74, 1981, pp. 672-677.
[58] Back, C., Boisvert, J., Lacoursiere, O., and Charpentier, G., *Canadian Entomologist*, Vol. 117, 1985, pp. 1523-1534.
[59] De Moor, F. C. and Car, M., *Onderstepoort Journal of Veterinary Research*, Vol. 53, 1986, pp. 43-50.

[60] Voshell, J. R., Jr., "Trophic basis of production for macroinvertebrates in the New River below Bluestone Dam," Report submitted to West Virginia Department of Natural Resources, Charleston, WV, 1985.
[61] Virginia Polytechnic Institute and State University, "An Ecological Investigation of the New River and Bluestone Lake," report submitted to U.S. Army Corps of Engineers, Huntington, WV, 1985.
[62] Voshell, J. R., Jr., Layton, R. J., and Hiner, S. W. in *Proceedings of the New River Symposium,* J. Wild and N. DeJong, Eds., National Park Service, Oak Hill, WV, 1987, pp. 201–212.
[63] Krueger, C. C. and Martin, F. B., *Limnology and Oceanography,* Vol. 25, 1980, pp. 773–777.
[64] Benke, A. C. in *The Ecology of Aquatic Insects,* V. H. Resh and D. M. Rosenberg, Eds., Praeger, New York, 1984, Chap. 10, pp. 289–322.
[65] Flossner, D., *Limnologica (Berlin),* Vol. 10, 1976, pp. 123–153.
[66] Uutala, A. J. in *Effects of Acidic Precipitation on Benthos,* R. Singer, Ed., North American Benthological Society, Springfield, IL, 1981, pp. 139–154.
[67] Layton, R. J., "Production of Aquatic Macroinvertebrates in a River Used for Commercial Navigation: Kanawha River, West Virginia," M.S. thesis, Virginia Polytechnic Institute and State University, Blacksburg, VA, 1985.
[68] Heisey, D. M. and Hoenig, J. M., *Limnology and Oceanography,* Vol. 31, No. 1, 1986, pp. 211–215.
[69] Resh, V. H. and Price, D. G., *Environmental Management,* Vol. 8, No. 1, 1984, pp. 75–80.
[70] Sheldon, A. L., *Hydrobiologia,* Vol. 111, 1984, pp. 147–152.
[71] Reger, S. J., Brothersen, C. F., Osborn, T. G., and Helm, W. T., *Journal of Freshwater Ecology,* Vol. 1, No. 5, 1982, pp. 451–465.
[72] Wrona, F. J., Culp, J. M., and Davies, R. W., *Canadian Journal of Fisheries and Aquatic Sciences,* Vol. 39, No. 7, 1982, pp. 1051–1054.
[73] Pollard, J. E. and Melancon, S. M., *Journal of Freshwater Ecology,* Vol. 2, No. 4, 1984, pp. 383–391.
[74] Roff, J. C. and Hopcroft, R. R., *Canadian Journal of Fisheries and Aquatic Sciences,* Vol. 43, No. 10, 1986, pp. 2044–2048.
[75] Levin, S. A. and Kimball, K. D., Eds., *Environmental Management,* Vol. 8, No. 5, 1984, pp. 375–442.
[76] Matthews, R. A., Buikema, A. L., Jr., Cairns, J., Jr., and Rodgers, J. H., Jr., *Water Research,* Vol. 16, 1982, pp. 129–139.
[77] Perry, J. A., Troelstrup, N. H., Jr., Newsom, M., and Shelley, B., *Water Science and Technology,* Vol. 19, No. 11, 1987, pp. 55–71.
[78] Barnthouse, L. W., O'Neill, R. V., Bartell, S. M., and Suter, G. W., II, *Aquatic Toxicology,* Vol. 9, 1986, pp. 82–96.
[79] Hurlbert, S. H., *Ecological Monographs,* Vol. 54, No. 2, 1984, pp. 187–211.
[80] Allan, J. D. in *The Ecology of Aquatic Insects,* V. H. Resh and D. M. Rosenberg, Eds., Praeger, New York, 1984, Chap. 16, pp. 484–507.
[81] Stewart-Oaten, A., Murdoch, W. W., and Parker, K. R., *Ecology,* Vol. 67, No. 4, 1986, pp. 929–940.

Interpretation of Microcosm Data

Loveday L. Conquest[1] *and Frieda B. Taub*[2]

Repeatability and Reproducibility of the Standardized Aquatic Microcosm: Statistical Properties[3]

REFERENCE: Conquest, L. L. and Taub, F. B., **"Repeatability and Reproducibility of the Standardized Aquatic Microcosm: Statistical Properties,"** *Aquatic Toxicology and Hazard Assessment: 12th Volume, ASTM STP 1027,* American Society for Testing and Materials, Philadelphia, 1989, pp. 159–177.

ABSTRACT: A general problem faced by ecologists and mathematicians alike is how to statistically analyze data from dynamic biological systems observed for limited periods of time. Examination of biological and statistical properties of standardized aquatic microcosms (SAMs) can provide extra insights or artifacts; several examples will be used to support recommendations and cautions on the use of various statistical procedures and interpretation.

Within-experiment statistical analyses of the SAM data include analyses of variance (ANOVAs) on both biological and chemical data. The results of the ANOVAs on each sampling day are expressed as an "interval of nonsignificant difference" (IND) around the control mean. These IND graphics are essentially results of modified *t*-tests between the control and treatment means, are easily interpretable, and allow one to quickly view the results of the entire experiment for any response variable. Apparent statistical artifacts can be investigated by examining the replicate data, for example, cases in which the biological and statistical interpretations do not seem to agree. Several other statistical procedures have demonstrated shortcomings, including (1) classical time series, (2) other multivariate methods, and (3) nonparametric multiple comparisons.

Between-experiment statistical comparisons have proven more difficult. For the interlaboratory copper experiments, we use the days-weighted-by-variable (DWV). With the DWV, we have been able to come to grips with the problem that faces almost any ecological study—many measurements (and thus lots of data) but a limited number of statistical degrees of freedom. The DWV has allowed us to demonstrate where the similarities and differences are between different experiments. For instance, we have been able to state statistically that experiments agreed in the nature and direction of certain effects, but that the exact timing of such effects has differed. Rejected techniques for between-experiment comparisons include the following: (1) day-by-day two-way ANOVA; (2) nonparametric correlations of statistical differences; (3) comparison of fitted regression parameters via ANOVA; (4) differential equation modeling of individual microcosms; (5) integrated distances between microcosm time traces.

KEY WORDS: toxicity statistics, microcosms, statistics

The statistical properties of microcosm data share many of the same problems as ecological monitoring in general. However, microcosm data have several advantages for statistical studies: more replicates may be available, the data are more transparent—there are fewer opportunities for unknown happenstances (for example, a school of fish swam through and ate all the *Daphnia* just before the sample was taken), and there are fewer missing values. Also, in the development and testing of the SAM protocol, several related variables have been measured; experiments have been repeated many times within the host laboratory and performed in other

[1]Associate professor, Center for Quantitative Science, University of Washington, Seattle, WA 98195.
[2]Professor, School of Fisheries, WH-10, University of Washington, Seattle, WA 98195.
[3]School of Fisheries Contribution No. 771.

laboratories [1–7]. We therefore suggest that statistical analyses of microcosm data will provide insights that may be useful in ecological sampling and analyses.

Many of the problems of ecological sampling are (1) nonindependence of variables (for example, the algal biomass is related to pH) and (2) nonindependence of sequential sampling times (the repeated sampling problem). The nonindependence of variables may be complex, involving lags, competitive interactions, or predator-prey cycles; in these cases the variables may be related mathematically, but may not correlate in a simple manner. These are not simple problems, and the biological realities may not satisfy the mathematical assumptions of various statistical methods.

Another ecological-statistical problem is whether (1) the researcher can claim to have shown an effect if the statistical test does not indicate a statistical difference, and (2) the difference is biologically important if a statistical difference is shown. Others have raised questions whether hypothesis testing is appropriate for ecological tests or if regression of dose-response relationships would not be a more efficient use of experimental effort.

The problems mentioned above are generally involved in statistical analyses within an experiment; a new group of problems arises in considering comparisons between experiments. How can the similarity of two experiments be compared; how similar do the responses have to be to state that the experimental outcomes have agreed? The purpose of this paper is to display various statistical procedures that have either been successful in demonstrating ecological responses, have introduced serious artifacts, or, although correct and useful, are more complex than simpler statistics which served just as effectively.

Methods

Only a brief summary of the microcosm methods is included here, as they have been described extensively in a formal protocol [8], have been published in modified form [9], and are summarized elsewhere [1–7]. The microcosms consist of 3-L cultures of ten species of algae and five animals. The liquid is a chemically defined medium, and the sediment is sand with chitin and cellulose. The algae are inoculated on Day 0 and the animals on Day 4; the test chemical is added on Day 7. The algal nutrients, oxygen dynamics, pH, and organism abundances are monitored twice each week for 63 days, except for algal nutrients which are monitored only once per week after Day 28. A typical experiment consists of 24 microcosms with six replicates each of four treatment groups. The microcosm response to several chemicals have been studied [2,3,7], and an interlaboratory test using $CuSO_4$ has been conducted [4–7]; those experiments used control, 500, 1000, and 2000-μgL^{-1} Cu. The interlaboratory test included seven experiments, two (of many) performed at the host laboratory (University of Washington), two performed at Duluth—EPA with the University of Minnesota, two performed at Aberdeen Proving Ground, and one performed at Marine Bioassays. The Aberdeen experiments used control and 500-μgL^{-1} Cu, but not the other standard concentrations. Although more experiments were available from the host laboratory, we used only two so as not to have more than from the other laboratories.

Within Experiment Statistics

One-Way Analysis Of Variance (ANOVA)

These experiments were designed for analyses of variance on both biological and chemical data. This technique has been generally useful. We have been able to use statistical power equations to determine sample sizes and number of treatment groups [10], thus allowing us to obtain a high degree of statistical power to detect differences between means. Analysis of variance is commonly accepted, easily understood, and with equal sample sizes is fairly robust to violations of the assumptions of normality and equal variances [11]. The reporting of the data is best when both the treatment results and the statistical differences are shown simultaneously. This has been accomplished (Figs. 1 and 3) through the use of the IND graphic [*4–5*], allowing the reader to grasp the nature and magnitude of the effects and to simultaneously observe how large a difference is required to be statistically different from the control mean. The interval of nonsignificant difference is based upon a hypothesis test between the control mean and a treatment mean. It is the graphical equivalent of such a test and is not the same as an ordinary confidence interval about the control mean. It takes into account the sample sizes of the two means in the test, the pooled variance estimate from all the groups, and the difference between the two means.

Intervals Of Nonsignificant Difference

Intervals of nonsignificant difference are based upon parametric hypothesis tests for comparing means. To test the hypothesis that two population means are equal, compute the test statistic

$$(\bar{x}A - \bar{x}B)/\sqrt{s^2(1/nA + 1/nB)}$$

where

$\bar{x}A, \bar{x}B$ = observed means,
s^2 = pooled variance estimate, and
nA, nB = respective sample sizes.

and compare it to the appropriate tabled critical value with corresponding degrees of freedom (df), depending upon the chosen error rate and the kind of test being done. Sokal and Rohlf

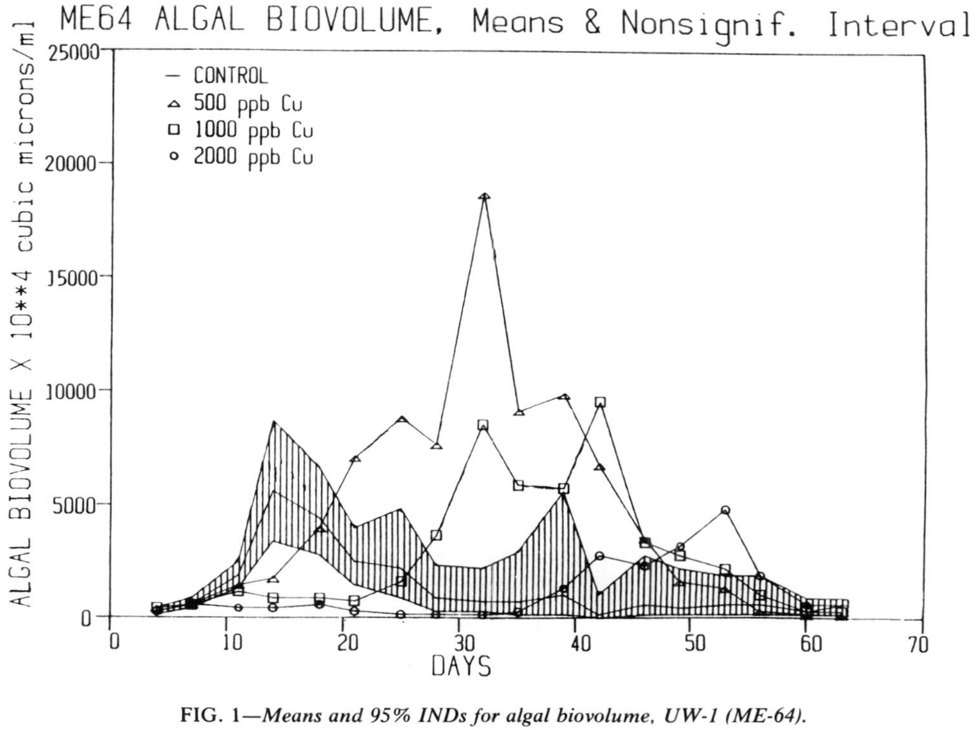

FIG. 1—*Means and 95% INDs for algal biovolume, UW-1 (ME-64).*

[12] and Steel and Torrie [13] are but two of the many statistics references which discuss the distinctions between the kinds of tests listed in Table 1. Some comments are noted beside each test. A "conservative" test is one which does not reject the null hypothesis of equality of means very often and hence lacks statistical power to detect differences when they are present.

$\bar{x}A$ and $\bar{x}B$ are considered to be nonsignificantly different from each other whenever the distance between the two does not exceed (tabled value) $\sqrt{(s^2(1/nA + 1/nB))}$, or equivalently, $\bar{x}B$ is nonsignificantly different from $\bar{x}A$ whenever $\bar{x}B$ falls in the interval $\bar{x}A \pm$ (tabled value) $\sqrt{(s^2(1/nA + 1/nB))}$. Hence the name interval of nonsignificant difference or IND. For graphical purposes, these are easy and efficient to use only when the length of the interval does not change with each comparison. LSD, Tukey's HSD, Scheffe's procedure, and Dunnett's tests can all satisfy this. Unequal sample sizes can cause problems, unless nA and nB are such that $(1/nA + 1/nB)$ does not change (for example, nA = sample size for the control group and nB = the equal sample size for all other groups). Andrews et al. [14] give some suggestions for finding a weighted average of all possible combinations of the sample sizes involved for highly unequal sample sizes and concede that unequal sample sizes do cause problems for doing exact multiple comparison tests via graphics alone. They state, "The principal benefit of these plots is the provision of a summary display that enables one to get a general picture of [the] nature and magnitude of the differences among the means of the different groups." When in doubt as to whether two means are significantly different, a formal test of significance should be performed in addition to any graphical procedures.

If $\bar{x}A$ is the control mean, $\bar{x}B$s are treatment means, and when we are looking at the results of several sampling days through time, we can use IND plots to see (1) where the treatment means are significantly different from the control mean, (2) where "strings" of significant differences occur, and (3) where the INDs get quite large due to large variance estimates, thus offering the potential for obscuring large differences between means.

How are INDs different from "ordinary" confidence intervals? The main distinction is that these are confidence intervals for the *difference* between two observed means. While the formula $(\bar{x}A - \bar{x}B) \pm t\sqrt{s^2(1/nA + 1/nB)}$ appears at least once in almost every elementary statistics book, what often appears in graphical form is $\bar{x} \pm t\sqrt{s^2/n}$, the confidence interval around $\bar{x}$, a single (often the control) mean, rather than the IND for the difference between two means. Then a variety of temptations arise. One is to declare an *observation* (a mean from a sample size of 1) different from a control mean if the observation falls outside the control mean's confidence interval. Another is to declare two means nonsignificantly different whenever one falls within the confidence interval of the other. If confidence intervals are shown around two means, the temptation is to declare them nonsignificantly different when the intervals overlap and significantly different when they do not. In extreme cases (that is, means very far apart, or very close together), the conclusion may be the same, but any single-mean procedure is *not valid* for comparing two means. An ordinary confidence interval around a single mean (or each of two means) fails to take into account the variability (and sample sizes) associated with the difference between two means, and it is this variability for the difference that must appear in the graphic.

TABLE 1—*Tabled critical values and kind of test.*

Student's "t"	Least significant difference (LSD) for planned comparisons between means.
Dunnett's "q"	Dunnett's test of the control mean to each treatment mean; for example, looking for the significant treatment means least different from the control.
Tukey's "q"	Tukey's honestly significant difference (HSD); too conservative for planned comparisons.
$(k - 1)F$	Scheffe's test for all possible linear contrasts among a set of k means (extremely conservative for planned comparisons of means).
S-N-K "q"	Student-Newman-Keuls multiple-range "q" to compare pairs of means; tabled value depends upon relative positions of means, unlike the others.

An example illustrating this phenomenon is shown in Figs. 2 and 3 for nitrate values from UW-1. Both plots show the means for the control and three treatment groups. The only difference between the two plots is the meaning attached to the shaded area. The shaded area in Fig. 2 is much narrower than that in Fig. 3. The former is a 95% confidence interval around the control mean. It should *not* be used to detect treatment effects from the control mean. Only Fig. 3, with its larger INDs, is appropriate for doing comparisons between the control and treatment means. The only legitimate test of significance available from Fig. 2 is a comparison of the control mean to a prespecified *constant* value, for example, a population mean with zero variability attached to it. This is *not* the same as comparing an observed control mean to another observed treatment mean, especially if the treatment mean is computed from a sample size of 1 (a single observation).

How much larger are the INDs than ordinary confidence intervals? The confidence interval around the control mean is based upon adding and subtracting the quantity $t\sqrt{s^2(1/nc)}$ to the control mean. The IND is based upon adding and subtracting the quantity $t\sqrt{s^2(1/nc + 1/nt)}$ to the control mean. The ratio of the latter to the former is $\sqrt{1 + nc/nt}$ which equals 1.414 whenever $nc = nt$. So under equality of sample size, one can expect a $\sqrt{2}$ increase on each side when doing an IND rather than a confidence interval around the control mean.

Student's t versus Dunnett's q—We have been using values from the Student's *t* distribution, appropriate for doing planned comparison of means. On any given sampling day, we do *not* compare every mean to every other mean, but only the control to each treatment mean. For the specific case of assessing which treatment is the least significantly different from the control (and then perhaps doing further experimentation with it), Dunnett [*15,16*] has devised a more appropriate set of table values. Franco et al. [*17*] and Giddings et al. [*18*] have used Dunnett-

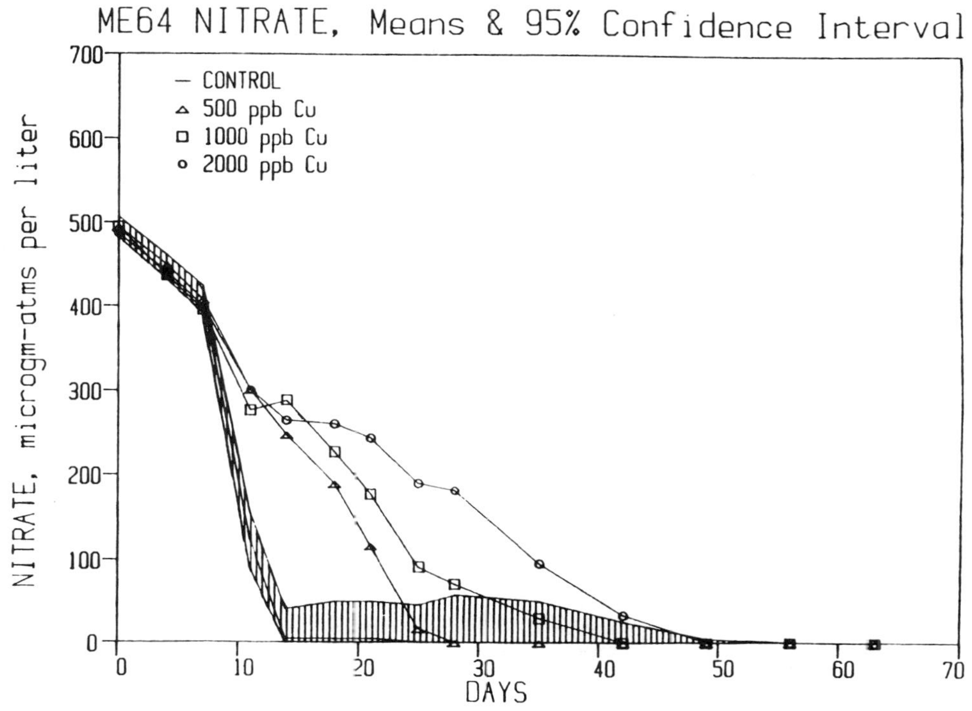

FIG. 2—*Means and 95% confidence interval for nitrate. UW-1 (ME-64).*

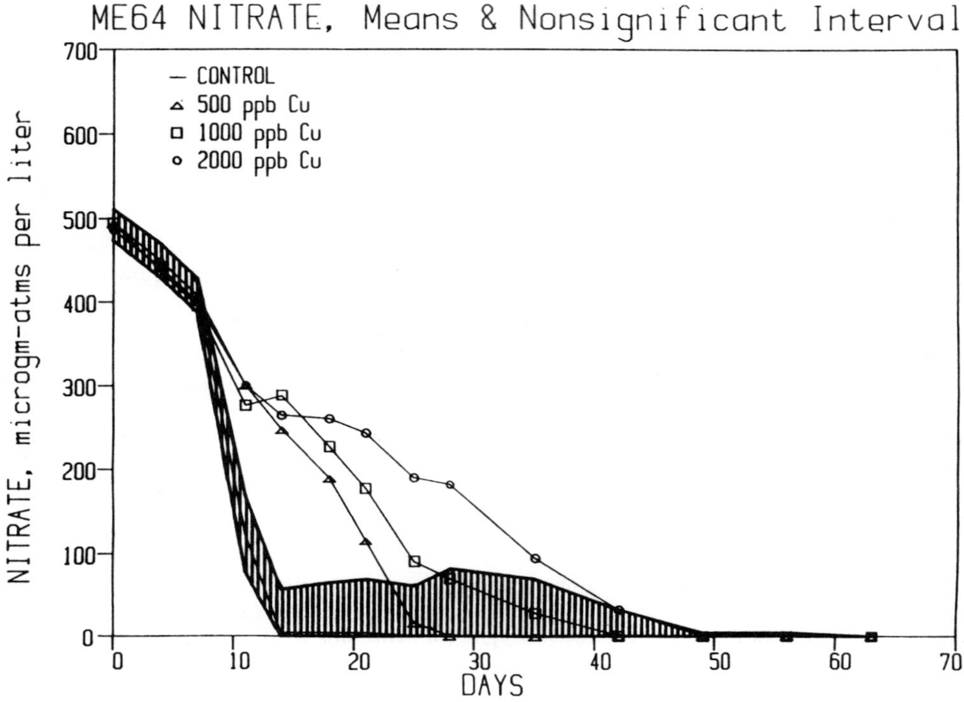

FIG. 3—*Means and 95% INDs for nitrate, UW-1 (ME-64).*

type values in certain comparisons. In deciding which values (Student's t or Dunnett's q) to use, it is important to keep in mind the treatments we are using, why we are doing the comparisons, and what happens afterwards. A situation where the t-test (least significant difference) would not be valid is the following: Consider a control group and k separate treatment groups (for example, k entirely different chemicals). The one chemical least significantly different from the control will be subjected to further testing. In a case like this, focusing on a treatment mean simply because it happens to be the one least significantly different from the control can easily invite a Type I error (declaring a difference when in fact none exists), so one would want to make sure that such a treatment mean was indeed significantly different, and that is what Dunnett's q values ensure. But we are not really in that kind of situation. We consider each control to treatment on its own merits and do *not* focus on a particular concentration simply because it displays the most significant results, nor do we focus on a particular concentration simply because it is the lowest to show significant effects. Furthermore, we always check that the trend shown by the increasing concentrations makes biological sense. Carmer and Walker [*19*] have presented further arguments, based on comparisonwise error rates, as to why a least significant difference approach is often in order when doing pairwise multiple comparisons. To see how a Student's t-based IND plots would compare with one generated from Dunnett's q-values, we recalculated the nitrate INDs for UW-1 (Fig. 4). The INDs created using Dunnett's q-values are slightly wider. There is virtually no difference in the number of days showing significant differences. This would clearly not alter any of the biological conclusions; in all three treatment groups, the decrease in nitrate is delayed according to the concentration. In Table 2 some comparisons for *Daphnia* and algal biovolume show essentially no differences with respect to the number of significant days via the t-test or Dunnett's test. So we remain assured in using least significant differences based on t-values and in making sure that any sequences of statistical differences

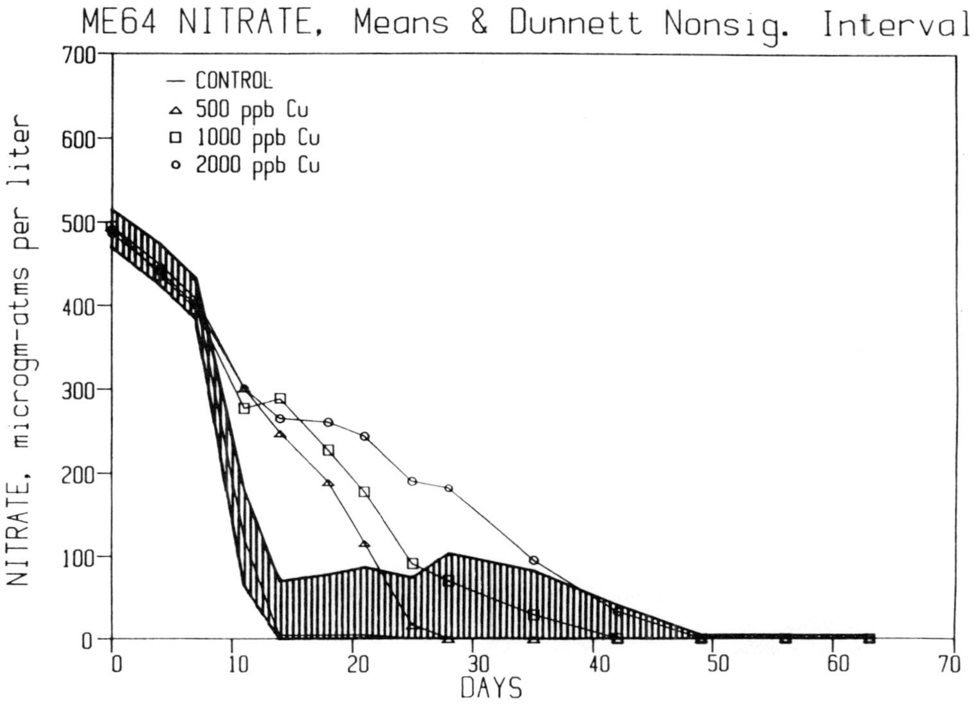

FIG. 4—*Means and 95% Dunnett IND for nitrate, UW-1 (ME-64).*

TABLE 2—*Number of days with at least one significant contrast, divided by number of sampling days in the experiment.*

	t-test, P	Dunnett, P	Dunnett, NP
Algal biovolume	12/18	11/18	7/18
Nitrate	7/14	7/14	7/14
Daphnia	15/17	15/17	15/17

NOTE: P = parametric test; NP = nonparametric test.

are *always* confirmed by biological conclusions and not just observed statistical artifacts due to chance.

Logarithmic Transformations of the Data—It is well established that the statistical behavior of certain variables follows the requirements of parametric tests (normality, equal variances) more closely if we use the logarithm of the variable. We are using log $(x + 1)$ for all enumerated variables (population densities and variables calculated from them). We have shown that the log-transformed variables do conform to the statistical requirements for parametric tests [*20*].

Arrow Diagrams—We have been using "arrow diagrams" (Fig. 5) as a convenient tool to get a quick summary of the statistical differences without the need of a graphics package. In one sense these arrow diagrams are not adequate because they do not indicate the magnitude of the effect, nor do they indicate whether no statistical differences could have been shown due to excessive variances. However, the arrow diagrams do tell one very quickly whether any statistical differences occurred at all. They are useful in telling an investigator what further variables

UW-1 NO3

Day	Contrast 1: Control vs. 500	Contrast 2: Control vs. 1000	Contrast 3: Control vs. 2000
00			
04			
07			
11	↑↑	↑↑	↑↑
14	↑↑	↑↑	↑↑
18	↑↑	↑↑	↑↑
21	↑↑	↑↑	↑↑
25		↑	↑↑
28			↑↑
35			↑↑
42			
49			
56			
63			

FIG. 5—*Arrow diagrams for nitrate, UW-1 (ME-64).*

(in addition to *Daphnia,* algal biovolume, nitrate, pH, and oxygen gain) one may want to display as IND plots.

Lack Of Independence Over Time for a Single Variable—It is readily acknowledged that measurements of a variable over time are correlated; hence the overall probability of a Type I error is greater than the specified alpha. To analyze our data via a more complex time-series approach could require at least 30 sampling days, which we do not have. Time series analysis also requires accompanying statistical assumptions, such as stability of the covariances through time. The lack of a sufficient number of data points and lack of the robustness of more sophisticated statistical procedures makes them less desirable for our use. It is better to use simpler techniques which are easily interpreted and which require far fewer assumptions.

Rejected Statistical Techniques for Within Experiment Statistics

Nonparametric Comparisons—We have tried and rejected nonparametric comparisons [*20*] as we have found that too much power is lost. Because a nonparametric approach deliberately ignores the magnitude of the data and concentrates only on the ranks, graphing mean ranks with nonparametric INDs around them can be noninformative at best and often misleading. We have seen several cases when a variable went down to zero in magnitude, but the graph of mean ranks went *up,* due to tied values among all the groups (Fig. 6). Means of zero may well occur, but *rank* values of zero will never occur. Despite the initially appealing nature of being able to forego the restrictive normality-and-equal-variances assumption by using a nonparametric approach, the resulting loss of power to detect differences and the occasionally misleading results due to tied values appear to be too much of a disadvantage.

Repeated Measures and Multivariate Techniques—Crow [*21*] has investigated a number of techniques for the through-time analysis of microcosm data, including one-way ANOVA (but without IND graphics), two-way factorial ANOVA with treatment groups and days as the two factors, repeated measures analysis, multivariate analysis of variance, and nonlinear curve fitting. For within-lab experiments, Crow concluded that the most reliable method of analysis appeared to be the univariate one-way ANOVA. "ANOVA never failed to show significant effects when more elaborate and difficult-to-interpret techniques did." Additionally ANOVA showed these effects much more clearly than the more elaborate methods. Multivariate tech-

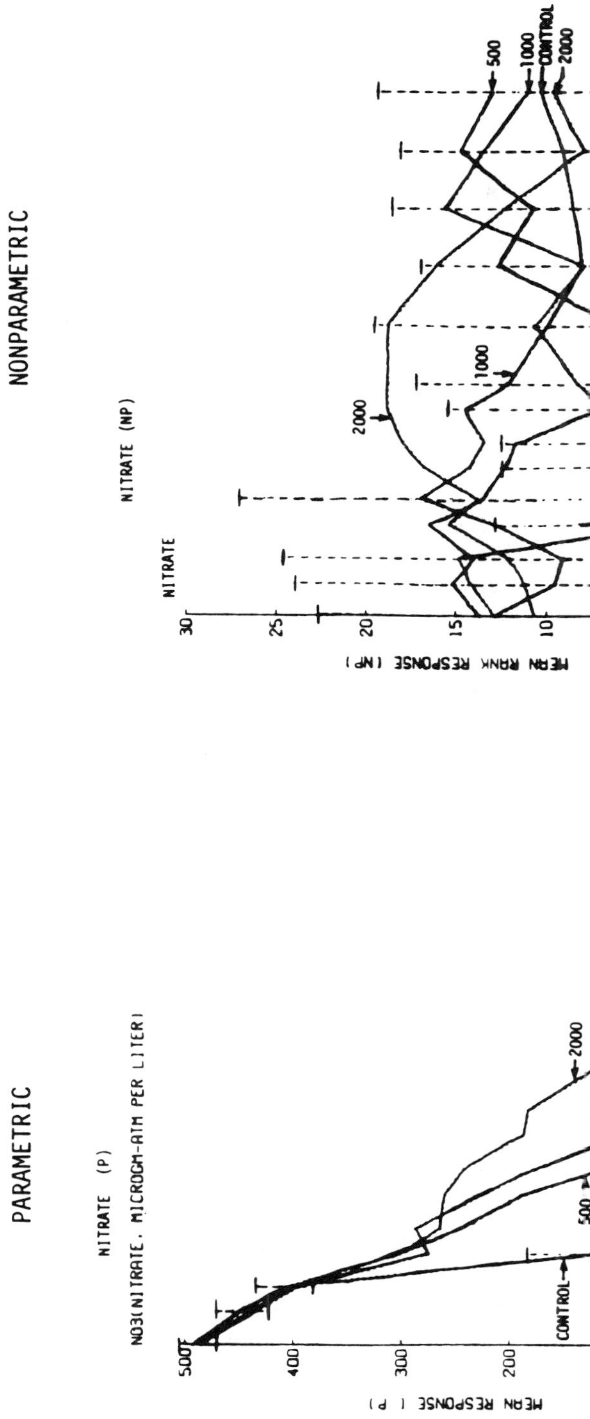

FIG. 6—*Example of parametric and nonparametric INDs for nitrate.*

niques tend to lack statistical power for low sample sizes, in addition to requiring much more strict assumptions to be satisfied (multivariate normality, regularity of the variance-covariance matrix among variables). The nonlinear curve fits were very time consuming; the dynamics of the microcosm behavior were too complex to be fitted by simple equations with low numbers of parameters.

Regression Analysis Approaches—We have seen in the literature some discussion as to a regression analysis approach and a differences-between-means approach for analysis of data within experiments [*22*]. For a given experiment, the analysis approach taken can certainly affect the design. For instance, if a clear trend on a single-output response variable is suspected, then the investigator may wish to include as many values of the predictor variable as possible. However this does *not* allow one an independent estimate of within-group error, as "error" here is not model independent; it is the "leftover" or residual from the particular model being fitted. One also has to think about what a regression approach involves for experiments where the experimental units are followed over time and where the relative positions of the group means may *shift*. For each sampling day, whether or not treatment groups are replicated, a regression analysis must be performed and a separate relationship statistically modeled. Such a procedure produces the following estimated parameters: intercept, slope, correlation coefficient, standard deviation of the residuals, etc., all of which must then be followed over the course of the experiment and interpreted. If the best-fitting relationship is linear at one sampling point and quadratic or cubic at another, the overall meaning is not particularly clear. The value of the correlation coefficient could vary considerably; this would make it difficult to interpret results. It is difficult to follow *complex* dynamic changes through time, especially if, due to horizontal shifts, the relationships between the controls and treatments (which are higher, which are lower, and by how much) change through time. These shifts are not artifacts of microcosm experiments; they are commonly shown in outdoor enclosures of natural aquatic communities treated with toxic chemicals [*23–25*]. A hypothetical example is displayed in Figs. 7 and 8. At sampling Day 10, the response is inversely related to level of concentration. At Day 20, the relationship has shifted to a parabolic one, while by Day 30 the response is again parabolic, but of a different nature with respect to concentration. All these changing relationships are largely due to the nature of the horizontal displacement between the four groups. The dynamics, graphed so clearly when viewing the treatment means, would be hard to infer from the graphs of the regressions or from a simple inspection of estimated regression parameters. Regression analyses are better for a dose-response relationship that is monotonic and essentially unchanging through time (for example, single species toxicity data). Regression analysis is a powerful statistical tool for appropriate data (for example, the data in Fig. 7 through Day 14), but the appropriateness of the regression model must be ascertained.

Between Experiment Statistics

The Days-Weighted-by-Variable (DWV) Statistic

To compare the outcomes of different experiments statistically, we desire to know if the experiments have yielded essentially the same results. It is unlikely that complex biological events will match on a day-by-day basis. This has led to the development of the days-weighted-by-variable, or DWV, statistic to compare results from different experiments [*5–6*]. Its principal use is to measure the amount of "horizontal shift" (for example, amount by which the *Daphnia* bloom is delayed, or the amount by which the nitrate depletion is delayed) between otherwise similar time traces. By not depending upon day-by-day comparisons of time traces, we avoid the time lag problems caused when two microcosms exhibit nearly identical behavior but for a slight time shift.

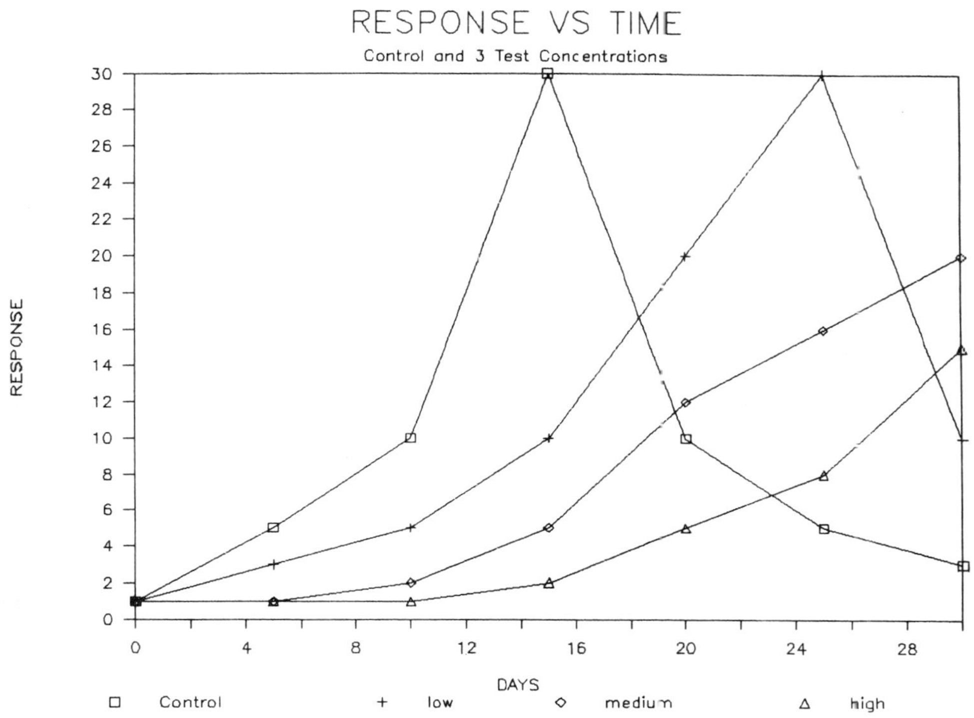

FIG. 7—*Hypothetical example of responses over time for control and three test concentrations.*

The DWV (in days) for a given microcosm is computed as follows:

$$\text{DWV} = \Sigma\,(\text{day} \times \text{value})/\Sigma\,(\text{value})$$

where value = organism count, nutrient measure, etc. and the summation is performed over all sampling days of a variable. By weighting each sampling day with its associated value, the DWV can reflect differences between an early low bloom (low DWV) versus a period of inhibition followed by an increased abundance (high DWV). This is quite different from simply taking an average value over the experiment. The DWV produces a single value for each microcosm, the experimental unit. By having the simplicity of a summary value to represent each microcosm's time trace, one can do interlaboratory comparisons using a standard two-way analysis of variance approach, with treatment group as one factor and experiment as the other.

We present some calculations of the DWV to show how it has been used to compare results from different experiments. Table 3 shows the replicate data and calculated DWVs for four microcosms from UW No. 1, one from each group (control, 500 ppb Cu, 1000 ppb Cu, 2000 ppb Cu). For both algal biovolume and *Daphnia,* the higher the concentration, the longer the delay in the peak, resulting in an increased DWV. In this way the DWV distinguishes between when particular time traces bloom or peak for different microcosms.

There are occasions where no organism recovery occurs, as is shown in Table 4. In this case, organism recovery clearly has been delayed well beyond Day 63, the end of the experiment. The DWV here is best characterized as simply having a value ">63." Such a value is not acceptable in a parametric ANOVA, but may still be used in a nonparametric analysis, as will be shown. First we show some DWV results for Algal Biovolume and *Daphnia* for various experiments.

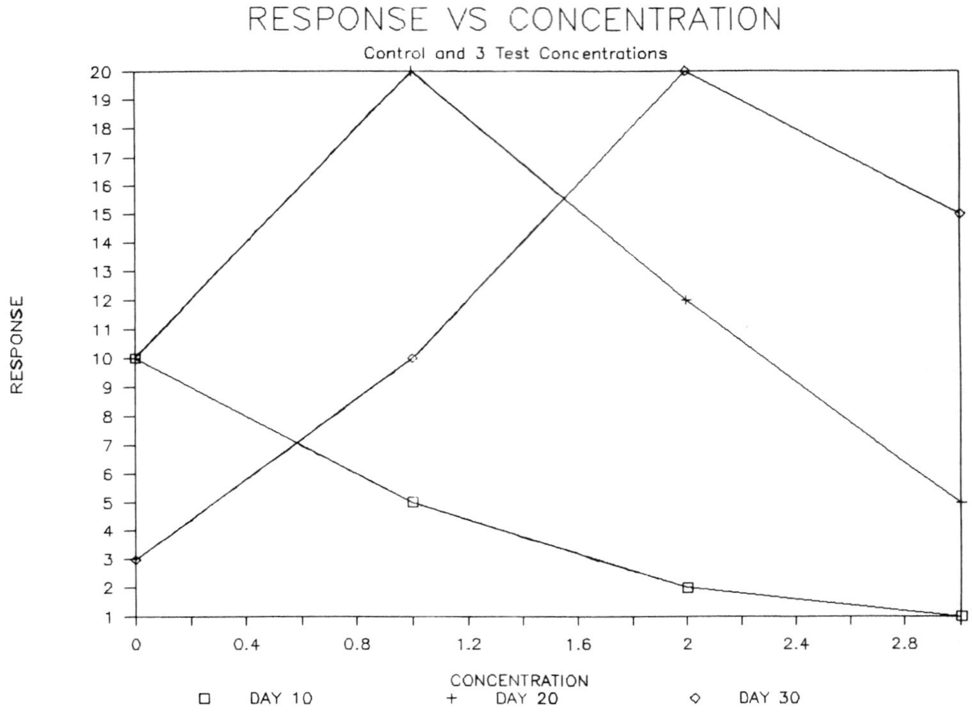

FIG. 8—*Regression responses for Days 10, 20, 30.*

These are displayed in Fig. 9. In each experiment, the effect of the increasing concentrations of Cu has been to delay the development of the algal and *Daphnia* populations. The greater the Cu concentration, the greater the delay, sometimes beyond the end of the experiment. Friedman's nonparametric two-way ANOVA by ranks [*11*] demonstrates that the events happened later as Cu concentration increased [*6*].

The DWV may also be used to detect shifts in trends for variables such as nitrate, which is depleted by the algal incorporation. As the Cu concentration increases and the nitrate depletion is further delayed, the DWV increases. Figure 10 shows mean DWVs for each experiment. The delay in nitrate depletion at the higher concentrations has been confirmed by Friedman's rank test [*6*]. The actual amount of the delay with increased concentration does vary enough between experiments to have resulted in a significant treatment-by-experiment interaction when subjected to a parametric two-way ANOVA [*6*]. By use of both parametric and nonparametric analyses for results between laboratories, we have been able to state statistically that all the experiments agreed in the nature and the direction of the effect, but that the exact timing of such effects has differed.

For changes that are more complex than simply a horizontal shift (for example, some change in the actual shape of the curve), it may be possible to generalize the DWV to second, third, fourth, etc. moments to uncover more information about shapes of the time traces.

Other Rejected Techniques for Between-Experiment Comparisons

Factorial Two-Way ANOVA Performed on Each Sampling Day—The two factors were treatment group and experiment; there were many problems with this approach. Small temporal lags between experiments in rapidly rising or falling variables resulted in tremendous variation on any given sampling day, despite identical patterns of change. The variables are not independent of each other, nor are different sampling days independent. These problems were only aggravated when a two-way ANOVA was performed on each day, resulting in an overwhelmingly large number of significant results.

Nonparametric Correlations of Arrow Diagram Results—Another approach used Spearman and Kendall rank correlations of arrow diagram results. For each sampling day, we used +1, 0, and −1, respectively, to indicate whether the treatment was statistically significantly greater than, not different from, or less than the control. These analyses also looked poor: results that were biologically quite similar appeared statistically different by this method. Again, the day-by-day comparison had distinct drawbacks. Furthermore, it was hard to get a biological sense of what the correlation values really meant.

Comparison of Fitted Regression Parameters—Various regression models were used to summarize the behavior of selected variables. The type of model was suggested by inspection of graphs of the data. This approach was found to depend greatly on the goodness of fit of the equation of the original data. Since the fit was so variable, it was not appropriate to test the statistical properties of the calculated regression equations as a measure of experiment reproducibility. Gurevitch and Chester [*26*] have done regression fits using orthogonal polynomials (suitable when a polynomial model is applicable and when the sampling days are equally spaced) to partition the trends into the linear, quadratic, and cubic portions. Correlations that assess the fit must be high to place any faith in further analyses upon the computed coefficients. Furthermore, this method would still compare controls to controls, treatments to treatments, and would not take into account the specific relationship between the controls and treatments within each experiment. It would also not give information as to *when* the treatments drifted away from the control values.

Differential Equation Modeling of Individual Microcosms—Another approach involved constructing a simple model of three differential equations with biologically meaningful parameters relating nitrate, algal biovolume, and *Daphnia;* this was a subset of the more complex MICMOD models [*27*]. By fitting this simple model to the SAM data for each microcosm and by employing multivariate ANOVA to the model parameters (for example, mortality, growth rates), it should have been possible to measure differences between experiments in a biologically meaningful way. This systems model would provide quantitative measures of interactions among variables. There is a *great* deal of time and effort involved in fitting by trial and error a set of complex differential equations to each microcosm. This modeling method is beyond our current resources due to the sheer enormity of the effort.

Distances Between Microcosm Time Traces—We obtained a single distance measure when comparing a control microcosm to a treatment microcosm (or comparing any two microcosms), instead of a set of correlated measurements for each sampling day in the time trace [*4*]. We considered mean absolute distance (averaged vertical distances for a given pair of microcosms), mean square distance, and euclidean distance [*28*]. For each distance measure considered and each control-to-treatment comparison, 36 measures from each experiment (6 controls × 6 treatments) were recorded for each variable. It was hypothesized that if the experiments exhibited the same relationships between the controls and the treatments, the 36 measures should show roughly the same values across the experiments. The probability distributions of the distance measures were intractable since the measures themselves were somewhat correlated. While it would be theoretically possible to produce the probability distributions once the correlation structure was specified, this would be a far from trivial exercise. Since the distance measures still depend upon making day-by-day comparisons between time traces, microcosms following the same pattern but just a few sampling days apart could still look very different. This method would also not distinguish between a time lag (a horizontal difference) and differences between the magnitude of, say, two different blooms on the same sampling day (a vertical difference). Thus, anomalous statistical results which did not match the biological patterns were often observed.

TABLE 3—*DWV calculations for* Daphnia *and algal biovolume from UW No. 1.*

	0 ppb Cu, Microcosm 21			
Day	*Daphnia*	Algal Biovolume	Day × *Daphnia*	Day × Algal Biovolume
4	Missing	278.182	Missing	1112.73
7	0.547	518.856	3.829	3631.99
11	1.719	1348.85	18.912	14837.3
14	12.0	3611.73	168.0	50564.2
18	41.0	4403.77	738.0	79267.8
21	50.0	2134.81	1050.0	44831.1
25	107.0	2521.81	2675.0	63045.2
28	97.0	1399.37	2716.0	39182.5
32	52.0	665.028	1664.0	21280.9
35	6.5	1021.40	227.5	35748.9
39	0.119	1093.82	4.633	42658.9
42	0.0396	359.844	1.663	15113.5
46	0.0400	564.0	1.840	25944.0
49	0.364	484.412	17.816	23736.2
53	0.816	711.089	43.264	37687.7
56	0.939	601.977	52.573	33710.7
60	6.667	195.706	400.002	11742.4
63	9.333	228.550	587.998	14398.6
Sum	386.0836	22143.204	10371.03	558494.62

DWV *(Daphnia)* = 10371.03/386.0836 = 26.86
DWV (algal biovolume) = 558494.62/22143.204 = 25.22

	500 ppb Cu, Microcosm 1		1000 ppb Cu, Microcosm 2		2000 ppb Cu, Microcosm 4	
Day	*Daphnia*	Algal Biovolume	*Daphnia*	Algal Biovolume	*Daphnia*	Algal Biovolume
4	Missing	308.722	Missing	293.076	Missing	429.895
7	0.305	631.892	0.271	535.846	0.513	852.648
11	0.0	1502.82	0.0	1081.29	0.0	339.346
14	0.0	1610.37	0.0	709.298	0.0	204.843
18	0.0	3944.58	0.0	741.259	0.0	542.663
21	0.0	6265.40	0.0	987.153	0.0	326.489
25	0.0	9301.65	0.0	2180.88	0.0	232.641
28	0.0	9902.10	0.0	7177.72	0.0	80.056
32	6.667	18904.1	0.755	21070.5	0.0	78.886
35	16.0	14934.3	0.457	11792.7	0.0	92.020
39	87.0	8387.47	20.0	11011.0	0.0	21.152
42	129.0	3900.96	24.0	10477.7	0.0	41.092
46	282.0	759.534	96.0	7229.73	0.119	732.853
49	211.0	64.441	199.0	2485.48	0.238	2770.39
53	89.0	132.373	197.0	1052.92	1.320	4709.67
56	63.0	131.336	132.0	97.034	39.500	1012.55
60	40.0	20.540	30.0	313.842	115.0	477.555
63	46.0	169.576	20.0	115.480	147.0	72.181
	DWV *(Daphnia)* = 47.88		DWV *(Daphnia)* = 35.47		DWV *(Daphnia)* = 60.80	
	DWV (algal biovolume) = 30.17		DWV (algal biovolume) = 51.27		DWV (algal biovolume) = 42.73	

TABLE 4—*Example of no* Daphnia *recovery from UW No. 2R.*

	1000 ppb Cu, Microcosm 9	
Day	*Daphnia*	Algal Biovolume
4	0.0	285.619
7	0.765	364.972
11	0.0	1556.09
14	0.0	1734.99
18	0.0	1463.95
21	0.0	1417.57
25	0.0	1130.32
28	0.0	1156.81
32	0.0	992.316
35	0.0	1063.69
39	0.0	488.983
42	0.0	1458.61
46	0.0	3341.38
49	0.0	5921.07
53	0.0	4735.99
56	0.0	12972.00
60	0.0	8437.74
63	0.0	10376.40

NOTE: DWV *(Daphnia)* is >63 days (no organism recovery by 63 days). DWV (algal biovolume) = 49.11.

Discussion

We have explored comparisons within and between similarly designed experiments, but have not handled the more difficult problem of statistical comparisons of different kinds of communities in their responses (this pond with that lake; Lake A with Lake B). The problems discussed here are not those of microcosms per se, but of ecological measurements which consist of related (nonindependent) variables and repeated samples within the same communities. Similar problems are encountered in many multispecies or community toxicity tests [*30,31*]. Statistical techniques are as yet inadequate to handle many of the ecologist's realistic problems.

Based on the experience with extensive microcosm data, we recommend that one-way ANOVAs with accompanying IND plots be used for analyzing data within experiments and that DWVs be used to compare results from different experiments. An alternative method to the IND graphic is shown by Giddings (this symposium); statistical power and replicability of multispecies toxicity test is reviewed by Giesy and Allred [*31*]. To facilitate data handling within an experiment, we are preparing a microcomputer data handling package (for both MacIntosh and IBM-AT compatibles) that will automate the calculations from laboratory measures through the statistical analyses and report tables and graphics.

It would be helpful if ecologists and toxicologists could test different statistical techniques on data sets from a variety of different experimental systems. To facilitate statistical experiments, some of our microcosm data has been prepared on microcomputer disks for limited distribution (through the International Joint Commission). It would be useful to have a workshop setting in which researchers could explore statistical techniques on their and other data sets to determine if the statistical approaches are valid on a variety of data sets.

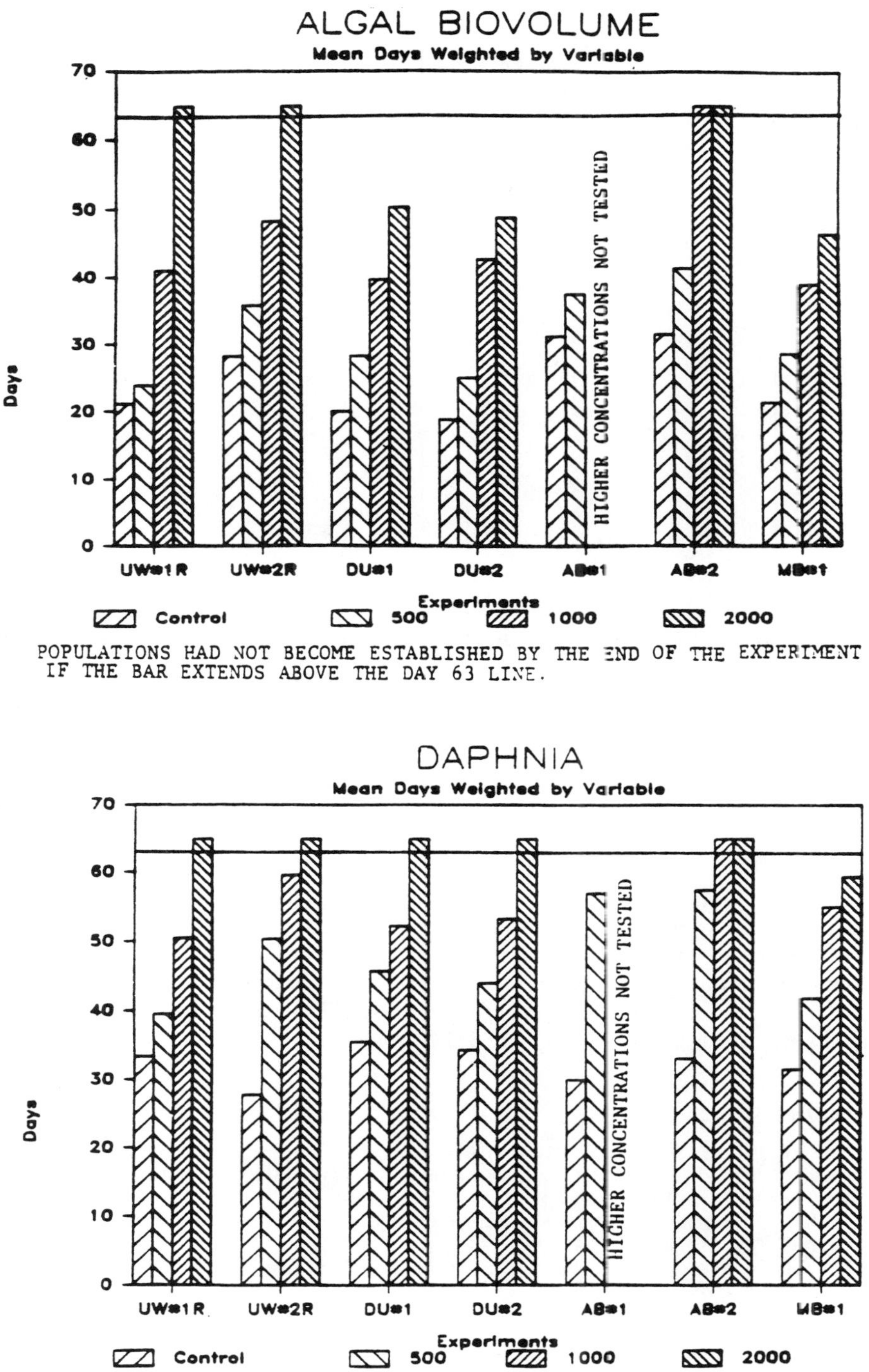

FIG. 9—*Comparison of interlaboratory experiments for algal biovolume and* Daphnia *(DWV).*

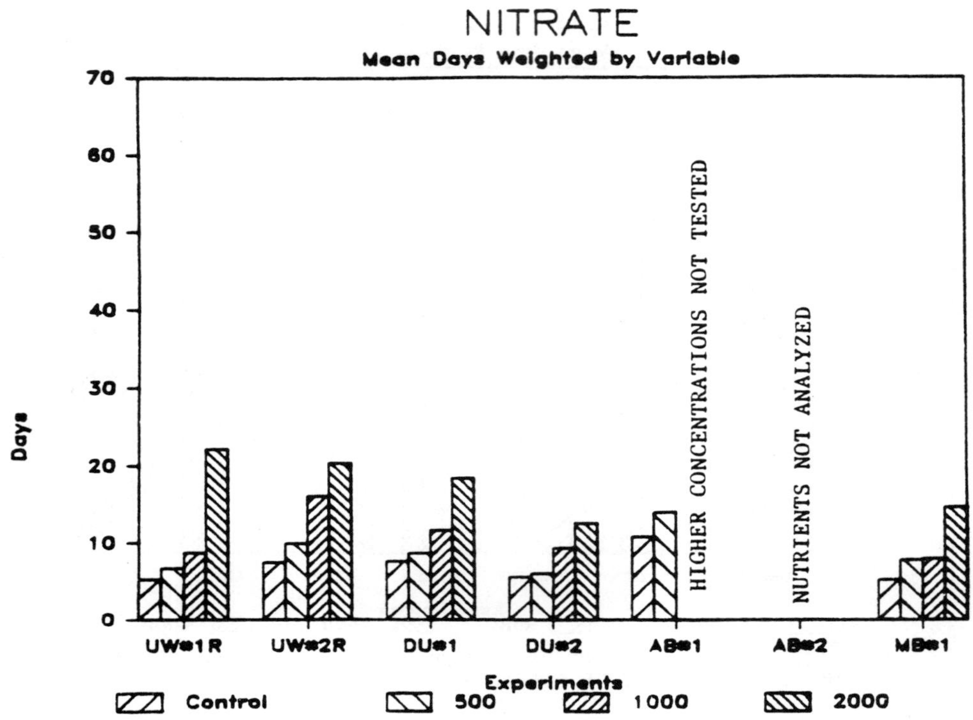

FIG. 10—*Comparison of interlaboratory experiments for nitrate (DWV).*

Acknowledgments

Special thanks are due to the staffs of the participating laboratories: Duluth—EOA, David Yount; University of the Michigan—Duluth, Lyle Shannon, Michael C. Harrass (during these studies), and Chuck Walbridge; U.S. Army—Aberdeen, Wayne Landis, Mark Haley, and Dennis Johnson; Marine Bioassay Laboratory—Allan Thum, Skip Newton, Sherrie Hamer, and John Hardin. The project has been supported by the U.S. Food and Drug Administration under contracts 223-76-8348, 223-80-2352, and 223-83-7000.

References

[1] Taub, F. B. and Crow, M. E., "Synthesizing Aquatic Microcosms," *Microcosms in Ecological Research*, J. P. Geisy, Ed., *DOE Symposium Series 52*, CONF781101, U.S. Department of Energy, Washington, DC, 1980, pp. 69–104.

[2] Taub, F. B. in *Algae as Ecological Indicators*, L. E. Shubert, Ed., Academic Press, New York, 1984, Chap. 13, pp. 363–394.

[3] Taub, F. B., Read, P. L., Kindig, A. C., Harrass, M. C., Hartmann, H. J., Conquest, L. L., Hardy, F. J., and Munro, P. T. in *Aquatic Toxicology and Hazard Assessment: Sixth Symposium, ASTM STP 802*, W. E. Bishop, R. D. Cardwell, and B. B. Heidolph, Eds., American Society for Testing and Materials, Philadelphia, 1983, pp. 5–25.

[4] Taub, F. B., Kindig, A. C., and Conquest, L. L., "Preliminary Results of Interlaboratory Testing of a Standardized Aquatic Microcosm," *Community Toxicity Testing, ASTM STP 920*, J. Cairns, Jr., Ed., American Society for Testing and Materials, Philadelphia, 1986, pp. 93–120.

[5] Taub, F. B., Kindig, A. C., and Conquest, L. L., "Interlaboratory Testing of a Standardized Aquatic Microcosm," *Aquatic Toxicology and Hazard Assessment: Tenth Volume, ASTM STP 971*, W. J.

Adams, G. A. Chapman, and W. G. Landis, Eds., American Society for Testing and Materials, Philadelphia, 1987, pp. 384–405.
[6] Taub, F. B., Kindig, A. C., Conquest, L. L., and Meador, J. P., "Results of Interlaboratory Testing of the Standardized Aquatic Microcosm Protocol," Eleventh ASTM Symposium on Aquatic Toxicology and Hazard Assessment. ASTM, in press.
[7] Taub, F. B., "Standardized Aquatic Microcosm—Development and Testing," *Aquatic Ecotoxicology*, A. Boudou and F. Ribeyre, Eds., CRC Press, Inc., in press.
[8] Taub, F. B. and Kindig, A. C., "Standardized Aquatic Microcosm Protocol," final report, FDA Contract No. 223-83-7000, Section II, Dept. of Health and Human Services, Rockville, MD, 1986.
[9] Environmental Protection Agency, "Toxic Substances Control Act Test Guidelines, Proposed Rule," *Federal Register*, Vol. 52, No. 187, pp. 36344–36352, September 28, 1987.
[10] Conquest, L. L., *International Journal of Environmental Studies*, Vol. 20, 1983, pp. 209–221.
[11] Zar, J. H., *Biostatistical Analysis*, 2nd ed., Prentice Hall, Inc., Englewood Cliffs, NJ, 1984.
[12] Sokal, R. R. and Rohlf, F. J., *Biometry*, 2nd ed., W. H. Freeman and Co., New York, NY, 1981.
[13] Steel, R. G. D. and Torrie, J. H., *Principles and Procedures of Statistics*, 2nd ed., McGraw-Hill, New York, NY, 1980.
[14] Andrews, H. P., Snee, R. D., and Sarner, M. H., *The American Statistician*, Vol. 34, 1980, pp. 195–199.
[15] Dunnett, C. W., *Journal of the American Statistician Association*, Vol. 50, 1955, pp. 1096–1121.
[16] Dunnett, C. W., *Biometrics*, Vol. 20, 1964, pp. 482–491.
[17] Franco, P. J. and Giddings, J. M., *Environmental Toxicology and Chemistry*, Vol. 3, 1984, pp. 447–463.
[18] Giddings, J. M. and Franco, P. J., *Environmental Toxicology and Chemistry*, Vol. 3, 1984, pp. 465–488.
[19] Carmer, S. G. and Walker, W. M., *Agronomy Journal*, Vol. 74, 1982, pp. 122–124.
[20] Conquest L. L., "Statistics Support Document," final report, Section III of V, "Interlaboratory Testing of an Aquatic Microcosm Protocol," USFDA Contract No. 223-83-7000, Dept. of Health and Human Services, Rockville, MD, 1986.
[21] Crow, M. E., "Evaluation of Techniques for the Through-Time Analysis of Aquatic Microcosm Data," 1979, pp. 105–142 in Conquest, L. L., [20].
[22] Chew, V., *HortScience*, Vol. 11, 1976, pp. 348–357.
[23] Thomas, W. H., Holm-Hansen, O., Seibert, D. L. R., Azam, F., Hodson, R., and Takahashi, M., *Bulletin of Marine Science*, Vol. 27, No. 1, 1977, pp. 34–43.
[24] Zischke, J. A., Arthur, J. W., Hermanutz, R. O., Hedtke, S. F., and Helgen, J. C., *Aquatic Toxicology*, Vol. 7, 1985, pp. 37–58.
[25] Giesy, J. P., Kania, H. J., Bowling, J. W., Knight, R. L., Mashburn, S., and Clarkin, S., "Fate and Biological Effects of Cadmium Introduced into Channel Microcosms," EPA 600/3-79-039, EPA, Washington, DC, April 1979.
[26] Gurevitch, J. and Chester, S. T., Jr., *Ecology*, Vol. 67, 1986, pp. 251–255.
[27] Swartzman, G. L. and Rose, K. A., *Ecological Modelling*, Vol. 22, 1983/84, pp. 123–134.
[28] Zimmerman, G. M., Goetz, H., and Mielke, P. W., Jr., *Ecology*, Vol. 66, 1985, pp. 606–611.
[29] *Multispecies Toxicity Testing*, J. Cairns, Jr., Ed., Pergamon Press, New York, 1985.
[30] *Community Toxicity Testing, ASTM STP 920*, J. Cairns, Jr., Ed., American Society for Testing and Materials, Philadelphia, May 1985.
[31] Giesy, J. P. and Allred, P. M., "Replicability of Aquatic Multispecies Test Systems," *Multispecies Toxicity Testing*, J. Cairns, Jr., Ed., Pergamon Press, New York, 1985.
[32] Stephan, C. E. and Rogers, J. W. in *Aquatic Toxicology and Hazard Assessment, ASTM STP 891*, American Society for Testing and Materials, Philadelphia, 1985; pp. 328–338.

Patrick J. Sheehan[1]

Statistical and Nonstatistical Considerations in Quantifying Pollutant-Induced Changes in Microcosms

REFERENCE: Sheehan, P. J., **"Statistical and Nonstatistical Considerations in Quantifying Pollutant-Induced Changes in Microcosms,"** *Aquatic Toxicology and Hazard Assessment: 12th Volume, ASTM STP 1027,* U. M. Cowgill and L. R. Williams, Eds., American Society for Testing and Materials, Philadelphia, 1989, pp. 178–188.

ABSTRACT: Data from four generic microcosm experiments examining the effects of toxic chemicals were used to examine statistical and nonstatistical considerations in quantifying pollutant-induced changes in aquatic microcosms. The minimum limits of detectable change within the experimental design were assessed in terms of mean ecosystem dissolved oxygen production and consumption, and algal community primary productivity rate. Consistent reproducibility in hazard rankings based on the area of "relative impact" was found for response variables in microcosms treated with four toxic chemicals in repeat experiments. Based on changes in microcosm dissolved oxygen production, the response pattern for atrazine treatment was found to be similar among repeat experiments. Primary productivity response data for the four experiments could not be fitted to a single linear or curvilinear regression model, and, therefore, the similarity of repeat test results could not be analyzed statistically.

KEY WORDS: aquatic, microcosms, hazard ranking, response curve, statistical considerations, limits of detectable change

Microcosms provide a fundamental experimental model to study the effects of chemical perturbations on aquatic ecosystems. Microcosm studies, through the incorporation of replication in experimental design, provide data which can be analyzed statistically to determine significant changes in ecological structure or function in treated systems.

The problem in interpreting microcosm data is to decide whether an observed change in some system response variable represents a deviation caused by the presence of a toxic chemical or whether such changes as are seen are part of the natural fluctuations or cycles inherent in the microcosm system. The main purpose of replication is to supply an estimate of variability (error) by which significance of treatment and control comparisons can be judged.

Interpretation of the results of microcosm studies present some analytical difficulties. The analytical difficulties are associated with several factors:

1. There is a good deal of inherent biological variability in even simple aquatic microcosm models, and the level of variability influences the limits of detectable change as well as the degree of difference in patterns of system response to perturbation.
2. The response of microcosms to toxic substances represents the sum of direct toxic and indirect ecological effects to treatment.
3. Hazard is associated with both the magnitude and duration of induced change.
4. Strict standardization of microcosm testing procedures is not possible; therefore, consis-

[1]Environmental toxicologist, Aqua Terra Technologies, Inc., Walnut Creek, CA 94596.

tency in the reproduction of test results in repeat tests needs to be continuously evaluated to determine the utility of microcosm models in predicting response and ranking potential hazards.

5. Microcosm response to chemical perturbations is perhaps best represented as a curve of response versus exposure time; however, statistical procedures to compare response curves are not well developed and are in some cases difficult to apply to microcosm data.

This paper examines the statistical and nonstatistical considerations in quantifying chemical pollutant-induced changes in microcosms. Specifically, the manuscript addresses the following issues:

1. Determining the limits of detectable change in tests with a generic microcosm model.
2. Evaluating consistency in the ranking of potential ecological hazard based on microcosm data from repeat experiments.
3. Distinguishing significant differences in the patterns of response curves based on microcosm data from repeat experiments.

Background

In most microcosm experiments examining the effects of toxic chemicals, the test chemical is introduced into replicate test systems as a single dose at some point in the development of these dynamic systems. In the generic microcosm studies described by Sheehan et al. [*1*], the chemical was introduced either during a phase of rapid phytoplankton growth under "nutrient sufficient" conditions or at a later point after excess nutrients had been depleted under "nutrient deficient" conditions. In either case the systems received the chemical as a single well-mixed pulse. Prior to the introduction of the chemical and at regular intervals thereafter, microcosm response is monitored by measuring some combination of ecosystem, community, and population response variables describing system function (for example, net ecosystem production) and community structure (for example, zooplankton biomass or numbers).

The response of ecological variables in microcosm studies provides quantitative or graded data rather than the all or none quantal response information such as lethality often monitored in short-term single species toxicity tests. The magnitude of response measured in microcosm tests is the sum of direct toxic effects and indirect ecological effects. In most cases, the proportion of response associated with direct toxicity cannot readily be separated from that associated with ecological interactions. The magnitude and the duration of effects must be considered in hazard assessment.

Reported monitoring of microcosms exposed to toxic chemicals for several weeks or months results in a series of data with time. For each discrete sampling time, response variables are measured in control and treated systems. Because microcosms are dynamic systems and the conditions, even in control units, will change with time, pretreatment and posttreatment comparisons are generally not valid.

Comparisons must be made between treated microcosms and untreated controls. There is, however, little agreement among microcosm experimenters and statisticians about the best way to analyze data on treatment effects. One approach has been to make week-by-week comparisons of weekly or weekly averaged data using a one-way analysis of variance (ANOVA) to detect significant treatment effects (for example, Refs *1–4*). Data may be presented graphically with the area of nonsignificance indicated. Sheehan et al. [*1*] have taken this approach a step further to rank potential hazards by describing the response of treatment replicates in terms of:

1. *Maximum change* in treatment response range from the control response range.
2. *System resilience* defined as the time required for the treatment range once displaced from the normal range to return.

3. *Relative impact* measured as the integral difference between the perturbed trajectory of the treatment range and that of the control range.

Another approach has been to look at the rate of change of response variables over selected time intervals [*5*].

In microcosm data analysis, it is important to determine the magnitude of impact and the pattern of response as well as the relative differences in effects of concentrations of the same chemical or of different chemical treatments.

Neither of the described analytical approaches above addresses the comparison of response curves describing the patterns of disturbance and recovery with exposure time in treated versus untreated microcosms. A statistical analysis of response curves depends upon the fitting of data to linear or curvilinear regression models and then comparing regression coefficients for fitted curves using analysis of covariance techniques [*6*]. Recently, Pantone et al. [*7*], described an application of this technique to compare herbicide tolerance in weed species. This type of analysis has not been applied to microcosm data even though currently applied week-by-week ANOVA approaches are generally inadequate to statistically evaluate important questions such as the reproducibility of response patterns in repeat tests.

Methods

A generic aquatic microcosm model was used in four repeat experiments to assess the reproducibility of microcosm results. The generic test system is based on the premise that small aquatic microcosms, containing naturally coadapted communities of algae, protozoa, bacteria, and zooplankton, exhibit typical ecosystem properties and processes, but simulate no particular natural aquatic system.

Microcosms were set up by first diluting a parent culture inocula with distilled water ($\sim 1:10$) in a 50-L polypropylene carboy, enriching this mixture with inorganic medium, and then redistributing 3.5-L portions into 4-L glass jars. The culture media was described in detail previously [*1*].

Specific media differences for the four experiments are described below:

Experiment 1—Inoculum was MacKay Lake water collected under winter conditions from under ice; inoculum was maintained in the laboratory culture for two weeks prior to use.

Experiment 2—Inoculum was the same as Experiment 1 except that it was maintained in laboratory culture for more than three months prior to use.

Experiment 3—Inoculum was MacKay Lake water collected under summer conditions; inoculum was not maintained in laboratory culture prior to use.

Experiment 4—Inoculum was Heney Lake water collected under winter conditions; inoculum was maintained in the laboratory culture for more than two weeks prior to use.

Illumination for an experiment was at levels similar to that provided for parent cultures. Every second day (Monday, Wednesday, Friday) the jars were moved clockwise one quarter of the way around the laboratory bench to minimize effects due to uneven lighting. Each microcosm was stirred daily to resuspend particulate materials, and distilled water was added once weekly to maintain constant volume.

Microcosm response to toxic chemicals was analyzed in terms of the metabolic response variables: (1) diel production and consumption of dissolved oxygen; and (2) ^{14}C primary productivity.

1. *Diel production and consumption of dissolved oxygen* were estimated from three-point oxygen readings in each microcosm on two consecutive mornings just after lights were turned on, and approximately 9 h later on the intervening afternoon.

2. *^{14}C-Primary productivity rates* of ^{14}C-labelled inorganic carbon uptake into particulate matter were used to estimate rates of photosynthesis and primary productivity [*8*]. Assays were

conducted by placing 100-μL of individually ampoulated stock solutions of $Na_2^{14}CO_3$ (~10.7 μCi/mL) into 10 mL of microcosm water in 22-mL liquid scintillation counting (LSC) glass vials. After a 2-h incubation at 20°C under the same light system as for the microcosms, the samples were filtered onto 0.45-μm cellulose nitrate membranes in the dark. These filters were then digested with 1 mL of ethylene glycol monomethyl ether and their radioactivity determined by LSC using 10 mL of Aquasol II as a cocktail. Background levels of uptake were determined using a comparable set of samples kept in the dark at all times. Finally, rates of photosynthetic carbon uptake were calculated using the difference in activity between light and dark vials and then adjusted for measured dissolved inorganic (DIC) concentration (by alkalinity titration).

Four replicates of control and treated microcosms were monitored in all experiments. The toxic substances tested were copper (100 μg/L), methoxychlor (200 μg/L), atrazine (450 μg/L), and trifluralin (200 μg/L). The concentrations of test substance represent expected environmental levels. Microcosms were treated during a period of rapid phytoplankton growth. Primary productivity and dissolved oxygen production and consumption were monitored just prior to the introduction of the toxic substances and weekly thereafter. All experiments were run for four weeks to assure that observed impacts were not of a transient nature [*9*].

Data from four replicate experiments were analyzed in this presentation. Data were analyzed to determine minimum detectable treatment differences and to evaluate test reproducibility. The reproducibility of experimental results was examined in terms of: (1) the consistency of hazard ranking among experiments based on system response as relative impact (Fig. 1); and (2) the statistical differences in response curves among experiments based on curve fitting and analysis of covariance techniques comparing regression coefficients [*6*].

Results and Discussion

Levels of Detectable Change

The experimental design used in the generic microcosm experiment incorporated four replicate microcosm units per treatment. The selection of four replicates was based on experimental

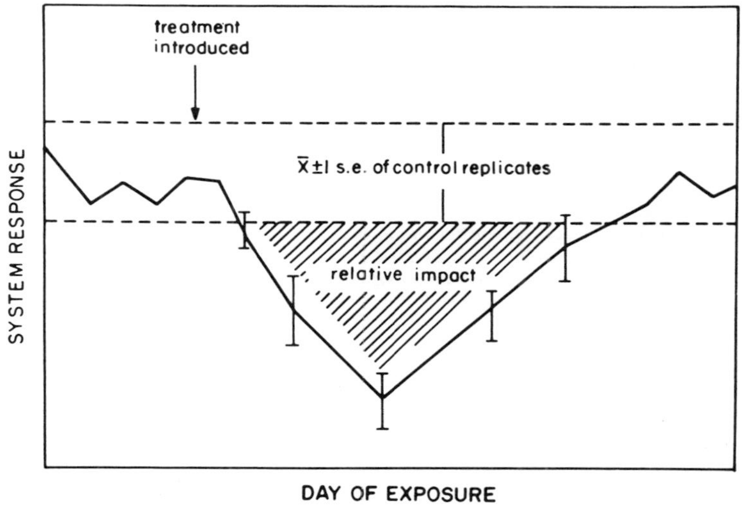

FIG. 1—*Hypothetical response model based on assessing impact in terms of changes for chemically treated systems as measured from the normal operating range of control systems.*

data that showed that four to six replicates provided maximum sensitivity within practical constraints on laboratory space. Based on data from four repeat experiments, the inherent variability in microcosm system metabolism expressed as the coefficient of variation (CV) of measured primary productivity rate and dissolved oxygen production and consumption is presented in Table 1. There is little variability associated with measurement techniques for either primary productivity or dissolved oxygen production and consumption. CV values for the primary productivity rate in replicate untreated microcosms varied from 4 to 74% based on 20 data collections from four experiments. The median CV value was 28%. Based on the median CV, the minimum detectable statistical difference ($\alpha = 0.05$) as a percent of mean treatment response is 52%.

CV values for dissolved oxygen production and consumption ranged from 1 to 23%, and 2 to 36%, respectively. Based on median CV values of 9 and 11% for dissolved oxygen production and consumption, the minimum detectable differences as percent mean response were 17 and 20%, respectively. As measures of system metabolic function, dissolved oxygen production and consumption are more than twice as sensitive in detecting treatment effects as primary productivity rate. The replicibility of dissolved oxygen metabolism in microcosms, as well as the ease and low cost associated with monitoring dissolved oxygen, suggest that this variable is important to detecting and quantifying microcosm response to toxic chemicals.

Reproducibility of Hazard Rankings

The reproducibility of hazard rankings in repeat experiments is essential to an acceptable toxicity screening procedure. In the generic microcosm model, we assume that functional response of treated microcosms is relatively uninfluenced by the taxonomic composition of the community inocula. Unfortunately, there is no statistical procedure to prove the hypothesis which states that "there is no influence of taxonomic structure on functionally based hazard ranking."

Hazard ranking, however, may be compared within treatments in an experiment and between experiments with the graphical approach first introduced by Leffler [*10*] and modified by Sheehan et al. [*1*] (Fig. 1). This approach defines an area of "relative impact" which is measured directly with a planimeter from graphical presentation of the data. Relative impact is the integral difference between the treatment range and the control range.

This graphical approach was applied to analyze treatment reproducibility in data sets (Figs. 2 and 3) for dissolved oxygen production and primary productivity from four experiments representing microcosms with different inocula. Table 2 shows the hazard ranking comparisons based on the dissolved oxygen production data. The hazard ranking order was consistent among the four experiments with a rank order from high hazard to low or no hazard: atrazine,

TABLE 1—*Replication in measurement techniques and among control microcosm units during four toxicity testing experiments. Variability is expressed as the coefficient of variation. Minimum detectable difference is calculated from the CV median.*

Response Variable	Technique Variability ($N = 4$), CV	Microcosm Variability, $N = 4$: CV Median	Microcosm Variability, $N = 4$: CV Range	Minimum Detectable Difference,[b] Percent Mean Response
^{14}C primary productivity rate	7	28 (20)[a]	4 to 74	52
Dissolved oxygen production	2	9 (20)	1 to 23	17
Dissolved oxygen consumption	2	11 (20)	2 to 36	20

[a]Number of replicate sample collections during the four experimental studies.
[b]Minimum detectable difference is based on the median CV assuming $a = 0.05$ and $P = 0.80$.

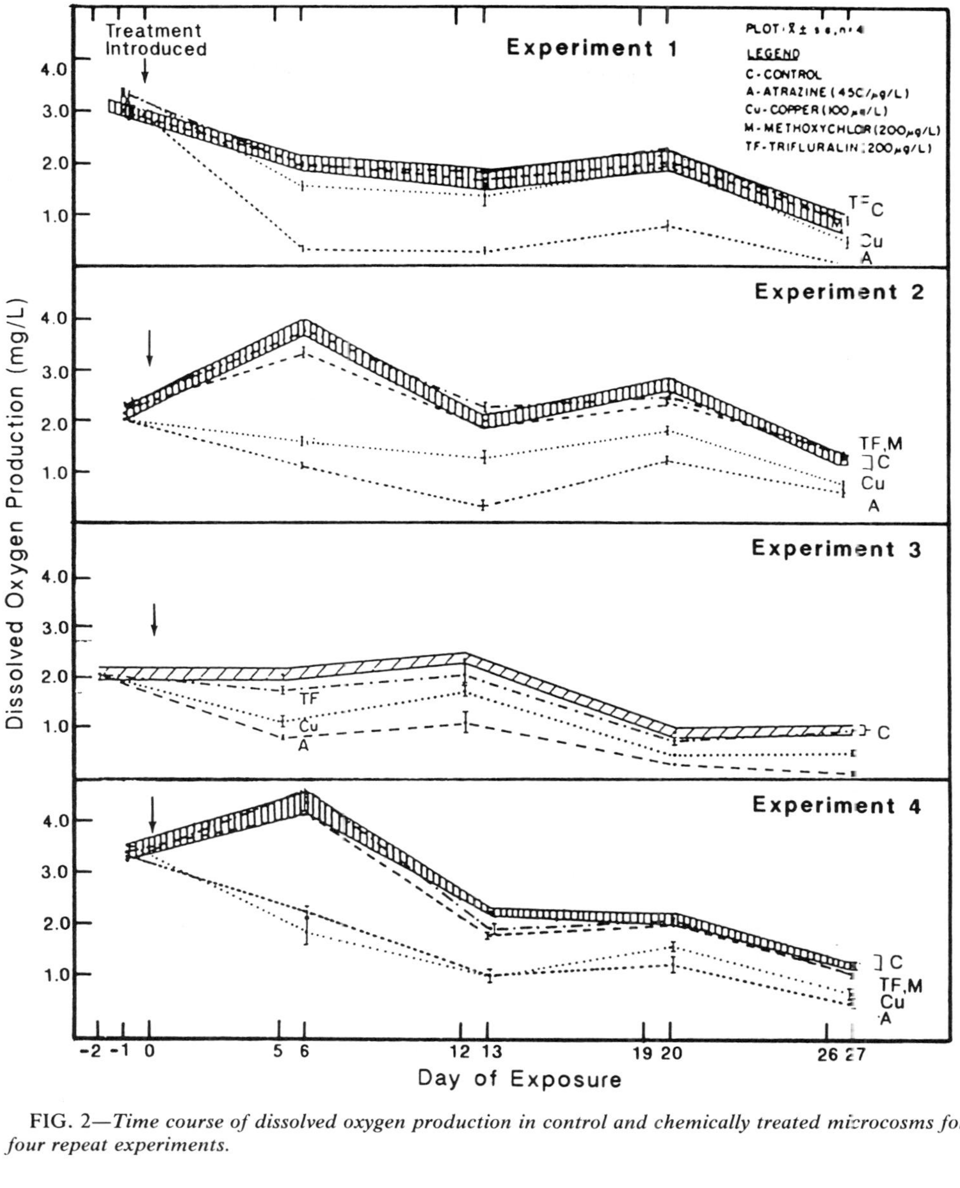

FIG. 2—*Time course of dissolved oxygen production in control and chemically treated microcosms for four repeat experiments.*

copper, trifluralin, and methoxychlor. In Experiment 1, trifluralin and methoxychlor treatments response was indistinguishable from that of untreated systems.

Table 3 displays hazard ranking comparisons based on algal community primary productivity rate. Again, hazard ranking order was consistent among the four experiments and was the same as predicted from system dissolved oxygen production. The methoxychlor treatment results in Experiments 1 and 4 were indistinguishable from those of the control systems.

The consistency in hazard rankings between the response variables was not unexpected as algal primary productivity is a major input to ecosystem production. The consistency in hazard ranking among the four experimental data sets is encouraging to advocates of a microcosm test

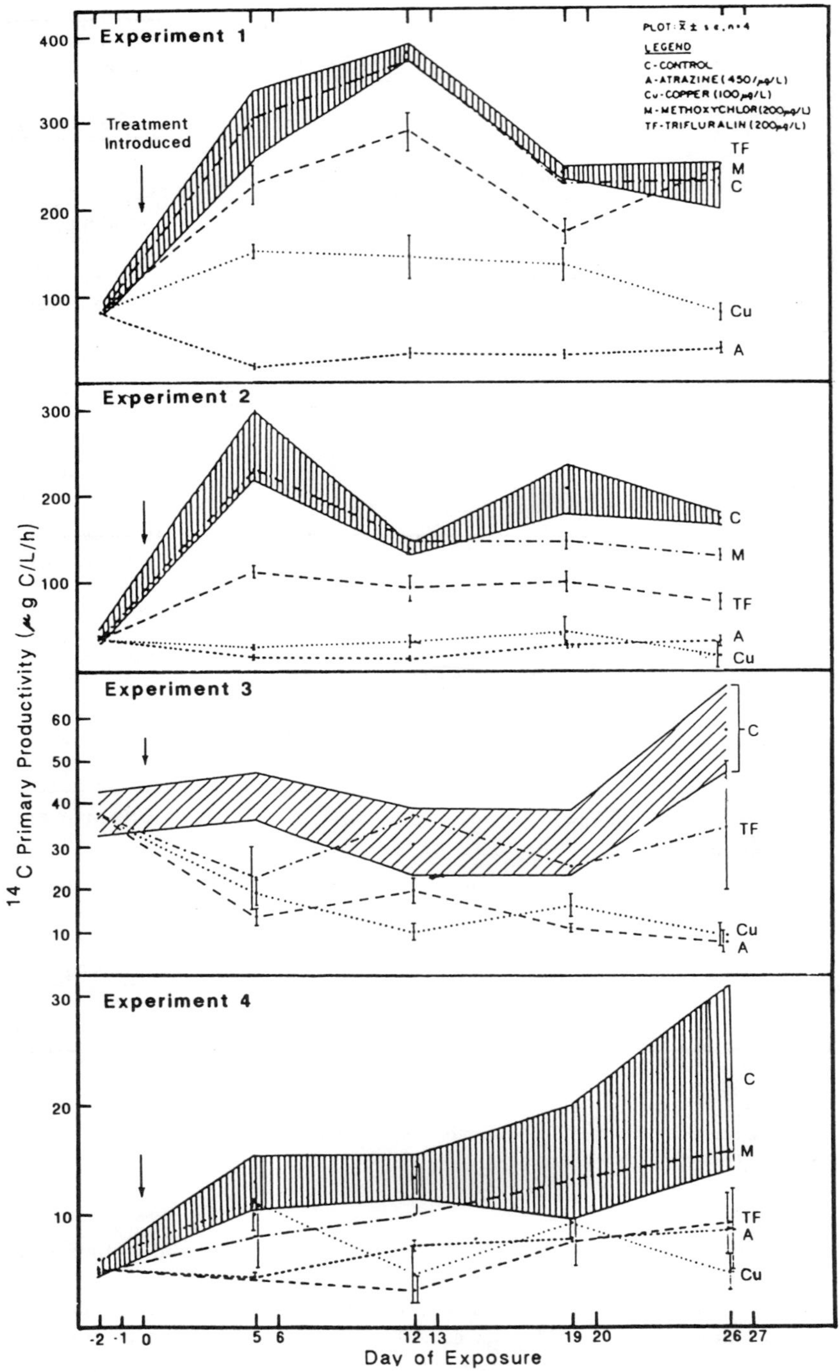

FIG. 3—*Time course of* [14]*C-primary productivity in control and chemically treated microcosms in four repeat experiments.*

TABLE 2—*Comparison of hazard rankings based on the area of the zone of relative impact for dissolved oxygen production as measured from four repeat microcosm experiments. Rankings are from high hazard (top) to low or no hazard (bottom).*

Hazard Ranking Based On Dissolved Oxygen Production[a]			
Experiment 1	Experiment 2	Experiment 3	Experiment 4
A	A	A	A
Cu	Cu	Cu	Cu
[TF][b]	TF	TF	TF
[M][b]	M	. . .[c]	M

NOTE: A = atrazine; Cu = copper; TF = trifluralin; M = methoxychlor.
[a]Data plotted in Fig. 2.
[b]Response not distinguishable from that of the control systems.
[c]Methoxychlor was not tested in Experiment 3.

TABLE 3—*Comparison of hazard rankings based on the area of the zone of relative impact for primary productivity rate from four repeat microcosm experiments. Rankings are from high hazard (top) to low or no hazard (bottom).*

Hazard Ranking Based On Primary Productivity[a]			
Experiment 1	Experiment 2	Experiment 3	Experiment 4
A	A	A	A
Cu	Cu	Cu	Cu
TF	TF	TF	TF
[M][b]	M	. . .[c]	[M][b]

[a]Data plotted in Fig. 3.
[b]Response not distinguishable from that of the control systems.
[c]Methoxychlor was not tested in Experiment 3.

screening procedure for toxic chemicals. The available data supporting consistency is, however, limited, and further studies of generic microcosm hazard ranking reproducibility are warranted.

Evaluation of Response Patterns

The results from a series of four hazard ranking experiments are shown in Figs. 2 and 3. In Fig. 2, treatment response as dissolved oxygen production is plotted in four separate panels representing the four experiments. In all experiments, atrazine treatment reduced dissolved oxygen production with respect to that of untreated systems. Atrazine is a potent inhibitor of photosynthesis [*11*]. In all cases, the recovery of atrazine-treated systems was incomplete within the experimental period and dissolved oxygen production in control systems declined with exposure time during the four weeks of postexposure observation.

For the four weeks of exposures, the shape of response curves was similar for dissolved oxygen production in the four repeat experiments. Linear regression analysis was used to represent the response-exposure relationship for differences between the control microcosm and atrazine-treated microcosm response. Regression lines (response = $b + m$ duration) for the four experimental data sets are plotted in Fig. 4. The four data sets were analyzed for homogeneity of variance with Bartlett's test [*6*], and no statistically significant difference was detected ($P \leq$

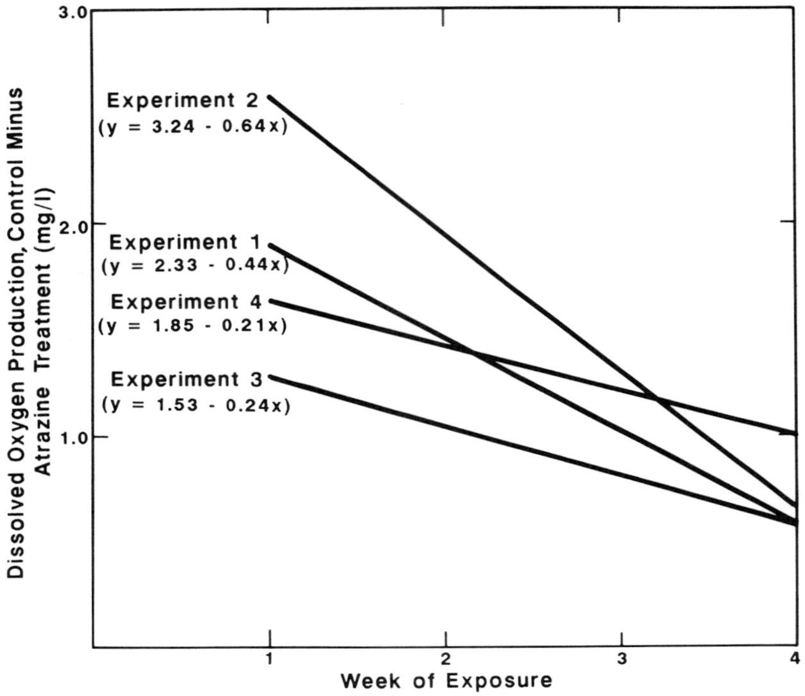

FIG. 4—*Plot of regression lines for the difference in dissolved oxygen production in control versus atrazine-treated microcosms in four repeat experiments.*

0.01). Calculated F values for all regression coefficients (b) were significantly different from zero at $P \leq 0.01$. Analysis of covariance [*6*] was used to compare the slopes (m) of response lines. The comparison of response lines for the four experiments showed that there was no statistically significant difference in slopes ($F = 0.4$, $df = 3.8$; $P \geq 0.10$).

In this rather simple case where linear regressions nicely fit dissolved oxygen production response data ($r = 0.86$, 0.96, 0.86, and 0.79 for regressions of data from Experiments 1, 2, 3, and 4, respectively), the statistical procedure described above can be used to demonstrate similarity in response curves. However, where response curves are more complex and not easily represented by linear or curvilinear regression models, statistical comparison of the reproducibility of response patterns is not possible. Such is the case for primary productivity response data collected in the four repeat experiments. Figure 3 presents primary productivity data for the four experiments. In all experiments, the rate of primary productivity in untreated microcosm replicates fluctuated substantially during the four-week exposure period. The rate of primary productivity in atrazine-treated systems was severely depressed relative to that of untreated systems, and there was very little weekly fluctuation in the response rate of atrazine-treated microcosms.

Figure 5 presents the response patterns for primary productivity data sets as the difference between control replicates and atrazine-treated replicates. Because untreated system response in the four experiments followed distinctly different patterns, response based on control-treatment difference also followed disparate patterns. In Experiments 1 and 2, data appear to follow an oscillating pattern apparently reflecting increases and decreases in untreated system productivity associated with nutrient availability. The primary productivity rates measured in Experiments 3 and 4 were substantially lower than those measured in Experiments 1 and 2, and

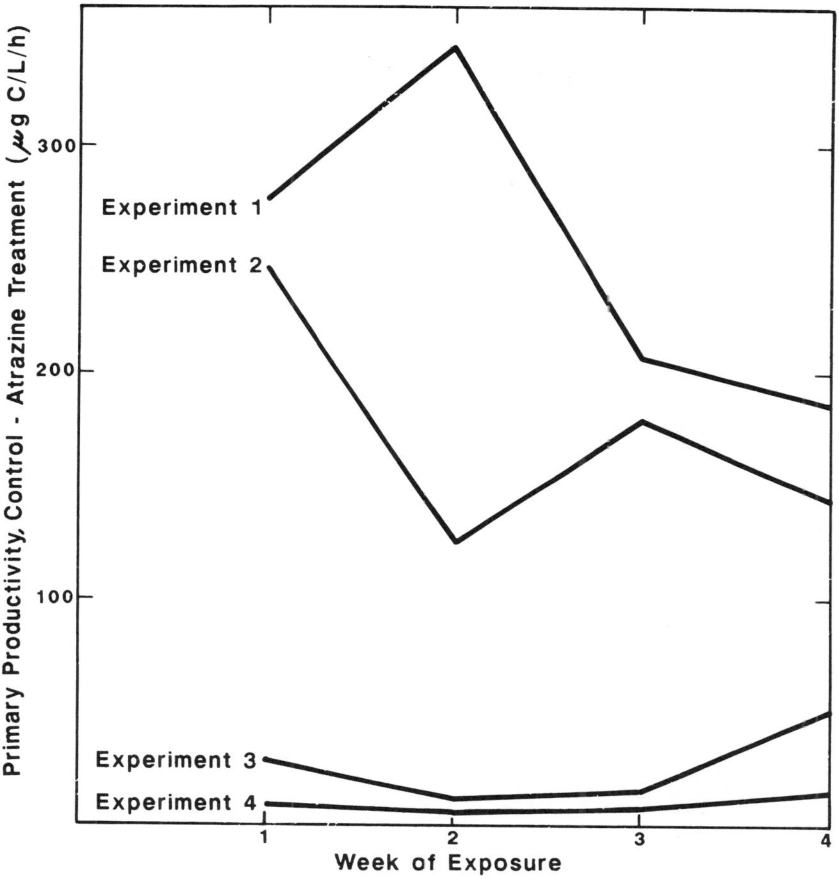

FIG. 5—*Plot of response pattern for the difference in primary productivity rate in control versus atrazine-treated microcosms in four repeat experiments.*

as such, were less responsive to treatments. Primary productivity rates in Experiments 3 and 4 do not appear to follow patterns of oscillation within the four weeks of exposure. It is possible that such a pattern might emerge from a longer period of time series observations.

No single curvilinear regression model adequately fits data from the four experiments. Therefore, no statistical procedure could be employed to test the reproducibility of response pattern. This problem is recurrent with response with exposure time microcosm data. Experimenters often have no statistical option to analyze similarities in response curve data from similar biological experiments. This deficiency presently dampens arguments for the more extensive use of microcosm models to predict response patterns in treated aquatic ecosystems.

Acknowledgments

Specific thanks for financial support go to W. K. Marshall of the National Wildlife Research Centre and D. R. Miller and L. van der Berg of the National Research Council Canada. Thanks go to Jan Hale-Levin of Aqua Terra Technologies for manuscript review and to Daisy Hoellwarth for word processing assistance.

References

[1] Sheehan, P. J., Axler, R. P., and Newhook, R. C., "Evaluation of Sample Generic Aquatic Ecosystem Tests to Screen the Ecological Impacts of Pesticides," *Community Toxicity Testing, ASTM STP 920,* J. Cairns, Jr., Ed., American Society of Testing and Materials, Philadelphia, 1980, pp. 58–179.

[2] Giddings, J. M., "A Microcosm Procedure for Determining Safe Levels of Chemicals in Shallow-Water Communities," *Community Toxicity Testing, ASTM STP 920,* J. Cairns, Jr., Ed., American Society of Testing and Materials, Philadelphia, 1986, pp. 121–134.

[3] Shannon, L. J., Harrass, M. C., Yount, J. D., and Walbridge, C. T., "A Comparison of Mixed Flask Culture and Standardized Laboratory Model Ecosystems for Toxicity Testing," *Community Toxicity Testing, ASTM STP 920,* J. Cairns, Jr., Ed., American Society for Testing and Materials, Philadelphia, 1986, pp. 135–157.

[4] Taub, F. B., Kindig, A. C., and Conquest, L. L., "Preliminary Results of Interlaboratory Testing of a Standardized Aquatic Microcosm," *Community Toxicity Testing, ASTM STP 920,* J. Cairns, Jr., Ed., American Society for Testing and Materials, Philadelphia, 1986, pp. 93–120.

[5] Schindler, J. E., Waide, J. B., Waldron, M. C., Hains, J. J., Schreiner, S. P., Freedman, M. L., Benz, S. L., Pettigrew, D. R., Schissel, L. A., and Clark, P. J., "A Microcosm Approach to the Study of Biogeochemical Systems. 1. Theoretical Rationale," *Microcosms in Ecological Research,* CONF-781101, J. P. Giesy, Jr., Ed., Technical Information Center, U.S. Department of Energy, NTIS, Springfield, VA, 1980, pp. 192–203.

[6] Snedecor, G. W. and Cochran, W. G., *Statistical Methods,* The Iowa State University Press, Ames, IA, 1973.

[7] Pantone, D. S., Larsen, L. C., Williams, W. A., *Journal of Agronomy and Crop Science,* Vol. 160, 1988, pp. 54–59.

[8] *A Manual on Methods for Measuring Primary Production in Aquatic Environments,* 2nd ed., IBP No. 12, Vollenweider, R. A., Ed., Blackwell Scientific Publications, London, 1974, p. 225.

[9] Stockner, J. G. and Anita, N. J., *Journal of the Fisheries Research Board of Canada,* Vol. 33, 1976, pp. 2089–2096.

[10] Leffler, J. W., "Microcosmology: Theoretical Applications of Biological Models," *Microcosms in Ecological Research,* J. P. Geisy, Ed., DOE Symposium Series 52, Technical Information Center, U.S. Department of Energy, 1980, pp. 14–29.

[11] Esser, H. O., Dupuis, G., Ebert, E., Vogel, C., and Marco, G. J., "S-Triazines," *Herbicides, Chemistry, Degradation and Mode of Action,* 2nd ed., P. C. Kearney and D. D. Kaufman, Eds., Marcel Dekker, New York, 1975, pp. 129–208.

Frank S. Stay,[1] *Terry E. Flum,*[2] *Lyle J. Shannon,*[2] *and J. David Yount*[1]

An Assessment of the Precision and Accuracy of SAM and MFC Microcosms Exposed to Toxicants

REFERENCE: Stay, F. S., Flum, T. E., Shannon, L. J., and Yount, J. D., **"An Assessment of the Precision and Accuracy of SAM and MFC Microcosms Exposed to Toxicants,"** *Aquatic Toxicology and Hazard Assessment: 12th Volume, ASTM STP 1027,* U. M. Cowgill and L. R. Williams, Eds., American Society for Testing and Materials, Philadelphia, 1989, pp. 189-203.

ABSTRACT: The results of 30 mixed flask culture (MFC) and four standardized aquatic microcosm (SAM) microcosm experiments were used to describe the precision and accuracy of these two protocols. Coefficients of variation (CV) for chemical measurements (DO, pH) were generally less than 7%, for ecosystem processes (primary productivity, respiration) less than 20%, and for structure (organism density) between 20 and 200%. Microcosm age had no effect on the CV; however, treatments causing a high magnitude of response resulted in significantly higher CV values. Coefficients of variation, statistical power, and minimum detectable differences for several treatment-replicate combinations are presented to aid in designing future MFC experiments. Accuracy was evaluated for the SAM and MFC microcosms by comparison to results reported for field studies of fluorene, atrazine, and chlorpyrifos. Comparisons were based primarily on the lowest effect level of short-term direct effects and the relative hazard ranking of these chemicals. The results of comparisons suggest that these microecosystems have properties common to natural systems, and their responses are more than an artifact of the microcosm.

KEY WORDS: aquatic ecosystems, microcosms, coefficients of variation, accuracy, power

Recognition of the importance of toxicity testing at the community and ecosystem levels [*1-3*] has increased. A number of protocols have been developed to address this need [*4-6*]. Although these tests have demonstrated that such procedures are capable of detecting effects, the question of their precision and accuracy remains largely unanswered.

Three aspects of precision, replicability, repeatability, and reproducibility were defined by Giesy and Allred [*7*] in their extensive review of precision of multispecies tests. Replicability, which is a function of the variance among experimental units within a treatment at a given time, is the aspect of precision described in this paper.

Microcosm protocols have been designed to assess a variety of ecological and toxicological questions. The experimental design and selection of measured variables can greatly influence the precision of the protocol. Measured variables generally are grouped into three categories, which include those dealing with the chemistry of the system, those which estimate ecosystem processes such as primary productivity and respiration, and organism densities which reflect structural characteristics of the microcosm community. Crow and Taub [*8*] compared the repli-

[1] Research aquatic biologist, Duluth Environmental Research Laboratory, U.S. EPA, Duluth, MN.
[2] Junior scientist, University of Minnesota—Duluth, Duluth, MN.
[3] Research fellow, University of Minnesota—Duluth, Duluth, MN.
[4] Life scientist, Duluth Environmental Research Laboratory, U.S. EPA, Duluth, MN.

cability of response variables commonly used in microcosm protocols. A higher coefficient of variation (CV, ratio of the standard deviation to the mean) was associated with structural measurements than with process measurements. This might suggest, as proposed in the Leffler mixed flask culture (MFC) protocol [*6*], that process measurements alone can be used to assess the effect of toxicants. This is not always sufficient, however, since the questions asked often deal with changes in structural characteristics of the system, and process variables appear to give little insight into major structural changes [*9*]. In order to answer questions about the effects of a toxicant on ecosystem structure, inferences must be made from measurements of structural characteristics rather than from process variables.

Knowing the statistical properties of each measured variable is of great value in development of experimental designs. The relationship between (1) the CV, (2) the number of replicates, (3) Type I, alpha, error (the probability of falsely concluding that a toxicant had an effect), (4) Type II, beta, error (the probability of falsely concluding that a toxicant had no effect), and (5) the power of the test ($1-\beta$, the probability of correctly concluding a toxicant had an effect) should be used to design cost-effective experiments capable of detecting toxic effects with the desired level of inference error [*7,10*]. The CV is commonly used to compare the variation of two populations, independent of their means. However, when comparing response variables in treatments with means approaching zero, the CV values can be highly variable without a change in the standard deviation. Because of this, the CV should be used cautiously when comparing CV values of response variables approaching zero [*10*].

Ideally, experiments should be designed to discern the smallest possible difference for the selected variables with an affordable number of replicates. Because the consequences of a Type II error are undesirable, we would like to keep the power of the test high.

The objectives of many microcosm tests are to predict the lowest concentration of a toxicant which might cause harm or dysfunction in the natural environment and to determine which components of the ecosystem are affected. In order to assess accuracy, it is necessary to determine how closely measurements from laboratory tests agree with responses in natural systems.

Relative accuracy [*5*] is an alternative method for assessing microcosm performance. Relative accuracy is determined by comparing the hazard ranking of a series of toxicants by laboratory tests with the ranking of actual impacts on natural systems. Although it would be useful to be able to infer the magnitude of change and the indirect effects caused by a toxicant, it is unlikely that we will have much success without an understanding of the structural and limiting characteristics of both the laboratory test and the natural system for which the inferences are made [*11*]. Unfortunately, there are usually insufficient field data available for robust evaluation of the hazard ranking in natural systems.

The purpose of this report is to evaluate the precision and accuracy of the MFC and SAM microcosm protocols. Over the past six years, the authors have conducted a large number of MFC and several SAM microcosm experiments with several different toxicants. The data used here include experiments conducted at the U.S. Environmental Protection Agency, Environmental Research Laboratories at Duluth, MN and Corvallis, OR.

The specific objectives of this paper were: (1) to describe the mean CV for each response variable used in the MFC and SAM protocols and to determine whether microcosm age or magnitude of effect caused by a toxicant affected these CV values; (2) to describe the relationship between CV, statistical power, and minimum detectable difference for use in the design of MFC microcosm experiments; and (3) to describe the accuracy and, to a lesser extent, the relative accuracy of these microcosm protocols.

Materials and Methods

Data from 26 MFC [*6*], 7 modified MFC [*14*], and 4 SAM [*4*] microcosm experiments were analyzed to evaluate the replicability and accuracy of these procedures.

Treatment groups of all MFC experiments contained five replicate microcosms and all SAM treatment groups contained six replicates.

Microcosms Description

For brevity, a summary only of the MFC and SAM protocols is presented. The MFC microcosms were 1-L systems inoculated with 50 mL of a stock culture originally derived from natural systems. Although the same major groups were present in all tests, species composition varied. In the seven modified MFC experiments, 50 mL of sand was added to enhance the benthic portion of the systems [*12*].

The SAM systems contained a similar medium [*13*], but were larger (3 L) than the MFC, and organisms were introduced from single species cultures at defined densities rather than as a mixed community. These cultures were not axenic, so bacteria and minor populations of other organisms were also introduced. Both the MFC and SAM systems contained species of algae, cladocerans, ostracods, amphipods, protozoans, and rotifers.

Variables Measured

Dissolved oxygen (DO) and pH were measured in all experiments. Redox potential adjusted to pH 7 (E7) was also measured in some of the MFC experiments. The DO measurements were taken on a three-point schedule to allow calculation of daytime oxygen gain and nightly oxygen loss, which were used as estimates of net primary production and community respiration, respectively. From these the production to respiration (P/R) ratio was calculated.

Data Base Assembly

CV values were calculated for all chemical, ecosystem process, and structural measurements analyzed in the MFC and SAM microcosm experiments. The number of experiments for which each variable was available are presented in Table 1.

In order to determine whether the magnitude of change of a response variable affected the precision of that measurement in the MFC microcosm, each treatment group was assigned to an

TABLE 1—*Chemical, ecosystem process, and community structural variables for which the CV was analyzed and the number of experiments included for each variable are presented for MFC and SAM microcosm experiments.*

	Number of Experiments			Number of Experiments	
Variable	MFC	SAM	Variable	MFC	SAM
Morning pH	26	4	Oxygen gain	33	4
Afternoon pH	33	4	Oxygen loss	33	4
Morning E7	18	4	P/R ratio	33	4
Afternoon E7	18	4	Small and single cell algae	6	4
Diel pH change	26	4	Total zooplankton	7	4
Diel E7 change	18	4	Ciliate protozoa	7	4
Morning DO	33	4			
Afternoon DO	33	4			

effect level category. The definition of each effect level category and the number of MFC experiments included in each category follow:

Controls—The reference microcosm used in determining statistically significant effects. In some cases, this was an untreated set of microcosms; in others this was a solvent control group treated with 0.1 mL of either acetone or methanol (33 MFC experiments).

No-Effect Level (NEL)—No consistent statistically significant effect was observed for any variable measured (17 MFC experiments).

Lowest-Effect Level (LEL)—The lowest treatment level producing a statistically significant effect for two or more consecutive measurement days (30 MFC experiments).

Medium-Effect Level (MEL)—A treatment level producing an effect generally more pronounced than the LEL yet not as severe as the HEL (24 MFC experiments).

High-Effect Level (HEL)—A treatment level producing a deviation from the control of at least 1 pH unit or 3 mg/L DO in the afternoon measurements for a period beyond the first two days after dosing (16 MFC experiments).

In some cases, more than one treatment level could be assigned to the same effect level category. In such instances, the highest treatment level was assigned to that category. In all cases, the LEL included measured parameters which were on the margins of the statistically significant ($P < 0.05$) envelope produced by Dunnett's procedure [*14*]. The effects were quite low in relative magnitude. The HEL was defined to exclude short-term transient effects to which the community could quickly adjust. It was then generally possible to assign an intermediate level of effect (MEL).

Statistical Analysis

Coefficients of variation were calculated for each response variable measured in each treatment group for every measurement day in all experiments.

Statistical analyses of factors affecting the CV values were conducted using SPSS PC+ [*15*]. A one-way ANOVA was conducted on the CV values of each response variable to determine significant ($P < 0.05$) differences among experiment days and to determine significant ($P < 0.05$) differences among treatment categories. Where differences were significant, CV values of treatments were contrasted to those of controls and to other effect categories using the Student-Neumann-Kuehls method [*14*].

Calculation of Minimum Detectable Differences and Power

The approach suggested by Conquest [*10*] was applied to MFC data set to show the relationship between number of treatments and replicates, CV, alpha error, statistical power, and minimum detectable difference. According to this approach, a minimum detectable difference between population means (MDD) in units of pooled standard deviations based on the Student t distribution was calculated as follows

$$[\text{MDD}] = [\overline{X}_1 - \overline{X}_2/Sp] = \sqrt{2/n}[\text{t}_{(0.05)\nu} + \text{t}_{2(1-p)\nu}]$$

and

$$[\text{MDD}] = [Sp/\overline{X}_c] = [\overline{X}_1 - \overline{X}_2/\overline{X}_c]$$

where

n = the number of replicate microcosms per treatment,
ν = degrees of freedom = $k(n - 1)$ where k is the number of treatment groups,

p = power = (1 − beta),
Sp = pooled standard deviation, and
$\bar{X}_c$ = control mean.

The minimum detectable difference as a proportion of the control mean for any variable is then the product of its CV and the MDD value. In all cases, an alpha of 0.05 was used. Calculations were made for powers of 80 and 95%. The CV values used in these calculations were based on a pooled standard deviation from control, NEL, LEL, and MEL over all days. The calculations are based on the assumption of normality and homogeneity of variance and should be considered approximations.

Accuracy of MFC and SAM Microcosm Results

Accuracy of microcosm results were evaluated through comparisons with published descriptions of the responses of experimental ponds exposed to fluorene (2-2′-methylenebiphenyl) [*16*], experimental ponds exposed to atrazine (2-chloro-4-ethylamino-6-isopropylamino-s-triazine) [*17,18*], and enclosures exposed to chlorpyrifos (0,0-diethyl 0-(3,5,6-trichloro-2-pyridyl) phosphorothioate) [*19*]. Evaluation of accuracy was primarily based on a comparison of the lowest concentration of the toxicant which caused a discernible effect in both the field and microcosm experiments. The general type of direct effects on the phytoplankton and zooplankton communities in both field and microcosms experiments were also compared.

Results and Discussion

Replicability of MFC and SAM Microcosms

Mean CV values for seven different response variables are presented for the MFC microcosm in Fig. 1 and for the SAM microcosm in Fig. 2. These values were calculated for control microcosm measurements from all experiments. The CV of response variables presented in the two figures are generally representative of other chemical, ecosystem process, and structural variables not shown.

In the MFC systems, the lowest variability was associated with the chemical measurements (pH and DO). Throughout the experimental period, CV values ranged from about 3 to 7% for DO and from about 2 to 3% for pH. The CV values of the chemical measurements were also less variable through time than the other measurements. The CV values of the ecosystem process estimates (primary productivity and respiration) were intermediate between the chemical and structural variables and ranged between about 9 and 22% during the experiments. Structural measurements had the highest CV values. The CV values associated with total zooplankton and single-cell algae ranged between 20 and 60%. The CV values of the ciliate counts generally were between 60 and 200%.

The CV values of the response variables measured in the SAM microcosms (Fig. 2) were similar to those of the MFC microcosms, except that the CV of DO measurements, total zooplankton, and single-cell algae counts were higher in the SAM microcosms. This may be due to the *Daphnia magna* bloom, which commonly occurred between Days 20 and 40. This bloom was not always synchronous among replicate microcosms, resulting in higher variation among the zooplankton densities. The magnitude of the zooplankton bloom is also sufficient to affect algal abundance and increase the CV associated with these counts.

The Effect of Age and Magnitude of Response on Replicability

The possibility that the variability between replicate microcosms becomes greater with age has been a concern in long-term experiments. In both the MFC and SAM microcosms, micro-

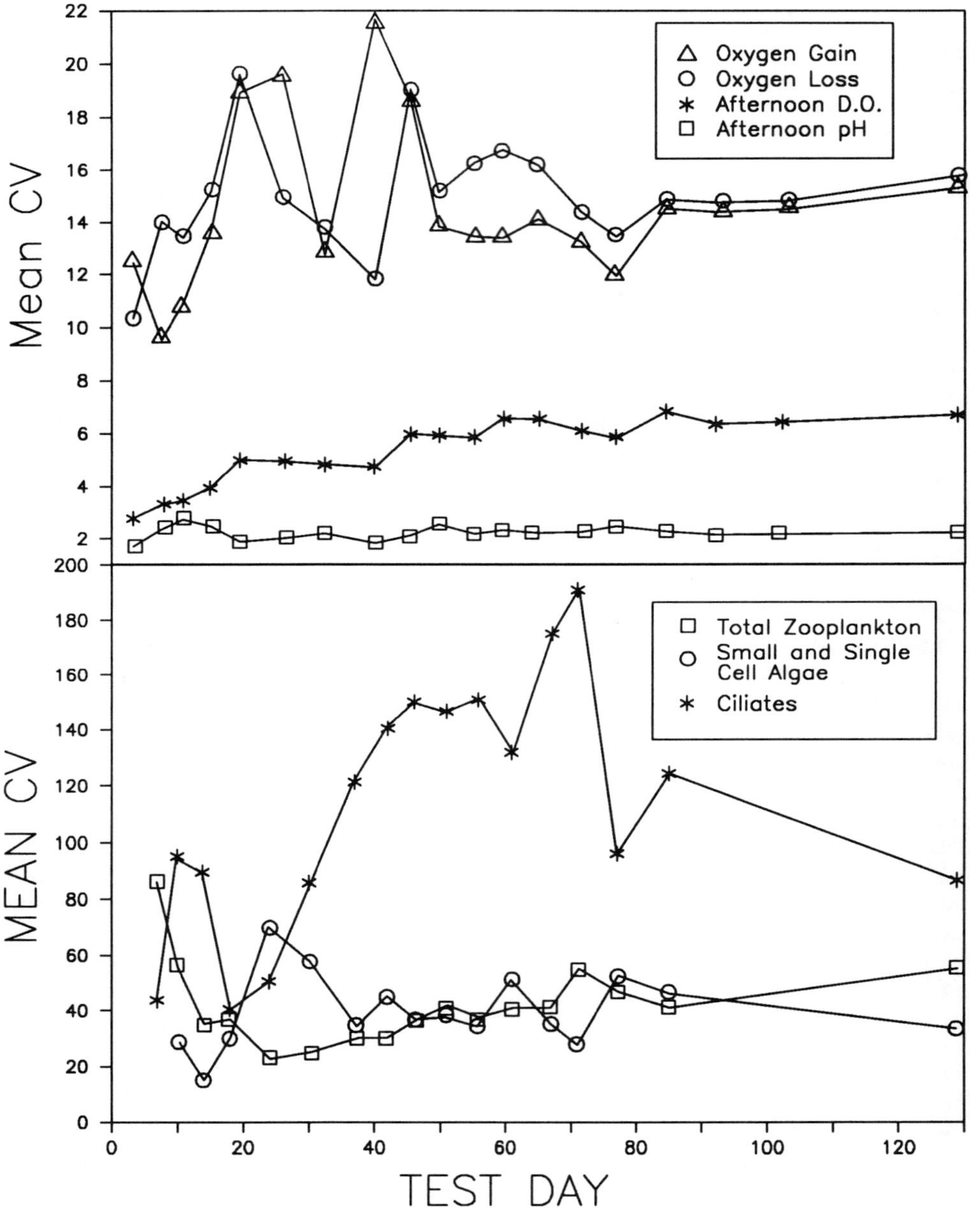

FIG. 1—*Mean coefficients of variation (CV) for chemical (DO, pH), process (primary productivity, respiration), and structural (zooplankton, algal, and ciliate densities) variables measured in control microcosms of all MFC experiments.*

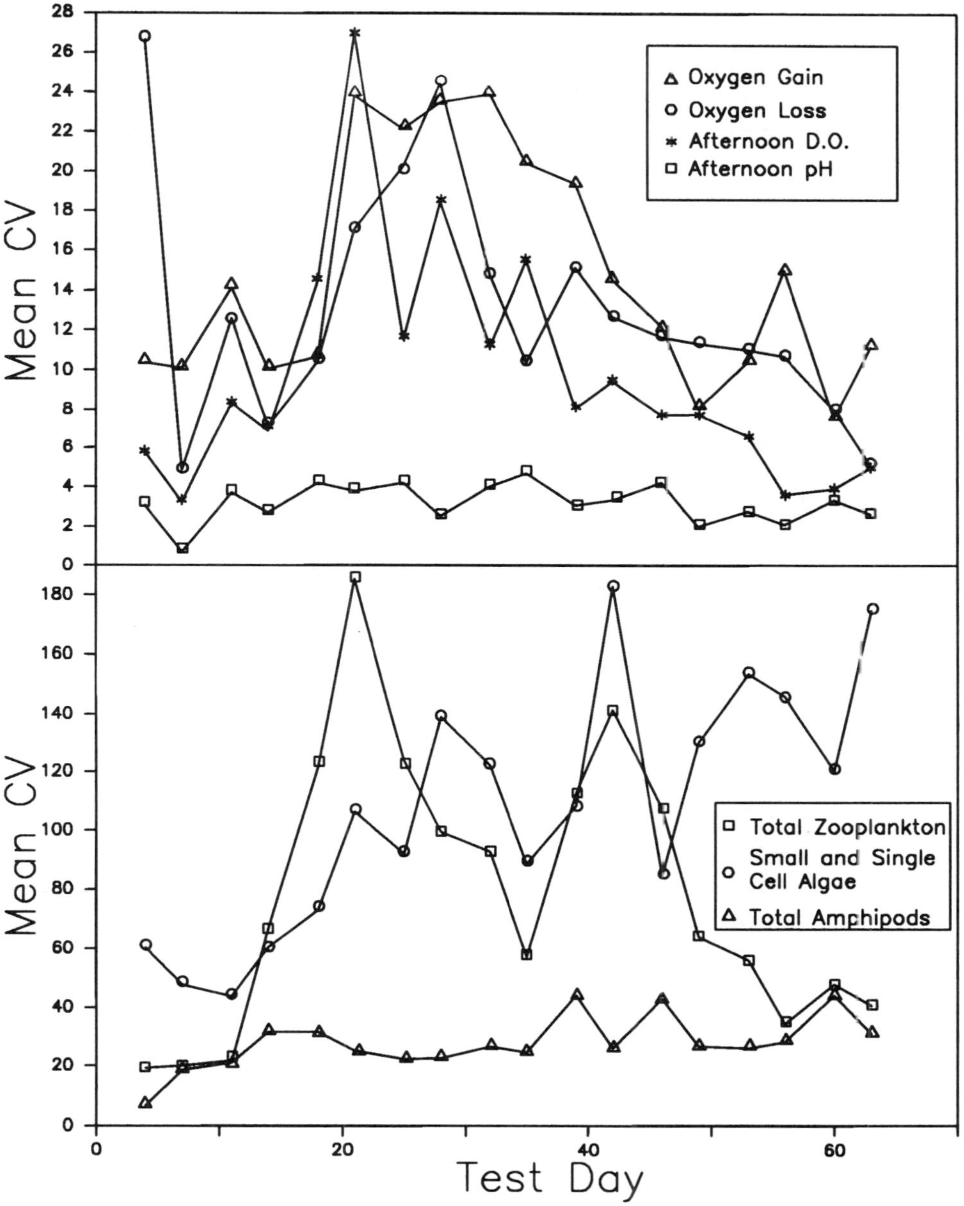

FIG. 2—*Mean coefficients of variation (CV) for chemical (DO, pH), process (primary productivity, respiration), and structural (zooplankton, algal, and ciliate densities) variables measured in control microcosms of all SAM experiments.*

cosm age had little effect on replicability. ANOVA of CV values in control microcosms showed no significant difference among days for any variable in either MFC or SAM microcosms. Therefore, microcosm age or successional stage was not a significant factor influencing the replicability of any of the measured variables.

Another concern was that the effect level or magnitude of change caused by a toxicant might affect the variability of a treatment group within a microcosm experiment. Mean CV values of chemical and process variables of the MFC microcosm experiments are presented in Fig. 3 for each response category. The mean CV values of response variables in each effect category were compared by ANOVA. There was generally no significant difference between categories before toxicant addition. After toxicant addition, CV values of both chemical and process variables were significantly different ($P < 0.05$). Additional analysis using the Student-Neumann-Kuels method indicated that these differences were due to significantly higher CV values in the HEL group than in the controls or other response categories. The CV values of the other categories were not significantly different from those of the controls.

The relationship between the level of perturbation and the CV values of structural measurements (Fig. 4) was not as clear as it was with the chemical and process variables. This was due to the greater variation of the CV values associated with these measurements, which may have resulted from low organism counts at these treatment levels. Only the CV values of the small- and single-cell algae were significantly greater in the HEL than in the controls.

Design of Experiments

Prior knowledge about ecosystem components most likely to be affected by a toxicant can be useful in selecting variables for measurement. The objectives to be addressed in a particular experiment may also influence which variables are measured. Each of these situations could require a different level of importance to be applied to the selected response variables.

Figure 5 shows an estimate of the relationship between power and minimum detectable difference for representative chemical, process, and structural variables. These curves were generated from CV values for each variable measured in the MFC microcosms, alpha of 0.05, and seven treatment groups with five replicates each. The relative precision with which the three types of variables can be measured becomes clear in this figure. These curves indicate that the MFC protocol, requiring seven treatments and five replicates, can be used to detect differences as low as 10% for chemical variables and 25% for process variables with 90% power. However, objectives requiring 90% power and relying exclusively on measurement of structural variables might not be appropriate because the minimum detectable change would have to be greater than 70% of the controls.

Table 2 was prepared to provide information which could be used in the development of alternative MFC experimental designs. The information in Table 2 was calculated for different numbers of treatments and replicates combinations using CV values calculated from our data base for each of the variables shown. Examples are presented for powers of 80 and 95%. We have also limited the total number of microcosm to about 35. The range of treatment and replicate numbers was selected to provide minimal detectable differences that would be useful for chemical, process, and structural variables.

Others [*7,8,10*] have described how increasing the number of replicates (n) results in a reduction of the minimum detectable distance. As shown in Table 2, the benefit or reduction of the minimum detectable distance from increasing n decreases as n gets larger for each category of response variables. Increasing the number of treatments (k) results in only a slight reduction in the minimum detectable distance. However, because one of the objectives in ecotoxicology is to determine the LOEL, one must balance the need for treatments with that for replicates. When there is little prior information about probable effects on the ecosystem, more treatments are needed in order to estimate the LOEL within a reasonable concentration range. Other objectives can require different decisions. Table 2 provides information for making those decisions.

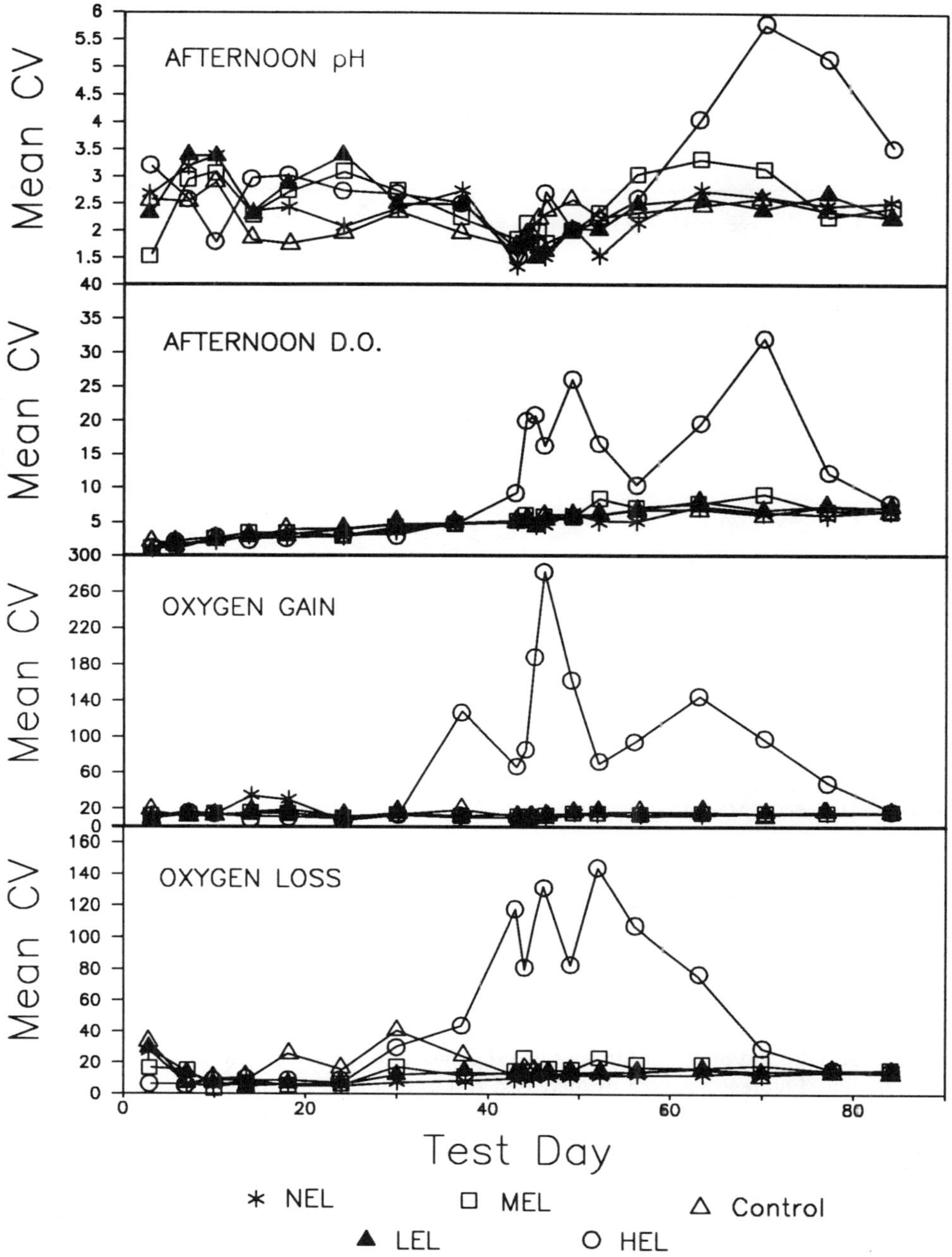

FIG. 3—*Mean coefficients of variation (CV) for chemical (DO, pH) and process (primary productivity, respiration) variables in control, no-effect level (NEL), low-effect level (LEL), medium-effect level (MEL), and high-effect level (HEL) response categories before and after treatment (Day 41) of MFC microcosms.*

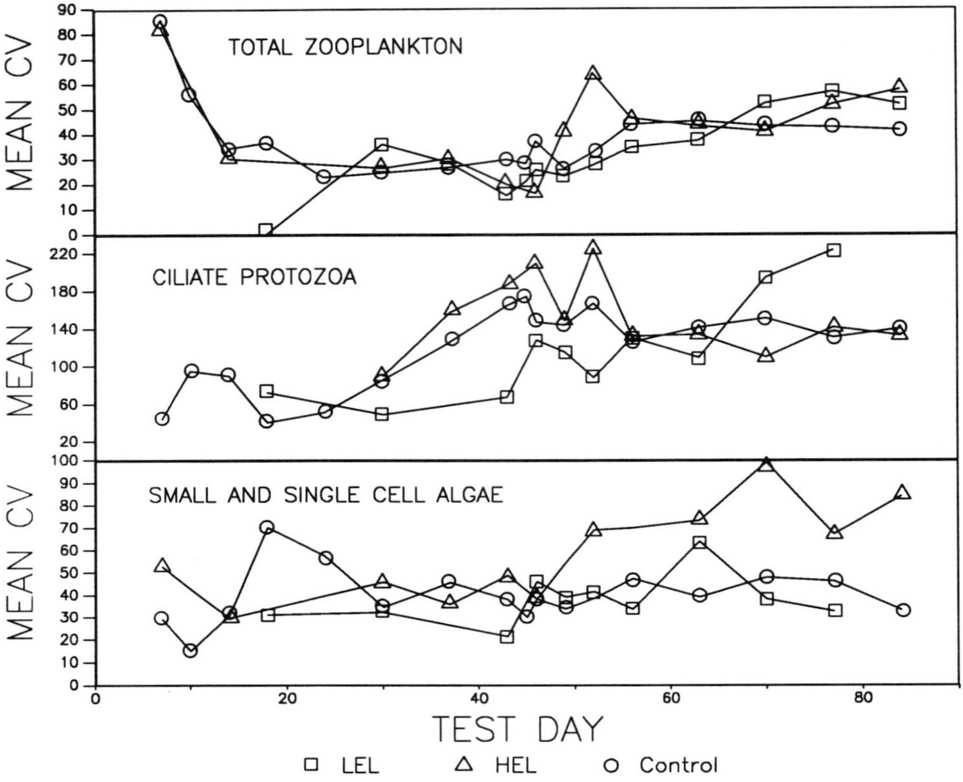

FIG. 4—*Mean coefficients of variation (CV) for structural (zooplankton, algal, and ciliates counts) variables in control, low-effect level (LEL), and high-effect level (HEL) response categories before and after treatment (Day 41) of MFC microcosms.*

Accuracy of MFC and SAM Microcosm Experiments

The responses of microcosm and field systems exposed to three different toxicants are presented in Table 3. Accuracy was assessed through comparison of the lowest observed effect level (LOEL) and general types of changes in microcosm and the three field experiments.

The types of changes which occurred in both the experimental pond studies conducted by Boyle et al. [*16*] and in MFC microcosms [*9*] exposed to the same concentrations of fluorene were similar. Fluorene resulted in increased respiration and elimination of the zooplankton communities in both the pond and MFC microcosm. Fluorene concentrations resulting in increased respiration were 120 μg/L and greater in the MFC. In the ponds, respiration increased at all concentrations except 10 000 μg/L [*16*]. The zooplankton community was almost completely eliminated at 2000 μg/L in the microcosms and at 5000 μg/L in the ponds. Autotrophic activity was not affected in the MFC, and gross primary productivity was reduced only in the 10 000 μg/L treatment pond. In both the pond and microcosm experiments, the LOEL was 120 μg/L. In the microcosms, the LOEL was based on respiration and *P/R* ratio, and in the ponds it was based on respiration and fish survival.

Similar concentrations of atrazine, a herbicide, were used in the MFC and SAM [*20*] microcosms and experimental ponds studies conducted by deNoyelles et al. [*17,18*]. Atrazine caused a significant reduction in primary productivity at 100 μg/L in the MFC, SAM, and pond plank-

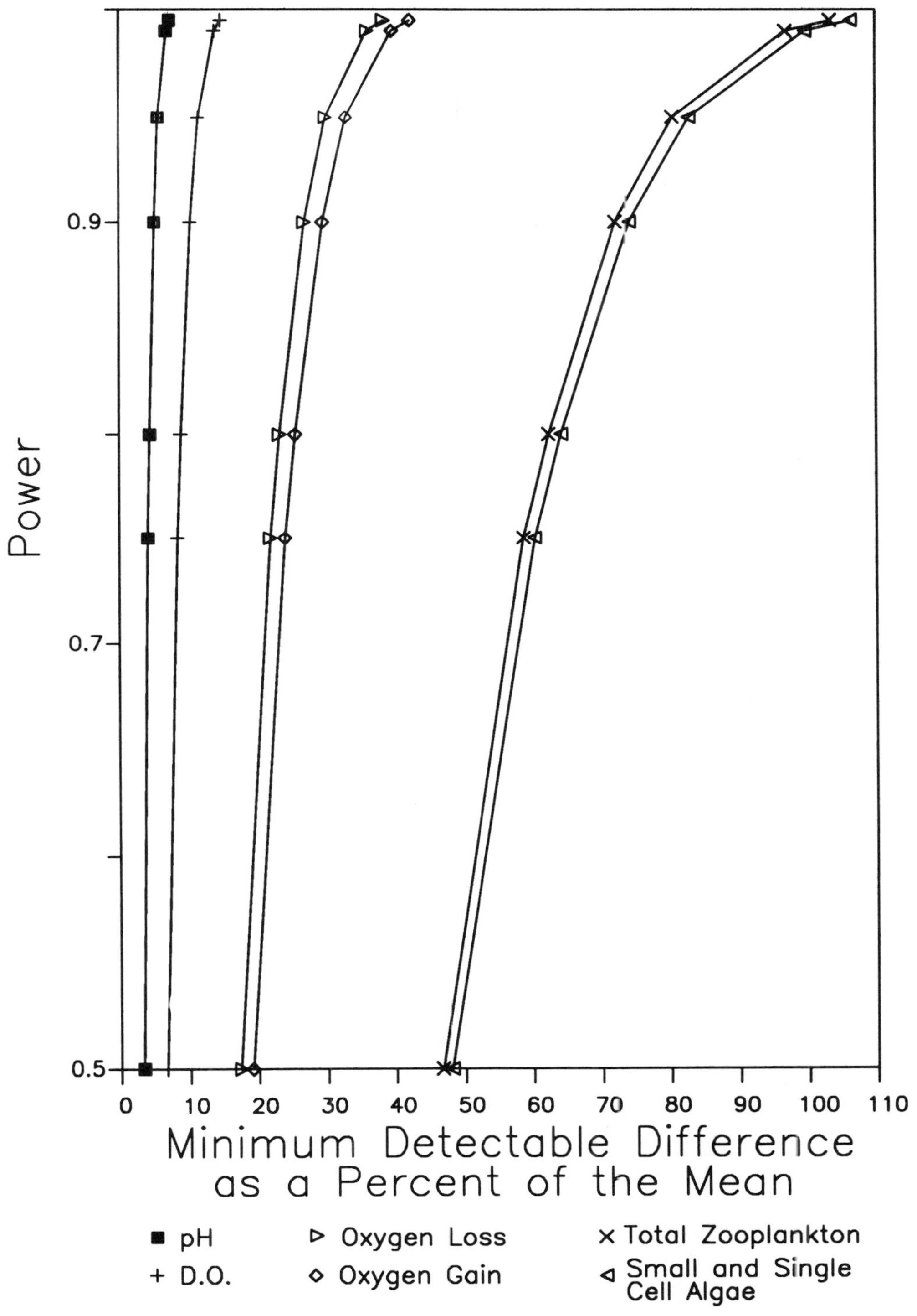

FIG. 5—*Power curves showing the relationship between power (1-beta) and minimum detectable distance (percent of the control mean) for DO, primary productivity, and algal densities in MFC microcosms. Curves are based on alpha equal 0.05, seven treatments, and five replicates.*

TABLE 2—*Relationship between number of treatment groups* (k), *replicates* (n), *and statistical power of 80 and 95% on minimum detectable differences expressed as a percentage of the control means.*

			Minimum Detectable Differences					
			Afternoon Dissolved Oxygen, CV = 4.85		Oxygen Gain, CV = 13.9		Small and Single Cell Algae, CV = 35.0	
k	*n*	Power	80%	95%	80%	95%	80%	95%
7	5		9	11	26	33	64	83
7	4		10	13	29	37	73	94
7	3		12	15	34	44	86	111
6	6		8	10	23	30	58	75
6	5		9	12	26	33	65	83
6	4		10	13	29	38	73	95
6	3		12	16	35	45	87	113
5	7		8	10	22	28	54	70
5	6		8	11	23	30	59	76
5	5		9	12	26	34	65	84
5	3		12	16	35	46	89	115
4	9		7	8	19	25	47	61
4	7		8	10	22	28	55	70
4	6		8	11	24	31	59	76
4	5		9	12	26	34	66	85
4	4		10	14	30	39	75	98
4	3		13	17	37	48	93	121
3	12		6	7	16	21	41	53
3	10		6	8	18	23	45	59
3	8		7	9	20	26	51	66
3	5		9	12	27	35	67	88
3	3		13	17	38	50	96	125

ton systems. Primary productivity in 60 μg/L treatment of the SAM was also significantly less than the controls for a short period. Little or no direct effect of atrazine on the zooplankton community was observed at concentrations of <500 μg/L in the ponds and SAM microcosms. At concentrations of 500 μg/L and greater, zooplankton decreased in response to the reduced algal biomass. The LOEL for atrazine in the MFC and pond plankton communities was 100 μg/L, and in the SAM, the LOEL was 60 μg/L. The LOEL in the ponds and microcosms was based on a reduction of autotrophic activity.

Siefert et al. [*19*] added chlorpyrifos, an insecticide, to enclosures containing natural littoral communities. The responses noted in these experiments were compared to MFC microcosms exposed to the same treatment concentrations. In both systems, chlorpyrifos caused a major reduction in invertebrate populations at 0.5 μg/L, the lowest treatment concentration, and no discernible change in the autotrophs at the highest treatment of 20 μg/L. In both the enclosures and microcosms, cladocerans and amphipods were the most sensitive, followed by copepods and ostracods. The LOEL of chlorpyrifos in both the MFC microcosms and enclosures was 0.5 μg/L. The LOELs in the microcosms were based on reduced invertebrate abundance and in the ponds reduced invertebrate abundance and fathead minnow growth.

Relative accuracy of these microcosm systems was evaluated by comparing the ranking of their LOEL with the ranking of field experiments. The relative ranking of the toxicants by both the laboratory and field studies was the same. The relative hazard ranking based on the LOEL was chlorpyrifos < atrazine < fluorene. The LOEL was not necessarily determined by the same response variables in both laboratory and field systems.

The similarity between the responses of microcosms and controlled field systems in the exper-

TABLE 3—*Comparison of the effects of fluorene, atrazine, and chlorpyrifos in MFC and SAM microcosms and field sites exposed to similar concentrations of each chemical.*

Field Results	Microcosm Results
FLUORENE	
Concentrations tested (μg/L), 0, 120, 500, 2000, 5000, 10 000	Concentrations tested (μg/L), 0, 120, 500, 2000, 5000, 10 000; only MFC microcosms tested
1. Increased community respiration at 120 to 500 μg/L. 2. Near elimination of zooplankton at 5000 μg/L. 3. Reduced survival and growth of fish at 120 μg/L. 4. Reduced gross primary production at 10 000 μg/L.	1. Increased community respiration, reduced *P/R* and pH at 120 μg/L. 2. Reduced zooplankton at 2000 μg/L. 3. No effect on net primary production at 10 000 μg/L.
ATRAZINE	
Concentrations tested (μg/L), 0, 20, 100, 200, 500	Concentrations tested (μg/L), 0, 20, 100, 200, 500, 1000, 5000 MFC; 0, 60, 100, 200, 500, 1000, 5000 SAM
1. Reduced phytoplankton primary productivity at 100 μg/L. 2. No direct effect on zooplankton at <500 μg/L. 3. Primary productivity recovered in 2 to 3 weeks at 100 μg/L.	1. Reduced primary production, *P/R*, and pH at 100 μg/L for MFC, SAM. 2. No direct effect on *Daphnia magna* at <500 μg/L in SAM. 3. MFC *P/R* recovered to control levels at 100 μg/L. Primary production did not recover. 4. Recovery to control levels in SAM at 200 μg/L.
CHLORPYRIFOS	
Concentrations tested (μg/L), 0, 0.5, 5.0, 20.0	Concentrations tested (μg/L), 0, 0.5, 5.0, 20.0; only MFC microcosms tested
1. Reduced invertebrate abundance and growth of fathead minnow at 0.5 μg/L. 2. Reduced bluegill sunfish survival 62% at 5.0 μg/L. 3. No mortality of fathead minnow or plants at 20.0 μg/L.	1. Reduced cladoceran abundance at 0.5 μg/L. 2. Reduced amphipod abundance and eliminated cladocerans at 5.0 μg/L. 3. No effect on primary production at 20.0 μg/L. 4. Reduced copepods and ostracods at 20.0 μg/L.

iments presented here appear more than fortuitous. The three toxicants used caused widely different direct effects; however, each of the three microcosm experiments accurately reflected the type of responses which occurred in the field experiments when exposed to the same toxicant. Also, concentrations which caused perturbations in the microcosm experiments were generally within a factor of two of those causing the same perturbation in the field experiments. These assessments are based primarily on direct, short-term effects. In order to predict the effects of these chemicals in more detail, over longer periods, or over more diverse ecosystems, a better understanding is needed of the characteristics which control the indirect effects.

Similarities between direct effects in microcosms and field studies were also noted by Harrass and Taub [*11*] when comparing the effects of copper in several unrelated field studies to those in the SAM microcosms, and by Giddings and Franco [*21*] when comparing the effects of a phenolic compound in their littoral microcosms to those in experimental ponds. The authors of both studies concluded that extrapolation of indirect effects from microcosm ecosystems to natural ecosystems, or from natural ecosystem to other natural ecosystems, can be done only with an understanding of the structure and function of both systems. The conclusions of the comparisons presented here agree with those of these two studies.

Summary and Conclusion

More than 30 different MFC and 4 different SAM microcosm experiments were used to describe the level of precision (CV) associated with chemical, ecosystem process, and structural measurements. The effect of microcosm age and magnitude of response on the CV values of each of these three categories of measurements was evaluated. The approach suggested by Conquest [*12*] was used to show how CV values can be used to determine the appropriate number of treatment groups and replicate microcosms per treatment in the design of MFC experiments. Accuracy was assessed through comparisons of the responses of MFC and SAM microcosms with those reported for field studies exposed to fluorene, atrazine, and chlorpyrifos. Conclusions drawn from these evaluations follow:

1. Three general levels of precision were noted. The CV values associated with chemical measurements (DO and pH) were generally less than 7%, process variables (primary productivity and respiration) were generally less than 20%, and structural variables (organism counts) were generally greater than 50% and less than 200%.
2. Microcosm age or successional stage did not appear to affect the precision of these two microcosm protocols. Large changes in response to toxicant stress, however, caused a significant increase in the CV of most response variables.
3. The CV values associated with response variables measured in over 30 MFC experiments were used to develop information on the relationship between replicate numbers, statistical power, and minimum detectable differences. This information can be used by others to design MFC microcosm experiments.
4. The LOEL and general types of responses noted with the SAM and MFC microcosm protocols exposed to atrazine, fluorene, and chlorpyrifos accurately reflected those of experimental field sites exposed to similar concentrations of the same toxicants. The types of responses and effect concentrations noted in the microcosm experiments would have been useful in estimating the potential hazard of these chemicals to the field sites. However, until factors which control the indirect and longer term effects of natural ecosystems are better understood, predictions must be made with caution.

Acknowledgments

The authors would like to thank the following people for their contributions: Delli Byers for data entry, Sandra Beder-Miller and Steven Hedtke for providing technical review, and Evelyn Hunt for providing editorial review of this manuscript.

References

[*1*] Cairns, C., *Hydrobiologia*, Vol. 100, 1984, pp. 47–57.
[*2*] Levin, S. A. and Kimball, K. D., "New Perspectives in Ecotoxicology," *Environmental Management*, Vol. 8, 1984, pp. 375–442.
[*3*] National Academy of Sciences, *Testing for Effects of Chemicals on Ecosystems*, National Academic Press, Washington, DC, 1981.
[*4*] Taub, F. B. and Read, P. L., "Standardized Aquatic Microcosm Protocol," final report, Vol. 2 of FDA Contract 223-80-2352, Food and Drug Administration, Washington, DC, 1982, p. 55.
[*5*] Giddings, J. M. in *Methods for Ecological Toxicology, A Critical Review of Laboratory Multispecies*

Tests, A. S. Hammons, Ed., Ann Arbor Science Publishers Inc., Ann Arbor, MI, 1981, Chap. 3, pp. 23-91.

[6] Leffler, J. W. in *Concepts in Marine Pollution Measurements*, H. H. White, Ed., University of Maryland, College Park, MD, 1984, pp. 139-157.

[7] Giesy, J. P. and Allred, P. M. in *Multispecies Toxicity Testing*, J. Cairns, Ed., Pergamon Press, New York, NY, 1985, Chap. 14, pp. 187-247.

[8] Crow, M. E. and Taub, F. B., *International Journal of Environmental Studies*, Vol. 13, 1978, pp. 141-147.

[9] Stay, F. S., Katko, A., Rohm, C. M., Fix, M. A., and Larsen, D. P., *Society of Environmental Toxicology and Chemistry*, Vol. 7, 1988, pp. 635-644.

[10] Conquest, L. L., *International Journal of Environmental Studies*, Vol. 20, 1983, pp. 209-221.

[11] Harrass, M. C. and Taub, F. B. in *Validation and Predictability of Laboratory Methods for Assessing the Fate and Effects of Contaminants in Aquatic Ecosystems, ASTM STP 865*, T. P. Boyle, Ed., American Society for Testing and Materials, Philadelphia, PA, 1985, pp. 57-74.

[12] Shannon, L. J., Flum, T. E., and Yount, J. D., "Draft Protocol for a Mixed Flask Culture Aquatic Microcosm Toxicity Test and Support Document," EPA Project Report No. 7767A, U.S. EPA, Environmental Research Laboratory, Duluth, MN, 1987, p. 59.

[13] Taub, F. B. and Dollar, A. M., *Limnology and Oceanography*, Vol. 9, 1964, pp. 61-74.

[14] Steel, R. G. D. and Torrie, J. H., *Principles and Procedures of Statistics*, McGraw-Hill Book Company, New York, NY, 1960, Chap. 7, pp. 110-115.

[15] Norusis, M. J., "SPSS/PC+: SPSS for the IBM PC/XT/AT," SPSS INC., Chicago, IL, 1986, p. 268.

[16] Boyle, T. P., Finger, S. E., Paulson, R. L., and Rabeni, C. F. in *Validation and Predictability of Laboratory Methods for Assessing the Fate and Effects of Contaminants in Aquatic Ecosystems, ASTM STP 865*, T. P. Boyle, Ed., American Society for Testing and Materials, Philadelphia, PA, 1985, pp. 134-151.

[17] deNoyelles, F., Kettle, W. D., and Sinn, D. E., *Ecology*, Vol. 63, 1982, pp. 1285-1293.

[18] deNoyelles, F. and Kettle, W. D. in *Validation and Predictability of Laboratory Methods for Assessing the Fate and Effects of Contaminants in Aquatic Ecosystems, ASTM STP 865*, T. P. Boyle, Ed., American Society for Testing and Materials, Philadelphia, PA, 1985, pp. 91-103.

[19] Siefert, R. E., Lozano, S. J., Brazner, J. C., and Knuth, M. L., *Miscellaneous Publication Series*, Entomological Society of America, Lanham, MD, 1988, in press.

[20] Stay, F. S., Larsen, D. P., Katko, A., and Rohm, C. M. in *Validation and Predictability of Laboratory Methods for Assessing the Fate and Effects of Contaminants in Aquatic Ecosystems, ASTM STP 865*, T. P. Boyle, Ed., American Society for Testing and Materials, Philadelphia, PA, 1985, pp. 75-90.

[21] Giddings, J. M. and Franco, P. J. in *Validation and Predictability of Laboratory Methods for Assessing the Fate and Effects of Contaminants in Aquatic Ecosystems, ASTM STP 865*, T. P. Boyle, Ed., American Society for Testing and Materials, Philadelphia, PA, 1985, pp. 104-119.

Michael C. Harrass[1] *and Philip G. Sayre*[2]

Use of Microcosm Data for Regulatory Decisions

REFERENCE: Harrass, M. C. and Sayre, P. G., **"Use of Microcosm Data for Regulatory Decisions,"** *Aquatic Toxicology and Hazard Assessment: 12th Volume, ASTM STP 1027,* U. M. Cowgill and L. R. Williams, Eds., American Society for Testing and Materials, Philadelphia, 1989, pp. 204-223.

ABSTRACT: Microcosm data provide new kinds of information which go beyond currently used regulations and procedures to evaluate environmental fate and effects of chemicals. Microcosm data provide a basis for considering ecological questions such as the extent of effects at several trophic levels, the breadth of effects on competitive species, the extent of apparent detoxification, and effects on the expected system recovery profile. Several models for applications of microcosm data are discussed: as a model ecosystem in a high-level tier test; as a means of integrating fate and effects processes to validate inferences and estimates; as a low-level tier or screening test when specific concerns require evaluation in an ecosystem context; as a means to anticipate patterns of ecosystem magnification and amelioration; and as a means to rank and compare environmental impacts based on community- or ecosystem-level characteristics. Besides applications in evaluating the impact potential of chemicals, microcosms may be useful in evaluating the potential environmental impact of genetically engineered microorganisms. As these new methods become available and credible, regulatory agencies such as EPA and FDA will need to consider how to incorporate the resulting data into their evaluation schemes.

KEY WORDS: microcosm, regulatory agency, environmental impact assessment, tier testing, multispecies tests, ecosystem-level endpoints

Protection of an environment requires assumptions and information about how chemical and physical agents affect the ecosystems in that environment. As more information about an ecosystem is available, presumably fewer or simpler assumptions need to be made. The study of the effects of chemical and physical agents on ecosystems has been termed ecotoxicology, with a primary focus on populations and communities and their interactions, in contrast to toxicology, with a focus on organisms of a single species [*1,2*]. However, the lack of adequate ecotoxicological test methods has resulted in the reliance on toxicological tests on organisms assumed to be representative of those in the environment as indicators of potential ecosystem impact.

Development of ecotoxicity test methods, such as microcosms, has been seen as one way to reduce the burden of assumptions required to navigate from test data to decision making for environmental protection [*3*]. By microcosm we mean an experimental unit containing multiple species, usually of a size suitable for testing in laboratory settings, designed to contain the interacting biotic and abiotic components that characterize a natural ecosystem.

Federal regulatory agencies are major consumers of toxicity data, as well as developers and practitioners of the field of ecotoxicology. Federal agencies must work out with care and detail

[1]Environmental scientist, Center for Food Safety and Applied Nutrition, U.S. Food and Drug Administration, Washington, DC 20204.

[2]Microbiologist, Office of Toxic Substances, U.S. Environmental Protection Agency, Washington, DC 20460.

how test data will be used in support of regulatory decision making. Responsibilities for enforcement of specific legislation are distributed among several federal agencies. Under the Federal Insecticide, Fungicide and Rodenticide Act (FIFRA), the U.S. Environmental Protection Agency (EPA) regulates the registration of pesticides [*4*]. Under the Toxic Substances Control Act (TSCA), EPA regulates all chemicals already in commerce or intended for entry into commerce that are not specifically covered by other regulatory authorities [*5*]. Under the Federal Food, Drug and Cosmetic Act, the U.S. Food and Drug Administration (FDA) regulates foods, food additives, human and animal drugs, human biological products (for example, vaccines), cosmetics, and medical devices. FDA's responsibility for evaluating the potential environmental impact of its actions derives from the National Environmental Policy Act (NEPA). Other regulatory authorities exist; however, we will discuss only some of the activities of EPA and FDA which derive from the legislation identified above.

At this time, regulatory experience with microcosm data is very limited, and no consensus as to the role of microcosm testing has developed [*6*]. The purpose of this discussion is to consider how microcosm testing has developed and how it may prove useful to regulators. We address in particular what aspects of microcosm testing might be especially useful (for example, identifying appropriate endpoints) and what kinds of regulatory situations might benefit from microcosm test results.

Current Practices

Ecological Toxicity Testing and the EPA

Federal regulatory agencies require several types of toxicity data for evaluation of environmental hazard under the statutes they administer. For example, EPA has presented information about the role of toxicity testing in evaluating premanufacture notice (PMN) chemicals under Section 5 of TSCA or in pesticide registration under FIFRA [*5, 7*]. EPA's Office of Toxic Substances (OTS) may also require testing of a chemical or mixture which is already in commerce under Section 4 of TSCA or for significant new uses under Section 5e of TSCA [*8*]. The discussion here focuses on the evaluation of PMN chemicals in OTS.

The evaluation schemes used by EPA are described as tiered testing, and include two different elements: toxicity tests and triggers (the decision-making logic which invokes the tests). The three-tiered testing scheme used by OTS is shown in outline form in Fig. 1 [*9*]. For PMN chemicals, basic information is received with the initial PMN submission (for example, chemical structure, water solubility, melting temperature, and projected market volume) and is used to estimate environmental fate and concentrations of the new chemical. Information on chemical analogs and structures likely to exhibit toxicity is sought in the literature and in available quantitative structure-activity relationship (QSAR) data bases. If a PMN chemical is potentially toxic, but there are insufficient data to evaluate the risk, or if the predicted environmental concentration is high, then the level of concern increases and testing may be initiated under Section 5 of TSCA. If this information raises sufficient concern (that is, activates the first trigger), then Tier I testing is requested.

For evaluation of PMN chemicals under Section 5 of TSCA, the first trigger is a comparison between the predicted environmental concentration and a concern level. The concern level is determined by the nature of available toxicity information (for example, whether based on QSAR or toxicity of analogous chemicals) and is calculated as the quotient of a toxicity value and an assessment factor. The assessment factor used depends on the extent and quality of the information available: the greater the certainty that the estimates reflect the toxicity of the premanufacture chemical, the smaller the assessment factor. For example, an assessment factor of 1000 would be applied if one LC_{50} estimate was available from a QSAR or an analog chemical, while an assessment factor of 10 would be applied if a maximum allowable toxicant

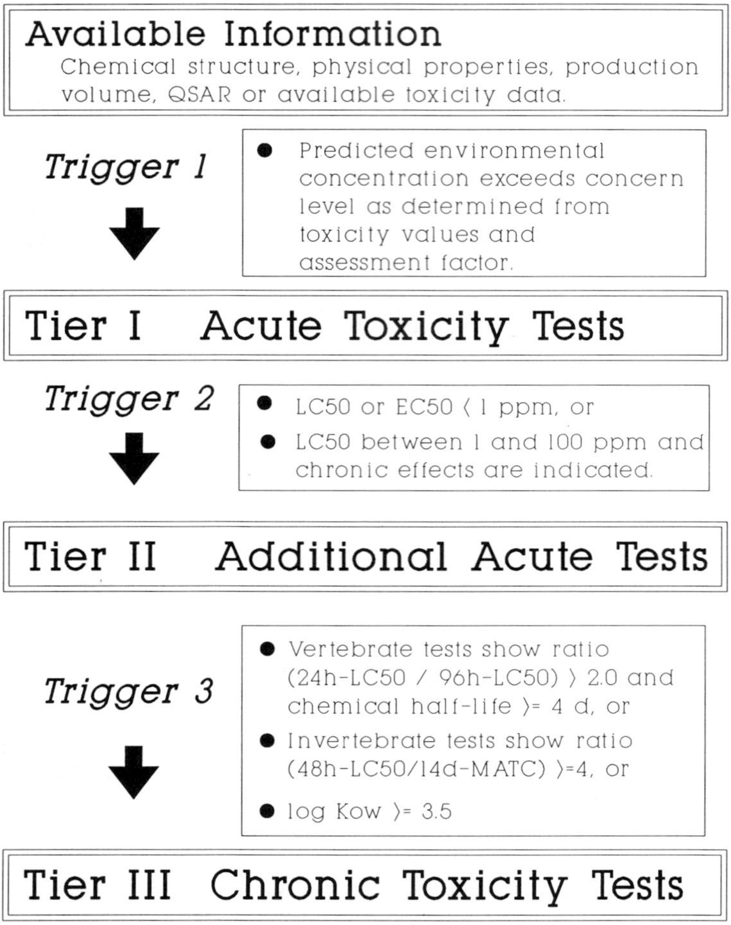

FIG. 1—*Example of a tiered testing scheme. The tier testing scheme used by EPA's Office of Toxic Substances is shown. Tier I tests are used to determine if a chemical exhibits any ecotoxicological activity and to estimate the inherent activity. Tier II tests are used to identify the more sensitive species. Tier III includes tests designed to investigate potential chronic or cumulative toxicity and bioconcentration potential. The triggers shown are examples of the major criteria used to determine the need for higher-tier testing to evaluate impacts to the aquatic environment. Additional specific criteria are discussed by EPA* [9].

concentration (MATC) estimate was available. If the concern level equals or exceeds the predicted environmental concentration, then Tier I testing is necessary.

The tiered testing scheme for TSCA Section 5 PMN chemicals involves less testing (in practice) relative to the tiered testing for chemicals on the TSCA inventory done under Section 4 authority. Testing of PMN chemicals under Section 5 is simplified because the PMN regulatory mechanism is a screening process that identifies only those chemicals which are reasonably expected to have adverse environmental impacts and have not yet been introduced into commerce. Testing done under Section 4 authority involves more detailed testing because these chemicals are being manufactured and sold and therefore have an economic base to support testing.

Tier I testing is designed to determine if a chemical has potential toxicity to representative species. Tier I testing under Section 4 authority examines the acute toxicity of a chemical using five species: an aquatic vertebrate, an aquatic invertebrate, an aquatic plant, a terrestrial vertebrate, and a terrestrial plant. For a PMN chemical, only aquatic tests are requested in Tier I: a 96-h fish EC_{50} test, a 48-h invertebrate test (usually *Daphnia*), and an algal toxicity test [*9*]. If the Tier I tests indicate potential toxicity, then Tier II tests may be requested for a chemical under Section 4 authority (that is, the second trigger is activated). Tier II testing examines whether an additional species from each organism group is significantly more sensitive than the five species previously tested for Tier I. Once the sensitive species are identified, the third trigger considers the acute toxicity values, persistence of the chemical, and indications of chronic toxicity. Common practice with PMN chemicals under Section 5 of TSCA is to bypass Tier II testing and request Tier III tests if the third trigger considerations indicate the potential for chronic toxicity. Tier III testing is performed to determine chronic effects and uptake of the chemical and consists of the following five test classes for a Section 4 chemical: an aquatic vertebrate chronic toxicity test, an aquatic invertebrate chronic toxicity test, an aquatic bioconcentration test, a terrestrial vertebrate reproduction test, and a terrestrial plant uptake test. When only aquatic exposure is involved for a PMN chemical, only a chronic fish and a chronic daphnid test are requested for Tier III testing.

Aside from the above testing for Section 5 PMN chemicals and Section 4 inventoried chemicals, a chemical on the TSCA inventory may have to undergo testing under Section 5e of TSCA. This testing uses a tiered approach similar to PMN chemicals if the inventoried chemical is to be used for a significant new use which had not been listed on the TSCA inventory.

Another tiered testing scheme has been developed by EPA's Office of Pesticide Programs (OPP) for pesticide registration under FIFRA [*7,10*]. While the FIFRA scheme resembles the above-described OTS scheme in that it uses a similar quotient-based comparison of exposure and toxicity to trigger additional testing, it incorporates additional criteria such as the anticipated use pattern of the pesticide. For example, Tier I testing for OPP's effects assessment for wildlife and aquatic organisms requires three acute aquatic tests: coldwater fish, warmwater fish, and freshwater invertebrate. Tier II testing includes fish early life stage or aquatic invertebrate life-cycle and can be triggered by finding an acute LC_{50} less than 1 mg/L, or an LC_{50} equal to or less than 1/100 of the estimated environmental concentration [*10*]. Simulated or actual field testing (Tier IV) may be required on a case-by-case basis depending on the results of lower tier studies [*10*].

Ecological Toxicity Testing and the FDA

Less familiar to many ecotoxicologists is the use of environmental fate and effects data by the FDA in review of, for example, food additive petitions and new animal drug applications. FDA's involvement in ecotoxicity testing derives from its responsibilities under the National Environmental Policy Act (NEPA), which requires that Federal agencies consider the environmental impact of their actions. While NEPA does not supersede FDA's other statutory duties or require FDA decisions to favor environmental protection over other relevant factors, NEPA does provide the FDA with supplementary authority to base substantive decisions on environmental considerations.

To implement NEPA and the regulations of the Council on Environmental Quality, FDA has established regulations describing its environmental review process [*11*]. FDA actions to approve products such as food additives and new animal drugs may result in environmental introductions of chemicals; thus, FDA's review includes consideration of the potential for significant environmental impact resulting from FDA approval. Under NEPA, FDA may find that the action will have no significant impact on the environment, in which case the agency publishes a "Finding of No Significant Impact." If FDA finds that a proposed action may significantly affect the quality of the human environment, then an environmental impact statement (EIS) is prepared. The preparation of an EIS follows procedures established by FDA and Council on Environmental Quality (CEQ) regulations, which ensure public participation in the decision-making process and consideration of alternative actions.

FDA approves the use of a wide variety of food additives, human and animal drugs, and devices. Some of these products are intentionally biologically active (for example, drugs), while some are as inactive as possible (for example, food packaging). Under FDA's environmental regulations, the review process typically requires the industry sponsor of a petition or application to prepare an environmental assessment of the proposed action. To complete the assessment, appropriate testing of the environmental fate and effects of chemicals entering the environment may be required. To assist petitioners and applicants, FDA has described its environmental review process in some detail [*12*].

FDA has therefore maintained an interest in ecotoxicity testing and in organizations like ASTM's Committee E47 on Biological Effects and Environmental Fate. FDA has also, in a limited way, supported research on new methods for multispecies testing. Development by Robert L. Metcalf (University of Illinois) of a model ecosystem for evaluating the environmental fate of chemicals was supported by FDA [*13*], as well as by EPA. Development of the standardized aquatic microcosm (SAM) for evaluating environmental effects by Frieda B. Taub (University of Washington) was also supported by FDA [*14*]. This research has also been supported by EPA and the U.S. Army. Support for this type of research has resulted from recognition of the gaps and uncertainties which remain under current approaches to estimating environmental fate and effects of chemicals, as noted in the literature [*15,16*].

Acceptance of Ecotoxicity Testing in Regulatory Decision Making

As has been pointed out previously [*17*], regulatory decision making is constrained by what information is available. Agencies typically are also constrained by deadlines imposed by statute or regulation in making a decision. As a result, decisions cannot wait for development of new methods of testing, and regulations now in place reflect existing methods of ecotoxicity testing. This means that the procedures and endpoints of single species tests are used in the current schemes, while other methods, including microcosm testing, have been relegated to future consideration.

Concern has been expressed about the utility of microcosm tests to provide information useful in making a regulatory decision [*17,18*]. Microcosm responses were seen as difficult to use in establishing limitations, as potentially impossible to interpret and as of doubtful replicability or potential for extrapolation [*17*]. Further, these authors were skeptical of the relevance of microcosm endpoints to protecting or managing ecosystems [*17,18*]. Since these comments were made (1983), additional experience with microcosm protocols, along with further identification of parameters reflecting ecosystem stress and recognition of the usefulness of ecosystem-level data in addressing specific concerns, as discussed below, permits us to suggest that microcosms can provide information useful in regulatory decision making.

In considering what features of a proposed microcosm test would advance its potential for regulatory use, we suggest that three areas may be evaluated: credibility, interpretability, and applicability. The credibility of a protocol, that is, its believability, develops from the history and experience with a test. The interpretability of a protocol is a function of the endpoints it provides and how those endpoints relate to the appropriate theory and to real-world situations. The applicability of a protocol is determined by how the test results contribute to the decision-making process. Acceptance of a protocol for use in regulatory activities would appear feasible if these three features were demonstrated.

1. *To be credible for use in regulatory decision making, a protocol should be backed by adequate experience.* The protocol should have developed past a research stage to where a recognizably consistent procedure is used. Experience with the protocol by several staffs should be available to demonstrate that the protocol is feasible under different test situations and that

hidden problems in implementing the test are minimal. Test results for a variety of standard chemicals (benchmarks) should be available for toxicity test protocols to demonstrate how the test systems respond, allowing comparison with other toxicity test protocols. Experience with a protocol should show that test results are repeatable and reproducible when tests are conducted at different times and by different workers. Endorsement of the protocol by members of a standards-setting organization such as ASTM also contributes to the credibility of a protocol.

The SAM protocol [*14*] provides one example of how background experience might support the credibility of a protocol. As noted above, the protocol was developed by Frieda Taub and associates beginning in 1976. Alternative designs were considered during the developmental research [*19–21*]. Tests with several chemicals (for example, streptomycin sulfate, malathion, diflubenzuron, copper sulfate) were conducted and described the range of responses of the test system [*22–26*]. Preparation of a revised protocol document [*14*] allowed other laboratories to conduct tests using the protocol, both as a test of the repeatability of the protocol (using copper as a reference chemical [*27–31*]) and as a test of other chemicals [*32*]. SAM test results were favorably compared with the effect of field tests of copper [*25*] and atrazine [*32*]. The protocol is currently under review by a task group within Committee E47 of ASTM and has been included in a proposed EPA guideline for multispecies toxicity testing [*33*]. In addition, SAM data were used to develop a computer model of nutrient-phytoplankton-zooplankton interactions and ecosystem responses to toxicants [*34*].

2. *To be interpretable for use in regulatory decision making, a protocol should produce results which relate to meaningful and reasonably measurable endpoints.* While it is easy to agree that an appropriate endpoint should be measured, less agreement exists as to what parameters are reliable indicators of ecosystem health [*35,36*]. Ecosystem endpoints may be broadly distinguished as either structural or functional elements. Structural elements of an ecosystem are those based on populations such as species presence, organism densities, biomass, or relative abundance patterns [*37*]. Functional elements are those based on material and energy movement throughout an ecosystem [*38*], such as primary productivity, and decomposition and nutrient cycling.

Selection of appropriate endpoints must reflect, to some degree, the public perception of what constitutes meaningful endpoints [*17,18*]. Many structural ecosystem endpoints result from processes encountered in the more traditional toxicity endpoints (for example, mortality, reproduction, growth) and so may be more easily accepted. Public acceptance of functional ecosystem endpoints (for example, primary productivity) may depend on the perceived relationship of the endpoint to accepted values, such as the role of primary production in maintaining a sound fishery.

Historical data can be helpful to identify what responses characterize ecosystems under stress. Rapport et al. [*39*] reviewed available information on the behavior of ecosystems under stress and presented a number of "indicator signs." Similar compilations have been published by others [*40–43*]. The endpoints reported in these reviews to be indicative of ecosystem responses to stress are listed in Table 1. The traits commonly described include changes in primary productivity, respiration, nutrient cycling rates and loss, species diversity and richness, biomass, size distribution of species, dominance of species, amplitude of population fluctuations, disease incidence, and successional patterns (including seasonal succession patterns, [*44*]). Of note is the observation that these indicators involve structural, as well as functional, effects on the ecosystem. Also listed in the table are the typical types of data used to characterize the ecosystem response.

We believe that a substantial proportion of these ecosystem endpoints are measurable using microcosm tests. Table 2 presents the endpoints reported in four selected aquatic microcosm protocols, including the SAM protocol [*14*], a mixed-flask culture (MFC) protocol [*27,45*], a generic aquatic ecosystem test [*46*], and a pond-derived microcosm [*47*]. All the protocols presented include estimates of photosynthesis and respiration. Three include enumeration of species, which permits estimation of interactions, numerical indexes, age and size-structure, as

TABLE 1—*Endpoints indicative of ecosystems responses to stress.*

Observation of how ecosystems have responded to stress provides one way to identify what types of data should be obtained and what endpoints might be most useful from microcosm data. The lists below represent the endpoints suggested by several reviewers [*37–44*].

Type of Response	Typical Data Used
Ecosystem Function	
Photosynthesis	^{14}C uptake, oxygen production, chlorophyll concentration, fluorescence
Respiration	Oxygen consumption, CO_2 production
Ratio of photosynthesis to respiration	Oxygen gain/oxygen loss
Decomposition rate	^{14}C acetate degradation
Nutrient uptake, turnover, loss	Levels of nitrogen (ammonia, nitrate, and nitrite), phosphate and silicate present as free or bound, lost to system.
Trophic regulation of function	Predator, grazer, or competitor-mediated control of functions
Ecosystem Structure	
Species abundances and biomass	Enumeration of individuals within species
Trophic interactions (competition, predation, grazing)	Presence of indicator or key species, dominance patterns
Numerical indexes of diversity, richness and similarity	Calculation of index value using enumeration data
Size structure of community	Size distribution of individuals in species or within community
Age structure of population	Fractions of individuals of specified age or size
Gene pool makeup	Genetic composition of population, allele frequency
Reproductive rate and success	Eggs per female, recruitment patterns
Ecosystem Dynamics	
Population fluctuations	Amplitude of population peak and low densities
Successional or developmental patterns	Nature of species replacements over time
Stability descriptors (resilience, elasticity, amplitude, inertia)	Sensitivity and recovery of ecosystem-level parameter following perturbation

well as population dynamics. In several cases, the methods used are the same as techniques used in field studies: extracted chlorophyll, ^{14}C fixation, nutrient concentration, and organism sampling and preservation. Comparison of Tables 1 and 2 suggests that both structural and functional endpoints, which have been of interest in field studies of pollutant effects, can be monitored in laboratory-scale microcosm tests.

Although no terrestrial microcosm protocols have been included in Table 2, protocols such as the soil-core microcosm include many of the same types of ecosystem endpoints. For example, the soil-core microcosm protocol [*48,49*] provides information on vegetation growth and reproduction and nutrient uptake and cycling and can provide information on bioaccumulation and transport to groundwater.

A significant omission from microcosm-scale aquatic tests is the absence of top predators such as fish, predaceous insects, and amphibians. For example, because fish are both economically and ecologically important, the inability to readily include populations of fish in laboratory microcosms because of size factors clearly is a limitation in the use of microcosms as surrogate ecosystems [*50,51*]. Given the current procedures for regulatory review, however, toxicity data for fish are likely to be obtained independently. Consequently, what remains missing are procedures for evaluating the indirect effects on fish of community-level impacts (and vice versa).

Heretofore, we have addressed only regulatory decision making concerning chemicals. Ecosystem testing has also been suggested as a necessary part of regulatory review of the products

TABLE 2—*Values measured in selected aquatic microcosm protocols.*

The endpoints measured and reported by selected aquatic microcosm protocols are marked "+" below. Parentheses indicate that data are collected as an optional procedure or could be obtained without substantially revising the published protocol. "P" indicates that the endpoint can be calculated using data collected as part of the published protocol.

Type of Ecosystem Endpoint	Variable Measured	Microcosm Protocol[a]			
		SAM	MFC	Pond	GAE
Chemical and Fate Information	pH	+	+	+	+
	Total alkalinity	(+)		+	+
	Redox potential		+		
	Toxicant level	(+)	(+)	+	+
Ecosystem Function					
Photosynthesis	Dissolved oxygen	+	+	+	+
	Chlorophyll *a*	(+)		+	
	In vivo fluorescence	+			+
	^{14}C fixation	(+)			+
Respiration	Dissolved oxygen	+	+	+	+
Ratio of Photosynthesis to Respiration		+	+	+	+
Decomposition rate	Electron-transport system				+
	^{14}C-acetate uptake				+
Nutrient uptake, turnover, loss					
	Nutrient concentration	+		+	
Trophic regulation of function		P	P	P	P
Ecosystem Structure					
Species abundances and biomass	Phytoplankton counts	+		+	+
	Macrophyte weight	NA[b]	NA	+	NA
	Zooplankton counts	+		+	+
Trophic interactions (competition, predation, grazing)		P		P	P
Numerical indexes of diversity, richness, and similarity		P		P	P
Size structure of community		P		P	P
Age structure of population		P		P	P
Reproductive rate and success		P		P	P
Ecosystem Dynamics					
Population fluctuations		P		P	P
Successional or developmental patterns		+	P	P	P
Stability descriptors (resilience, elasticity, amplitude, inertia)		P	(+)	P	(+)
Descriptive Information	Volume contained, L	3	1	80	3.5
	Sediment[c]	+		+	
	Chemically defined medium	Yes	Yes	No[d]	Yes

[a]Microcosm protocols: SAM, standardized aquatic microcosm [*14*]; MFC, mixed flask culture [*27*]; Pond, pond-derived microcosms [*47*]; GAE, generic aquatic ecosystem [*46*]. The MFC protocol, as cited in Refs *27, 45,* and *61*, was developed by John Leffler, Ferrum College, VA, and did not include any quantitative enumeration of organisms; the recent modifications to the MFC protocol reported elsewhere in this volume [*83*] are not reflected in these tables.

[b]NA—not applicable, since macrophytes are not included in microcosm.

[c]Sediments included are: SAM—mix of sand, chitin, and cellulose; Pond—ca. 10 L of sediment obtained from shallow ponds, littoral zones of lakes, slow-moving rivers, or other similar habitats.

[d]Water for pond microcosms is collected with the sediments, as above [*47*].

and processes of genetic engineering [*35*]. As pointed out elsewhere, the nature of potential impacts of genetically engineered microorganisms (GEMs) requires different approaches to evaluating hazard [*52,53*]. Questions about the environmental fate of GEMs cannot be resolved with the same procedures used for estimating fate of chemicals, leading to difficulty in determining exposure estimates and in relating exposure to environmental effects. GEM impact assessment may follow a different set of triggers than chemical impact assessments. TSCA Section 5 test guidelines are currently under revision to incorporate new reporting requirements for microbial products of biotechnology. Microcosm tests are frequently included in discussions of GEM impact assessment [*35,52–54*].

3. *To be applicable to regulatory decision making, a protocol should provide effective and efficient information.* Tier-testing schemes attempt to delineate approaches to obtaining information in the most cost- and time-efficient method feasible. Other schemes for determining appropriate testing may involve case-by-case review of the specific uses and properties of a substance [*12*]. Under either approach, the goal is to select from the many available test protocols those which may best contribute to evaluation of potential impact. When the potential impacts of concern can be expressed only where ecological interactions are present, microcosm test methods may prove a more efficient approach to data collection than other available methods. When available information suggests that a wide variety of ecotoxicity information will ultimately be needed, microcosm test methods might provide an efficient means to obtain data on a large number of species, tested under the same conditions of exposure.

Microcosm testing may provide appropriate information without some of the limitations associated with field testing. Although the questions addressed in microcosms could be examined in field tests, practical considerations such as logistics, financing, and administrative problems often prevent field testing of chemicals [*55*]. Obtaining clearance to conduct field testing illustrates the administrative problems which may be encountered. Obtaining local and state clearance for field testing of a potentially toxic substance can be especially difficult when there is potential for release of the toxic test substance to navigable waterways. Consequently, tests such as the Shayler Run, Ohio stream dosed with copper for 33 months [*56*] are difficult to implement. Obtaining clearance for estuarine field tests can also be problematic. EPA's Gulf Breeze Laboratory in Florida attempted to conduct field testing of a pesticide on a portion of a bay. However, the state blocked the testing. Eventually, EPA monitored a state-planned application of the pesticide elsewhere [*57*].

Costs and logistics may also pose problems. Field test sites must be cleaned up after completion of the test, a process which becomes more burdensome, both technically and financially, as the size of the test site increases. Cleanup or other steps are usually done to remove the toxicant so that the site can be used again. For example, experimental ponds require draining and flushing, usually interspersed with fallow periods, to ensure that toxicants are degraded between tests. In addition, many sites do not permit field testing chemicals with long half-lives, that is, which resist degradation. Although removal and disposal of sediments and liners is seen as possible, the expense involved and the need to obtain sediment dredged from a lentic site discourage the practice [*58*].

The Role of Microcosm Testing: A Niche in the Regulatory Scheme

If the credibility, interpretability, and applicability of one or several microcosm protocols are judged acceptable, then it becomes helpful to address how microcosm protocols might fit into the processes of regulatory decision making. Discussion of how decisions should be made is difficult, especially because suggestions eventually may attain the status of policy and regulations. We carefully note that neither EPA nor FDA currently requires microcosm tests (although EPA has proposed test guidelines which may be utilized in developing chemical-specific TSCA Section 4 rules under 40 CFR Part 799 [*33*]).

Regulatory use of new types of data is likely to begin with approaches paralleling current models. Regulatory agencies are conservative in their acceptance of new methods; convincing a reviewer that a new method is valid often requires first convincing the reviewer that the new method is equivalent to the old method. The acceptance of a new practice is enhanced by showing how the practice is already used in other applications. For example, the LD_{50} concept has a long history of use in agriculture and pharmacology [*59,60*]. Adapting this model to single species toxicity testing (that is, LC_{50} and EC_{50}) gave credibility to the analysis of acute toxicity tests. In a similar fashion, MFC tests on aniline and three derivatives have been compared with results from fathead minnow acute tests [*61*].

A clear question which follows is what models are available to suggest roles for microcosm tests in regulatory decision making. We suggest below several models for situations where microcosm test data would seem particularly useful.

Microcosm Testing as a Model Ecosystem in a High-Level Tier Test

The suggestion which is most widely considered at this time is that microcosm testing be fit into the tiered testing scheme as a preliminary step to field testing (Fig. 2). Microcosms would be considered as a model ecosystem, that is, a minimal test system which can demonstrate ecosystem-level effects. The accepted three-tiered testing scheme in OTS might reasonably be revised to include two additional tiers. Tier III testing would presumably include full life-cycle chronic toxicity tests. Tier V would consist of field tests to fully determine the effects of a chemical. In such a scheme, microcosm tests could occupy the Tier IV position between chronic toxicity testing and field tests.

Using the analogy of a microcosm as a model ecosystem, microcosms would permit evaluation of ecosystem endpoints in a laboratory setting. Decision rules triggering microcosm testing would be based upon acute and chronic single-species tests: if individuals may be affected at the estimated exposure, then toxic effects at the community or ecosystem level are likely. In such a case, a microcosm test could address issues such as the representativeness of the tested

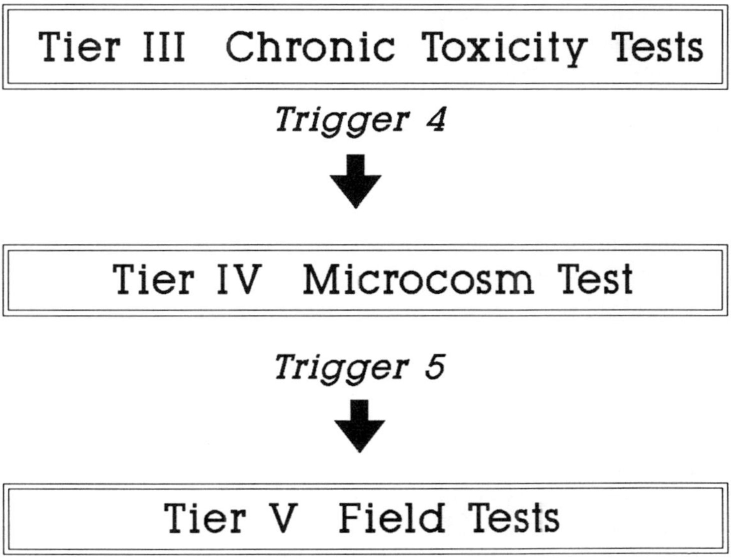

FIG. 2—*Microcosms in a tiered testing scheme. A widely considered role for microcosm testing is as a test triggered by the results of Tier III testing for chronic effects, prefatory to requiring field testing.*

species, the likelihood of replacement species being able to maintain ecosystem function, and the potential for cascading or indirect effects [*62*]. Used as a precursor of field testing, microcosm tests would reduce the logistic, financial, and administrative problems present with field testing as described above.

A critical assumption of single-species testing is the representativeness of the surrogate species used. For some categories of chemicals, a mode of action may be suspected from structure-activity relationships, but the breadth of activity between different species (selectivity) is not known. In some cases, testing of a single species may not adequately represent the responses of related species. Such is the case with the antibiotic streptomycin, where minimum inhibitory concentrations to several species of green algae ranged over two orders of magnitude (0.66 to 66 mg/L) [*63*]. Microcosm tests might demonstrate the range of species that are susceptible as well as the degree of species replacement and ecosystem-level functional impact. Of course, any specific microcosm test will include only a limited number of species, limited either by design (for example, SAM) or by the number in the source inocula (for example, MFC, GAE). If single-species tests were conducted on an equivalent taxonomic spectrum, an improved chance of observing "representative" responses would exist. As discussed earlier, however, current practice generally relies on tests of a single species within a group (for example, Tier I requirements) or two species (for example, Tier II requirements).

Microcosm Testing to Integrate Fate and Effects Estimations

A second model of appropriate microcosm application would be where the exposure is expected to approach or exceed a criterion concentration established by tests from lower tiers, or where fate processes are expected to significantly alter exposure. As discussed earlier, current assessment schemes most often rely on a comparison of an exposure estimate and a toxicity estimate (Fig. 3A). The exposure estimate is the result of expected use levels or loadings and of several estimates of partitioning, transformation, and persistence, while the toxicity estimate is the result of one or several isolated tests. As described earlier, decision making is based on numerical combinations of these estimates and test results. Microcosm designs have incorporated various features to include fate processes as well as effects (for example, [*20*]) and so provide a system which integrates fate and effects processes. To the extent that these processes are representative, a microcosm test functions as a model ecosystem, an ecological integrator to demonstrate the presence or absence of effects resulting from a specified chemical loading (Fig. 3B).

Microcosms could be used to determine whether acute and chronic effects identified in earlier tier tests actually exhibit themselves in more environmentally realistic surroundings. For example, microcosms simulating the Illinois River [*64*] were used to determine the fate of butyl benzyl phthalate (BBP). The results indicated that the toxicity of BBP was mitigated by the degradability of the compound in water and sediments. This site-specific microcosm study was submitted in response to a proposed test rule under Section 4 of TSCA and was patterned after a marine microcosm test protocol.

Another example of integrating fate and effects information is the marine microcosm system used by Perez et al. [*65*] to test di(2-ethylhexyl) phthalate (DEHP). Their microcosms contained a sediment subsystem and were run to simulate Narragansett Bay, Rhode Island under summer and winter conditions. Perez et al. concluded that tests of DEHP in the water column and on a few associated organisms conducted only under summer conditions would have underestimated the potential environmental hazard of DEHP. Perez et al. found significant increases in the winter body burdens of DEHP in the filter-feeding bivalves *Pitua morhuana* and *Mulinia lateralis.* The observed increases apparently resulted from interactions of other seasonal patterns: DEHP concentrations in the water column were, on the average, higher in winter because of reduced DEHP breakdown rates; algae, which adsorb DEHP and serve as food for the benthic bivalves, were at higher densities under winter conditions. Perez et al. concluded that the biotic complexity of their test system was crucial in demonstrating additional routes of potential human exposure and in observing any ecological effects of DEHP.

The chemical nonylphenol provides another example where integration of fate and effects processes would be useful in evaluating environmental impact. Nonylphenol is used in the preparation of other chemicals. About 70% of the 75 000 metric tons of nonylphenol produced in the U.S. is incorporated into ethoxylates for nonionic surfactants [*66*]. Although nonylphenol is not released directly into the environment, hydrolysis and bacterial action on the surfactants and detergents (ethoxylates and carboxylic acids) degrades these materials back to nonylphenol [*67–69*]. Measurable concentrations have been observed in wastewater effluents (0.04 to 4.0 ppm [*66, 70*]), sludges (280 to 1,000 ppm [*68*]), rivers, and groundwaters [*66, 68, 71, 72*]. Data on degradation of nonylphenol is somewhat contradictory: little degradation was reported in a biological oxygen demand (BOD) test (0 to 3% [*66*]), while some degradation was observed in a pilot wastewater treatment plant (25% in about 8 h [*73*]), and disappearance in stream and pond waters was extensive (50% in 16 to 25 d, 80% in 71 d [*74*]). Acute toxicity studies report LC_{50} values to invertebrates of 0.14 to 0.48 ppm [*66, 75*] and to fish of 0.14 to 1.4 ppm [*66, 75, 76*]. Physical-chemical data such as an octanol/water partition coefficient of 13 000 suggest a strong tendency to partition to sediments [*66*]; but measured bioconcentration factors ranged from 1.4 to 280 [*66, 75*], suggesting limited bioaccumulation.

The combination of factors seen in this case suggest that a microcosm study would be useful in evaluating whether environmental introductions of nonylphenol-based materials might have environmental effects. In a microcosm test the several fate processes (degradation, partitioning, transformation) as well as the effects processes could be evaluated. Although nonylphenol was the subject of a recent EPA call for additional testing under Section 4 of TSCA [*77*], microcosm testing was not suggested. Our use of nonylphenol is only as an example of a situation where microcosm data could help integrate fate and effects processes; it is not to suggest that microcosm testing be required for nonylphenol.

Microcosm Testing to Identify Potential Ecosystem-Level Impacts

Microcosm testing might precede or be concurrent with other testing in cases where the critical questions in the evaluation inherently require testing in a community or ecosystem matrix, for example, species persistence and displacement (Fig. 4). In the case of evaluating the potential impact of GEMs, these ecological endpoints may be precisely the endpoints that are of concern. For example, Regal concluded that the evaluation of potential impact of recombinant organisms may rely on understanding what factors affect the success or failure of introduced species [*78*]. Specific microcosm systems have been used, for example, to monitor the maintenance and stability of introduced genotypes in groundwater material [*79*], plasmid transfer between river bacteria [*80*], and competition between genetically engineered and parental strains of *Pseudomonas* [*81*]. For example, in microcosms simulating a sewage treatment plant, growth of chlorobenzoate-degrading recombinant *Pseudomonas* populations was similar to populations of the parental strain in the absence of chlorobenzoates. In the presence of chlorobenzoates, however, the genetically engineered strain grew better than the parental strain.

The National Institutes of Health (NIH) proposed endpoints to consider in evaluating environmental release of organisms containing recombinant DNA [*82*]. These endpoints are presented in Table 3 in the framework used in Tables 1 and 2. Table 3 illustrates that a major concern of GEMs evaluation is the survival and reproduction of the recombinant organism, typically bacteria.

Neither naturally occurring nor recombinant bacteria have been regularly identified or enumerated in most microcosm tests, although microbially mediated functions have been measured (for example, acetate metabolism, electron-transport system (ETS) function in GAE microcosms [*46*]). However, minor additions to microcosm protocols such as those listed in Table 2 would permit bacterial enumeration; in some cases such modifications have already been tried.

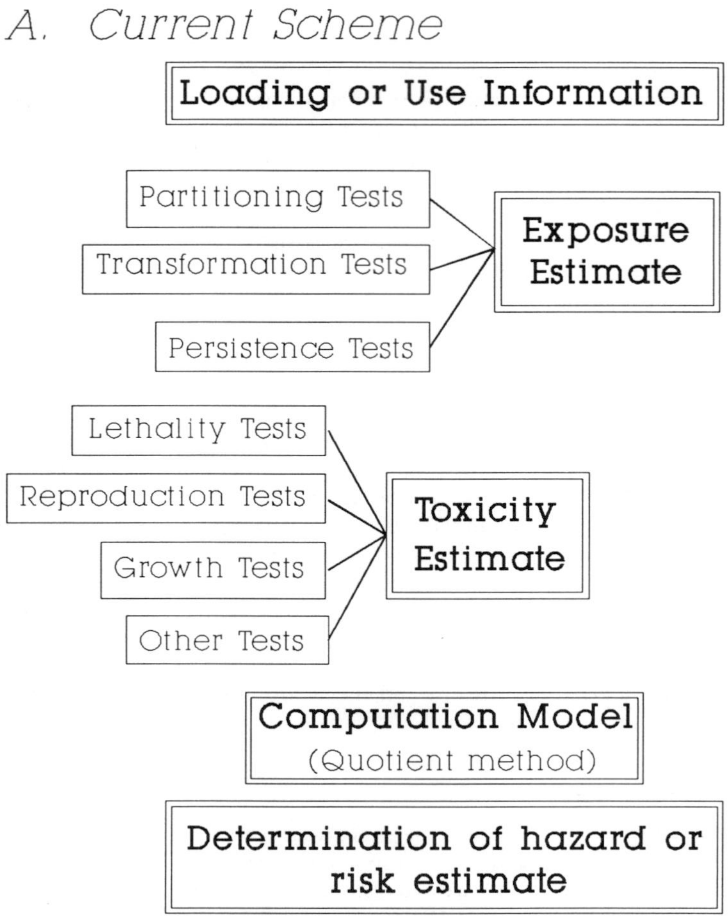

FIG. 3—*Potential role of microcosm testing in an assessment scheme:* (A) *Currently, a determination of hazard or risk assessment is based on the quotient of a predicted environmental exposure to an estimate of what concentration causes toxic effects. Test data obtained from isolated tests are used to compute these estimates, which are in turn used (sometimes with a safety factor) to evaluate the potential for environmental impact.* (B) *Microcosm systems may provide an alternative to the computational approach because the fate and effects processes are interacting within the test system. Considering a microcosm as a model ecosystem, an assessment scheme could use the presence or absence of significant response to chemical loading of a microcosm as a basis for evaluating potential environmental impact.*

For example, Shannon, Flum, Anderson, and Yount reported in this symposium on their work with the microbial control agent *Bacillus thuringensis* introduced into a mixed flask culture microcosm [*83*]. Tests of streptomycin in the SAM system have also included bacterial enumeration [*22*]. Consequently, we expect that such modifications could be accomplished reasonably easily in many microcosm systems. Therefore, we included in Table 3 an identification of several aquatic and terrestrial protocols which, with minor modifications, would seem capable of providing data on the endpoints mentioned by NIH [*82*]. Of course, additional modifications would be required to distinguish and enumerate recombinant bacteria.

Table 3 includes one type of endpoint not included in Table 1 or 2, reflecting the specific

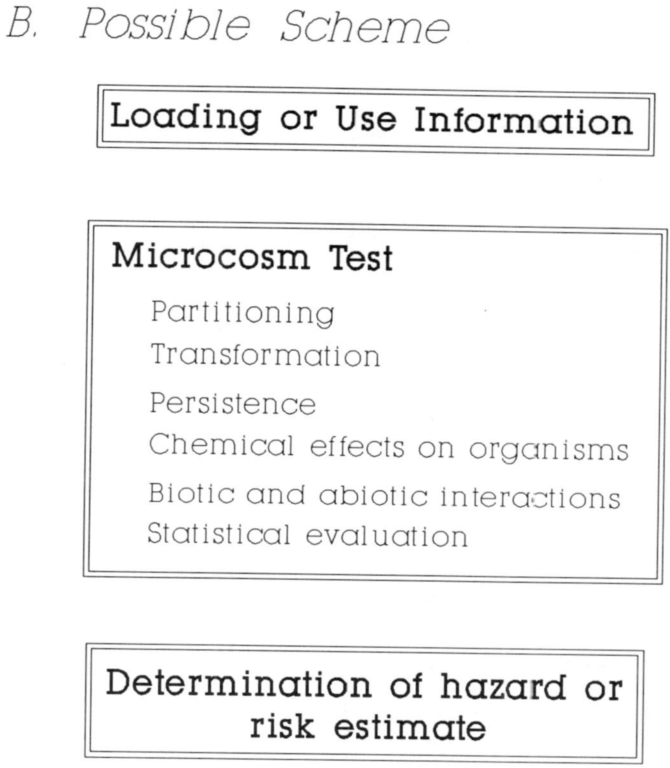

FIG. 3—*Continued.*

nature of concerns over GEM impacts. Data on pathogenicity, infectivity, toxicity, virulence, and vectors reflect concerns specific to microbial ecology. Similarly, data on hosts and symbiotic relationships were not previously considered. As mentioned earlier, the survival and reproduction of the recombinant organism, as well as of related wild-type organisms, are a major concern in GEMs evaluation. Bacterial identification and enumeration will not be adequate, however, to address other concerns related to GEM survival, such as the fate of introduced gene sequences or efficacy of gene markers. GEM evaluation also focuses on biogeochemical cycling processes and the degradation of important biological chemicals (for example, cellulose and lignin), as reflected in the endpoints on nutrient cycling processes and decomposition.

EPA expects that the majority of new GEM applications will involve agricultural chemicals, biomass conversion, industrial chemical production, metal recovery, nitrogen fixation, polymer and macromolecule production, enhanced oil recovery, and waste and pollutant degradation [*54,84*]. Rissler [*54*] noted that OPP-regulated microorganisms are developed specifically to be pathogenic or toxic to pest species, while OTS-regulated microorganisms are designed to establish themselves in a given environment and perform functions such as those listed above.

Since several of these GEM applications involve intentional environmental introduction, environmental fate and effects testing is expected to be necessary. OTS-regulated microorganisms will be examined for environmental expression, including survival, multiplication, exposure, and hazard identification. OPP-regulated organisms will be initially examined for their toxicity and pathogenicity to nontarget organisms, and secondarily examined for environ-

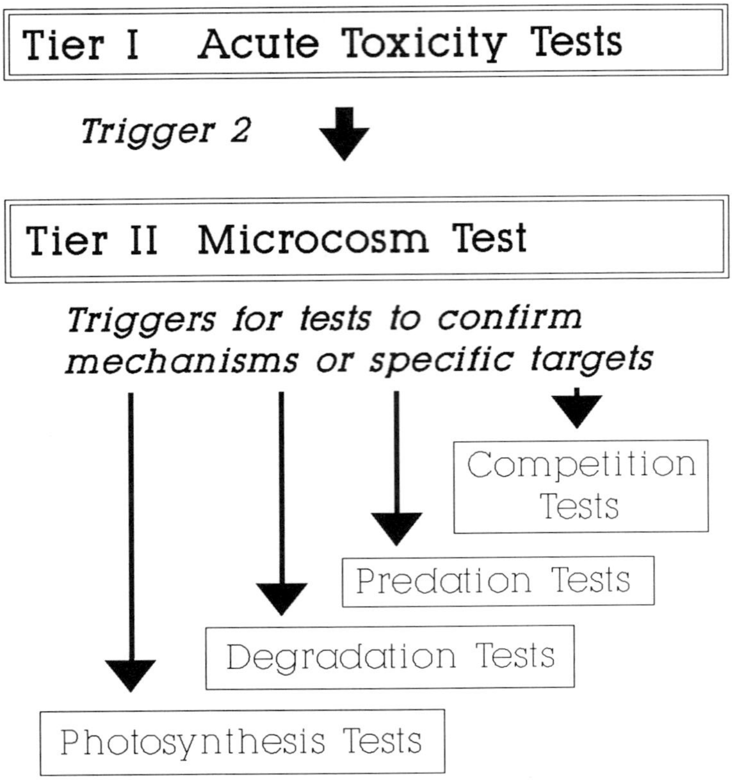

FIG. 4—*Alternative model for microcosms in tiered testing scheme. Microcosms might prove useful at a low level in tiered testing, where test results would be used to address ecological endpoints of concern. For example, the competitive success of an introduced species in a microcosm test might trigger specific subsequent testing to identify competitive mechanisms or strategies.*

mental expression. Because it is difficult to contain microorganisms or restrict reproduction during a field test, it is reasonable to expect that, in some situations, microcosm tests under strict containment procedures will be justified prior to field testing.

Other Uses of Microcosm Testing

One aspect of evaluating impact for exposures of short or limited duration is the potential for amelioration, recovery, or resiliency of an exposed ecosystem. Where such a situation might affect regulatory decision making, microcosm testing might demonstrate the pattern of recovery. An example of this was shown by Flum and Shannon [*45*] for the three related amides, diuron, 2-(octyloxy)-acetanilide, and salicylanilide. This type of data might prove especially useful in evaluating impacts of spills or mixed materials by suggesting what pattern ecosystem recovery might take.

Separate regulatory review processes exist for ranking the relative hazard of materials entering or already in an ecosystem, often to help prioritize subsequent work. Microcosm data may help provide a ranking system, as was done by Sheehan et al. for a series of toxic substances [*46*].

TABLE 3—*Endpoints of concern for testing potential impact of GEMs.*

Endpoints for testing the potential environmental impact of GEMs are listed below in the sequence used in Tables 1 and 2. The specific types of data called for in the NIH "Points to Consider" document [*82*] are identified. Microcosm protocols which include data on these endpoints, or which could be modified slightly to include such data, are identified.

Type of Ecosystem Endpoint	Data Suggested by NIH Points to Consider	Protocols Which Might Provide This Data	
		Aquatic	Terrestrial
Ecosystem Function			
Photosynthesis		B[a], C, D	E, F, G
Respiration		B, C, D	E, F, G
Decomposition rate	Cellulose, lignin and pesticide degradation	B, C, D	E, F, G
Nutrient uptake, turnover, loss	Nitrogen fixation and mineral cycling	A, B, C, D	E, F, G
Ecosystem Structure			
Species abundances and biomass	GEM survival, wild-type survival	A, B, C, D	G
Trophic interactions	Competition, predation, hosts, symbionts, parasites, pathogens	A, B, D	
Biological Toxicity[b]	Pathogenicity, infectivity, toxicity, virulence, vector		
Gene pool makeup	Likelihood or frequency of gene exchange, gene stability		
Reproductive rate and success	GEM reproduction	A, B, C, D	G
Ecosystem Dynamics			
Population fluctuations			
Successional or developmental patterns		A, B, D	
Stability descriptors (resilience, elasticity, amplitude, inertia)	Shifts in ecological characteristics	A, B, D, C	

[a]Protocols referenced are:

A Site-specific aquatic microcosm test, Subpart 797.3100 of Ref *33*, patterned after Perez et al. [*65*].

B Standardized aquatic microcosm test, Subpart 797.3050 of Ref *33*, patterned after Taub and Read [*14*].

C Mixed flask culture microcosm test, Subpart 797.3050 of Ref *33*, patterned after Shannon et al. [*27*].

D Pond-derived microcosm test [*47*].

E Soil microbial community microcosm test, Subpart 797.3700 of Ref *33*.

F Rhizobial-legume chronic toxicity test, Subpart 797.2900 of Ref *33*.

G Soil-core microcosm test, Subpart 797.3775 of Ref *33*, patterned after Van Voris et al. [*49*].

[b]Not included in Table 1 or 2, but is of specific interest for GEMs evaluation.

Summary

Because current regulatory evaluation schemes are designed to rely on single-species testing, they may require revision before community or ecosystem-level concerns can be included. Incorporation of a new approach into a regulatory agency's evaluation schemes will depend on the extent of experience in using the new approach, its ability to yield meaningful and measurable endpoints, and its efficiency to provide useful information. We have termed these aspects the credibility, interpretability, and applicability of a protocol. We have presented several models for the use of microcosm test data by regulatory agencies: (1) as a model ecosystem, a high-level tier test, preparatory to field testing; (2) as a test procedure to integrate environmental fate and effects processes; (3) as a way to identify problems for further evaluation by pinpointing specific ecosystem-level effects; (4) as a procedure for demonstrating ecosystem recovery following a stress; and (5) as a method to rank chemicals based on ecosystem-level impacts. Microcosm tests have been suggested for evaluation of the potential environmental impact of GEMs. Use of microcosm tests for such a purpose seems appropriate in that the microcosms provide an ecological matrix for observation of GEM fate and effects, although most microcosm protocols would require revisions to meet many of the concerns of microbial ecologists. We believe that as the credibility, relevance, and applicability of microcosm protocols are further demonstrated, regulatory agencies will better be able to fit microcosm data into their decision making processes.

Acknowledgments

Discussions with Jim Gilford and Vince Nabholz of EPA's Office of Toxic Substances were most helpful in considering the use of microcosms in tiered testing schemes. Thomas LaPoint of the U.S. Fish and Wildlife Service's Columbia, Missouri, laboratory provided helpful comments on the pond systems. We also thank Sylvia Matson of FDA for help in preparing the manuscript and Buzz Hoffmann and Lisa Nowell of FDA for their useful comments.

Disclaimer

This report has been reviewed by the Office of Toxic Substances, U.S. Environmental Protection Agency, and approved for publication. Approval does not signify that the contents necessarily reflect the views and policies of the U.S. Environmental Protection Agency.

References

[1] *SCOPE 12: Principles of Ecotoxicology,* G. C. Butler, Ed., John Wiley & Sons, New York, 1978, p. xx.
[2] Moriarty, F., *Ecotoxicology: the Study of Pollutants in Ecosystems,* Academic Press, New York, 1983, p. 1.
[3] *Methods for Ecological Toxicology, A Critical Review of Multispecies Tests,* A. S. Hammons, Ed., Ann Arbor Science Publishers, Inc., Ann Arbor, MI.
[4] U.S. Environmental Protection Agency, *Federal Register,* Vol. 49, No. 207, 24 Oct. 1984, pp. 42856–42903.
[5] Stern, A. M., *Environmental Management,* Vol. 10, No. 4, 1986, pp. 453–462.
[6] Dickson, K. L., Duke, T., and Loewengart, G. in *Multispecies Toxicity Testing,* J. Cairns, Jr., Ed., Pergamon Press, New York, 1985, Chapter 15, pp. 248–253.
[7] Urban, D. J. and Cook, N. J., "Standard Evaluation Procedure: Ecological Risk Assessment," EPA 540/9-85-001, U.S. Environmental Protection Agency, Washington, DC, June 1986.
[8] *Code of Federal Regulations,* Title 40, Part 721.
[9] "Testing for Environmental Effects under the Toxic Substances Control Act" and attached support document, unpublished manuscript, U.S. Environmental Protection Agency, Ecological Effects Branch, Office of Toxic Substances, Washington, DC, 9 Sept. 1983.
[10] *Code of Federal Regulation,* Title 40, Part 158.145.
[11] Food and Drug Administration, "National Environmental Policy Act; Policies and Procedures," *Federal Register,* Vol. 50, No. 81, 26 April 1985, pp. 16636–16669.
[12] Food and Drug Administration, *Environmental Assessment Technical Handbook,* PB87-175345/AS, National Technical Information Service, Springfield, VA, 1987.
[13] Metcalf, R. L., "Evaluation of the Utility of the Model Ecosystem for Determining the Ecological Fate of Substances Subject to FDA Regulatory Authority," final report for Contract No. 74-127, Food and Drug Administration, Washington, DC, December 1976.
[14] Taub, F. B. and Kindig, A. C., "Standardized Aquatic Microcosm Protocol," Vol. II of final report for Contract No. 223-83-7000, Food and Drug Administration, Washington, DC, 31 Dec. 1986.
[15] Cairns, J., Jr., *Hydrobiologia,* Vol. 100, 1983, pp. 47–57.
[16] Kimball, K. D. and Levin, S. A., *BioScience,* Vol. 35, No. 3, 1985, pp. 165–171.
[17] Tebo, Jr., L. B. in *Multispecies Toxicity Testing,* J. Cairns, Jr., Ed., Pergamon Press, New York, 1985, Chap. 3, pp. 19–26.

[18] Mount, D. I. in *Multispecies Toxicity Testing*, J. Cairns, Jr., Ed., Pergamon Press, New York, 1985, Chap. 2, pp. 13-18.
[19] Crow, M. E. and Taub, F. B., *International Journal of Environmental Studies*, Vol. 13, 1979, pp. 141-147.
[20] Taub, F. B. and Crow, M. E. in *Microcosms in Ecological Research*, J. P. Geisy, Jr., Ed., U.S. Department of Energy, Symposium Series 52, CONF-781101, National Technical Information Service, Springfield, VA, 1980, pp. 69-104.
[21] Taub, F. B. in *Algae as Ecological Indicators*, L. W. Shubert, Ed., Academic Press, New York, 1984, pp. 363-394.
[22] Taub, F. B., Read, P. L., Kindig, A. C., Harrass, M. C., Hartmann, H. J., Conquest, L. L., Hardy, F. J., and Munro, P. T. in *Aquatic Toxicology and Hazard Assessment: Sixth Symposium, ASTM STP 802*, W. E. Bishop, R. D. Cardwell, and B. B. Heidolph, Eds., American Society for Testing and Materials, Philadelphia, 1983, pp. 5-25.
[23] Kindig, A. C., Conquest, L. L., and Taub, F. B. in *Aquatic Toxicology and Hazard Assessment: Sixth Symposium, ASTM STP 802*, W. E. Bishop, R. D. Cardwell, and B. B. Heidolph, Eds., American Society for Testing and Materials, Philadelphia, 1983, pp. 192-203.
[24] Taub, F. B. in *Concepts in Marine Pollution Measurement*, H. H. White, Ed., University of Maryland, College Park, MD, 1984, pp. 159-192.
[25] Harrass, M. C. and Taub, F. B. in *Validation and Predictability of Laboratory Methods for the Assessment of Fate and Effects of Contaminants in Aquatic Ecosystems, ASTM STP 865*, T. P. Boyle, Ed., American Society for Testing and Materials, Philadelphia, 1985, pp. 57-74.
[26] Taub, F. B. in *Aquatic Toxicology*, A. Boudou and F. Ribeyre, Eds., CRC Press, in press.
[27] Shannon, L. J., Harrass, M. C., Yount, J. D., and Walbridge, C. T. in *Community Toxicity Testing, ASTM STP 920*, J. Cairns, Jr., Ed., American Society for Testing and Materials, Philadelphia, 1986, pp. 135-157.
[28] Taub, F. B., Kindig, A. C., and Conquest, L. L. in *Community Toxicity Testing, ASTM STP 920*, J. Cairns, Jr., Ed., American Society for Testing and Materials, Philadelphia, 1986, pp. 93-120.
[29] Taub, F. B., Kindig, A. C., and Conquest, L. L. in *Aquatic Toxicology and Hazard Assessment: Tenth Volume, ASTM STP 971*, American Society for Testing and Materials, Philadelphia, 1988, pp. 384-405.
[30] Landis, W. G., Chester, N. A., Haley, M. V., Johnson, D. W., and Muse, W. T., Jr. in *Aquatic Toxicology and Hazard Assessment: Eleventh Volume, ASTM STP 1007*, American Society for Testing and Materials, Philadelphia, in press.
[31] Taub, F. B., Kindig, A. C., Conquest, L. L., and Meador, J. P. in *Aquatic Toxicology and Hazard Assessment: Eleventh Symposium, ASTM STP 1007*, American Society for Testing and Materials, Philadelphia, in press.
[32] Stay, F. S., Larsen, D. P., Katko, A., and Rohm, C. in *Validation and Predictability of Laboratory Methods for Assessing the Fate and Effects of Contaminants in Aquatic Ecosystems, ASTM STP 865*, T. P. Boyle, Ed., American Society for Testing and Materials, Philadelphia, 1985, pp. 75-90.
[33] U.S. Environmental Protection Agency, *Federal Register*, Vol. 52, No. 187, 28 Sept. 1987, pp. 39334-36371.
[34] Swartzman, G. and Rose, K. A., *Ecological Modelling*, Vol. 22, 1983/84, pp. 123-134.
[35] Cairns, J., Jr. and Pratt, J. R., *Water Resources Bulletin*, Vol. 22, No. 2, 1986, pp. 171-182.
[36] Miller, D. R. in *SCOPE 22: Effects of Pollutants at the Ecosystem Level*, P. J. Sheehan, D. R. Miller, G. C. Butler, P. Bourdeau, and J. M. Ridgeway, Eds., Wiley, New York, 1984, pp. 15-22.
[37] Sheehan, P. J. in *SCOPE 22: Effects of Pollutants at the Ecosystem Level*, P. J. Sheehan, D. R. Miller, G. C. Butler, P. Bourdeau, and J. M. Ridgeway, Eds., Wiley, New York, 1984, pp. 51-99.
[38] Sheehan, P. J. in *SCOPE 22: Effects of Pollutants at the Ecosystem Level*, P. J. Sheehan, D. R. Miller, G. C. Butler, P. Bourdeau, and J. M. Ridgeway, Eds., Wiley, New York, 1984, pp. 101-145.
[39] Rapport, D. J., Regier, H. A., and Hutchinson, T. C., *American Naturalist*, Vol. 125, No. 5, 1985, pp. 617-640.
[40] Moore, C. G., "The Use of Community Structure in Pollution Monitoring," *Environmental Toxicology and Ecotoxicology*, World Health Organization, Ed., WHO Regional Office for Europe, Copenhagen, 1986, pp. 231-253.
[41] Couture, P., Thellen, C., Thompson, P. A., and Auclair, J.-C., *Canadian Journal of Fisheries and Aquatic Science*, Vol. 44, 1987, pp. 167-175.
[42] American Management Systems, Inc., "Review of the Literature on Ecological Endpoints," Work Assignment WA-87-45 for the Science-Policy Integration Branch, Regulatory Integration Division, Office of Policy, Planning and Evaluation, U.S. Environmental Protection Agency, Contract No. 68-01-7002, 30 Sept. 1987.
[43] Mittelman, A., Settel, J., Plourd, K., Fulton, R. S., III, Sun, R. S., Chaube, S., and Sheehan, P.,

"Ecological Endpoint Selection Criteria," Technical Resources, Inc., Contract No. 68-02-4199, Exposure Assessment Group, Office of Research and Development, U.S. Environmental Protection Agency, Washington, DC, 30 Sept. 1987 (draft).

[44] Sommer, U., Gliwicz, Z. M., Lampert, W., and Duncan, A., *Archiv fur Hydrobiologia,* Vol. 106, No. 4, 1986, pp. 433–471.

[45] Flum, T. F. and Shannon, L. J., *Ecotoxicology and Environmental Safety,* Vol. 13, 1987, pp. 239–252.

[46] Sheehan, P. J., Axler, R. P., and Newhook, R. C. in *Community Toxicity Testing, ASTM STP 920,* J. Cairns, Jr., Ed., American Society for Testing and Materials, Philadelphia, 1986, pp. 158–179.

[47] Giddings, J. M. in *Community Toxicity Testing, ASTM STP 920,* J. Cairns, Jr., Ed., American Society for Testing and Materials, Philadelphia, 1986, pp. 121–134.

[48] U.S. Environmental Protection Agency, *Federal Register,* Vol. 52, No. 187, 28 Sept. 1987, pp. 36363–36371.

[49] Van Voris, P., Tolle, D. A., Arthur, M. F., and Chesson, J. in *Multispecies Toxicity Testing,* J. Cairns, Jr., Ed., Pergamon Press, New York, 1985, pp. 117–141.

[50] Harrass, M. C. and Taub, F. B. in *Aquatic Toxicology and Hazard Assessment (Seventh Symposium), ASTM STP 854,* R. D. Cardwell, R. Purdy, and R. C. Banner, Eds., American Society for Testing and Materials, Philadelphia, 1985, pp. 117–133.

[51] Harte, J., Levy, D., Rees, J., and Saegebarth, E. in *Microcosms in Ecological Research,* J. P. Geisy, Jr., Ed., U.S. Department of Energy Symposium Series 52, CONF-781101, National Technical Information Service, Springfield, VA, 1980, pp. 105–137.

[52] Fiskel, J. R. and Covello, V. T. in *Biotechnology Risk Assessment: Issues and Methods for Environmental Introductions,* J. Fiskel and V. T. Covello, Eds., Pergamon Press, New York, 1986, pp. 1–34.

[53] Omenn, G. S. in *Biotechnology Risk Assessment: Issues and Methods for Environmental Introductions,* J. Fiskel and V. T. Covello, Eds., Pergamon Press, New York, 1986, pp. 144–163.

[54] Rissler, J. F., *Recombinant DNA Technical Bulletin,* Vol. 7, No. 1, 1984, pp. 20–30.

[55] Pritchard, P. H. and Bourquin, A. W., *Advances in Microbial Ecology,* Vol. 7, 1984, pp. 133–215.

[56] Geckler, J. R., Horning, W. B., Neiheisel, T. M., Pickering, Q. H., Robinson, E. L., and Stephan, C. E., "Validity of Laboratory Tests for Predicting Copper Toxicity in Streams," EPA-600/3-76-116, U.S. Environmental Protection Agency, Washington, DC, December 1976.

[57] Personal communication, Jim Gilford, Office of Toxic Substances, U.S. Environmental Protection Agency, Washington, DC.

[58] Personal communication, Thomas LaPointe, U.S. Fish and Wildlife Service, Columbia, MO.

[59] Bliss, C. I., *Science,* Vol. 79, 1934, pp. 409–410.

[60] Finney, D. J., *Probit Analysis,* 2nd ed., University Press, Cambridge, 1952.

[61] Yount, J. D. and Shannon, L. J., *Environmental Toxicology and Chemistry,* Vol. 6, 1987, pp. 463–468.

[62] Carpenter, S. R., Kitchell, J. F., and Hodgson, J. R., *BioScience,* Vol. 35, 1985, pp. 634–639.

[63] Harrass, M. C., Kindig, A. C., and Taub, F. B., *Aquatic Toxicology,* Vol. 6, 1985, pp. 1–11.

[64] Adams, W. J., Renaudette, W. J., Doi, J. D., Stepro, M. G., Tucker, M. W., Kimerle, R. A., Franklin, B. B., and Nabholz, J. V. in *Aquatic Toxicology and Hazard Assessment: Eleventh Volume, ASTM STP 1007,* American Society for Testing and Materials, Philadelphia, in press.

[65] Perez, K. T., Davey, E. W., Lackie, N. F., Morrison, G. E., Murphy, P. G., Soper, A. E., and Winslow, D. L. in *Aquatic Toxicology and Hazard Assessment: Sixth Symposium, ASTM STP 802,* W. E. Bishop, R. D. Cardwell, and B. B. Heidolph, Eds., American Society for Testing and Materials, Philadelphia, 1983, pp. 180–191.

[66] Office of Toxic Substances, U.S. Environmental Protection Agency, Chemical Hazard Information Profile, Draft Report, Nonylphenol, 25154-52-3, 17 Sept. 1985.

[67] Ahel, M., Conrad, T., and Giger, W., *Environmental Science and Technology,* Vol. 21, No. 7, 1987, pp. 697–703.

[68] Giger, W., Brunner, P. H., and Schaffner, C., *Science,* Vol. 225, 10 Aug. 1984, pp. 623–625.

[69] Marcomimi, A. and Giger, W., *Analytical Chemistry,* Vol. 59, 1987, pp. 1709–1715.

[70] Wagener, S. and Schink, B., *Water Research,* Vol. 21, No. 5, 1987, pp. 615–622.

[71] Stephanou, E. and Giger, W., *Environmental Science and Technology,* Vol. 16, 1982, pp. 800–805.

[72] Sheldon, L. S. and Hites, R. A., *Environmental Science and Technology,* Vol. 12, 1978, pp. 1188–1194.

[73] Kravetz, L., Chung, H., Guin, K. F., Shebs, W. T., and Smith, L. S., "Primary and Ultimate Biodegradation of an Alcohol Ethoxylate and a Nonylphenol Ethoxylate Under Average Winter Conditions in the U.S.," Technical Bulletin SC: 779-83, Shell Chemical Co., Houston, TX, 1983.

[74] Sundaram, K. M. S. and Szeto, S., *Journal of Environmental Science and Health, Series B,* Vol. 16, 1981, pp. 767–776.

[75] McLeese, D. W., Zitko, V., Sargean, D. B., Burridge, L., and Metcalfe, C. D., *Chemosphere,* Vol. 10, 1981, pp. 723–730.

[76] Yoshimura, K., *Journal of American Oil Chemists Society,* Vol. 63, 1986, pp. 1590-1596.
[77] U.S. Environmental Protection Agency, *Federal Register,* Vol. 52, No. 203, 21 Oct. 1987, pp. 39273-39274.
[78] Regal, P., *Recombinant DNA Technical Bulletin,* Vol. 10, No. 3, 1987, pp. 67-85.
[79] Jain, R. K., Sayler, G. S., Wilson, J. T., Houston, L., and Pacia, D., *Applied and Environmental Microbiology,* Vol. 53, 1987, pp. 996-1002.
[80] Bale, M. J., Fry, J. C., and Day, M. J., *Journal of General Microbiology,* Vol. 133, 1987, pp. 3099-3107.
[81] Dwyer, D., "Behavior and Functioning of a GEM in Model Ecosystems," Proceedings of the First International Conference on the Release of Genetically Engineered Microorganisms (REGEM 1), Cardiff, Wales, 5 April 1988, Academic Press, New York.
[82] National Institutes of Health, *Federal Register,* Vol. 50, No. 60, 28 Mar. 1985, pp. 12456-12459.
[83] Shannon, L. J., Flum, T., Anderson, R., and Yount, J. D., p. 224, this volume.
[84] Office of Toxic Substances, Information Management Division, U.S. Environmental Protection Agency, 1987.

Lyle J. Shannon,[1] *Terry E. Flum,*[1] *Richard L. Anderson,*[2] *and J. David Yount*[2]

Adaptation of Mixed Flask Culture Microcosms for Testing the Survival and Effects of Introduced Microorganisms

REFERENCE: Shannon, L. J., Flum, T. E., Anderson, R. L., and Yount, J. D., **"Adaptation of Mixed Flask Culture Microcosms for Testing the Survival and Effects of Introduced Microorganisms,"** *Aquatic Toxicology and Hazard Assessment: 12th Volume, ASTM STP 1027,* U. M. Cowgill and L. R. Williams, Eds., American Society for Testing and Materials, Philadelphia, 1989, pp. 224–239.

ABSTRACT: A microcosm test was used to evaluate the survival and effects of *Bacillus thuringiensis* var. *israelensis* (Bti) in aquatic systems. Two commercial formulations (Vectobac from Abbott; Mosquito Attack from Reuters Laboratories) and a laboratory preparation of Bti spores were tested. In all cases, Bti spores persisted in the microcosms, showing no change in abundance over six weeks of observation. The Bti preparations killed larval mosquitoes *(Aedes aegypti)* and midges (*Tanytarsus* sp.) in microcosms within 24 to 48 h. Other effects were limited to minor increases in primary production and community respiration (as measured by changes in pH and dissolved oxygen). These effects appeared to be a response to the loss of grazing insect species, the decomposition of dead insects, and the addition of a biodegradable compound (the formulation) with the formulated products. Bioassays of samples from the microcosms against *Aedes aegypti* larvae showed that Bti lost its toxicity within seven days.

The microcosm procedure proved to be a useful method for evaluating the survival and effects of microbial spores. The results obtained with this procedure were similar to those reported from field tests of Bti.

KEY WORDS: microcosms, microorganisms, genetically engineered organisms, microbial pest control, *Bacillus thuringiensis* var. *israelensis,* ecological effects, statistical analysis

Recent advances in the commercialization of genetic engineering technology have resulted in the creation of many new and exotic microorganisms. Any of these genetically engineered microorganisms (GEMs), whether through intended or accidental means, could be released into the environment. The potential negative environmental effects resulting from such releases have become a source of concern for both regulators and the general public.

Concern over the environmental hazards presented by GEMs is based on numerous examples of invading species colonizing new ecosystems and displacing existing species. Instances of relatively violent disruptions have been widely reported in the literature. They include such disasters as the introduction of the Sea Lamprey *(Petromyzon marinus)* into the Great Lakes, the loss of elm trees across the United States through the rapid progress of the Dutch Elm fungus *(Ceratocystis ulmi),* the displacement of native herbs by the purple loose-strife *(Lythrum salicaria)*

[1]Research fellow and junior scientist, respectively, Biology Department, University of Minnesota-Duluth, Duluth, MN 55803.

[2]Research entomologist and life scientist, respectively, USEPA Environmental Research Laboratory, Duluth, MN 55804.

across many areas of the East, and the extensive defoliation of forests by the gypsy moth *(Lymantria dispar)*.

Simberloff [*1*], in a summary of 854 introductions reported in the literature, concluded that in 678 cases there was no observed effect on the ecosystem. While other authors have disputed this interpretation and insisted that the frequency of disruption was actually much higher [*2*], it is clear that introduced species do not always persist or disrupt an ecosystem. As Herbold and Moyle [*2*] note, many unsuccessful introductions are never reported in the literature. The probability of survival may be particularly low for an exotic microorganism. There are many reports (for example, Refs *3–5*) of rapid declines in populations of introduced microbes. In many cases they quickly disappeared while in others low densities persisted for short periods of time. Nonetheless, the release of GEMs into the environment raises a legitimate concern, especially since some of these organisms have been developed for their ability to survive in the particular environment in which they are to be used. For example, the first genetically engineered organism to be patented was a *Pseudomonas* developed to survive in oil spills and degrade crude petroleum [*6*].

It is not yet possible to accurately predict the survival or ecological effects of a microorganism released into the environment. Consequently, each GEM developed for release to the environment should be thoroughly evaluated to assess its ability to survive and its potential for causing adverse ecological effects.

A tiered testing approach is currently employed, using single species laboratory exposures to estimate acute and chronic or life-cycle effects on several invertebrate and fish species. Although these tests can provide information on direct toxicity or pathogenicity of a microorganism toward a given species, they provide no information on the critical questions of survival and ecological effects of the microorganism. Single species laboratory tests are not useful for understanding the complex interactions which occur among species in a natural environment [*7*]. As a number of authors have noted, alternative tests are needed which can measure effects at the community and ecosystem level [*8–10*].

The need for environmental data presents a dilemma, since an experiment in outdoor systems could result in the escape of the microorganism. Laboratory microcosms containing self-sustaining populations of microorganisms, plants, and invertebrate species are a rational alternative to outdoor tests. Such systems have been successfully used for testing a variety of chemicals [*11–17*].

While microcosms have been used in many shapes and sizes, not all are well suited for testing microorganisms. The ideal system should contain functioning populations of producers, consumers, and decomposers, yet be small enough to permit the use of a large number of replicates to insure adequate statistical power. A system allowing some flexibility in community structure is desirable in that it could be used either as a generic or site-specific test. The method should allow for measurement of both structural and functional characteristics of the ecosystem through simple cost-effective procedures. Microcosms must have containment features which prevent the accidental release of test microorganisms. Finally, the systems should have acceptable levels of precision and accuracy.

We have developed a system based on these criteria for testing the survival and effects of microorganisms in aquatic environments. Our microcosm protocol [*18*] is a modification of the procedure originally described by Leffler [*19*]. This procedure was used successfully in chemical hazard assessment and appeared to have considerable potential for evaluating risk presented by microorganisms. The system is currently constructed to serve as a generic surrogate of a small eutrophic pond. The protocol is flexible, however, and permits the development of site-specific microcosms. Their simplicity and small size permits up to 40 microcosms to be maintained and monitored in a small, contained growth chamber. This allows a large number of replicates and enhances the statistical power of the test.

In testing this system, our objectives were to determine how well it could assess the fate and effects of an introduced microorganism. The test organism was *Bacillus thuringiensis* var. *israelensis* (Bti), a registered microbial pest control agent (MPCA) widely used for mosquito and blackfly control. Bti is a naturally occurring soil microorganism which produces a proteinaceous parasporal crystal during sporulation. This crystal has been shown to be toxic to mosquitoes (Culicidae), blackflies (Simuliidae), and some midges (Chironomidae). Other organisms appear to be unaffected.

Bacillus thuringiensis varieties are of particular interest because their genes for toxin production are now being inserted into a variety of other bacteria and higher plants. The species has been well studied, resulting in a considerable body of literature on its effects on both "target" (host) and nontarget organisms. It is one of the few bacterial species whose survival and effects have been studied in the field. This provided an opportunity to determine how well this microcosm protocol was able to monitor the survival and detect effects of an introduced microorganism, and to evaluate how closely microcosm responses correlated with reported field results.

Methods

The Microcosm System

The microcosm protocol described by Shannon et al. [*18*] was used in these experiments. These microcosms are biological models of small autotrophic ponds, containing a variety of bacterial and fungal decomposers, protozoa, algal primary producers, and herbivorous zooplankton and insects. This community performs the ecological functions necessary for a self-sustaining system.

Microcosms were constructed in 1-L beakers covered with a large petri dish to prevent the accidental contamination of control systems with the test organism. The lip on the beaker allowed the free exchange of gases with the atmosphere. Each microcosm consisted of 50 mL of acid washed sand sediment and 900 mL of Taub No. 82 medium [*20*], into which 50 mL of inoculum was introduced. The inoculum was taken from a mature stock community containing cladocerans, copepods, ostracods, amphipods, chironomids, species of unicellular and filamentous algae, and an assortment of protozoans, fungi, and bacteria.

A test consisted of four treatment groups (a control and three manipulated groups), each composed of five replicate microcosms. Microcosm vessels were incubated at 20°C in a Percival growth chamber on a 12-h light-dark cycle. To distribute any effect of shelf position over all treatment groups, the microcosm beakers were arranged in five randomized blocks (by replicate) that were moved to a new shelf location twice weekly. To simulate naturally occurring immigration and permit the reintroduction of extirpated species, all microcosms were reinoculated each week with 10 mL of the stock community.

Measurement Regime—Microcosms were allowed to mature for six weeks prior to treatment with Bti and followed for six to twelve weeks after exposure. During the exposure period, monitoring was done on a schedule of decreasing frequency. Measurements were taken twice weekly for the first two weeks after dosing, and then weekly for the next four weeks. Some microcosm tests were allowed to run beyond the standard six week exposure and were monitored a final time after twelve weeks.

Variables Measured—Both functional and structural (population) variables were measured to determine effects on the system. Structural measurements provided an indication of the effects on individual taxa, while functional variables gave an integrated measure of overall effect on the system. The variables measured are listed in Table 1 and described below.

Functional Variables—Dissolved oxygen (DO) and pH were used as indicators of effect at the ecosystem level. Both are easily measured quantities which reflect trophic interactions within the biogeochemical matrix of the ecosystem [*21*]. By measuring oxygen at three times, just before lights on (predawn), just before lights out (afternoon), and just before lights on the following morning (second predawn), calculations of diel oxygen gain and loss could be calcu-

TABLE 1—*Functional and structural variables measured in the aquatic microcosm protocol.*

- A. Functional Variables
  - 1. Measured Variables
    - Predawn pH
    - Afternoon pH
    - Predawn dissolved oxygen (on two successive days)
    - Afternoon dissolved oxygen
  - 2. Computed Variables
    - Primary production (oxygen gain)
    - Community respiration (oxygen loss)
    - Production/respiration ratio
- B. Structural Variables
  - 1. Measured Variables
    - Bacteria
      - Total bacterial density
    - Protozoa
      - Ciliate protozoan density
      - Flagellate protozoan density
      - Amoeboid protozoan density
    - Rotifers
      - Total rotifer density
    - Algae
      - Density of each genus
    - Zooplankton
      - Cladoceran density
      - Copepod density
      - Ostracod density
    - Macroinvertebrates
      - Amphipod density
      - Chironomid density
  - 2. Computed Variables
      - Total protozoan density
      - Single-celled algae—biovolume
      - Filamentous algae—biovolume
      - Nitrogen-fixing blue green algae—biovolume
      - Nonnitrogen fixing blue green algae—biovolume
      - Total algae—biovolume
      - Total zooplankton density

lated. Oxygen gain is a measure of net primary production, while oxygen loss is a measure of community respiration. From estimates of production and respiration, an oxygen gain/oxygen loss (analogous to a production/respiration) ratio was computed as a further indicator of the status of the system. Measurements of pH were taken at the time of the predawn and afternoon dissolved oxygen measurements.

Dissolved oxygen concentrations were measured with a Beckman meter and polarographic oxygen electrodes. Separate electrodes were used for control and treated groups to avoid contamination. Measurement of pH was done with a Fisher Model 52 pH meter equipped with four glass combination electrodes and a junction box for electrode selection. Separate electrodes were used for each treatment group.

Structural Variables—Populations of major taxa were determined by the procedures outlined below.

1. *Bacteria.* Bacterial density was determined by counting the number of colony-forming units (CFU) appearing on standard pour plates. A large bore pipette was used to collect 13-mL combined sediment-water samples from the microcosms. These samples were vigorously shaken in sterile water. Appropriate dilutions were made and plated on tryptic soy agar (TSA). Plates were incubated at 30°C for 24 to 48 h, after which colonies were counted.

This method was used to monitor both Bti and other bacteria. The spores of Bti are highly heat resistant, so they could be separated from vegetative bacteria by pasteurizing samples at 80°C for 3 min prior to diluting and plating. Colonies showing typical Bti colony color and morphology (that is, cream color, irregular shape, rough edge) were counted. Since colonies of this type rarely appeared in pasteurized samples from untreated microcosms, it was assumed that all such colonies were Bti. As a check on this assumption, bacteria from typical Bti colonies were regularly tested for their ability to kill mosquito larvae *(Aedes aegypti)*. Colony counts of plates from pasteurized aliquots were compared to counts from nonpasteurized samples to determine the extent of Bti spore germination in the microcosms.

2. *Algae.* Samples were withdrawn from microcosms with a Pasteur pipette and transferred to a Palmer-Maloney cell. Algal genera were counted at X400 magnification using the method of Wetzel and Likens [*22*]. The biovolume of each taxon was estimated according to the procedures of Wetzel and Likens [*22*]. Individual taxa were categorized into single-celled algae, colonial algae, clumped algae, filamentous algae, and periphyton. These categories were chosen to reflect the availability of taxa to planktonic grazers.

3. *Zooplankton, amphipods, and larval insects.* Amphipods were counted as the total number per microcosm. Zooplankton and insects were usually too abundant for total counts and must be subsampled by removing 50-mL aliquots. The organisms in each subsample were grouped into major categories (Cladocera, Copepoda, Ostracoda, Chironomidae, Culicidae) and counted. Aliquots were returned to the system. Sterile subsampling devices were used to avoid contamination. These materials were autoclaved after use to prevent contamination of the lab.

4. *Protozoa and rotifers.* Protozoa and rotifers were counted by withdrawing 0.5 mL of stirred sample from each microcosm and distributing it among 5 to 8 droplets on a glass plate. Each droplet was examined under a X40 dissecting microscope. The total number of rotifers and protozoa were counted. Protozoan counts were categorized into flagellated, ciliated, and amoeboid groups.

The Test Microorganisms

Three different preparations of Bti spores were tested in three microcosm experiments. Two of these, Vectobac (Abbott Laboratories lot No. 52-009 BR) and Mosquito Attack (Reuters Laboratories), were commercial products formulated as wetable powders. The third Bti preparation was a pure culture of washed spores grown in our laboratory from the Mosquito Attack.

Bioassays of larval *Aedes aegypti* were used to evaluate the potency of Bti preparations. Five second-instar larvae were transferred to 50-mL beakers containing a range of Bti dilutions. Mortality data were recorded at 15-min intervals for the first 2 h and at 24 h.

Bioassays were also performed on samples periodically withdrawn from microcosms to monitor the changes in the toxicity of Bti. Microcosms were stirred and 4 mL of mixed sample withdrawn. Samples from each replicate of a treatment group were pooled into a single 20-mL aliquot. Five second-to-third-instar *Aedes aegypti* larvae were added to each beaker and mortality recorded.

Experimental Design

We conducted three experiments to evaluate the ability of the microcosm protocol to measure survival and ecological effects of introduced microorganisms. In the first experiment Vectobac was added to three microcosm groups at densities of 10^3, 10^4, and 10^6 spores/mL. This experiment was performed to determine whether inoculation density influenced survival of Bti or its effects on the microcosm community.

The second and third experiments tested whether the presence of a target organism in the microcosm would affect the survival and growth of introduced Bti or alter its effect on the system. While the microcosms in the first experiment contained chironomids of the genus *Tanytarsus,* these organisms were only moderately susceptible to Bti. To insure that a suitable, highly susceptible host was present, 40 second-instar mosquito larvae *(Aedes aegypti)* were added to control and treatment microcosms in the second and third experiments. The presence of a target organism served as a positive control, that is, a check on the toxicity of Bti under microcosm conditions.

Four treatment groups were used in the second and third experiments: a group with mosquito larvae and Bti, a group with mosquito larvae and no Bti, a group with heat-killed mosquito larvae and no Bti, and a group with no mosquito larvae and no Bti. Larvae were added to the systems at the time of dosing. The Mosquito Attack Bti formulation was used in Experiment 2 and the laboratory Bti preparation was used in Experiment 3.

All Bti exposures in Experiments 2 and 3 were at 10^4 spores/mL. This density was selected on the basis of a bioassay of Mosquito Attack against *Aedes aegypti.* It was the minimum Bti density that was 100% effective in killing mosquito larvae. It was also sufficiently low to allow enumeration of bacteria other than Bti on pour plates. At the 10^6 spores/mL test concentrations used in the first experiment, Bti outnumbered the native bacterial community by a factor of 1000. This difference made it impossible to determine the densities of non-Bti microorganisms by the plate count method. When samples were sufficiently diluted to permit plate counts of Bti, the native microbial community had been diluted to numbers below detectable levels.

Statistical Procedures

Treatment groups were compared for differences in every measured variable using Dunnett's procedure [*24*]. All counted variables (that is, population densities) were transformed to logarithms prior to the analysis. The results of the analysis were plotted on figures indicating the temporal variation of each treatment group in terms of either the mean or the deviation from the control. A shaded area around the control in such figures indicates Dunnett's minimum significant difference at $p \leq 0.05$. Points lying outside the shaded area are significantly different from the control.

In some cases, comparisons of treatments and controls were done using a t-test at a significance level of $p \leq 0.05$.

The sensitivity of the measured variables was determined by calculating a minimum detectable difference using the method given by Snedecor and Cochran [*23*]. According to this method, the detectable difference (*dd*) is

$$dd^2 = \frac{(10.5)\, s^2}{n}$$

where n is the sample size, s^2 is the variance, and 10.5 is based on a paired, two-tailed design with a power of 90% and a significance level of $p \leq 0.05$.

Results

Survival of Bti in Microcosms

The pasteurization/pour plate method worked well for monitoring Bti density. Bti colonies were clearly distinguishable on the plates and were confirmed through bioassays with larval *Aedes aegypti*. Occasionally non-Bti colonies were present on plates from pasteurized samples, but they were easily distinguished from Bti on the basis of color and morphology. In fact, Bti colonies were sufficiently distinctive to be recognized and counted on plates from nonpasteurized samples. This differentiation was possible because there were relatively few naturally occurring *Bacillus* sp. in these microcosm systems.

In all experiments, Bti persisted in test microcosms at densities near the initial inoculation density. There was no significant increase or decrease in spore concentrations. This was most clearly shown in Experiment 1 where microcosms treated with 10^3, 10^4, and 10^6 spores per mL maintained those densities throughout the experiment (Fig. 1). This lack of population growth was confirmed in Experiments 2 and 3 by comparison of Bti densities on pasteurized and unpasteurized plates. There was no significant difference in Bti density between pasteurized and nonpasteurized samples, indicating that little germination had occurred and most of the Bti remained in the spore stage.

Persistence of Toxicity—In Experiment 1 we observed an apparent loss of Bti toxicity. Within a week after Vectobac was added to the microcosms, populations of the chironomid *Tanytarsus* sp. showed major declines at all test concentrations. However, with continued reinoculation, the populations quickly increased, and numbers returned to control levels within two weeks (Fig. 2). It appeared that Bti, while still present, was no longer toxic.

We used bioassays in Experiments 2 and 3 to systematically test the persistence of Bti toxicity. Both Mosquito Attack (Experiment 2) and unformulated Bti (Experiment 3) showed a rapid decrease in toxicity after application to microcosms. Mosquito Attack toxicity persisted for at least seven days following inoculation but did not last more than 14 days. Test mosquito larvae died within 48 h when placed into samples withdrawn from microcosms treated with Bti, while no mortality occurred in microcosm samples from any other treatment group. The unformulated Bti showed a similar loss of toxicity, although, because the mosquito colony failed, bioassays could be performed only on Days 3 and 28 after treatment. Larvae died when exposed to the Day 3 samples from Bti-treated microcosms, but survived in the Day 28 samples.

Ecological Effects

Target organisms (mosquito larvae) were killed within 24 to 48 h and related dipteran species (*Tanytarsus* sp. larvae) were killed within one week, but no other species were lost from the system. No other direct toxic effects were seen. Some variables were affected indirectly through the removal of grazers from the system and the addition of their carcasses to the detritus. These effects, which were similar in all experiments, are summarized in Table 2.

Effects were most pronounced on functional variables. Typical of these were shifts in oxygen gain (primary production) and oxygen loss (community respiration). In each of the experi-

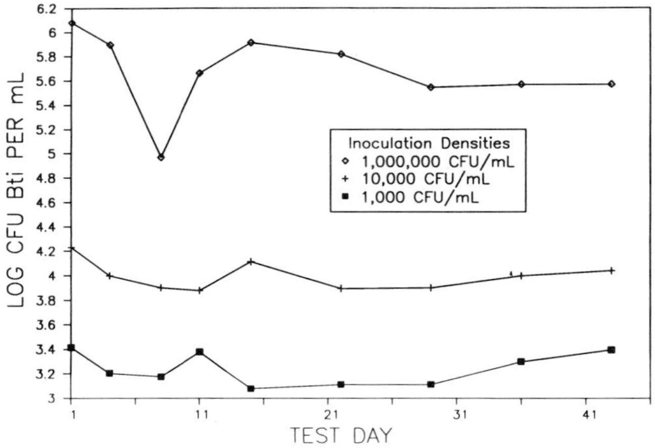

FIG. 1—*Survival of Bti spores (Experiment 1—Vectobac) at different inoculation densities.*

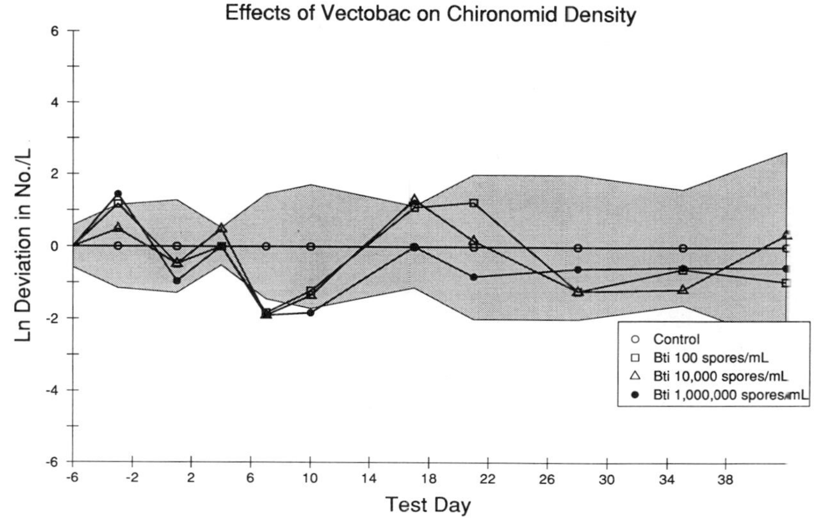

FIG. 2—*Effects of different densities of Bti (Experiment 1—Vectobac) on chironomid (*Tanytarsus *sp.) abundance. Shaded area represents Dunnett's significant difference interval about the control at* P $\leq$ *0.05.*

TABLE 2—*Summary of microcosm responses to Bti at* 10^4 *spores/mL.*

1. Target mosquito species *Aedes aegyptii* killed.
2. Nontarget chironomid *Tanytarsus* sp. killed.
3. Increase in the indigenous bacterial community.
4. Increase in primary production (oxygen gain).
5. Increase in community respiration (oxygen loss).
6. Changes in algal community structure.

ments, there was an increase in both primary production and respiration following Bti exposure. The Mosquito Attack test data were used to illustrate these patterns of effect (Fig. 3). Other functional variables (pH, production/respiration ratio) showed similar responses.

The increases in primary production suggest an increase in algal densities. Indeed, algal counts were higher in Bti-treated systems, but because of their variability, they were not statistically different from controls. The increase in primary production may have resulted from the removal of grazing mosquito and chironomid larvae, or the addition of nutrients contained in the Bti preparations or mosquito carcasses.

The initial increases in respiration (within the first week) were at least partially a reflection of increased microbial activity. In Experiments 2 and 3, we measured the indigenous bacterial community and found a significant population increase after Bti killed mosquito larvae in the systems (Fig. 4). The brief increase in bacterial density may have been in response to mosquito carcasses, or may have indicated the biodegradation of materials in the Bti preparations. Later increases in respiratory activity were probably related to increases of total biomass in the treated microcosms.

Other variables showed few statistically significant responses to Bti exposure. Protozoan populations increased in all microcosms, but none of the Bti-treated groups showed consistent differences from controls. Cladocerans, copepods, ostracods, and amphipods all followed cyclic patterns of abundance, but rarely differed significantly from control systems.

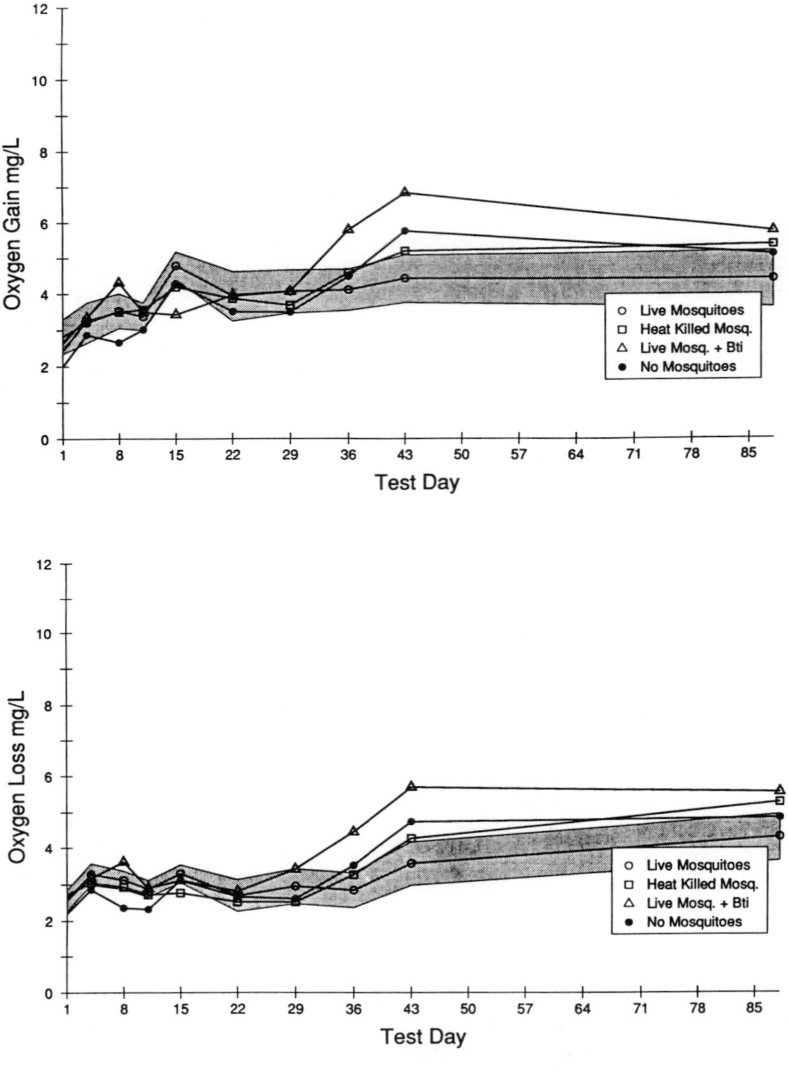

FIG. 3—(top) *Effects of Bti (Experiment 2—Mosquito Attack at 10^4 spores/mL) on oxygen gain. Comparisons are relative to controls with live mosquito larvae added. Shaded area represents Dunnett's significant difference interval about the control at* $P \leq 0.05$. (bottom) *Effects of Bti (Experiment 2—Mosquito Attack at 10^4 spores/mL) on oxygen loss. Comparisons are relative to controls with live mosquito larvae added. Shaded area represents Dunnett's significant difference interval about the control at* $P \leq 0.05$.

Performance of the Protocol

Even though microcosm species composition differed among experiments, especially within the algal community, control microcosms behaved similarly in each of the three tests. For most of the test period, there was no significant difference ($p \leq 0.05$) among experiments with respect to control populations of total algae or zooplankton (Fig. 5). Afternoon dissolved oxygen concentrations (a measure of primary production) in the second experiment averaged about 20% lower than in the first and third experiments, however. Microcosms responded in a consis-

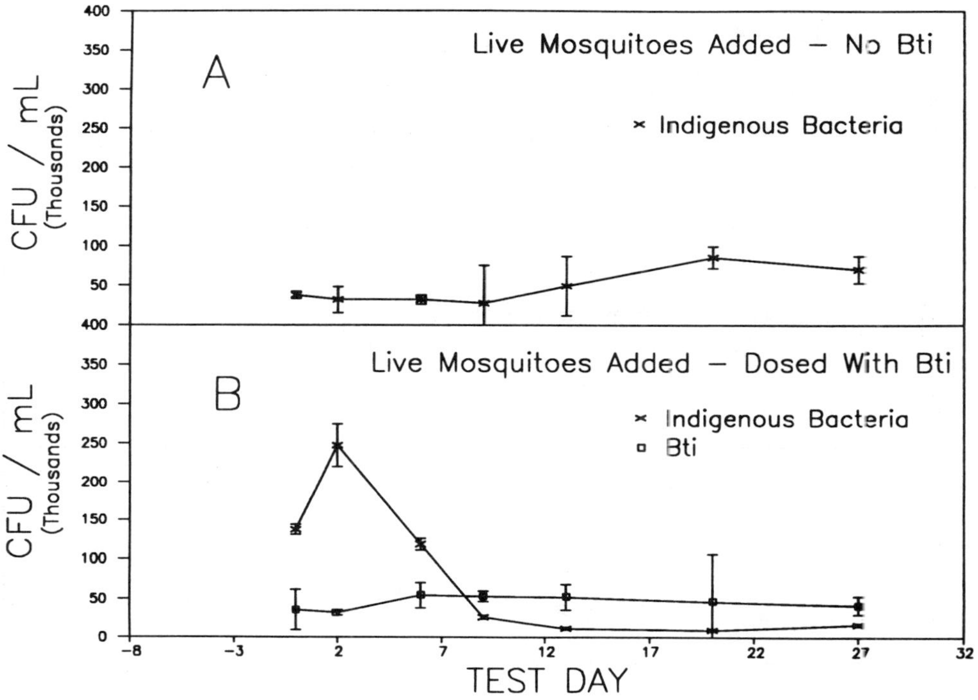

FIG. 4—*Changes in the bacterial community as a result of Bti (Experiment 2—Mosquito Attack at* 10^4 *spores/mL) addition to microcosms. Error bars represent 95% confidence intervals.*

tent and predictable manner to the Bti preparations, that is, the same variables were affected in each of the three runs of the procedure and the magnitude of response was similar. Three different Bti preparations were tested, so identical microcosm responses would not necessarily be expected.

Structural and functional measures differed widely in their sensitivity in a manner related to their background variability. As summarized in Table 3, functional variables were by far the most sensitive with average minimum detectable differences for all tests ranging from 3 to 19%. Structural variables were considerably less sensitive. For the broadest taxonomic groupings (where the number of individuals was large and the sampling variability low) detectable differences were approximately 50%. Detectable differences for rare taxa were much higher, ranging well over 100%.

Addition of Mosquitoes to Test Communities—Mosquito larvae survived well in control microcosms (Fig. 6), although pupation success was poor in Experiment 2 (Mosquito Attack). This appeared to be related to food quality since most of the mosquitoes in the control groups survived and pupated in Experiment 3. An analysis of the algal community structure in these two experiments indicated important differences (Fig. 7). In Experiment 2 the algal community was dominated by colonial bluegreen species (primarily *Gloeocapsa* sp.), while in Experiment 3 filamentous bluegreen species (chiefly *Oscillatoria* sp.) predominated. Mosquitoes may have been unable to develop properly when restricted to a diet consisting primarily of colonial bluegreen taxa, although further work is needed to confirm this.

The differences in algal community structure in Fig. 7 indicate the effects of two weeks of larval mosquito grazing in the microcosms. It was apparent the *Aedes* larvae utilized filamen-

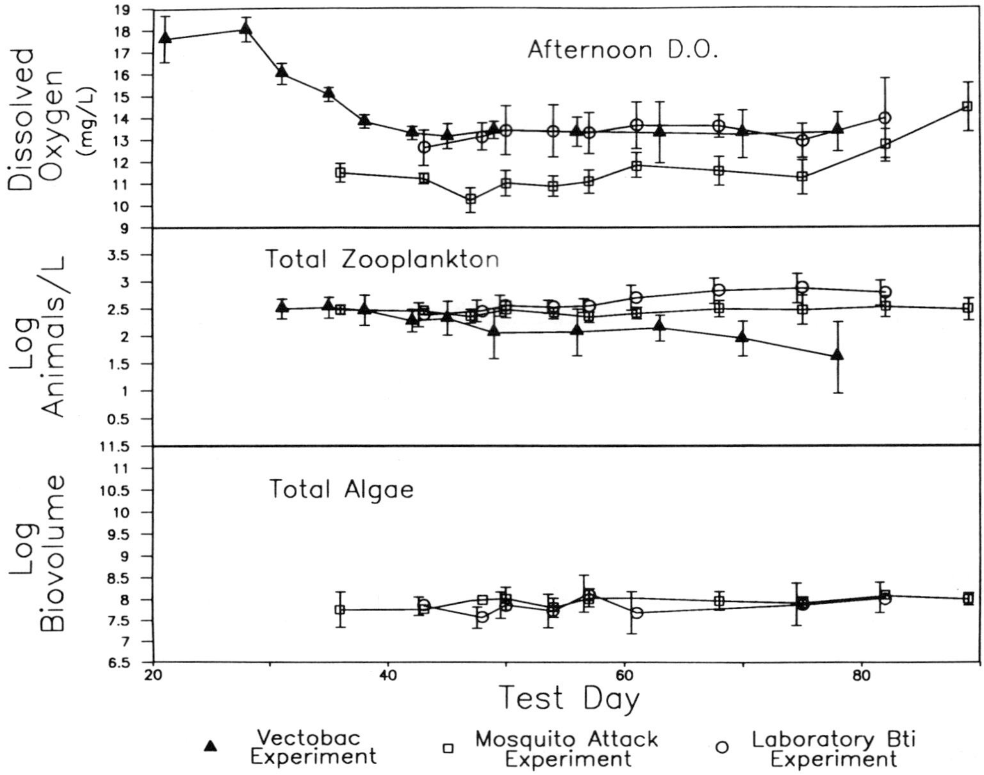

FIG. 5—*Development of control microcosms from Experiments 1, 2, and 3.*

TABLE 3—*Average minimum detectable differences (*p ≤ *0.05 and 90% power) for microcosm variables measured in tests of* Bacillus thuringiensis *var.* israelensis *preparations.*

	Minimum Detectable Difference, percent
FUNCTIONAL VARIABLES	
Predawn pH	3.45
Afternoon pH	2.06
Predawn dissolved oxygen	5.59
Afternoon dissolved oxygen	6.75
Primary production (oxygen gain)	18.97
Community respiration (oxygen loss)	18.09
Production/respiration ratio	11.13
STRUCTURAL VARIABLES	
Total zooplankton density	55.26
Cladoceran density	71.11
Copepod density	68.79
Amphipod density	64.99
Single-celled algae	61.26
Total protozoa	130.31

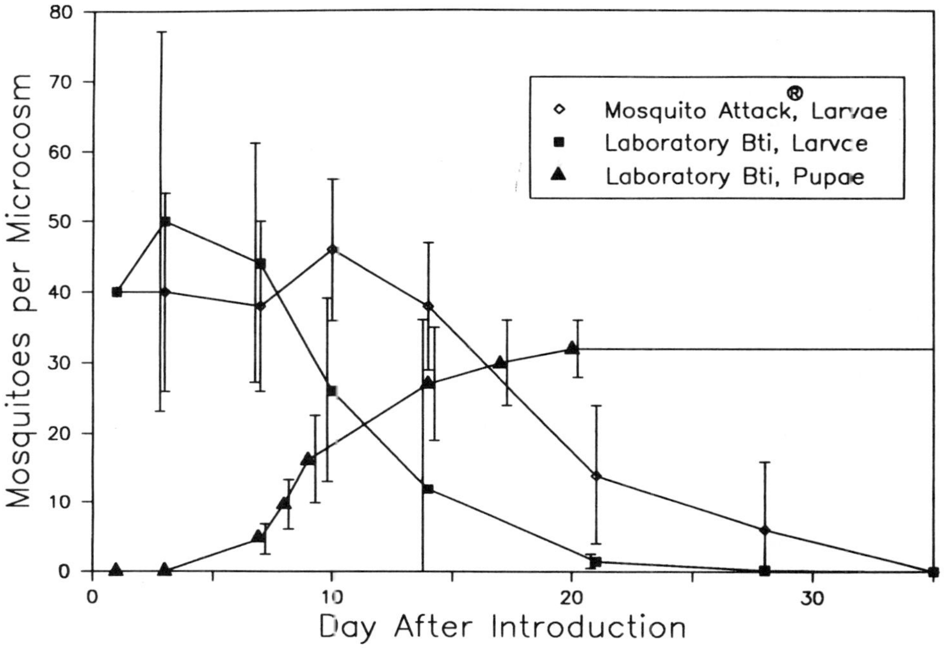

FIG. 6—*Larval and pupal mosquito density in control microcosms of Experiments 2 (Mosquito Attack) and 3 (Laboratory Bti). Numbers of pupae in Experiment 2 were too low to present.*

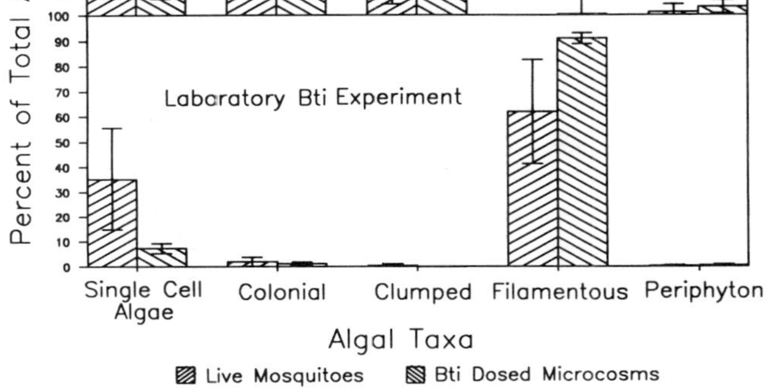

FIG. 7—*Comparison of the algal community structure for Experiments 2 (Mosquito Attack) and 3 (Laboratory Bti). Both control and Bti treated systems are shown.*

tous species in Experiment 3. The abundance of filamentous species declined rapidly in control systems containing mosquito larvae relative to microcosms without larvae. Single-celled algae, however, showed an increase in response to the presence of mosquitoes. It appears that mosquitoes were effectively recycling nutrients from filamentous algal species into smaller species less susceptible to grazing pressure.

Discussion

The release of microbes into the environment raises important risk assessment concerns. In contrast to chemical risk assessment, where the primary concerns involve the effect, fate, and transport of introduced materials, microbial risk assessment must also consider a microbe's ability to survive, multiply, colonize new habitats, and transfer genetic information. The mass balance measurements so often used in chemical testing are generally inappropriate or inadequate for describing the effects of microorganisms. Effects are likely to depend more on the ability of a microorganism to survive and multiply than on initial release densities, although there may be some baseline density necessary for survival. The possibility of gene transfer from an introduced species to indigenous species further complicates the hazard assessment process. Alexander [*25*] has grouped these concerns into five categories: the ability to survive, the potential to increase in abundance, the potential for causing deleterious effects, the ability to transfer deleterious genetic information, and the potential for transport or dissemination. All of these need to be considered in making informed judgments on the risk posed by a given microbe. With the exception of the dissemination question, these concerns can be appropriately addressed in a microcosm system.

The probability of survival for a microorganism depends in part on its life-style. Microbes may operate as either free-living organisms, opportunistic pathogens, symbionts, or obligate pathogens. Any of these groups may survive in a given ecosystem, although the probability of survival decreases as requirements for growth become more fastidious. That is, a free-living organism would be much more likely to survive than an obligate pathogen dependent on the presence of a specific host. Yet any introduced organism *may* survive in a given system. Designing a microcosm procedure to adequately address the survival question for each of these varying life-styles presents a unique challenge.

The ability of any introduced organism to survive depends on its ability to exploit a niche in the environment into which it is released. An introduction may fail for a number of reasons, including the absence of a suitable substrate or host, competitive exclusion, abiotic limitation, and predation or direct toxic action (for example, antibiotic production) by another organism. Test systems, in order to provide a reasonable assessment of an organism's survival potential, must contain a habitat suitable for the microorganism being tested. For free-living organisms, the availability of a complex pool of organic matter increases the probability of an organism encountering a suitable substrate. For obligate pathogens the inclusion of a suitable host is a necessity.

In designing a microcosm system it is neither possible nor desirable to develop a system where all checks on the growth of an invading organism are removed. The aim is rather to develop a realistic environmental surrogate which is sufficiently complex to offer a variety of microbial niches. Controlled temperature and lighting conditions are used to minimize the potential for abiotic limitation through stochastic disturbances, although sampling itself obviously disturbs the systems. This disruption is unavoidable, but it is at least imposed uniformly on all microcosms. Tests with Bti indicated that these systems can measure both structural and functional ecological effects caused by introduced microorganisms. The sensitivity of the response variables was sufficiently high to detect effects. This series of experiments also showed that repeated experiments develop similar levels of producer and consumer populations and respond to similar perturbations in a similar manner.

The second and third experiments included a control containing mosquito larvae. This group served as a positive control to demonstrate that Bti is an effective larvicide under microcosm conditions. It is important to demonstrate this in testing any MPCA; otherwise a finding of "no effect" applies only to an environment lacking target organisms. Many MPCAs (Bti included) do not grow or reproduce well outside a host organism and might remain in an inactive state in the absence of a host. Mosquitoes survived in all the systems and appear to be a logical addition for tests where mosquito control agents are evaluated. The addition of mosquitoes, or any host organism, may require manipulation of the microcosm's community structure to insure the presence of an appropriate and adequate food supply.

Aside from its obvious effects on target and closely related species, Bti appears to present little risk to aquatic ecosystems of the type modeled by these microcosms (that is, small eutrophic ponds). No other species were lost from the community as a result of Bti exposure. These responses are in agreement with literature reports of the host specificity of Bti [*26,27*]. The minor ecological changes observed were indirect responses to the loss of a large grazing organism, and the decomposition and recycling of nutrients from dead target carcasses. There may also have been some direct stimulation of the algal community through the addition of nutrients associated with the Bti formulations.

Although Bti spores persisted in the microcosms, they quickly lost their toxicity. Similar results have been reported from field tests of Bti. This rapid reduction in toxicity has been attributed to inactivation on exposure to ultraviolet light [*28*] and adsorption onto soil particles [*29,30*]. Ohana et al. [*31*] found larvicidal activity decreased rapidly when Bti was adsorbed onto sediments but the process was reversible. Vigorous shaking of the sediments released the spore/crystal and renewed activity. No evidence of reversibility was found in the present study. Bti toxicity may not always decrease with time. Dupont and Boisvert [*32*] have suggested that low water temperature may prolong toxicity and even stimulate spore germination.

The microcosms used in these experiments contained a diverse biotic community and a pool of organic matter sufficient to support a variety of microbial species. Rates of decomposition and nutrient cycling were sufficient to support a self-sustaining system. Species richness in the microbial community could probably be increased through the addition of a more complex sediment, however. The sand sediment used in the current protocol remains highly aerobic and supports a bacterial community whose primary functions are oxidative. In preliminary experiments with finer sediments, we have developed anaerobic layers within the microcosm sediment and found corresponding increases in anaerobic microorganisms (for example, sulfur-reducing bacteria). Clearly further work needs to be done in evaluating the role of sediment type in determining the survival and effects of an introduced microbe.

Conclusions

Small microcosms were used in testing the survival and effects of an MPCA. Bti spores released in the microcosms survived but did not increase in abundance. The inclusion of mosquito larvae in the microcosm community served as a positive control and confirmed that Bti was active against target organisms under microcosm conditions.

Bti caused statistically significant but minor changes in the structure and function of micro-ecosystems. Functional measures of primary production and respiration were most affected while populations of plants and animals, except for the mosquito larvae and the chironomid *Tanytarsus*, were unaffected. There was a brief increase in the density of indigenous microbial flora, presumably in response to an increase in available nutrients.

The data from these three experiments show that microbes introduced in spore form can be successfully monitored. Since spores are inactive, data from this experiment are not necessarily applicable to the survival and effects of microorganisms in vegetative form. With modifications in the methods for monitoring microorganisms, the procedure should be widely adaptable to testing any introduced microorganism. While the protocol produced important information on ecological effects, additional modifications could provide valuable enhancements to the procedure. Of particular interest are: (1) procedures to monitor the distribution of a microbe within trophic levels (that is, a more precise measure of the fate of the organism); (2) procedures to more closely monitor populations of indigenous microbial species; and (3) investigation of the effect of sediment type on the behavior of introduced microorganisms.

Acknowledgments

We gratefully acknowledge the valuable laboratory assistance provided by Eric Mead. Laboratory space, supplies, and equipment were largely supplied by the U.S. Environmental Protection Agency. Support of non-EPA personnel was funded through EPA cooperative agreement CR812799-02.

References

[*1*] Simberloff, D. S., testimony before U.S. Senate Subcommittee on Toxic Substances and Environmental Oversight of the Senate Committee on Environment and Public Works, 27 Sept. 1984.

[*2*] Herbold, B. and Moyle, P. B., "Introduced Species and Vacant Niches," *American Naturalist,* Vol. 128, No. 5, 1984, pp. 751–760.

[*3*] McFeters, G. A., Bissonette, G. K., Jegeski, J. J., Thompson, C. K., and Stuart, D. G., "Comparative Survival of Indicator Bacteria and Enteric Pathogens in Well Water," *Applied Microbiology,* Vol. 42, 1974, pp. 823–829.

[*4*] Chandler, D. S. and Craven, J. A., "Relationship of Soil Moisture to Survival of *Escherichia coli* and *Salmonella typhimurium* in Soils," *Australian Journal of Agricultural Research,* Vol. 31, 1980, pp. 547–555.

[*5*] Sinclair, J. L. and Alexander, M., "Role of Resistance to Starvation in Bacterial Survival in Sewage and Lake Water," *Applied and Environmental Microbiology,* Vol. 48, No. 2, 1984, pp. 410–415.

[*6*] Chakrabarty, A. M., "Microorganisms Having Multiple Compatible Degradative Energy-Generating Plasmids and Preparation Thereof," U.S. Patent 4,259,444, 31 Mar. 1981, Cl. 435-172.000 (filed 6 June 1972), patent to General Electric Co., NY, 1981.

[*7*] Cairns, J., "The Case for Simultaneous Toxicity Testing at Different Levels of Biological Organization," *Aquatic Toxicology and Hazard Assessment: Sixth Symposium, ASTM STP 802,* W. E. Bishop, R. D. Cardwell and B. B. Heidolph, Eds., American Society for Testing and Materials, Philadelphia, 1983, pp. 111–127.

[*8*] Gillett, J. W., Levin, S. A., Harwell, M. A., Alexander, M., Androw, D. A., and Stern, A. M., *Potential Impacts of Environmental Release of Biotechnology Products: Assessment, Regulation, and Research Needs,* ERC-075, Ecosystems Research Center, Ithaca, NY, 1984.

[*9*] Levin, S. A. and Kimball, K. D., "New Perspectives in Ecotoxicology," *Environmental Management,* Vol. 8, 1984, pp. 375–442.

[*10*] Odum, E. P., "The Mesocosm," *Bioscience,* Vol. 34, 1984, pp. 558–562.

[*11*] Bourquin, A. W., Garnas, R. L., Pritchard, P. H., Wilkes, F. G., Cripe, C. R., and Rubenstein, N. I., "Interpendent Microcosms for the Assessment of Pollutants in the Marine Environment," *International Journal for Environmental Studies,* Vol. 13, 1979, pp. 131–140.

[*12*] Hendrix, P. F., Langner, C. L., Odum, E. P., and Thomas, C. L., "Microcosms as Test Systems for the Ecological Effects of Toxic Substances: An Appraisal with Cadmium," EPA-600/S3-81-036, USEPA, Washington, DC, 1981.

[*13*] Van Voris, P., Arthur, M. F., and Tolle, D. A., "Terrestrial Soil Core Microcosm Test Protocol and Technical Support Document," draft document submitted to the Office of Toxic Substances, U.S. Environmental Protection Agency, Washington, DC, 1984.

[*14*] Giddings, J. M., "Microcosms for the Assessment of Chemical Effects on the Properties of Aquatic Ecosystems," *Hazard Assessment of Chemicals: Current Developments,* Vol. 2, 1983, pp. 45–94.

[*15*] Sheehan, P. J., Axler, R. P., and Newhook, R. C., "Evaluation of Simple Generic Aquatic Ecosystem Tests to Screen the Ecological Impacts of Pesticides," *Community Toxicity Testing, ASTM STP 920,* J. Cairns, Jr., Ed., ASTM, Philadelphia, 1986, pp. 158–178.

[*16*] Taub, F. B., Kindig, A. C., and Conquest, L. L., "Preliminary Results of Interlaboratory Testing on a Standardized Aquatic Microcosm Protocol," *Community Toxicity Testing, ASTM STP 920,* J. Cairns, Jr., Ed., ASTM, Philadelphia, 1986, pp. 158–178.

[*17*] Yount, J. D. and Shannon, L. J., "Effects of Aniline and Three Derivatives on Laboratory Microecosystems," *Environmental Toxicology and Chemistry,* Vol. 6, 1987, pp. 463–468.

[*18*] Shannon, L. J., Flum, T. E., and Yount, J. D., "Draft Protocol for a Mixed Flask Culture Aquatic

Microcosm Toxicity Test," draft document submitted to the Office of Toxic Substances, U.S. Environmental Protection Agency, Washington, DC, Output No. 7667A, 1987.

[*19*] Leffler, J. W., "The Use of Self-Selected, Generic Aquatic Microcosms for Pollution Effects Assessment," *Concepts in Marine Pollution Measurements,* Harris H. White, Ed., Maryland Sea Grant College, University of Maryland, College Park, MD, 1984.

[*20*] Taub, F. B. and Read, P. L., "Model Ecosystems: Design, Development, Construction and Testing, Vol. II: Standard Aquatic Microcosm Protocol," FDA Contract Report No. 223-80-2352, Food and Drug Administration, Washington, DC, 1982.

[*21*] Schindler, J. E. and Waide, J. B., "Theoretical Rationale," *Microcosms in Ecological Research,* J. P. Giesy, Ed., DOE symposium series 52, CONF-781101, 1980, pp. 192–203.

[*22*] Wetzel, R. G. and Likens, G. E., *Limnological Analyses,* W. B. Saunders Co., Philadelphia, 1974.

[*23*] Snedecor, G. W. and Cochran, W. G., *Statistical Methods,* Iowa State University Press, Ames, Iowa, 1976.

[*24*] Steel, R. G. D. and Torrie, J. H., *Principles and Procedures of Statistics,* McGraw-Hill, New York, NY, 1960.

[*25*] Alexander, M., testimony before U.S. Senate Sub-committee on Toxic Substances and Environmental Oversight of the Senate Committee on Environment and Public works, 26 Sept. 1984.

[*26*] Ali, A., "*Bacillus thuringiensis* serovar. *israelensis* (ABG-6108) against Chironomids and Some Nontarget Aquatic Invertebrates," *Journal of Invertebrate Pathology,* Vol. 38, 1981, pp. 264–272.

[*27*] Mulla, M. S., Fedreici, B. A., and Darwazeh, H. A., "Larvicidal Efficacy of *Bacillus thuringiensis* Serotype H-14 Against Stagnant-Water Mosquitoes and Its Effects on Nontarget Organisms," *Environmental Entomology,* Vol. 11, 1982, pp. 788–795.

[*28*] Mulligan, F. S., Schaefer, C. H., and Wilder, W. H., "Efficacy and Persistence of *Bacillus sphaericus* and *B. thuringiensis* H. 14 Against Mosquitoes Under Laboratory and Field Conditions," *Journal of Economic Entomology,* Vol. 73, 1980, pp. 684–688.

[*29*] Van Essen, F. W. and Hembree, S. C., "Simulated Field Studies with Four Formulations of *Bacillus thuringiensis* var. *israelensis* Against Mosquitoes: Residual Activity and Effect of Soil Constituents," *Mosquito News,* Vol. 42, 1982, pp. 66–72.

[*30*] Margalit, J. and Bobroglo, H., "The Effect of Organic Materials and Solids in Water on the Persistence of *Bacillus thuringiensis* var. *israelensis* serotype H-14.," *Zeitschrift für Angewandte Entomologie,* Vol. 97, 1984, pp. 516–520.

[*31*] Ohana, B., Margalit, J., and Barak, Z., "Fate of *Bacillus thuringiensis* Subsp. *israelensis* Under Simulated Field Conditions," *Applied and Environmental Microbiology,* Vol. 53, 1987, pp. 828–831.

[*32*] Dupont, C. and Boisvert, J., "Persistence of *Bacillus thuringiensis* serovar. *israelensis* Toxic Activity in the Environment and Interaction with Natural Substrates," *Water Air and Soil Pollution,* Vol. 29, 1986, pp. 425–438.

Quality Assurance in Ecotoxicity Testing

Alfred L. Korndoerfer, Jr.[1]

The New Jersey Toxicity Test Laboratory Certification Program

REFERENCE: Korndoerfer, A. L., Jr., **"The New Jersey Toxicity Test Laboratory Certification Program,"** *Aquatic Toxicology and Hazard Assessment: 12th Volume, ASTM STP 1027*, U. M. Cowgill and L. R. Williams, Eds., American Society for Testing and Materials, Philadelphia, 1989, pp. 243-250.

ABSTRACT: It is critical for the New Jersey Department of Environmental Protection to obtain scientifically valid and defensible data in order to successfully fulfill its environmental protection responsibilities. To that end, the Department developed and on 6 Aug. 1981 adopted "Regulations Governing Laboratory Certification and Standards of Performance," NJAC 7:18 et seq. Laboratories receiving certification in the bioassay category of the regulations must have complied with specific quality control criteria. This paper will discuss the regulations, our experiences in administering them, and the process of regulation development.

KEY WORDS: aquatic toxicology, acute toxicity, toxicity, aquatic biology, effluents, waste water, experimental design, sampling, environmental tests, performance standards, biological laboratories, quality assurance, quality control, LC_{50}, standardization, bioassay, standard operating procedures, tests, environmental regulations, water analysis, performance evaluation

To accomplish the legislative goals of the Federal Water Pollution Control Act[2] and the New Jersey Pollutant Discharge Elimination System Regulations,[3] the New Jersey Department of Environmental Protection (NJDEP), Division of Water Resources has since the early 1970s placed increasing importance upon the use of effluent toxicity bioassays for the identification and control of acutely toxic wastewater discharges. However, the Department recognized that in order to make valid environmental assessments it must have data of known quality; data of unknown quality cannot be used for setting permit limits, developing planning strategies, or demonstrating compliance with an existing standard. This paper will discuss the Department's efforts to manage the quality of toxicity bioassay data submitted to it for the aforementioned purposes.

Discussion

Background

In 1979, the Department conducted a quality assurance pilot study and determined that the quality of chemical data submitted to New Jersey under the National Pollutant Discharge Elimination System (NPDES) self-monitoring program was unacceptable.

Although the performance of toxicity bioassay laboratories was not included in this study, several years of experience with this program had demonstrated to the Department that, as with

[1]Acting section chief, Bureau of Monitoring Management, Division of Water Resources, New Jersey Department of Environmental Protection, Trenton, NJ 08625.
[2]As amended 1987, Section 101(a)3.
[3]New Jersey Administrative Code (NJAC) 7:14-A1.1, October 1984.

the chemical and microbiological analyses, a self-administered quality assurance program had not been reliable. Inspection and evaluation of toxicity bioassay laboratories conducting NPDES compliance monitoring revealed, as with the chemical laboratories, a substantial variability in minimum personnel qualifications, sample collection and handling procedures, laboratory facilities, testing methodology, data handling and reporting procedures, and laboratory quality control.

The Department realized that before it could have confidence in the toxicity bioassay or chemical monitoring data it was receiving from the permittees, it would have to ascertain if a permittee's laboratory was following sound scientific procedures when performing analyses, if it was operated by sound management and proficient supervisors, if it possessed and utilized proper equipment, and if it maintained and used accurate records [*1*]. There are two approaches to achieve a high level of comparability of testing data among laboratories [*2*]:

1. Develop a program in which the standards are tightly controlled, leaving little flexibility to the testing procedures.
2. Develop a performance-based "method equivalency" program. Such a program is dependent upon the development and routine use of standard toxicants and standard methods. Both standards must be sufficiently developed so that responses are well known and predictable.

In 1980 the Department chose the former path due both to a lack of well-established, predictable standard toxicant materials and to the lack of sufficiently controlled, predictable, and reproducible reference standard methods. Accordingly, the Department developed, and on 6 Aug. 1981 adopted, "Regulations Governing Laboratory Certification and Standards of Performance," NJAC 7:18 et seq. Those regulations were amended in July 1984 and currently are under review for additional amendments this year.

This discussion of the laboratory certification regulations will be confined to the bioassay laboratory certification program. The bioassay testing category for certification includes any acute toxicity test analyses required to determine compliance with the Water Pollution Control Act or the NJPDES regulations or both. The Bureau of Monitoring Management in the Division of Water Resources, in coordination with the Office of Quality Assurance, is responsible for managing the Department's bioassay laboratory certification program.

The Regulations

The New Jersey laboratory certification regulations (NJAC 7:18) are organized into a series of subchapters. These subchapters are summarized in Table 1. Subchapter 6 specifically covers the toxicity bioassay category of analyses. This paper will discuss Subchapter 6 and Subchapters 1 and 2 as they relate to toxicity bioassay analyses.

In Subchapter 1 the regulations incorporate, by reference, the "Guidelines Establishing Test Procedures for the Analysis of Pollutants," as amended, 40 CFR 136. To date, no toxicity bioassay test procedures have been included in this federal regulation.

This subchapter also specifies that all laboratories wishing to analyze water samples for compliance with regulations adopted or orders issued under the authority of the Water Pollution Control Act, NJSA 58:10A-1 et seq., or the Safe Drinking Act, NJSA 58:12A-1 et seq., shall follow the procedures established in the regulations in order to obtain and maintain certification.

The administrative requirements for the regulations are delineated in Subchapter 2, "Program Procedures and Requirements." All water analyses performed for compliance with the requirements of the New Jersey Pollutant Discharge Limitation System Regulations, NJAC 7:14 A-1, must be performed in laboratories certified for the specific analytical category. A laboratory may apply for one or more than one category.

To become certified a laboratory must submit a completed application along with the annual certification fee. The application is reviewed for completeness, especially as regards the labora-

TABLE 1—*Summarization of the New Jersey laboratory certification regulations.*

Subchapter	Content
1. "General Provisions"	Scope and authority State purpose of regulations Establish program requirements Pertinent definitions
2. "Program Procedures and Requirements"	Administrative procedures Establish certification categories and parameters Establish standards for laboratory personnel, facilities, and equipment Establish enforcement procedures Application procedures for in-state and out-of-state laboratories Establish fee schedule Establish proficiency testing requirements
3. "Criteria and Procedures for Microbiological Testing and Analysis"	Establish procedures and criteria for microbiological testing
4. "Criteria and Procedures for Chemical Testing and Analysis"	Establish procedures and criteria for limited chemistry, atomic absorption, and gas chromotography analyses
5. "Criteria and Procedures for Radiological Testing and Analysis"	Establish procedures and criteria for radiological testing
6. "Criteria and Procedures for Bioassay Testing and Analysis	Establish procedures and criteria for bioassay toxicity testing

tory supervisor's qualifications. The application will be rejected should the supervisor specified not meet the minimum qualifications.

Every certified laboratory, and laboratories seeking certification, must employ sufficient properly qualified personnel commensurate with the workload and types of tests being performed. Every laboratory must designate one individual to be the laboratory manager who is responsible for administering the operations of the lab. The laboratory must also have one or more supervisors who meet the education and experience qualifications summarized in Table 2. The laboratory manager may also serve as laboratory supervisor providing he or she meets the stated qualifications.

Once the application is accepted, the laboratory is placed in the "Applied" category and an on-site inspection for the evaluation of the laboratory is scheduled. During this inspection the laboratory must demonstrate compliance with the requirements of the regulations. Consideration is given to competence and attitude of staff; adequacy of facilities, equipment, and supplies; efficient organization of the laboratory; testing methods used; quality control procedures; and maintenance of all required records.

TABLE 2—*Bioassay laboratory supervisor qualifications.*

Education	Required Experience
B.S. or B.A. in biological science or chemistry	Subsequent to graduation at least one year laboratory training or experience in toxicity bioassay analyses.
M.S. or Ph.D. in biological or environmental sciences	Demonstrate competency in discipline by providing evidence of six successfully completed definitive bioassay analyses.

Following the inspection, the laboratory is furnished with an inspection report listing any deficiencies found by the inspector. The laboratory has at least 15 days but not more than 30 days from the date of the inspection report to correct said deficiencies. Once all deficiencies, if any, have been corrected, the laboratory is awarded certification in the acute toxicity bioassay category. The Department may, for due cause, suspend or revoke a laboratory's certification.

Subchapter 6, "Criteria and Procedures for Bioassay Testing and Analysis," is segregated into seven sections, not counting the scope and definitions sections; these sections are summarized in Table 3.

Since in toxicity bioassay analyses a living organism is used as the analytical reagent, many of the quality control measures contained in the methodology section of Subchapter 6 concern the test organisms. The test organisms specified are those with known sensitivities to toxicants that can also be reliably obtained or cultured. The standardized test organisms are listed in Table 4. Criteria for the size and age of each test organism are also established.

In addition to the criteria designed to standardize the biological "reagent," the "Methodology" section of the subchapter also contains criteria designed to standardize the experimental design for both range-finding and definitive toxicity bioassays. The criteria cover such aspects as test duration, number of test concentrations, number of replicate concentration series, number of test organisms exposed to each effluent concentration and the control, the type of test (static, renewal, or flow-through) permitted, the frequency of observations and the criterion for the effect being measured. Also specified are the approved statistical methods for the calculation of EC_{50}s and LC_{50}s, and the conditions which invalidate test results.

TABLE 3—*Listing of standardized bioassay test organisms.*

1. Laboratory Facilities and Safety
   - Ample workbench space per analyst
   - Adequate electrical and plumbing services
   - Adequate temperature and humidity controls
2. Equipment and Instrumentation
   - Listing of required equipment and instrumentation
   - Quality control specifications on all equipment and materials
   - Design, calibration, and operation criteria for flow-through bioassay diluter systems
3. Sample Collection, Handling, and Preservation
   - Standardization of procedures to assure representative samples
   - Establishment of required chain-of-custody procedures
4. Methodology
   - Standardization of test organisms
   - Criteria for test organism holding, handling, and culturing
   - Standardization of bioassay test temperatures
   - Establishment of minimum allowable dissolved oxygen concentrations
   - Establishment of maximum test organism exposure chamber loading
   - Establishment of approved statistical methodology
   - Establishment of conditions which invalidate bioassay results
5. General Laboratory Practices
   - Quality control criteria for laboratory water supplies
   - Standardization of test organism care and feeding procedures
   - Standardization of laboratory equipment cleaning procedures
6. Quality Control
   - Requirement for a written quality control procedures manual
   - Establishment of minimum acceptable bioassay analytical precision
7. Records and Data Reporting
   - Standardization bioassay results report
   - Requirement to keep all records for at least five years

TABLE 4—*Listing of standardized bioassay test organisms.*

Receiving Water[a]	Test Organism Species
Freshwater	Fathead Minnow *(Pimephales promelas)*
	Bluegill *(Lepomis macrochirus)*
Saltwater	Atlantic Silverside *(Menidia menidia)*
	Sheepshead Minnow *(Cyprinodon variegatus)*
	Grass Shrimp *(Palaemonetes pugio)*
	Opossum Shrimp *(Mysidopsis bahia)*

[a]Less than or equal to 1 ppt is designated as freshwater; greater than 1 ppt is designated as saltwater.

Every laboratory is required to develop and have on file a quality control manual. This manual must outline the laboratory's quality control procedures and describe how these procedures are utilized to ensure the generation of valid data. The quality control manual would include, for example, the laboratory's procedures for the calibration and maintenance of flow-through bioassay diluter systems, ph meters, and other instrumentation. The regulations specify the minimum requirements for said quality control practices. The regulations also specify a minimum acceptable degree of precision for toxicity bioassays. An acceptable degree of precision requires the 95% confidence intervals on the 96-h LC_{50} to be within $\pm 30\%$. A laboratory's quality control program can be more stringent.

In order for a state to evaluate a laboratory's quality assurance program and verify implementation of same, certified laboratories are required to keep adequate records of all toxicity bioassay results. These records are to include all raw data, quality control data, chain of custody forms, and laboratory reports. Also a daily log of feeding, behavioral observations, and mortality of organisms during holding and acclimation must be kept by laboratories. This log would include air and water temperatures of holding/culturing areas and analysis of laboratory grade water. All records must be retained by the laboratory for a minimum of five years. The regulations also specify how the toxicity bioassay test results are to be reported.

Pitfalls and Problems

The past nearly seven years of experience with New Jersey's laboratory certification program have revealed a number of unexpected problems. A discussion of our experiences with some of those problems may help others to avoid some of the pitfalls.

The potential problems one may encounter when conducting a laboratory certification program can be segregated into three categories: organization (format), technical content, and administration:

Organization—The format you choose for the regulations is very important. To avoid misinterpretations the format must be clear and logical; preferably, it should be progressive. A progressive format places the more general information at the beginning of the regulations with the more specific requirements following. This format is easier for the average reader to comprehend. For example, in our existing regulations we made the mistake of placing the "General Laboratory Practices" section in the last third of the subchapter. When laboratories refer to the document they often fail to realize that important criteria, say on disease treatment of test organisms, is contained in the "General Laboratory Practices" section. It would have been better to place this section nearer the beginning of the subchapter. For example, one could organize the sections of the New Jersey toxicity bioassay regulations as follows:

1. Scope.
2. Definitions.
3. Laboratory safety.
4. Laboratory facilities and equipment.
5. Sample collection, handling, and preservation.
6. General laboratory procedures.
7. Toxicity testing methodology.
8. Calculating, analyzing, and reporting of test results.
9. Laboratory quality control.

Whatever the format, it is critical to use clear, easily understood English. At all costs, avoid using "legalize" or bureaucratic double talk. If the words are not in "Webster" do not use them. It is far better to spend the written word explaining the meaning of regulation than to spend time later trying to correct misunderstandings. Furthermore, avoid the use of the "referral," that is, defining one section of the regulation by referring to another section of the same regulation. A certain amount of "referring" is unavoidable, but overuse of the technique proves to be much too confusing to the reader. Repeat rather than refer.

Technical Content—The byline when writing laboratory certification regulations must be "DO NOT TRUST ANY ONE REFERENCE." Even standard protocols, such as those given in either *Standard Methods for the Examination of Water and Wastewater* (published by the American Public Health Organization) or published by the USEPA or the American Society for Testing and Materials (ASTM), are not completely reliable. Methods often contain internal discrepancies, vague language resulting from compromise, and out-of-date procedures due to rapid developments in the field. Since the available standard methods are written as guidelines or recommendations, many contain procedures which have not been substantiated and which often are written too loosely to be used in a certification context. Also, agreement between standard methods is often not very good.

It is necessary, therefore, for the regulation writer to review as many references as possible, comparing the protocols point by point. The writer then edits those points, for which there is a consensus among the references, into a protocol for inclusion in the certification regulation. When a procedure or criterion is critical to the quality of the analytical results, compliance with it must be made mandatory. Only make a criterion mandatory when there is a clear consensus for it in the references. Caution, many of the available toxicity bioassay methods contain procedures that may not be practical, and if implemented might make a laboratory certification program unmanageable.

The use of standardized toxicants for laboratory performance evaluation is still an unresolved issue. While there are several promising compounds that also are relatively safe to use in the laboratory, the data bases on their toxicity to aquatic organisms are not always adequate. Before a compound can be used as a standardized toxicant, the regulatory agency must be reasonably sure of its toxicity to the test organisms to be used. The most direct method for collecting missing toxicity data is to conduct round-robin testing on the material(s). Recently, this approach has been utilized by the Department to begin generating standardized toxicant data bases.

Finally, one must recognize that, by their nature, laboratory certification regulations are temporal; they are near obsolete almost as soon as they are promulgated. To lessen the pain, try to retain flexibility in laboratory certification regulations, being careful not to exceed a level consistent with preserving the necessary data quality.

Administration—The administration of your toxicity bioassay laboratory certification program, good or bad, will determine how successful the program will be. The program must have the support of upper management before you even begin. The top level managers of your agency must, from the start, be made aware of the necessary financial and manpower commitments a certification program will require.

As of March 1988, there were 18 certified laboratories in New Jersey's program, with 4 laboratories in the "Applied" status. We average about 4 new laboratories applying to the certification program per calendar year. On the average, each certified laboratory conducts about 150 acute toxicity bioassays per calendar year. Under those conditions, operating New Jersey's program has required an annual commitment of at least one man year, preferably two man years, of effort. The funding source, or fee schedule, for comparably scaled programs must be sufficient to support same.

Besides the support of top management, it is necessary through education and coordination to receive the support of middle management. The various program administrators must appreciate and support the certification program's goals if they are to be attained. It is especially critical to establish and maintain good lines of communication between the certification program and the state or federal NPDES discharge permit writers and enforcement staff. Even with the best planning and coordination, unexpected problems come to light and invariably explode upon you at the worst possible moment. In New Jersey's program, for example, we discovered that the certified laboratories were sporadically encountering serious problems with the quality and quantity of some of their test organism supplies. This occurred despite assurances from major suppliers to the contrary. The regulatory agency must, therefore, retain administrative flexibility in order to creatively respond to such "land mines." New Jersey responded by instituting both intra- and interagency biomonitoring committees to expedite solutions to both unexpected and anticipated problems.

The intraagency committee consists of toxicity bioassay laboratory certification staff, toxicity bioassay compliance monitoring staff, and staff from the NJPDES permit writing/review programs. The interagency committee adds representatives from the local USEPA region and representatives from the NJDEP Office of Quality Assurance to the intragency core group. This approach to maintaining good lines of communication has worked well for problem solving in New Jersey's certification program.

Conclusions and Recommendations

The Department's experience over nearly seven years with the toxicity bioassay laboratory certification program has justified the wisdom in originally establishing it. While this program, like most, is not perfect in its design and execution, without it the Department would not be able to effectively enforce the NJPDES compliance biomonitoring regulations.

Significant improvements to New Jersey's toxicity bioassay program could be achieved if three problems received national attention. First, the standards setting organizations need to direct their energies towards attaining a higher level of both acute and chronic toxicity test standardization. Second, these same organizations need to accelerate their efforts to develop effective, practicable chronic toxicity testing methods which are useful for discharge compliance monitoring. Third, a national program, administered by the USEPA, should be created for the certification of test organism suppliers. This program should enforce strict criteria on aspects such as organism identification verification, organism feeding, and organism handling.

In conclusion, states currently including biomonitoring requirements in their discharge permits, or those contemplating instituting such monitoring, should seriously consider creating their own toxicity bioassay laboratory certification programs. Only through the efforts of a well-staffed aggressive certification program can the quality of monitoring data be ensured.

Acknowledgments

I wish to thank George G. McCann, Director, Division of Water Resources, for permission to publish this account. I also acknowledge the cooperation of J. Reading, who contributed to the general development of the paper and provided critical review.

References

[1] Bicking, C., Olin, S., and King, P., "Procedure for the Evaluation of Environmental Monitoring Laboratories," Research Report EPA-600/4-78-017, Office of Research and Development, U.S. Environmental Protection Agency, Cincinnati, OH, March 1978.

[2] Williams, L. R., "Harmonization of Biological Testing Methodology: A Performance-Based Approach," *Aquatic Toxicology and Hazard Assessment (Eighth Symposium), ASTM STP 891*, R. C. Bahner and D. J. Hansen, Eds., American Society for Testing and Materials, Philadelphia, 1985, pp. 288–301.

Krzysztof M. Jop[1]

Acute and Rapid-Chronic Toxicity of Hexavalent Chromium to Five Marine Species

REFERENCE: Jop, K. M., "**Acute and Rapid-Chronic Toxicity of Hexavalent Chromium to Five Marine Species,**" *Aquatic Toxicology and Hazard Assessment: 12th Volume, ASTM STP 1027,* U. M. Cowgill and L. R. Williams, Eds., American Society for Testing and Materials, Philadelphia, 1989, pp. 251–260.

ABSTRACT: Five different rapid-chronic toxicity tests were evaluated using hexavalent chromium as a reference toxicant. Rapid-chronic values were derived with hexavalent chromium for the fertilization test with the sea urchin *Arbacia punctulata,* the sexual reproduction test with the red macroalga *Champia parvula,* the larval growth test with the sheepshead minnow *Cyprinodon variegatus,* the reproduction test with the mysid *Mysidopsis bahia,* and the reproduction test with the polychaete *Dinophilus gyrociliatus.*

The results from rapid-chronic tests with mysids, sheepshead minnows, and the polychaete were compared with acute LC_{50} values. The acute values with hexavalent chromium were similar to rapid-chronic values for these species. Therefore, to evaluate the health of the testing animals and assess the variability of their response to a reference toxicant through time, less expensive short-term acute toxicity tests should be considered.

Among five species used in rapid-chronic toxicity tests with hexavalent chromium, *Champia parvula* was the most sensitive and the most variable. The least sensitive species tested was *Arbacia punctulata,* with a mean EC_{50} value of 341.8 mg Cr^{+6}/L.

The toxic effect of hexavalent chromium was consistent in repeated experiments for all species with the exception of two-day rapid-chronic tests with *Champia parvula.* The results of this evaluation clearly indicated that before introducing new species or a new test method, several toxicity tests should be conducted under standardized conditions using one reference toxicant in order to evaluate the suitability of the test method in aquatic toxicological studies.

KEY WORDS: hexavalent chromium, rapid-chronic toxicity tests, acute tests, *Arbacia punctulata, Champia parvula, Mysidopsis bahia, Cyprinodon variegatus, Dinophilus gyrociliatus,* reference toxicant

The need for a standardized test method to enhance comparison of toxicological data was recognized as early as 1951 [*1*]. Standardization can be achieved by adopting detailed test protocols that can be verified using a reference toxicant. A completely standardized method with a reference toxicant should answer the question of species sensitivity and the reliability of particular test designs. However, at present the definition of a standard test method varies substantially among laboratories. What is "strict" to one laboratory can be "loose" to another. Data generated from round-robin studies indicate that currently used test protocols are far from being standardized due to inconsistencies in interpretation and application of the test procedures and high interlaboratory variation in the results. Obviously, the current understanding of test procedures by personnel conducting these tests has not been sufficient to provide reliable toxicity data that will be meaningful to regulatory agencies. Recently Cowgill [*2*] identified factors

[1]Research scientist, Battelle Ocean Sciences, Duxbury, MA 02332.

that might affect the sensitivity of test zooplankton and the reproducibility of results obtained from repeated toxicity tests. She recommended conducting static acute toxicity tests through time to obtain an acceptably small coefficient of variation among the results with the same toxicant.

The suitability of hexavalent chromium as a reference toxicant has been established in previous research [*3–5*]. Hexavalent chromium can be readily available in a highly pure form; the chromate salt is preferable because the potential for alteration of the initial pH, and hence subsequent changes in chromium speciation, would be minimized [*6*].

Traditionally, the toxicity of wastewaters to aquatic organisms has been determined using 48-h or 96-h static acute toxicity tests. Although these methods are valuable and still are part of the effluent testing program, the U.S. Environmental Protection Agency (EPA) is recommending rapid-chronic toxicity tests as a part of biomonitoring for National Pollutant Discharge Elimination System (NPDES) permits. Many states have already incorporated rapid-chronic tests into their NPDES permits. The sublethal effects data derived from rapid-chronic tests often are considered to be better predictors of the potential environmental impact of pollutants than are the results of acute lethal toxicity tests. However, rapid-chronic tests are technically more difficult and therefore more expensive.

The purpose of the present study was to evaluate rapid-chronic methods with five marine species by using hexavalent chromium as a reference toxicant and to examine the suitability of short-term acute and rapid-chronic tests for assessing the health of test organisms through time.

Materials and Methods

Control Water

Natural seawater from Duxbury Bay, a high-salinity (28 to 33 ‰) inlet of Massachusetts Bay, was used in culturing and holding animals and as test dilution water. Seawater was Millipore-filtered (0.45 μm) and aerated. Water quality parameters of the dilution water were maintained within the following ranges: salinity 30 $\pm$ 3 ‰ (except for the *Dinophilus* test where salinity was 25 $\pm$ 3 ‰), pH 7.8 $\pm$ 0.3, and the dissolved oxygen concentration was always greater than 60% saturation.

Sea Urchin Fertilization Test Procedures

Mature sea urchins *Arbacia punctulata* were obtained from the Marine Biological Supply House, Woods Hole, Massachusetts, and held at Battelle Ocean Sciences in continuously flowing seawater at a temperature of 15 $\pm$ 3°C. The sea urchins were fed field-collected kelp, *Laminaria sp.*

Urchins were induced to spawn using electrical stimulation according to the procedure of Nacci et al. [*7*]. Eggs and sperm were collected from the gonophore by means of a disposable syringe fitted with a large-gauge blunt-tipped needle. The concentration was adjusted to 2000 eggs/mL of seawater. The density of sperm was estimated spectrophotometrically.

Test conditions included a temperature of 20 $\pm$ 1°C, salinity 30 ‰, and dark conditions during exposure. Glass scintillation vials with a volume of 20 mL were used as test chambers. Tests were conducted in triplicate with five chromium concentrations in a 50% dilution series and a seawater control. A predetermined density of sperm was dispensed to each test vial and incubated in the test solutions for 1 h. Eggs were then added to each vial. After 20 min, formalin was added to terminate fertilization. Fertilization, defined as the presence of a fertilization membrane, was measured microscopically by examining a 1-mL aliquot of eggs from each vial.

Sexual Reproduction Test with Red Macroalga

Separate cultures of male and female plants of the red macroalga *Champia parvula* were obtained from the EPA Environmental Laboratory in Narragansett, Rhode Island. Haploid male and female gametophytes were maintained in separate unialgal cultures. Unialgal stock cultures of males and females were maintained in 2000-mL aerated Erlenmeyer flasks containing natural seawater and GP2 culture medium [*8*]. Culture flasks were illuminated with cool white fluorescent light (500 fc) on a 16-h light, 8-h dark cycle [*9*].

One test was conducted with two replicates, and two tests were conducted with five replicates. Test conditions included a temperature of 22 ± 2°C and salinity 30 ± 2 ‰. Sterile polypropylene cups (110 mL) that had tight-fitting lids and contained 100 mL of test solution were used as test chambers during the exposure period. Each replicate contained five female branch tips and one male branch. To each cup, 1 mL of GP2 culture medium (25% of the full strength) was added. Test chambers were randomly placed on a gyratory shaker table. At the end of the two-day exposure period, female plants were removed to 250-mL Erlenmeyer flasks containing full-strength GP2 culture medium. Flasks were repositioned on shaker table trays and aerated through sterile, cotton-plugged 1-mL pipets. The number of cystocarps per female branch tip was determined microscopically at the end of a nine-day recovery period. The culture medium was renewed once after three days during the recovery period. Tests were conducted at a temperature of 22 ± 2°C.

Larval Growth Test with Sheepshead Minnows

Sheephead minnows were cultured at Battelle. Fish were cultured, held, and tested at a temperature of 25 ± 2°C. Ten days prior to the scheduled start of the test, two spawning trays were placed in a brood stock holding tank. Each tray was 20 by 45 by 5 cm with a 0.5-mm Nitex screen-meshed bottom covered with a 2-cm layer of dolomite. After three days of spawning, trays were removed, and the dolomite was transferred in seawater to a plastic tub. The dolomite was swirled in the tub to free the eggs, and the seawater containing the eggs was decanted over a 0.5-mm-mesh screen. Approximately 600 embryos were pipetted into 1-L glass jars containing 800 mL of continuously aerated seawater. Twenty-four hours prior to the start of the toxicity test, all hatched fish larvae were removed from incubation jars and transferred to the 1-L holding jars. The fish were fed ≤24-h-old brine shrimp *Artemia salina* nauplii.

Tests were conducted in triplicate with a sewater control and five chromium concentrations in 50% serial dilutions. Test conditions included a temperature of 25 ± 1°C, salinity 30 ± 2 ‰, and a 16:8-h light and dark photoperiod. Glass jars (1 L) containing 800 mL of test solution and covered with plastic Petri plates were used as test chambers. Chambers containing 15 posthatched fish larvae ≤24-h old were randomly placed in a plexiglass rack in a water bath. Mortality was recorded daily before test solution renewal.

Approximately 80% of the test solution was renewed daily using the renewal device constructed especially for this test. The renewal device was a Tygon siphon hose with a 4-cm screened funnel attached on the inlet to prevent siphoning fish larvae. A hose clamp on the outlet prevented loss of suction between solution renewal. Prior to renewal, the temperature of each test solution was adjusted to 25 ± 2°C. From each chamber 640 mL of test solution was siphoned daily. At the same time unconsumed *Artemia* nauplii and debris were pipetted from the bottom. The fish larvae were then counted, and 640 mL of freshly prepared test solution was gently decanted into the chamber. The fish larvae were fed 2 mL of a concentrated suspension of ≤24-h-old *Artemia* nauplii. The *Artemia* suspension was prepared by diluting in 200 mL of seawater 10 g of wet nauplii weight on Day 0, 1, and 2 and 15 g of wet weight of *Artemia* nauplii on Day, 3, 4, 5, and 6.

At the termination of each test on Day 7, minnows from each chamber were collected on a sieve, rinsed with deionized water, and transferred to preweighed aluminum pans. The fish larvae were dried to constant weight for 24 h at 60°C and then weighed in groups (all fish from each replicate were weighed concurrently) on a Cahn Model 25 electrobalance.

Reproduction Test with Mysids

Mysids used in the rapid-chronic toxicity test (Test A) and acute tests were cultured at Battelle. The second rapid-chronic test (Test B) was conducted using mysids obtained from a commercial supplier. The juvenile test organisms (less than 24-h old) were incubated at 26 ± 1°C in 10-L glass aquaria containing approximately 7 L of seawater for an eight-day period. During the incubation period, the mysids were fed brine shrimp *Artemia* nauplii, and approximately 80% of the seawater was replaced every other day.

Disposable plastic cups (200 mL), each containing 150 mL of test solution and five mysids, were loosely covered with plastic Petri plates and randomly placed in a plexiglass rack in a water bath. Test conditions included a temperature of 26 ± 1°C, salinity of 30 ± 2 ‰, and a 16:8-h light and dark photoperiod. Five replicates were used in Test A and eight in Test B. The renewal of approximately 80% of the test solution was conducted daily according to the procedure of Horning and Weber [*10*]. During the renewal process, each test chamber was examined over a light table and the number of surviving mysids recorded. Salinity, temperature, and dissolved oxygen were measured in one replicate of each treatment before and after the renewal. The mysids were fed 24-h-old brine shrimp nauplii after each renewal procedure. At the termination of each test on Day 7, surviving mysids in each replicate chamber were counted and then examined alive microscopically for sex determination. After sex determination, the mysids were rinsed with deionized water and separated into two groups (males and females). The mysids from each of these two groups for each treatment were dried for 24 h at 60°C and then weighed on a Cahn Model 25 electrobalance.

Reproduction Test with Dinophilus gyrociliatus

The culture of *Dinophilus gyrociliatus* has been maintained at Battelle for more than four years. The polychaetes were cultured according to procedures described by Carr et al. [*11*].

The tests were conducted in 20-mL glass Stender dishes with ground glass lids containing 10 mL of solution. Test conditions included a temperature of 20 ± 2°C, salinity of 25 ± 3 ‰, and a 16:8-h light and dark photoperiod. Five chromium concentrations in 50% serial dilutions plus a seawater control were used in each test. Static tests were conducted in triplicate with five 1 to 2-day-old juveniles per dish.

The animals were fed once at the beginning of each test with two drops of a 0.5% (weight/volume) suspension of dried spinach. Observations were performed daily using a dissecting microscope. Daily observations included number of live animals and number of eggs present.

Acute Toxicity Test with Mysids and Sheepshead Minnows

The mysids and sheepshead minnows used in acute tests were cultured at Battelle. Juvenile mysids and fish ≤48-h old were used for testing.

Static toxicity tests were conducted according to standard procedures [*12*]. Glass crystallizing dishes (200 mL) covered with plastic Petri plates with ten mysids per dish and 1-L glass jars with 15 sheepshead minnows per jar were used for the tests. Tests were conducted in triplicate with five chromium concentrations and a seawater control. Each static test with mysids was conducted in a temperature of 20 ± 2°C, salinity 30 ± 2 ‰, and a 16:8-h light and dark photoperiod. Each static test with sheepshead minnows was conducted in a temperature of 25 ± 1°C, salinity 30 ± 2 ‰, and a 16:8-h light and dark photoperiod. Mysids were fed twice daily with live ≤48-h-old posthydration *Artemia* nauplii. Feeding volume and frequency were adjusted based on the amount of live brine shrimp remaining in each test chamber from the previous feeding. Fish were not fed during the tests.

Analysis of Hexavalent Chromium

Stock solutions of hexavalent chromium solution were prepared from reagent-grade potassium chromate supplied from Fisher Scientific Co. in natural seawater. Each chromium concentration was prepared independently in a separate chamber and then delivered to the test chambers. The initial hexavalent chromium concentrations were measured in one replicate from all treatments. At the beginning of each test, subsamples were taken and acidified with nitric acid to a pH of 2. Hexavalent chromium was measured colorimetrically by reaction with diphenylcarbazide in acid solution according to Standard Methods [*13*]. The sample was measured at 540 nm using a Perkin-Elmer spectrophotometer.

Statistical Analyses

The percent mortality of mysids, sheepshead minnows, and polychaetes after the specified exposure period was used to calculate median lethal concentration (LC_{50}) values. The median effective concentration (EC_{50}) values for the sea urchin test were calculated based on a percentage of unfertilized eggs. The Trimmed Spearman-Karber [*14*] method with Abbott's correction was applied to the data to calculate LC/EC_{50} values with respective 95% confidence intervals. An analysis of variance followed by Duncan's multiple range test [*15*] was used to determine the no-observed effect concentration (NOEC) and the lowest observed effect concentration (LOEC) from rapid-chronic test data. The maximum acceptable toxicant concentration (MATC) was calculated as the geometric mean of the NOEC and LOEC. Statistical differences between LC_{50} values were determined by the method suggested by Sprague [*16*].

Results

The mean measured concentrations of hexavalent chromium during the toxicity tests were generally within 94% of the nominal concentrations. According to thermodynamic calculations, theoretically only the hexavalent form should be present in natural oxygenated waters. Although hexavalent chromium may be reduced to the trivalent form by organic material present in seawater, this reaction was of little consequence because measured and nominal concentrations were similar.

Natural seawater from Duxbury Bay was used in all tests as a dilution water. The chemical analysis of selected metals in dilution water conducted in October 1987 revealed dissolved and particulate concentrations of 0.08 and 0.06 μg/L for copper, 0.03 and 0.004 μg/L for cadmium, 0.39 and 0.04 μg/L for nickel, 0.12 and 0.08 μg/L for lead, and 0.0 and 0.04 g/L for chromium (unpublished data, 1987, C. Hunt, Battelle Ocean Sciences). Water quality was similar in the controls and in the highest chromium concentrations, with dissolved oxygen >6.2 mg/L, pH from 7.7 to 8.2, and salinity of 27 to 31 ‰. Salinity was adjusted to 25 ‰ with milli-Q deionized water only in dilution water used for tests with *D. gyrociliatus*.

A total of sixteen toxicity tests were conducted with five marine species using hexavalent chromium as a reference toxicant. Of the five species tested, the most sensitive to hexavalent chromium was the alga *Champia parvula*. In three tests, the mean number of cystocarps produced in the seawater controls varied from 11 to 19, but was always above 10, the minimum required for acceptance (Table 1). Two tests (A and C) produced well-defined concentration-response curves; in one test (B) four chromium concentrations produced very similar responses. The MATC values differed substantially due to a wide range (from 0.1 ppb to 1.0 ppm) of hexavalent chromium concentrations used in those tests (Table 2). Therefore, to compare the response

TABLE 1—*Effect of hexavalent chromium on production of cystocarps during two-day reproduction test with* Champia parvula.

Hexavalent Chromium Concentrations, mg/L	Number of Cystocarps TEST A (n = 10) Mean	±SD	TEST B (n = 25) Mean	±SD	TEST C (n = 25) Mean	±SD
Control	11.4	3.9	19.1	6.0	13.4	3.3
Nominal[a]						
0.0001	...	...	20.1	7.0	...	...
0.001	3.5	3.6	13.6	5.4	...	...
0.005	...	...	13.9	4.1	10.4	4.8
0.01	7.8	5.1	9.1	5.9	12.5	5.3
Measured						
0.05	4.2	2.9	9.2	5.4	8.7	7.5
0.1	3.0	3.0	10.8	4.7	8.7	5.3
0.2	...	...	10.5	4.8	...	...
0.5	0.1	0.3	...	...	4.8	5.6
1.0	0.0	0.0	...	...	3.4	2.5

[a]These concentrations were not measured; they were below the colorimetric method detection limit (0.05 mg Cr^{+6}/L).

of the red macroalga to hexavalent chromium, EC_{50} values were calculated using a graphical interpolation method because linear regression analysis could not be applied to these data. The EC_{50} values for those tests where 30μg Cr^{+6}/L (Test A), 60 μg Cr^{+6}/L (Test B), and 240 μg Cr^{+6}/L (Test C).

The second most sensitive species to hexavalent chromium was the polychaete *Dinophilus gyrociliatus*. The results of the seven-day test with *D. gyrociliatus* are summarized in Table 3. Both end points investigated, that is, survival and number of eggs/female, were well reproduced in the two tests. The 96-h LC_{50} values were only slightly higher (20 to 30%) than the 168-h LC_{50} values (Table 2). The MATC (reproduction end point) for *D. gyrociliatus* was more sensitive than survival in both tests by a factor of 10 or more.

The results of the seven-day static-renewal rapid-chronic tests with *Mysidopsis bahia* are summarized in Table 4. Mysid mortality after 96 and 168 h was similar in both tests. In support of this observation, the LC_{50} values were also similar (Table 2). The survival test data indicated that MATC values were similar to LC_{50} values. Survival was significantly reduced ($p > 0.05$) in 3.9 mg Cr^{+6}/L when compared with the survival in the controls. Reproduction and survival were equally sensitive in Test A, whereas survival was a more sensitive end point in Test B. In contrast, growth and reproduction were equally sensitive in Test B, whereas growth was more sensitive in Test A. The LC_{50} values from acute tests were similar to LC_{50} values obtained from rapid-chronic tests. Similar responses were observed during the acute tests when juvenile (48-h or less old) mysids were exposed to chromium compared to the seven-day-old mysids used in the rapid-chronic test. It was apparent that the size and age of mysids did not affect their sensitivity to hexavalent chromium.

Analytical results from the verification of the hexavalent chromium concentrations during the seven-day rapid-chronic toxicity test showed good agreement between nominal and measured concentrations [coefficient of variation (CV) averaged 6.6 percent]. The initial chromium concentrations in each treatment were measured daily.

All sheepshead minnow larvae survived (100%) in the seawater controls, indicating that the fish used for testing were healthy (Table 5). The results from both tests showed that growth (weight) was a more sensitive end point than survival (Table 2). Reduction of fish growth occurred at the chromium concentration with little or no mortality. The confidence limits for the 96-h acute LC_{50} value (19.5 to 26.6 mg Cr^{+6}/L) and for the 96-h rapid-chronic LC_{50} value (27.5

TABLE 2—*Rapid-chronic and acute values for five marine species exposed to hexavalent chromium. All values are expressed as mg Cr^{+6}/L.*

Species	Exposure Time, h	Acute LC_{50} Values			Rapid-Chronic Values			
		48 h	96 h	168 h	EC_{50}	NOEC	LOEC	MATC
Mysidopsis bahia	48	4.3						
	96	3.6	2.5					
	168	4.9	2.7	2.5		1.6	3.9	2.5[a]
						0.6	1.6	1.0[b]
						1.6	3.9	2.5[c]
	168	ND	ND	3.6		0.6	1.6	1.0[a]
						1.6	3.9	2.5[b]
						1.6	3.9	2.5[c]
Cyprinodon variegatus	96	22.8						
	168	32.1	31.6			14.1	29.2	20.3[a]
						<2.65	2.65	ND[b]
						ND	ND	ND[a]
	168					2.5	5.0	3.5
Arbacia punctulata	1				275.7	125	250	177[c]
	1				385.5	125	250	177[c]
	1				322.1	125	250	177[c]
	1				383.8	250	500	354[c]
Champia parvula	48				0.03	<0.001	0.001	ND[c]
	48				0.06	0.0001	0.001	0.0003[c]
	48				0.24	0.01	0.05	0.02[c]
Dinophilus gyrociliatus	168		3.3	2.2		1.0	5.0	2.2[a]
						0.1	0.5	0.2[c]
	168		3.0	2.4		1.0	5.0	2.2[a]
						<0.1	0.1	ND[c]

NOTE: Rapid-chronic values were calculated from: [a]survival data; [b]growth (weight); [c]reproduction; ND = not determined.

TABLE 3—*Percent mortality and reproduction effect for* Dinophilus gyrociliatus *exposed to hexavalent chromium during seven-day chronic toxicity tests.*

	Percent Mortality				Number of Eggs Produced		Number of Eggs/ Adult	
	Test A		Test B					
Cr^{+6}, mg/L	96 h	168 h	96 h	168 h	Test A	Test B	Test A	Test B
Control	0	0	0	0	58	58	3.8	3.4
0.1	0	0	0	0	55	33	3.2	2.4
0.5	0	0	0	0	20	2	1.3	0.1
1.0	0	0	0	0	2	0	0.1	0.0
5.0	73	83	67	100	0	1	0.0	0.1
10.0	100	100	100	100	0	0	0.0	0.0

TABLE 4—*Percent mortality and growth (weight) of mysids exposed to hexavalent chromium during seven-day rapid-chronic toxicity tests.*

	Percent Mortality				Weight, mg	
Cr^{+6}, mg/L	48 h	96 h	168 h	Percent Females with Eggs	Females Mean (n)	Males Mean (n)
				TEST A		
Control	4	4	12	64	0.435 (9)	0.436 (8)
0.25	8	12	16	73	0.494 (8)	0.514 (10)
0.65	4	4	12	100	0.499 (9)	0.480 (13)
1.60	4	4	12	80	0.408 (8)	0.324 (12)
3.90	24	88	96	0	...	0.167 (1)
8.40	100	100	100	0	...	...
				TEST B		
Control	0	2	5	89	0.499 (9)	0.279 (29)
0.25	0	0	0	100	0.428 (2)	0.277 (13)
0.65	0	0	0	67	0.412 (3)	0.333 (12)
1.60	0	13	20	100	0.428 (4)	0.368 (8)
3.90	7	33	53	50	0.379 (4)	0.293 (3)

NOTE: Test A was conducted from December 15 to 22, 1987 with mysids from a commercial supplier. Test B was conducted from January 6 to 13, 1988 with mysids from a Battelle culture.

TABLE 5—*Percent mortality and growth (weight) of sheepshead minnows exposed to hexavalent chromium during seven-day rapid-chronic toxicity test.*

Cr^{+6}, mg/L						
	Measured		Percent Mortality		Weight, mg	
Nominal	Mean	±SD	96 h	168 h	Mean	*n*
Control			0	0	1.37	29
3.0	2.65	0.41	0	0	1.15	30
7.0	6.31	0.69	0	0	1.08	23
15.0	14.11	0.48	3	3	0.66	29
30.0	29.20	0.37	40	47	0.20	15
60.0	60.20	1.17	93	93	0.15	12

to 37.6 mg Cr^{+6}/L) did not overlap, demonstrating a statistically significant difference. Newly hatched larvae from the same batch were used in both acute and rapid-chronic tests (tests were conducted simultaneously); therefore, feeding probably was the reason for this difference between LC_{50} values.

The least sensitive species tested was *Arbacia punctulata.* Four sea urchin sperm cell toxicity tests were conducted over a three-month period. The results of those tests showed that the gametes of sea urchins were not sensitive to hexavalent chromium, and successful fertilization was hindered only at relatively high hexavalent chromium concentrations (Table 2). The results also showed relatively consistent (average EC_{50} value of 341.8 mg Cr^{+6}/L with 15.5% coefficient of variation) responses of the sea urchin gametes to chromium.

Discussion

A wide variety of toxicity tests has been developed in recent years to provide rapid and accurate screening of chemicals and effluents for toxic effects. Although toxicity testing procedures have been developed recently for the application of such methods to characterize complex industrial mixtures [*10,12*], there has been little research dealing with the comparative effects of chemicals on various marine organisms using rapid-chronic test procedures. Nacci et al. [*7*] evaluated three rapid toxicity tests: sea urchin sperm cell test, sea urchin early embryo growth test, and a microtox test with eight organic chemicals and five metals. They concluded that these tests are useful in measuring toxicities of single compounds; however, the toxic effect of each material tested was based on a single experiment.

This study provides an opportunity for direct comparison of the sensitivities of five test methods to hexavalent chromium and also allows an assessment of the variability of each species response. Of the five tests methods used in this study, the sexual reproduction test with red macroalga was the most sensitive but less reproducible than the other test methods. Thursby et al. [*9*] in repeated experiments with seven chemicals using red macroalgae obtained similar toxic effects with only two chemicals, while cystocarp production in remaining tests differed at least by factor of 2. The reason for variability in cystocarp production is not known. The effluent testing program conducted at Battelle Ocean Sciences in 1987 (unpublished data, 1988, K. Jop, Battelle Ocean Sciences) demonstrated high variability in cystocarp production through time. During that testing program, the number of cystocarps produced in three seawater controls ($n = 45$) in one week averaged 30 (CV = 27%), whereas one week later cystocarp production dropped to 7 (CV = 84%). This variability in production of cystocarps could not be attributed to changes in water quality because the seawater for all tests came from one source. Therefore, measuring the health of isolated laboratory cultures before starting toxicity tests becomes important, and criteria for the assessment of health should be defined. The characteristics of healthy *Champia* culture should include a high reproductive rate, early appearance of the cystocarps, and weight of plants in relation to age. In addition, genetic changes due to long-term maintenance of laboratory populations should be defined. Finally, the toxicity tests should be conducted with reference toxicant to achieve a coefficient of variation not greater than 40%.

The rapid-chronic toxicity tests with the mysids, the sheepshead minnows, the sea urchin, and the polychaete appear to be very reproducible with hexavalent chromium. The toxicity of chromium changed slightly over time: the 96-h LC_{50} values for the mysids, the sheepshead minnow, and the polychaete were similar to 168-h LC_{50} values. The 96-h LC_{50} values obtained in this study for mysids and sheepshead minnows were similar to those previously reported [*4,17*]. The 96-h acute values reported for mysids [*17*] and for sheepshead minnows [*4*] were 2.0 and 23.2 mg Cr^{+6}/L, respectively.

The relationship between toxicant concentrations that produce acute and chronic effects can be used as a prediction of the chronic response to the same toxicant by other aquatic species when only acute data are available. This ratio should be similar or vary in a predictable manner as a function of ionic strength or other water quality parameters. The acute to rapid-chronic ratios for mysids, sheepshead minnows, and the polychaete ranged from 1.0 to 15.7. The narrow range of acute-chronic ratios for these species (1.0 to 2.6 for mysid; 1.5 to 9.0 for sheepshead minnow; 1.4 to 15.7 for polychaete) gave a good prognosis for applicability of the acute-chronic hypothesis [*18*] for hexavalent chromium.

In summary, acute toxicity data can be used for assessing the health of organisms through time. Acute toxicity tests provide sufficient information about organism health and should be adopted as a standard procedure to verify the sensitivity of test species.

Acknowledgments

I gratefully acknowledge J. Williams, A. Amaral, C. Fragata, and J. Anderson at Battelle Ocean Sciences for their laboratory assistance throughout the study.

References

[*1*] Doudoroff, P., Anderson, B. G., Burdick, G. E., Galtsoff, P. S., Hart, W. B., Pattrick, R., Strong, E. R., Surber, E. W., and Van Horn, W. M., *Sewage Industrial Wastes*, Vol. 23, 1951, pp. 1380–1397.
[*2*] Cowgill, U. M., *Water Research*, Vol. 21, 1987, pp. 1453–1462.
[*3*] Dorn, P. B., Rodgers, J. H., Jr., Jop, K. M., Raia, J. C., and Dickson, K. L., *Environmental Toxicology and Chemistry*, Vol. 6, 1987, pp. 435–444.
[*4*] Jop, K. M., Rodgers, J. H., Jr., Dorn, P. B., and Dickson, K. L., *Aquatic Toxicology and Environmental Fate: Ninth Volume, STP 921*, American Society for Testing and Materials, Philadelphia, 1986, pp. 380–403.
[*5*] Wang, W., *Environmental Toxicology and Chemistry*, Vol. 6, 1987, pp. 953–960.
[*6*] Jop, K. M., Parkerton, T. F., Rodgers, J. H., Jr., Dickson, K. L., and Dorn, P. B., *Environmental Toxicology and Chemistry*, Vol. 6, 1987, pp. 697–703.
[*7*] Nacci, S., Jackim, W., and Walsh, R., *Environmental Toxicology and Chemistry*, Vol. 5, 1986, pp. 521–525.
[*8*] Spotte, S., Adams, G., and Bubucis, P. M., *Zoo Biology*, Vol. 3, 1984, pp. 229–240.
[*9*] Thursby, G. B., Steele, R. L., and Kane, M. E., *Environmental Toxicology and Chemistry*, Vol. 4, 1985, pp. 797–805.
[*10*] Horning, W. B. and Weber, C. I., "Short-Term Methods for Estimating the Chronic Toxicity of Effluents and Receiving Waters to Marine and Estuarine Organisms," EPA-600/4-87-028, Office of Research and Development, U.S. Environmental Protection Agency, Cincinnati, OH, 1988.
[*11*] Carr, R. S., Curran, M. D., and Mazurkiewicz, M., *Environmental Toxicology and Chemistry*, Vol. 5, 1986, pp. 703–712.
[*12*] Peltier, W. H. and Weber, C. I., "Methods for Measuring the Acute Toxicity of Effluents to Freshwater and Marine Organisms," U.S. Environmental Protection Agency, EPA-600/4-85-013, Cincinnati, OH, 1985.
[*13*] *Standard Methods for the Examination of Water and Wastewater*, 16th ed., American Public Health Association, Washington, DC, 1985.
[*14*] Hamilton, M. A., Russo, R. C., and Thurston, R. V., *Environmental Science and Technology*, Vol. 11, 1977, pp. 714–718.
[*15*] "SAS User's Guide: Statistics," SAS Institute, Statistical Analysis System, Raleigh, NC, 1985.
[*16*] Sprague, J. B. and Fogels, A., "Watch the Y in Bioassay," Proceeding, 3rd Aquatic Toxicity Workshop, Halifax, NS, 2–3 Nov. 1976, Technical Report No. EPS-5-AR-77-1, Environmental Protection Service, Halifax, Canada, 1977.
[*17*] Lussier, S. M., Gentile, J. H., and Walker, J., *Aquatic Toxicology* Vol. 7, 1985, pp. 25–35.
[*18*] Mount, D. I., "An Assessment of Application Factors in Aquatic Toxicology," *Recent Advances in Fish Toxicology*, EPA-600/3-77-085, R. Tubb, Ed., 1977.

P. T. Heitmuller[1] *and James R. Clark*[2]

Bioaccumulation of 1,2,4-Trichlorobenzene from Food and Water Sources by Spot *(Leiostomus xanthurus)*

REFERENCE: Heitmuller, P. T. and Clark, J. R., "**Bioaccumulation of 1,2,4-Trichlorobenzene from Food and Water Sources by Spot** ***(Leiostomus xanthurus)*****,**" *Aquatic Toxicology and Hazard Assessment: 12th Volume, ASTM STP 1027,* U. M. Cowgill and L. R. Williams, Eds., American Society for Testing and Materials, Philadelphia, 1989, pp. 261-269

ABSTRACT: Contaminated food was prepared by exposing pink shrimp *(Penaeus duorarum)* to 10 μg/L of radiolabeled (^{14}C) 1,2,4-trichlorobenzene (TCB) for twelve days; whole body concentration of TCB in the exposed shrimp was 0.59 μg/g. Juvenile spot *(Leiostomus xanthurus),* a marine fish, were then fed the TCB-contaminated shrimp at a daily ration of 10% body weight for 28 days, and they accumulated <0.05 μg/g TCB (detection limits). Spot, exposed to nominal 10 μg/L TCB in water for 28 days and fed uncontaminated food, bioconcentrated TCB approximately 100 times the aqueous exposure concentration. Equilibrium was attained in these fish within seven days, and fish depurated TCB to concentrations <0.1 μg/g within 24 h after being placed in TCB-free water. Spot, exposed simultaneously to the contaminated food and water described above, bioaccumulated TCB equal to the aqueous exposure treatment.

Pharmacokinetic uptake and depuration rate constants were used to compare the potential for spot to bioaccumulate TCB, a moderately lipophilic compound, to that for chlordecone (Kepone), a highly lipophilic compound. Our results were also compared to those from a TCB bioaccumulation study with freshwater species; both studies indicated that TCB was moderately accumulated from contaminated water and that accumulation from contaminated food was negligible.

KEY WORDS: 1,2,4-trichlorobenzene, bioaccumulation, dietary accumulation, pharmacokinetics, *Leiostomus xanthurus*

Uptake of most xenobiotic chemicals by fishes is mainly by bioconcentration of the chemicals from water through the gill membrane [*1*]. However, uptake from food is an additional source of contaminants for aquatic organisms and may be significant for lipophilic compounds with high octanol-water partition coefficients (K_{ow}) [*2-3*]. Pizza and O'Conner [*4*] showed that striped bass *(Morone saxatilis)* bioaccumulated polychlorinated biphenyls (PCB) when fed successive, small doses. The authors suggested that the dietary PCB component may be of great significance to fishes, especially in heavily contaminated environments. A model developed by Connolly and Tonelli [*5*] to compute accumulation of chlordecone (Kepone) in the striped bass food chain of the James River estuary indicated that diet was the major route of accumulation for several high trophic-level fishes.

Macek et al. [*1*] stated that laboratory data generated for comparisons of uptake from aqueous and dietary sources can be ecologically relevant only when food organisms accumulate their body burden of the contaminant under conditions to which the consumer organisms are ex-

[1]Aquatic biologist, Technical Resources, Inc., Sabine Island, Gulf Breeze, FL 32561.
[2]Research aquatic biologist, U.S. EPA Environmental Research Laboratory, Gulf Breeze, FL 32561.

posed. They further stated that the relative contributions of aqueous and dietary sources must be quantitatively evaluated at steady-state (equilibrium) conditions and that feeding rates should be related to those expected in a natural ecosystem. Fisher et al. [6] incorporated these conditions into a study of bioaccumulation of aqueous and dietary sources of chlordecone by spot *(Leiostomus xanthurus)*. They found dietary-accumulated chlordecone to be additive to that bioconcentrated from water, and noted that the dietary contribution was significant, accounting for approximately 25% of the total body burden of chlordecone in fish exposed for 28 days.

The overall objective of this research program was to determine if a threshold level of K_{ow} of xenobiotics exists at which dietary contribution becomes significant to the total bioaccumulation of contaminants by spot. This can be tested by independently and simultaneously exposing spot to food and water sources of contamination in a simple, one-link food chain, maintaining conditions of equilibrium, and evaluating food and water sources of contaminant bioaccumulation by spot. For the initial chemical studied (reported here), we selected a test compound, 1,2,4-trichlorobenzene (TCB), that has a moderate K_{ow} of 4.16 and is not readily metabolized [7], and conducted exposures based on the above experimental format. Results are compared to those from some of the previously mentioned studies [1,6], and the potential for spot to bioaccumulate TCB from food and water is evaluated in terms of pharmacokinetic parameters of uptake and depuration.

The laboratory exposures and sample analyses were conducted at the U.S. EPA Environmental Research Laboratory, Gulf Breeze, FL (ERL/GB). The term "bioaccumulation," as used in the remainder of this paper, applies to uptake from aqueous and dietary sources, unless otherwise stated.

Experimental Procedures

Test Material

Radiolabeled and unlabeled TCB were used in the tests. The uniformly ring-labeled (^{14}C) TCB, specific activity 12.8 mCi/mmol and 98% purity, was purchased from Pathfinder Laboratories, Inc., St. Louis, Missouri. Radiolabeled TCB (1 mCi) was added to pesticide-grade acetone to produce 100 mL of primary ^{14}C stock solution. Unlabeled test material was purchased from Aldrich Chemical Co. (99% + purity, Gold Label). Stock solutions of unlabeled TCB were prepared by pipetting parent compound into analytical grade triethylene glycol (TEG) to obtain a stock solution of 360 mg/L. Stock solutions of radiolabeled TCB were prepared similarly; 0.24 mL of unlabeled TCB was pipetted into approximately 900 mL of TEG, 10 mL of primary radiolabeled stock added, and the final volume adjusted to 1 L with additional TEG. Stock solutions were kept in amber glass bottles with Teflon-lined caps and stored in a refrigerator. Radiolabeled and unlabeled stock solutions were comparable; all measured concentrations of TCB were $\geq$95% of the nominal 360 mg/L. The ^{14}C activity for stocks used in the shrimp (food-loading) and fish exposures was 2065 and 2095 disintegrations per minute (dpm)/μg TCB, respectively.

Test Organisms

The contaminated food source, adult pink shrimp *(Penaeus duorarum)*, 6 to 10 cm rostrum-telson length, were purchased from a local bait dealer. At the laboratory, shrimp were held in 60-L glass aquaria that contained a layer of clean sand 2 cm deep, and each tank received 15 L/h of flowing, filtered (20 μm) seawater (salinity, 20 $\pm$ 2‰; temperature 25 $\pm$ 2°C). Shrimp were acclimated to these conditions for five days prior to the start of the food-loading exposure and fed a combination of finely minced fish filets and commercial fish flake. After twelve days of exposure to radiolabeled TCB in flowing seawater, shrimp were harvested to use as TCB-contaminated food during bioaccumulation exposures with spot. The average measured concentration of TCB in seawater during exposure was 8.5 μg/L (±1.1); the average measured concentration of TCB in samples of food prepared from the shrimp was 0.59 μg/g (±0.08) (Table 1).

Test fish, juvenile spot *(Leiostomus xanthurus)*, were collected by seine from sand flats of Santa Rosa Sound, Florida during late spring. At the laboratory, the fish were held in 130-L glass aquaria that received >60 L/h of filtered seawater, 26 ± 1‰ and 25 ± 2°C. Ten days before testing, 40 fish were transferred to each of ten 60-L acclimation aquaria that received 15 L/h of filtered seawater under designated test conditions (20 ± 2‰ and 25 ± 2°C). During acclimation, fish were fed a combination of frozen brine shrimp and commercial fish flake. Fish ranged from 35 to 53-mm standard length, and the mean wet weight of fish one week before testing was 1.9 g (n = 80). Mortality was <1% during the last 48 h of acclimation.

Test Method

Test procedures were modified from those described in ASTM Practice for Conducting Bioconcentration Tests with Fishes and Saltwater Bivalve Molluscs (E 1022-84) [*8*] to include dietary and aqueous exposures. Test populations of spot were exposed to TCB for a maximum of 28 days, followed by a one-day depuration in TCB-free seawater. Fish were sampled periodically during the test and their tissues analyzed for TCB. A combination of radiolabeled and unlabeled TCB was used to differentiate aqueous and dietary sources of bioaccumulated TCB in fish that were exposed simultaneously to both. When appropriate, bioconcentration factors (BCF) were calculated by dividing mean whole-body residue of TCB by the mean concentration of TCB measured in exposure water at estimated steady state.

Test Systems and Conditions

Flow-through exposures to TCB were conducted in 60-L glass aquaria that contained 54 L of test water. The aquaria were positioned in a water bath with temperature controls, and calibrated glass siphons continuously delivered 15 L/h of filtered (20-μm) seawater. Aquaria were covered with plexiglass lids, and the seawater was gently aerated. A photoperiod of 11-h light: 13-h dark was maintained.

Syringe pumps metered appropriate stock solutions (10 mL/day) into seawater as it flowed through glass mixing tubes into the test aquaria. The nominal test concentration of TCB was 10 μg/L for all aqueous exposures.

For dietary exposures, TCB-contaminated food was prepared from pink shrimp contaminated with radiolabeled TCB by using the test system described above. After twelve days of exposure to 8.5 μg/L TCB, shrimp were netted from the tanks, dipped in chilled deionized water, blotted dry, weighed, and frozen. The frozen shrimp were coarsely minced (typical particle size, 3 mm) by using a blender, then packaged and returned to the freezer for storage (−20°C) until needed as food for test fish. TCB concentration in the contaminated food remained stable during storage; measured concentrations ranged from 0.58 to 0.65 μg/g in samples taken at four different times over a 28-day period.

Spot were exposed to four treatments during the 28-day uptake phase of the study. Test treatments were: (1) ^{14}C-TCB in water, no food (unfed); (2) ^{14}C-TCB in water, TCB-free food (water); (3) unlabeled TCB in water, ^{14}C-TCB food (water/food); and (4) TCB-free water, ^{14}C-TCB food (food). All TCB treatments were duplicated. A control aquarium received 15 L/h of TCB-free flowing seawater; a solvent-carrier control aquarium received 15 L/h of flowing seawater and 0.42 mL/h of TEG. These treatments were not duplicated because of limited space in the water bath.

The test system was operated for 48 h to allow test concentrations to equilibrate before test fish were placed in respective exposure tanks. During this equilibration period, fish were

TABLE 1—*Bioaccumulation of 1,2,4-trichlorobenzene (TCB) from aqueous and dietary sources by spot* (Leiostomus xanthurus). *Data from steady-state during uptake (7 to 28 days) followed by a 1-day depuration in TCB-free seawater.*

Treatment	Measured Concentration in Water, μg/L	Measured Concentration in Tissue, Uptake, μg/g	Measured Concentration in Tissue, for Depuration, μg/g	Bioconcentration Factor (BCF)	Neutral Lipids, % dry wt
Food loading: pink shrimp	8.50[a] (±1.13, n = 27)	0.59[b] (±0.08, n = 7)	...[c]	69	...
Fish exposures:					
TCB-water, no food	8.27 (±0.74, n = 25)	0.57[d] (±0.12, n = 15)	...[e]	69	4.6 (±2.4, n = 14)
TCB-water, clean food	8.51 (±0.87, n = 29)	1.15 (±0.66, n = 24)	<0.1	135	4.2 (±1.4, n = 23)
TCB-water, TCB-food	7.31 (±0.88, n = 19)	0.89 (±0.30, n = 24)	<0.1	122	3.6 (±1.4, n = 24)
Clean water, TCB-food	ND[f] (n = 4)	ND[g] (n = 24)	...	...	...

[a]Mean ± standard deviation, n = number of samples.
[b]12-day uptake.
[c]Not measured.
[d]21-day uptake.
[e]No fish remained.
[f]None detected, <0.05 μg/L.
[g]None detected, <0.05 μg/g.

weighed in groups of five to determine the animal loading of each test aquarium. After equilibrium was confirmed by water analysis, 40 fish were transferred from acclimation aquaria to each test aquarium to begin the uptake phase. Fish in each aquarium, except those unfed, received a daily ration of appropriate food equal to 10% of the total wet weight of fish that remained in the aquarium on that day.

Fish were sampled for TCB analysis after 0.5, 1, 2, 4, 7, 14, 21, and 28 days of exposure and after 1 day of depuration in TCB-free water. At the scheduled sampling intervals, three fish were netted from each exposure aquarium, rinsed with deionized water, blotted dry, and placed in an aluminum foil packet that was sealed and stored in a freezer for later weighing and analyses. Fish sampled at intervals during the estimated steady-state period (≥ 7 days) were also analyzed for neutral lipid content. Water samples were analyzed for TCB by radiometric and gas chromatographic (GC) analysis at least twice weekly. Samples were taken before daily feeding to eliminate the inclusion of food particles in the samples. Dissolved oxygen (DO) and pH were monitored at least once weekly; DO was measured with an oxygen meter (Yellow Springs Instrument Co.) and pH with a portable pH meter (Corning pH 101). Daily observations were made for survival, salinity, temperature, and function of the test system. DO remained $\geq 60\%$ saturation, pH ranged from 7.7 to 8.0, temperature ranged from 24 to 25°C, and salinity ranged from 20 to 21‰.

Analytical Methods

Water Analysis

One-liter samples from radiolabeled-TCB water treatments were extracted with 20 mL of hexane in a 2-L separatory funnel. The hexane extract was collected in a 25-mL concentration tube and the volume adjusted to the 20-mL mark with additional hexane. Usually, only 18 to 19 mL of the original hexane extract were recovered. A 5.0-mL aliquot of the extract then was pipetted into a 20-mL vial containing 10 mL of liquid scintillation cocktail (PCS, Amersham Corp.). One-liter samples of control seawater were spiked to 10 μg with ^{14}C-TCB and analyzed with each set of water samples to determine extraction efficiency; TCB recovery was $\geq 95\%$. The detection limit for radiometric analysis of TCB in seawater was 0.05 μg/L.

Test concentrations of unlabeled TCB in seawater were determined by GC analysis. One-liter seawater samples were extracted twice with 100 mL of petroleum ether. The extract was dried by passing through heat-treated (600°C) glass wool, concentrated on a steam table by using a Kuderna-Danish concentrator, and transferred to a 9-mm Chromaflex column. The column contained 3.0 g of Florisil topped with 2.0 g of anhydrous sodium sulfate that had been washed previously with 10 mL of hexane. TCB was eluted from the column with 20-mL hexane and assayed on a Hewlett-Packard Model 5710 electron-capture (^{63}Ni) gas chromatograph equipped with a glass column (182 cm by 2 mm ID) packed with 2% SP2100 on 100/120 mesh. The operating conditions were: oven temperature, 125°C; injector temperature, 200°C; detector temperature, 300°C; carrier gas, P-10 at 30 mL/min. The detection limit for GC analysis of TCB in seawater was 0.01 μg/L.

Tissue Analyses

The concentration of ^{14}C-TCB in samples of food prepared from shrimp was measured radiometrically. Samples of the food (200 to 500 mg) were weighed and then combusted to CO_2 by using a Packard Tri-Carb Sample Oxidizer. The $^{14}CO_2$ was counted on a Beckman LS 8100 liquid scintillation counter to measure activity.

TCB in the tissue of test fish was measured by radiometric analysis, GC analysis, or both, depending on the test treatment from which fish were sampled. TCB data are reported on a wet-weight analysis of tissue prepared from whole-fish samples. Tissue samples of control fish were spiked with 10 μg of TCB and analyzed with each set of fish samples to determine the extraction efficiency; recovery was ≥80% for all sets of samples. The detection limit of TCB in tissue was 0.05 μg/g for both radiometric and GC analysis.

Fish samples with only ^{14}C-TCB activity as an analytical endpoint were extracted in test tubes (125 by 50 mm) with 10 mL of acetonitrile for 30 s by using a PT 10-ST Brinkman Polytron. The emulsified sample, still in the test tube, was centrifuged for 10 min, and a 5-mL aliquot of the acetonitrile supernatant was pipetted into a scintillation vial containing 10 mL of PCS and counted on a Beckman LS 8100. Radioactivity data were converted to the amount of TCB and tissue residue concentrations calculated as μg TCB/g tissue (wet weight). Fish samples in which TCB was measured by the combination of GC and radiometric analyses and those in which neutral lipids were measured required additional processing. The samples were extracted twice with 10 mL of acetonitrile, as described above. Following centrifugation, the supernatants from both acentonitrile extractions were back extracted with petroleum ether by decanting the supernatants into a 250-mL separatory funnel containing 75 mL of 2% aqueous sodium sulfate and 20 mL of petroleum ether. The contents were shaken for 1 min and then allowed to sit for 20 min for the solvent phases to separate. The aqueous phase was drained and discarded, the ether phase was collected in a 25-mL Kuderna-Danish concentrator tube, and the volume was taken to 20 mL with additional ether. Subsamples were taken from the ether extract for analysis by GC, radiometric, or both methods. The remaining extract was retained for neutral lipid determination.

For radiometric analysis, 5 mL of ether extract were pipetted into a scintillation vial containing 10 mL of PCS and counted on a Beckman LS 8100. For GC analysis, 1 mL of ether extract, followed with 3 mL of ether, was eluted through a Florisil Sep-Pak fitted on a 5-cc disposable syringe. Analysis of TCB was performed on a Hewlett-Packard Model 5730A electron-capture (^{63}Ni) gas chromatograph equipped with a glass column (182 cm by 2 mm ID) packed with 2% SP 2100 on 100/120 mesh. The operating parameters were: oven temperature, 100°C; injector temperature, 200°C; detector temperature, 250°C; and carrier gas, P-10 at 30 mL/min. Neutral lipid content was determined for fish that had measurable TCB residues during the estimated steady-state period. The ether extract that remained after subsamples were taken for GC and radiometric analyses was rinsed from the 25-mL Kuderna-Danish concentrator tube with 1 to 2 mL of additional clean ether into a tared polystyrene weighing boat. After evaporating the contents to dryness, the boat and contents were weighed to determine the neutral lipid content. The lipid weight per fish was adjusted to compensate for the volumes of extract sampled for GC and radiometric analyses and was reported as percent of dry fish weight.

Results

Spot bioaccumulated TCB only from aqueous sources; no radiolabeled TCB was detected in fish that were fed ^{14}C-TCB-contaminated food (Table 1). Uptake of TCB from aqueous sources was rapid; in two of the aqueous treatments, maximum TCB concentration in tissue was reached by 0.5 day of exposure (Fig. 1). TCB concentrations then declined in fish during the first week of uptake and became constant for the remainder of the exposure. Mean concentrations of TCB bioaccumulated by spot during apparent steady-state (Days 7 to 28) were 0.57 (±0.12), 1.15 (±0.66), and 0.89 (±0.30) μg/g from test treatments (unfed, water, and water/food, respectively). BCFs of 69, 135, and 122 were calculated by using the mean exposure concentration of TCB measured for these treatments (Table 1). Measured concentrations of TCB in water from the three aqueous exposures during the 28-day uptake phase of the test ranged from 5.58 to 10.47 μg/L; the average concentration was 8.22 μg/L (±0.99, n = 73). No TCB was detected in water sampled from the treatment in which fish were exposed to a dietary source of TCB only (Table 1).

During the period from exposure Day 6 to 21, 45% mortality occurred in the unfed treat-

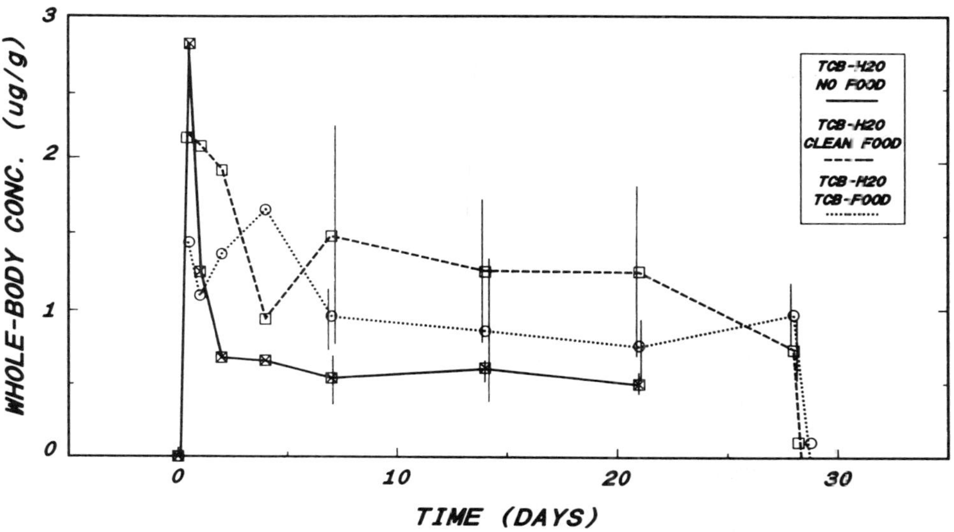

FIG. 1—*Bioaccumulation and depuration in whole-body tissue of spot* (Leiostomus xanthurus) *exposed to 8.2 μg/L 1,2,4-trichlorobenzene (TCB) in flowing seawater or to a daily 10% ration of TCB-contaminated food (0.59 μg/g) in combination with 8.2 μg/L TCB in water. Uptake was for 28 days followed by a 1-day depuration in TCB-free seawater. No fish survived in the unfed treatment after 21 days of uptake. Bars represent standard deviation for values plotted during estimated steady-state.*

ment. The remaining treatments appeared unaffected; only ten fish died among those treatments during the test. All surviving fish fed aggressively and appeared healthy.

Neutral lipid, based on dry weight, ranged from 2.1 to 9.8% for individually analyzed fish sampled during steady-state. When expressed as means for each treatment, the values were more comparable and ranged from 3.6% (±1.4) for the water/food treatment to 4.6% (±2.4) for the unfed treatment (Table 1).

Pearson's correlation coefficients [9], comparing body burden of TCB to neutral lipid content for individual fish within treatment groups, ranged from −0.08 to +0.27, indicating poor correlation between lipid content and TCB residues.

Discussion

The purpose of the TCB-water, no food treatment was for comparison with the TCB-water, clean food treatment to determine if uncontaminated food offered a means to enhance contaminant depuration as the food bolus traveled through the gut. Mortality among starved fish was not unexpected. The stress of starvation may have affected metabolism and longevity of individuals in this test population; however, we did not observe a decline in lipid values from Days 7 to 21 for starved fish. In fact, average lipid values for starved fish were greater than those for other treatments (Table 1). Because body burdens of TCB were less for starved fish compared to those fed clean food, it appears that enhanced depuration into the gut containing uncontaminated food was not a significant factor in our tests.

The increased body burdens of TCB observed in fish sampled at 0.5 days of exposure probably resulted from excitation and ventilation brought on during the transfer from acclimation aquaria to exposure aquaria. Such increases are not uncommon for chemicals that are rapidly bioconcentrated.

We used the BIOFAC computer program [10] to model the bioaccumulation kinetics of TCB. The nonlinear regression program generates uptake and depuration curves from first-order pharmacokinetic data, calculates rate constants for uptake and clearance, and estimates the bioconcentration factor, half time to depuration, and time to steady state. BIOFAC has been used widely, and the values it generates can be used, in addition to K_{ow}, for comparing bioaccumulation potential of different compounds. Depuration rate constants for the three aqueous TCB exposures with spot ranged from 2.7 to 3.2 per day, with depuration half-lives from 0.22 to 0.26 day (Table 2). Fisher et al. [6] used BIOFAC to model the bioaccumulation of chlordecone by spot for several aqueous treatments; depuration rate constants ranged from 0.02 to 0.04 per day, two orders of magnitude less than those for TCB, while the estimated BCF for chlordecone was approximately 10 000, compared to 100 for TCB.

The bioaccumulation data from our study with an estuarine species closely paralleled the freshwater study of Macek et al. [1], in which bluegill sunfish *(Lepomis macrochirus)* were exposed simultaneously to nominal 3 μg/L TCB in water and fed daphnia previously exposed to nominal 3 μg/L TCB. Bluegill bioaccumulated 0.57 μg/g of TCB (BCF, 190), of which at least 93% was from the aqueous source. In our study, spot apparently received ≥95% of their body burden from the aqueous fraction.

Lack of accumulation of TCB from dietary sources was expected. The BIOFAC-generated parameters and actual test measurements indicate that depuration of TCB is so rapid that fish depurate their body burdens within 12 to 24 h of placement in TCB-free water. In our food-only test treatment, TCB was not detected in any fish sampled during the 28-day exposure. These fish, fed a single daily ration of TCB-contaminated shrimp and held in TCB-free water, were routinely sampled for tissue analysis approximately 20 h after their previous feeding; consequently, they depurated their accumulated TCB by the time of sampling. Additional studies, with reduced intervals between feeding and sampling, are planned to assess the accumulation of TCB from food.

The moderate lipophilic nature of TCB (K_{ow}, 4.16) and its associated rapid depuration result in TCB being a transient xenobiotic in fish tissue. Other compounds with higher K_{ow}s are not depurated as readily and accumulate gradually over an extended time before attaining steady state. For such compounds, contributions from dietary sources may be accumulated. Pizza and O'Conner [4] showed that dietary PCB was sequentially accumulated by striped bass when fed multiple doses. Fish fed 287 ng PCB in food every 48 h cumulatively retained 67% of the predicted steady-state body burden (1592 ng PCB) by the fourth dose; steady state was predicted at Dose 17, or 32 days. At steady state, they calculated a 24% fluctuation as the result of PCB depuration between 48-h doses, resulting in a flux of 1210 to 1592 ng. For spot simultaneously exposed to quantitatively similar aqueous and dietary sources of chlordecone, Fisher et al. [6]

TABLE 2—*BIOFAC-modeled pharmacokinetic parameters for the bioaccumulation of 1,2,4-trichlorobenzene (TCB) by spot* (Leiostomus xanthurus). *The time-weighted average concentration of TCB in seawater was 8.5 μg/L; the contaminated diet was pink shrimp* (Penaeus duorarum) *previously exposed to TCB in water (body burden, 0.59 μg/g). Uptake was 28 days followed by a 1-day depuration in TCB-free seawater.*

Treatment	Uptake Rate Constant, mL/h/g	Depuration Rate Constant, h^{-1}	BCF	Depuration Half-Life, days	Time to 90% Steady-State, days
TCB-water/no food (unfed)	226[a] (±5.0)	2.67 (±0.02)	85 (±2)	0.26	0.86 (±0.01)
TCB-water/clean food (water)	456 (±34)	3.20 (±0.06)	143 (±11)	0.22	0.72 (±0.01)
TCB-water/TCB-food (water/food)	309 (±17)	2.96 (±0.07)	104 (±6)	0.23	0.78 (±0.02)

[a]Optimal parameter estimates and linearized standard deviations [10].

found that dietary chlordecone was additive and contributed up to 38% of the total body burden, depending on the ration size offered.

While the data from our study with TCB and spot corroborate the generalization by Macek et al. [1] that for most compounds dietary sources of contamination are relatively insignificant compared to aqueous contamination, the studies cited above indicate that some of the more environmentally persistent contaminants (that is, chlordecone, PCBs, and DDT) may be accumulated from dietary sources. TCB was the first compound we evaluated to determine the importance of dietary accumulation for compounds of different lipophilicity. We currently are conducting a similar test with hexachlorobenzene which is considerably more lipophilic (K_{ow} of 6.18) [7]. Data from this and other studies will contribute to evaluating the potential for chemical bioaccumulation from dietary sources based on K_{ow} values.

References

[1] Macek, K. J., Petrocelli, S. R., and Sleight, B. H. in *Aquatic Toxicology (Second Conference); ASTM STP 667,* L. L. Marking and R. A. Kimble, Eds., American Society for Testing and Materials, Philadelphia, 1979, pp. 251–268.

[2] Thomann, R. V., *Canadian Journal of Fisheries Aquatic Science,* Vol. 38, 1981, pp. 280–296.

[3] Bruggerman, W. A., Opperhuizen, W., Wijbenga, A., and Hutzinger, O., *Toxicological and Environmental Chemistry,* Vol. 7, 1984, pp. 173–189.

[4] Pizza, J. C. and O'Conner, J. N., *Aquatic Toxicology,* Vol. 3, 1983, pp. 313-327.

[5] Connolly, J. P. and Tonelli, R., *Estuarine Coastal and Shelf Science,* Vol. 20, 1985, pp. 349–366.

[6] Fisher, D. J., Clark, J. R., Roberts Jr., M. H., Connolly, J. P., and Mueller, L. H., *Aquatic Toxicology,* Vol. 9, 1986, pp. 161–178.

[7] "Water-Related Environmental Fate of 129 Priority Pollutants, Vol. II," EPA 440/4-79-029b, U.S. EPA, Washington, DC, 1979, pp. 76-1 to 77-13.

[8] *1985 Annual Book of ASTM Standards,* Vol. 11.04, American Society for Testing and Materials, Philadelphia, 1985, pp. 590–612.

[9] "SAS User's Guide: Statistics," SAS Institute, Inc., Cary, NC, 1982, pp. 501–512.

[10] Blau, G. E. and Agin, G. L., "A Users Guide for BIOFAC: A Computer Program for Characterizing the Rates of Uptake and Clearance of Chemicals in Aquatic Organisms," Dow Chemical Co., Midland, MI, 1978.

Kanth M. S. Sundaram[1]

Toxicity and Metabolism of Mexacarbate in Freshwater Crayfish Under Laboratory Conditions

REFERENCE: Sundaram, K. M. S., "**Toxicity and Metabolism of Mexacarbate in Freshwater Crayfish Under Laboratory Conditions,**" *Aquatic Toxicology and Hazard Assessment: 12th Volume, ASTM STP 1027*, U. M. Cowgill and L. R. Williams, Eds., American Society for Testing and Materials, Philadelphia, 1989, pp. 270–286.

ABSTRACT: Adult crayfish *(Orconetes limosus)* were exposed to 0.1, 1.0, 2.5, 4.0, 7.5, 8.0, 8.5, 9.5, 15.0, 20.0, and 25.0 μg/mL of mexacarbate [4-dimethylamino-3,5-xylyl N-methylcarbamate (Zectran)] in water at 15°C under laboratory conditions. Behavioral abnormalities were noted at levels ≥1.0 μg/mL, but as the duration of exposure increased the animals recovered and behaved similarly to the control group. Signs of acute toxicity and mortality were observed at concentration levels ≥2.5 μg/mL. The 96-h LC_{50} value was 8.8 μg/mL. Body weight depression was significant only in the three high dose groups (at 15, 20, and 25 μg/mL). The study was terminated 144 h after treatment. The parent compound (M) and seven metabolites, namely, 4-methylformamido mexacarbate (MFM), 4-formamido mexacarbate (FAM), 4-methylamino mexacarbate (MAM), 4-amino mexacarbate (AM), 4-dimethylamino-3,5-xylenol (DMAX), 4-methylamino-3,5-xylenol (MAX), and 4-amino-3,5-xylenol (AX) were detected in crayfish 144 h after exposure. The primary metabolites detected were MFM and FAM, which accounted for 30 and 20%, respectively, of the body residue. Other metabolites detected accounted for about 9% of the body residue altogether. About 41% of the body residue was contributed by the parent material M. The bioconcentration factor of mexacarbate in crayfish was very low, about 0.58, compared to 2000 to 3000 reported in the literature for fenvalerate and permethrin in fathead minnows.

KEY WORDS: mexacarbate, crayfish, 96-h LC_{50}, metabolic products, bioconcentration, LT_{50}, body weight depression, acute toxicity, behavioral abnormalities, rate of uptake

Mexacarbate (4-dimethylamino-3,5-xylyl *N*-methylcarbamate), which belongs to the carbamate group of insecticides, was first developed and marketed in 1961 by the Dow Chemical Co. under the trade name Zectran [*1*]. It was found to be effective against a wide spectrum of insects [*2–9*]. However, it showed only minor effects on various nontarget species [*10–11*] and was relatively nonpersistent in the environment [*12–14*].

Mexacarbate has a relatively high oral toxicity to mammals and to some types of fish [*11*]. Where tested, its toxicity to aquatic invertebrates was less than that of permethrin [3-phenoxybenzyl (±) cis-, trans-3-(2,2-dichlorovinyl)-2,2-dimethylcyclopropanecarboxylate], but slightly greater than that of aminocarb (4-dimethylamino-*m*-tolyl *N*-methylcarbamate) [*15*]. However, little is known about the toxicity of the chemical to freshwater crustaceans. Biotransformations of pesticides are known to alter their toxicity in the target and nontarget organisms [*16,17*]. The present study was designed to investigate the acute toxicity and in vivo metabolism of mexacarbate in freshwater crayfish, *Orconetes limosus*.

[1]Research scientist and project leader, Government of Canada, Canadian Forestry Service, Forest Pest Management Institute, Sault Ste. Marie, Ontario, Canada P6A 5M7.

Materials and Methods

1. *Crayfish Samples*

Adult crayfish with an average body weight of 5.3 g (SD = 0.7 g, n = 200) were trapped from the East Davignon Creek (physical properties and ionic contents of the river water are given in Table 1) near Sault Ste. Marie, Ontario, Canada. The animals were acclimatized in the laboratory for one week before experimentation in a 30-L glass aquarium filled with water collected from the same creek. They were fed shredded minnow until one day prior to the start of the experiment, and then starved for 24 h before exposure to the toxicant. This was done because starved animals could show a much greater initial response to toxic chemicals than unstarved ones [*18*].

2. *Acute Toxicity Study*

a. *Exposure System and Toxicant Preparation*
The experiment was conducted in an environmental chamber (2.80 by 1.90 by 2.44 m) maintained at a temperature of 15 ± 1°C, relative humidity of 85 ± 5%, and a photoperiod of 16 h light and 8 h darkness.

Thirty-six polypropylene tanks, each measuring 30 by 20 by 10 cm, were selected (that is, 33 for treatment and 3 for control) for exposure of the crayfish to the chemical. A 50 mg/mL stock solution of mexacarbate was made in absolute ethanol using crystalline Zectran (99% pure with a melting point of 85°C, obtained from Union Carbide Agricultural Products Co. Inc., North Carolina). Aliquots of 3.0-L stream water were taken in each exposure tank. Varying quantities of the stock solution, ranging from 0.006, 0.06, 0.15, 0.24, 0.45, 0.48, 0.51, 0.57, 0.90, 1.20, and 1.50 mL, were added to the treatment tanks to provide eleven different concentrations (in triplicate) of 0.1, 1.0, 2.5, 4.0, 7.5, 8.0, 8.5, 9.5, 15.0, 20.0, and 25.0 μg/mL, respectively. In the control tanks, 1.50 mL of absolute ethanol were added. Healthy crayfish, about 3 to 5 per tank (depending upon the body weight), were randomly assigned to each treatment group. A total of 15 crayfish were assigned to the control group at the rate of 5 per tank. The number used in each group varied from 10 to 15, and their body lengths and weights ranged from 44 to 66 mm and 4.4 to 6.7 g, respectively. The treated and untreated stream water was replaced every 24 h

TABLE 1—*Physical properties and ionic contents of river water used in the crayfish exposure study.*

Water temperature, °C	15
Turbidity, JTU	0.9
pH	6.4
Specific conductance, μmhos/cm	25.6
Alkalinity, mg/L $CaCO_3$	9.7
Hardness, mg/L $CaCO_3$	90
SO_4^{--}, mg/L	5.4
Cl^-, mg/L	4.0
Total nitrogen, mL/L as N_2	0.02
Cu^{++}, mg/L	0.01
Fe^{++}, mg/L	0.32
Mn^{++}, mg/L	0.03
Zn^{++}, mg/L	0.01
Pb^{++}, mg/L	0.001
Al^{+++}, mg/L	0.07

NOTE: JTU = Jackson turbidity units.

and was aerated during the experiment using a pump (Little Giant, Model No. 13152, Gelman Sciences, Inc., Ann Arbor, Michigan). The tanks were covered with perforated lids to ensure good air circulation. The animals were fed throughout the exposure period with shredded minnow [samples of which were monitored for mexacarbate residues and were found to have none, that is, levels below the detection limit (0.01 μg/g) of the analytical technique].

b. *Assessment of Toxicity*

i. *Behavioral patterns, mortality counts and LC_{50} determination*—Behavioral patterns of the crayfish were observed four times daily for a duration of 10 min each time. Simultaneously, the number of dead animals were noted, and they were removed from the tanks. Death was defined as complete immobilization and failure of the animal to respond to gentle prodding. Mortality occurred within 24 h after treatment only in tanks containing the highest concentration of the toxicant (that is, 25 μg/mL), and, therefore, these mortality counts were not used for further data treatment. Mortality data from triplicate tanks were combined, and the observed percentage mortality (P_o) was calculated at 48, 72, 96, 120, and 144 h after treatment (Table 2). The corrected percentage (P_t) was then computed using the Abbott's formula [*19*]

$$P_t = \frac{P_o - P_c}{100 - P_c} \times 100$$

where P_c represents the percentage mortality in the control group. The P_t values were then subjected to probit analysis [*20*] with mortalities expressed as probits and water concentrations converted to a logarithmic scale. The LC_{50} values, the median lethal concentration required to produce mortality in 50% of the test population over each period of exposure, that is, at 48, 72, 96, 120, and 144 h, respectively, were calculated using the SAS statistical computer package (SAS Institute Inc. [*21*]). Figures 1(*a*) and 1(*b*) provide the plots for 96-h LC_{50} value.

Aliquots of the untreated and treated water were taken from each tank immediately after fortification and twice every day (once every 12 h) until the end of the study (Table 2) for the determination of mexacarbate concentration by gas-liquid-chromatographic (GLC) analysis.

ii. *Changes in body weight*—Body weight depression is an important parameter in toxicity assessment [*22*]. To examine this symptom, each animal was weighed before treatment and the group mean ±SD values were calculated. This was accomplished by weighing the animal by difference in a 200-mL beaker, which was fitted with a lid and contained 40 mL of river water. During the course of the study, the dead animals were removed from the tank at each observation period (four times daily). The adsorbed water was wiped off using a paper towel, and the body weights and lengths were measured at once. This process was continued until the end of the study. Animals which were alive at 144 h were weighed, and the weights of the dead and live animals were combined together to calculate the mean ±SD for each treatment group (Table 2). From these data, the changes in body weight (Δw) were calculated.

iii. *Behavioral abnormalities and signs of intoxication*—Behavioral changes monitored at low concentrations of the toxicant were: locomotion (swimming and tail flipping), hyperactivity (fighting), and coordination. At high concentrations, immobilization of the animal and fluttering of the gills were also monitored.

c. *Water Concentrations*

Two-hundred and fifty mL of water, collected at the middepth of the tank, were extracted twice with 50 mL of dichloromethane. All the organic extracts corresponding to one treatment group were pooled and dried by passing through a 3-cm diameter by 5-cm length column of granular anhydrous sodium sulfate to remove all traces of moisture. The extract was then flash evaporated gently at 30°C to dryness, and the residue was taken in ethyl acetate for GLC analysis of mexacarbate without any further column cleanup. The data are given in Table 2.

d. *Residues in Crayfish*

i. *Extraction*—The crayfish samples that were dead during the course of the study were removed from the tank, rinsed twice with acetone to get rid of the adsorbed mexacarbate,

TABLE 2—*Toxicant concentrations and percentage mortality observed at different time intervals during the course of the study, and final body weights and body weight changes.*

Treatment Group No.	Toxicant Concentration, μg/mL (mean ± SD)[a]	Observed Mortality, %[b]					Body wt at 144 h, g, (mean ± SD)[c]	Body wt Changes, Δw (g)[d]
		48 h	72 h	96 h	120 h	144 h		
G-0.1	0.08 ± 0.02	0.0	0.0	0.0	0.0	0.0	5.2 ± 0.4	+0.2
G-1.0	0.83 ± 0.07	0.0	0.0	0.0	0.0	0.0	5.5 ± 0.6	+0.4
G-2.5	2.43 ± 0.32	0.0	0.0	0.0	13.3	13.3	5.5 ± 0.7	+0.2
G-4.0	3.74 ± 0.42	0.0	6.7	6.7	40.0	40.0	5.2 ± 0.5	−0.2
G-7.5	6.38 ± 1.07	0.0	8.3	16.7	48.3	48.3	5.2 ± 0.5	0.0
G-8.0	7.04 ± 1.40	0.0	27.3	38.6	55.5	65.5	4.8 ± 0.6	−0.6
G-8.5	6.97 ± 0.99	0.0	30.8	45.5	60.0	80.0	4.4 ± 0.4	−1.4
G-9.5	8.35 ± 1.40	0.0	36.7	60.0	71.9	85.0	4.4 ± 0.5	−0.6
G-15.0	13.45 ± 1.50	9.1	45.5	72.7	78.9	93.0	3.6 ± 0.6[e]	−1.6[e]
G-20.0	19.22 ± 1.40	25.3	51.2	87.5	90.0	100.0	3.8 ± 0.8[e]	−1.7[e]
G-25.0	24.11 ± 1.90	40.0	66.6	93.3	93.3	100.0	4.0 ± 0.8[e]	−1.6[e]
G-cont.	0.00 ± 0.00	0.0	0.0	0.0	0.0	0.0	6.1 ± 0.6	+0.3

[a]Mean ±SD values represent toxicant concentrations in water samples collected from the tanks every 12 h for a period of six days (n = 12 for each treatment group).

[b]The observed percentage mortality (P_o) is equal to the corrected percentage mortality (P_t) because mortality was nil in the control group.

[c]Mean ±SD values were calculated using body weights of animals which were alive and dead at 144 h. However, each dead animal was weighted at once and not at 144 h. This implies that when an animal died at, say, 72 h, its weight at 72 h was combined with the weights of the living animals (at 144 h) before the mean ±SD was calculated.

[d]Positive Δw represents weight gain, while negative Δw represents weight loss.

[e]These body weight changes are significantly different from the value observed in the control group, as indicated by the Student-Newman-Keuls test. See Table 5 for the relevant parameters.

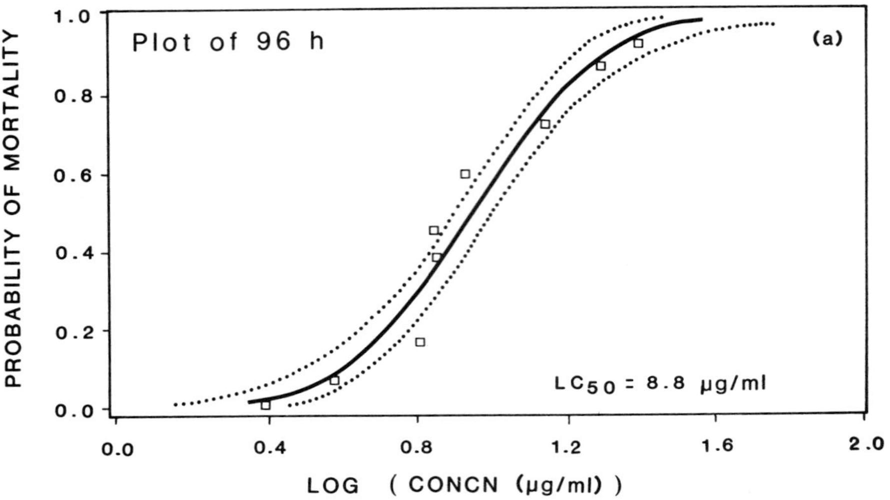

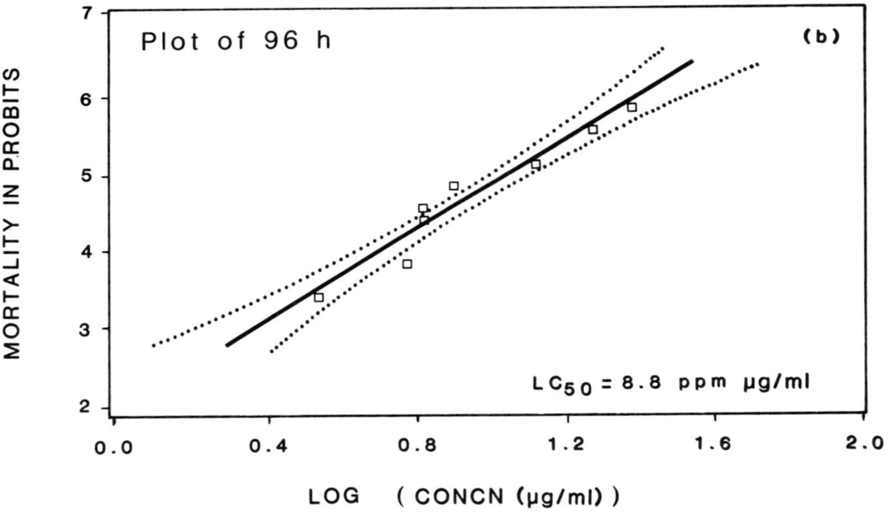

FIG. 1—*Log*$_{10}$ *of mexacarbate concentrations in water* versus (a) *probability of mortality,* (b) *mortality in probits, and their 95% confidence limits. The lines were determined by "SAS" statistics* [21].

weighed, and then stored at −20°C until GLC analysis. At test termination, controls and surviving mexacarbate-exposed crayfish were handled in a similar manner.

The crayfish samples of each tank were pooled so that three replicate samples could be analyzed in each treatment group. Each composite sample was chopped separately into small pieces using a sharp knife and mixed thoroughly. About 10-g aliquots (wet wt) were homogenized thrice in a Polytron PT-20 at a speed setting of 3.5 with 50 mL of ethyl acetate each time. The supernatant extract was filtered through a sintered glass (Pyrex) Buchner funnel (ID, 60 mm; 10 to 15-µm pore size) containing a 3-cm layer of anhydrous Na_2SO_4. The extraction was repeated twice more using the same volume of ethyl acetate and filtered through the same funnel. The filter cake in the funnel was finally rinsed with 20 mL of ethyl acetate. All the organic extracts belonging to one composite sample were pooled, flash evaporated to dryness, and the residue taken in 30 mL of acetonitrile. The polar acetonitrile was partitioned twice, each time with 10 mL of hexane. The hexane layers were discarded and the acetonitrile layer was flash evaporated gently at 30°C to dryness. The residue was taken up in 5-mL ethyl acetate for column cleanup so that 1 mL of extract was equivalent to 2 g of wet wt of crayfish. When necessary, the extracts were evaporated under a steady stream of nitrogen gas to bring the mexacarbate concentration above the detection limit.

ii. *Thin layer chromatographic (TLC) separation of mexacarbate*—The residues of mexacarbate and its metabolites (Fig. 2) present in the crayfish extracts were separated by thin layer chromatography (TLC) using precoated high-performance (HP) silica gel plates containing a 254-nm fluorescent indicator (HP-7011-4, 10 by 10 cm, 200-μm thickness; J. T. Baker

No.	CHEMICAL STRUCTURE	NAME (IUPAC USAGE)	ABBR.	R_f VALUES
1	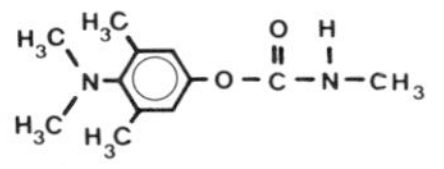	4-Dimethylamino-3,5-xylyl N-methylcarbamate	M	0.65
2	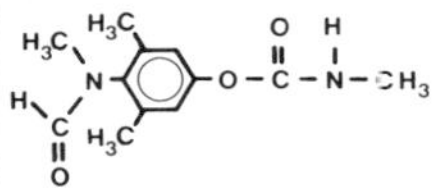	4-Methylformamido-3,5-xylyl N-methylcarbamate	MFM	0.26
3	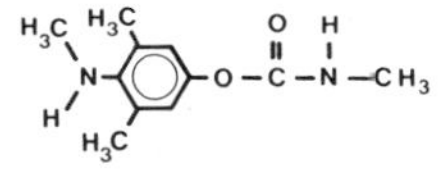	4-Methylamino-3,5-xylyl N-methylcarbamate	MAM	0.36
4	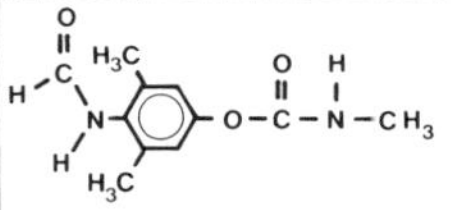	4-Formamido-3,5-xylyl N-methylcarbamate	FAM	0.12
5	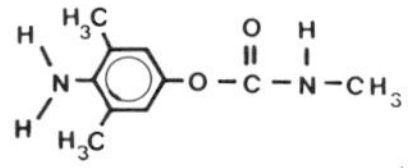	4-Amino-3,5-xylyl N-methylcarbamate	AM	0.31
6	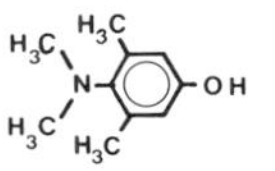	4-Dimethylamino-3,5-xylenol	DMAX	0.78
7	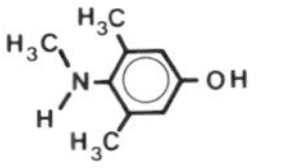	4-Methylamino-3,5-xylenol	MAX	0.39
8	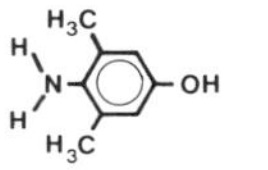	4-Amino-3,5-xylenol	AX	0.40

FIG. 2—*Chemical structures, IUPAC names and abbreviations used for mexacarbate and its metabolites, and their* R_f *values used in thin layer chromatography using a ether: hexane: ethanol (77:20:3 v/v%) solvent system.*

Chemical Co., Phillipsburg, NJ). The TLC plates were heated in an oven at 110°C for 1-h before using. One mL of the extract (2 g of crayfish/mL) was evaporated under a gentle stream of N_2 to a volume of 100 μL. This amount was spotted on the TLC plate about 1.5 cm above the lower edge. The authentic pure sample of mexacarbate acted as the reference standard. The R_f values obtained for the solvent system used in the study are given in Fig. 2. The spotted plates were developed in a glass tank saturated with the solvent system. The developed plates were removed from the tank when the solvent front was 8.0 cm from the origin. The plates were air dried and then viewed with the aid of a UV cabinet.

The spot corresponding to the R_f value of mexacarbate was scraped from the plate. The scrapings were transferred to a glass centrifuge tube containing 10 mL of acetonitrile, sonicated in a water bath for 15 min and then centrifuged at 2000 rpm for 10 min. The supernatant liquid was transferred quantitatively to another 15-mL glass centrifuge tube and evaporated to dryness under a gentle stream of N_2. The residue (corresponding to 2 g of crayfish) was taken in ethyl acetate, and the volume was adjusted volumetrically for GLC analysis.

iii. *GLC analysis of mexacarbate residues in tank water and crayfish tissues*—A Hewlett Packard (HP) model 5890A gas chromatograph fitted with a nitrogen-phosphorus flame-ionization detector, a HP3392A electronic integrator for measurement of peak areas, and a preconditioned J & W fused silica capillary column (15 m by 0.531 mm ID, liquid phase DB-17, film thickness 1.0 μm) was used. Two μL of the concentrated extract was injected each time onto the column using the splitless injection mode with a 30-s delay before the inlet was purged. Quantitation was by peak height. Carrier gas (He) and plasma gas (H_2 and air) flow rates (mL/min) were: He, 15; H_2, 4; air, 120; and He (makeup), 15. The operating temperatures (°C) were: detector, 220; injection port, 150. The oven temperature was programmed as listed in Table 3.

Prior to the analysis of mexacarbate and its metabolites in the crayfish, the solution of mixed standards was injected daily to establish detector sensitivity, linear range, and retention time [retention time (RT) in min] of the individual compounds. Under these conditions, the RTs of the different compounds were found to be those listed in Fig. 3.

Crayfish and water samples collected from the untreated tanks were subjected to the same extraction and analytical procedures to determine whether any coextracted impurities would interfere with the GLC analysis. Figure 3 provides, as an example, the chromatograms obtained from extracts of crayfish (*a*) exposed to mexacarbate at 15 μg/mL in water, and (*b*) untreated control.

iv. *Recovery*—Extracts from the control and pretreatment samples (both crayfish and water) were fortified with varying concentrations of mexacarbate and analyzed according to the above procedures to determine the extraction efficiencies of the experimental techniques. Measurements indicated excellent recovery efficiencies, ranging from 95 to 98% for the water samples and from 90 to 97% for the crayfish samples fortified with all concentration levels of mexacarbate used in the study (results reported throughout this paper are not corrected for recovery efficiency).

TABLE 3—*Oven temperature program used for the J & W fused silica capillary column for analyzing mexacarbate and its metabolites by GLC.*

Initial Oven Temperature, °C	Initial Time, min	Initial Ramp Rate, °C/min	Oven Temperature 1, °C	Time 1, min	Ramp Rate 1, °C/min	Final Oven Temperature, °C	Final Time, min
50	0.5	70	140	2.0	20	200	11.0

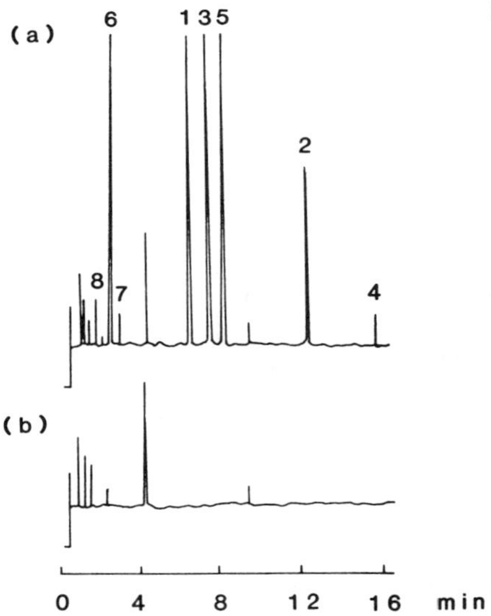

Peak no.	Compound abbr.	RT's (min)
1	M	6.6
2	MFM	12.3
3	MAM	7.5
4	FAM	15.7
5	AM	8.5
6	DMAX	2.5
7	MAX	3.0
8	AX	1.8

FIG. 3—*Chromatograms of crayfish extracts, indicating the peaks corresponding to different retention times (RTs) of mexacarbate and its metabolites, analyzed by GLC using the J & W fused silica capillary column:* (a) *crayfish exposed to 15 μg/mL in water, and* (b) *untreated control.*

3. Metabolism of Mexacarbate in Crayfish

The thin layer chromatographic treatment of the crayfish extract separated not only the intact chemical from its metabolites but also the different metabolites that were present (Fig. 2).

The spots corresponding to the R_f values of the different metabolites were scraped from the TLC plates, subjected to the same cleanup procedures as described earlier, and analyzed by GLC. Quantitation was done as before. The relative amounts of the eight metabolites found are given in Table 4.

Results and Discussion

1. *Acute Toxicity*

a. *Mortality Counts and LC_{50} Values*

The first mortality occurred at the highest exposure level of 25 μg/mL, within 24 h after treatment. No mortality was noted in the control group and at concentration levels ≤ 1.0 μg/mL (Table 2), and the first mortality at the 2.5 μg/mL level occurred within 120 h after treatment.

The LC_{50} values, calculated for the exposure periods of 48 to 144 h, are given in Fig. 4, which represents the plot of exposure period *versus* LC_{50}. The data indicate that the mortality counts were low in the animals during exposure for 24 to 48 h, but increased rapidly from 96 h onwards. There was a rapid decrease in the LC_{50} values (about five times) when the exposure increased from 48 to 144 h.

TABLE 4—*Residues of mexacarbate and its metabolites found in crayfish after 144 h exposure to different concentrations of the chemical in stream water.*

Treatment Group No.	Residues in Crayfish, $\mu g/g$ wet weight							
	M	MAM	AM	MFM	FAM	DMAX	MAX	AX
G-0.1	0.03	0.00	0.00	0.13	0.05	0.000	0.007	0.000
G-1.0	0.31	0.00	0.00	1.03	0.13	0.000	0.023	0.016
G-2.5	1.02	0.14	0.01	1.25	0.77	0.043	0.055	0.023
G-4.0	1.86	0.21	0.04	2.70	0.95	0.032	0.051	0.052
G-7.5	4.03	0.23	0.21	2.87	1.32	0.061	0.084	0.066
G-8.0	4.76	0.24	0.34	2.60	0.68	0.070	0.089	0.085
G-8.5	5.22	0.28	0.22	2.75	1.87	0.088	0.086	0.080
G-9.5	6.00	0.32	0.38	3.40	1.81	0.170	0.233	0.040
G-15.0	8.58	0.41	1.27	6.35	3.71	0.260	0.440	0.052
G-20.0	12.00	0.53	1.50	6.80	4.20	0.400	0.620	0.093
G-25.0	15.20	0.66	2.09	7.80	5.20	0.585	0.880	0.128
G-cont.	ND	ND	ND	ND	ND	ND	ND	ND

NOTE: ND = not detectable.

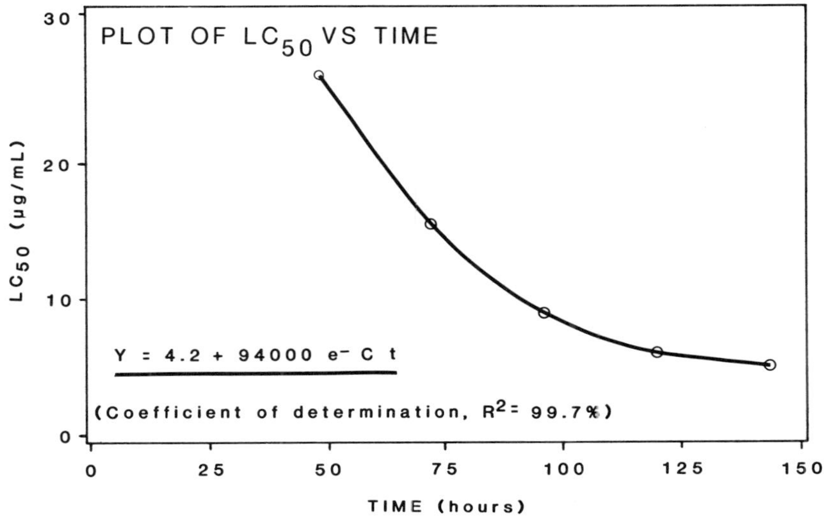

FIG. 4—*Exponential decrease of LC_{50} values of mexacarbate (in µg/mL of water) with time of exposure to crayfish.*

The decrease of the LC_{50} values with exposure time followed an exponential equation of the type

$$Y = A + Be^{-Ct} \tag{1}$$

from which Eqs 2 and 3 can be derived

$$Y = A \qquad (\text{when } t \rightarrow \infty) \tag{2}$$

$$Y = A + B \qquad (\text{when } t \rightarrow 0) \tag{3}$$

The constant A represents the LC_{50} value which would be encountered after exposure for a very long duration. Constants B and C provide the regression coefficient and the rate constant for the exponential relationship, respectively.

Curvilinear regression analysis of the data provided the numerical value of $A = 4.2$ µg/mL of water. The implication of this value is that, if exposure periods were very long, a relatively low water concentration of mexacarbate would be sufficient to cause 50% kill in the crayfish, provided that the metabolic processes would be similar at the low and medium exposure levels.

From the 96-h mortality data the LC_{50} and its 95% confidence limits were determined to be 8.8 and 7.4 to 10.5 µg/mL, respectively. The present 96-h LC_{50} is much lower than (about one fourth) that of aminocarb (32.7 µg/mL) toxicity to crayfish [*23*]. Similar observation was also made by Poirier and Surgenor [*15*], who demonstrated that mexacarbate was more toxic to certain stream invertebrates (black flies, dragon flies, mayflies, etc.) than aminocarb. The higher toxicity of mexacarbate than other insecticides was also shown in the western spruce budworm larvae, *Choristoneura occidentalis;* the 7-day LC_{50} for mexacarbate was 4.6 µg/g body weight as compared to those of DDT and pyrethrins, which were 24.4 and 12.2 µg/g, respectively [*11*].

b. *Body Weight Changes*

To compute the significance in body weight changes in crayfish due to exposure to mexacarbate, the following procedure was used. Initial average body weight was calculated prior to treatment for each of the three replicate tanks used for each dose group. This step provided three initial average body weight values for each dose group (that is, 3 × 12 = 36 average values for the 12 dose groups including the control group). Similarly, three final average body weight values were computed for each dose group after 144 h of exposure (for those animals which died before 144 h duration, the final body weights were obtained soon after death). The two sets of data (that is, 36 values before treatment and 36 after treatment) were subjected the Student-Newman-Keuls test. The relevant parameters are given in Table 5. No significant difference was noted in the exposure groups from 0.1 to 9.5 μg/mL and in the control group (Table 2). However, significantly lower body weights were noted in animals exposed to high levels of mexacarbate, that is, 15, 20, and 25 μg/mL in water. The data thus indicated that significant body weight depression occurred only at concentration levels greater than the 96-h LC_{50} value.

TABLE 5—*Results of the Student-Newman-Keuls test for body weight data [error degrees of freedom (EDF) = number of measurements* (K) × *number of replicates per dose group* (N) − *1] of crayfish prior to and after exposure to varying concentrations of mexacarbate in river water.*

Sample No.[p]	Toxicant Concentration,[q] μg/mL	Maximum Nonsignificant Ranges for Subsets a, b, and c[s]		
		1	2	3
1	0.1	a	b	c
2	1.0	a	b	c
3	2.5	a	b	c
4	4.0	a	b	c
5	7.5	a	b	c
6	8.0	a	b	c
7	8.5	...	b	c
8	9.5	a	b	c
9	15.0	a	b	c
10	20.0	a	b	c
11	25.0	...	b	c
12	0.00[r]	...	b	c
13	0.1	a	b	c
14	1.0	a	b	c
15	2.5	a	b	c
16	4.0	a	b	c
17	7.5	a	b	c
18	8.0	a	b	c
19	8.5	a	b	c
20	9.5	a	b	c
21	15.0	a	...	...[t]
22	20.0	a	...	...[t]
23	25.0	a	b	...[t]
24	0.00[r]	...	...	c

[p]Sample numbers 1 to 12 represent the dose groups of crayfish prior to exposure to the toxicant, and from 13 to 24, after exposure to the toxicant.

[q]The values refer to those added to water in the exposure tanks.

[r]These dose groups refer to the control group.

[s]These nonsignificant ranges were based on 95% confidence limits [EDF = 24 × (3 − 1) = 48, and error mean square value = 0.415].

[t]Only these three dose groups are significantly different from the control.

c. *Behavioral Abnormalities and Signs of Intoxication*

At the lowest exposure level, all animals were healthy and their behavioral patterns were comparable to those of the control animals. However, at 1.0 μg/mL the animals exhibited hyperactivity after 12 h of exposure, which was evident from the increased swimming activity (about 35% higher than that in control animals), fighting, and tail flipping, although all of these activities were performed with complete coordination. As the duration of exposure increased, the hyperactivity increased, but all animals recovered after 60-h posttreatment and were healthy afterwards similar to the control animals. At 2.5 μg/mL the hyperactivity increased and, as time progressed, the animals were found to swim near the surface of the water; this behavior was followed by darting and erratic bursts of swimming activity, which finally resulted in complete paralysis (or immobilization) of the animal. Once this level of intoxication was reached, mortality eventually occurred. At concentrations ≥8.0 μg/mL, paralysis occurred within 60 h. The fluttering of the gills was the last observable movement in the crayfish before death. The extent of paralysis and frequency of mortality was in proportion to the concentration of mexacarbate in water.

d. *Mexacarbate Concentrations in Water*

From the data in Table 2, it is evident that the measured water concentrations (mean ±SD) were only slightly lower than those the tanks were spiked with initially. This finding indicates that the 24-h replacement procedure for the tank water was quite adequate for meeting the requirements of the study.

e. *Concentration Factor, Time to 50% Mortality and Net Uptake Rate in Crayfish Tissues*

To describe mexacarbate uptake by crayfish in terms of the measured water concentration *C*, two types of parameters, namely, "concentration factor (*CF*)" and "net uptake rate," were derived. (i) The *CF* values were obtained by dividing the tissue residues (*M* in Table 4) by the mean water concentration (*C*) (Table 2) [*24*], and (ii) the net uptake rates were determined by dividing the tissue residues (in ng/g) by the "time to 50% mortality (LT_{50})." The latter was calculated from a nonlinear regression analysis [*25*] of the relationship between the corrected percentage mortality (P_t) and the duration of exposure [Y and X coordinates, respectively, in Fig. 5(*a*)]. The data are given in Table 6.

It is worthwhile to mention that the rate of uptake of mexacarbate by crayfish, as determined in a preliminary study, was very fast, and hence the fact that some of the tissue residue data were based on animals which died at varying times while some were based on those alive until 144 h did not significantly alter the net uptake rate values given in Table 6.

A plot of LT_{50} values versus *C* indicated a rapid decline between *C* values of 6 to 8, followed by a gradual decline from 8 until the highest value of 24 μg/mL [Fig. 5(*c*)].

The *CF* values provided a qualitative measure of the uptake of mexacarbate. A plot of *CF* versus *C* [Fig. 5(*b*)] indicated a steady increase at initial concentration levels, but with increasing mexacarbate amounts, the *CF* values showed a slight decrease, followed by an almost constant value for the concentration range of 12 to 25 μg/mL. This trend suggests an upper limit for the uptake, although it was due to earlier mortality and shorter exposure times in the high dose groups. In order to correct for this apparent anomaly the net uptake rate (in ng/g/h) was plotted against *C*. Figure 5(*d*) shows a steady increase in the uptake rate at all water concentrations.

The present findings are in agreement with those reported in the literature. Bradbury et al. [*26*] found a similar decline of the *CF* values at high water concentration levels for fenvalerate uptake in fathead minnows. The authors attributed the decline to early mortalities which occurred in the high dose groups, too soon to allow sufficient uptake. The net uptake, however, showed a steady increase from the low dose up to the highest dose group.

The *CF* values also provide a rough estimate of bioaccumulation [*26*]. The present data indicate low *CF* values at all dose levels (a mean value of 0.576 in Table 6) exhibiting very low bioconcentration, when compared to the *CF* values in fathead minnows of about 2000 for fen-

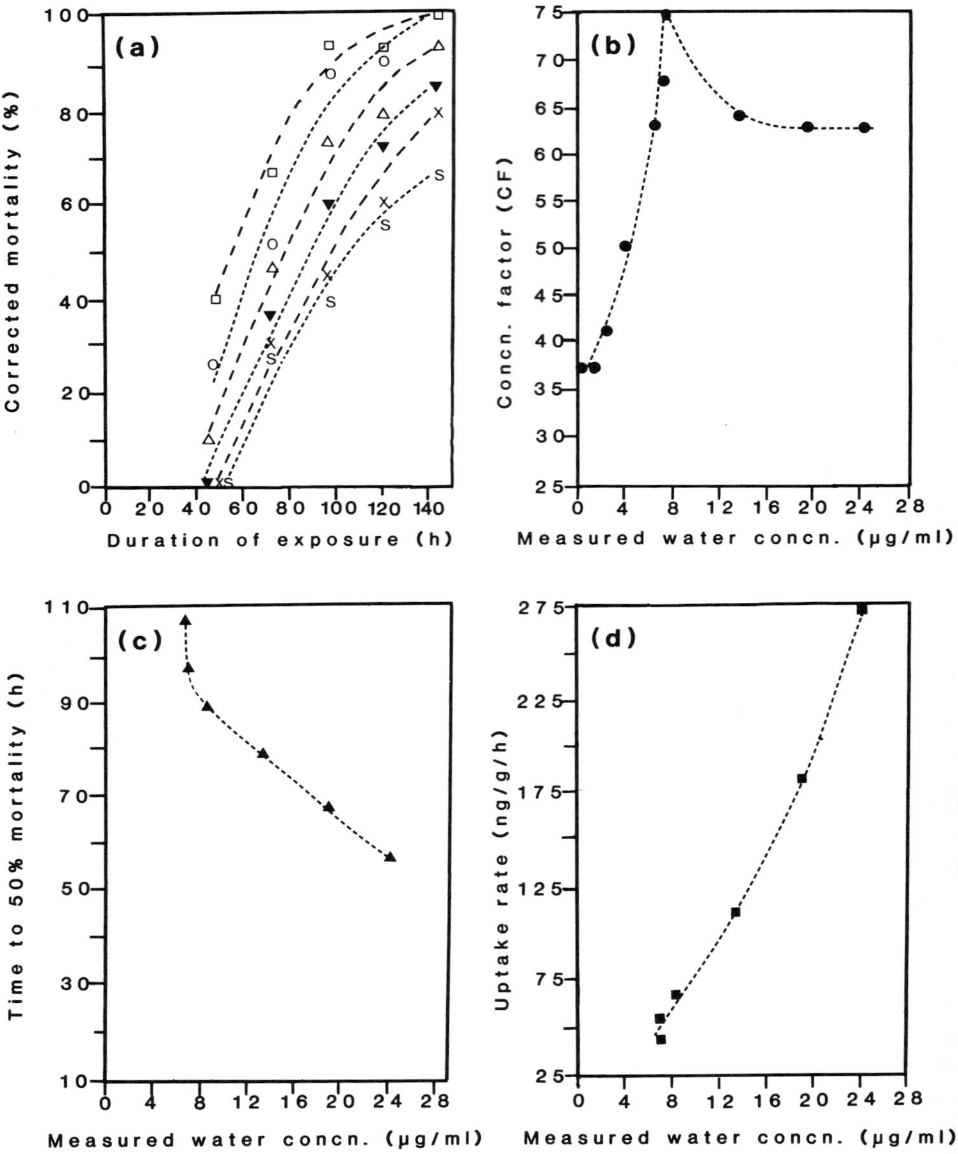

FIG. 5—*Plots of* (a) *corrected mortality* (P_t, %) versus *duration of exposure;* (b) *concentration factor,* CF, versus *water concentration;* (c) *time to 50% mortality* versus *water concentration;* (d) *uptake rate* versus *water concentration.*

valerate [*26*] and of about 2800 for permethrin [*27*]. The low value in the present study was due to the fact that crayfish do not have the high lipid content in their tissues as do fathead minnows, and also because carbamates generally do not bioaccumulate as much as other insecticides [*28*]. The present *CF* values are, however, slightly higher than those observed for aminocarb in crayfish (about 0.20, as compared to 0.576 for mexacarbate) [*23*], indicating that mexacarbate shows a slightly higher bioaccumulation capability in crayfish than aminocarb.

TABLE 6—*Concentration factor, time to 50% mortality, and uptake rate of mexacarbate in tissues of crayfish exposed to different concentrations in water.*

Treatment Group No.	Concentration Factor (*CF*) (M/*C*) × 10²[a]	Time to 50% Mortality,[b] h	Net Uptake Rate, ng/g/h[c]
G-0.1	37.5	...	...
G-1.0	37.3	...	...
G-2.5	42.0	...	...
G-4.0	49.7	...	...
G-7.5	63.2	...	...
G-8.0	67.6	106.0	45
G-8.5	74.9	97.0	54
G-9.5	71.9	89.0	67
G-15.0	63.8	79.0	109
G-20.0	62.4	67.0	179
G-25.0	63.0	56.5	269
G-cont.	0.00	...	...

[a,b,c] For explanation of these terms, see text under "Concentration Factor, Time to 50% Mortality, and "Net Uptake Rate in Crayfish Tissues." (The value of *CF* was multiplied by 100 to avoid handling fractions.)

2. *Metabolism of Mexacarbate in Crayfish*

The data in Table 4 indicate that the major identifiable constituents were the parent material (M) and the two metabolites MFM and FAM. The bulk of the tissue residues was due to intact mexacarbate. Figure 6 shows a gradual increase in mexacarbate and metabolite residues as the water concentration *C* increased. However, the plot was linear only in the case of M, since slight deviations were observed from linearity for all the seven metabolites. None of the eight plots showed any maximum or asymptote (Fig. 6). This behavior indicates that even at the highest exposure level of 25 ppm, the crayfish continued to absorb the chemical and convert it into the different metabolites.

The present data are in contrast with the metabolic pattern of aminocarb in crayfish [*23*]. The primary metabolite of aminocarb was found to be MA (4-methylamino-*m*-tolyl *N*-methylcarbamate), which accounted for 75 to 95% of the tissue residues. In the present study, however, the primary metabolites were MFM and FAM, which accounted for about 30 and 20% of the tissue residues, respectively. The reason for this dissimilarity appears to lie in the structural differences between aminocarb and mexacarbate.

Conclusions

The present findings indicate a "no acute-toxic-effect level" of 1.0 μg/mL (or 1000 ng/mL) in river water for freshwater crayfish exposed continuously to the toxicant for 6 days. In a recent aerial spray trial [*29*] conducted in the province of New Brunswick, Canada, using mexacarbate at a dosage rate of 70 g active material per ha, the maximum concentrations were 0.73 and 18.74 ng/mL in the stream and pond waters, respectively. The implications are that, under field application conditions, there is a minimum safety factor of 1000/18.74 (or about 53) for crayfish if they are exposed continuously to a pond water concentration of 18.74 ng/mL for 6 days. However, the field data also demonstrated a rapid decline of the maximum concentration to levels below the detection limit (0.01 ng/mL) within 3 days in the pond waters. Consequently, it can be concluded that there is sufficient safety margin for freshwater crayfish when exposed to mexacarbate after spray application under operational conditions over conifer forests.

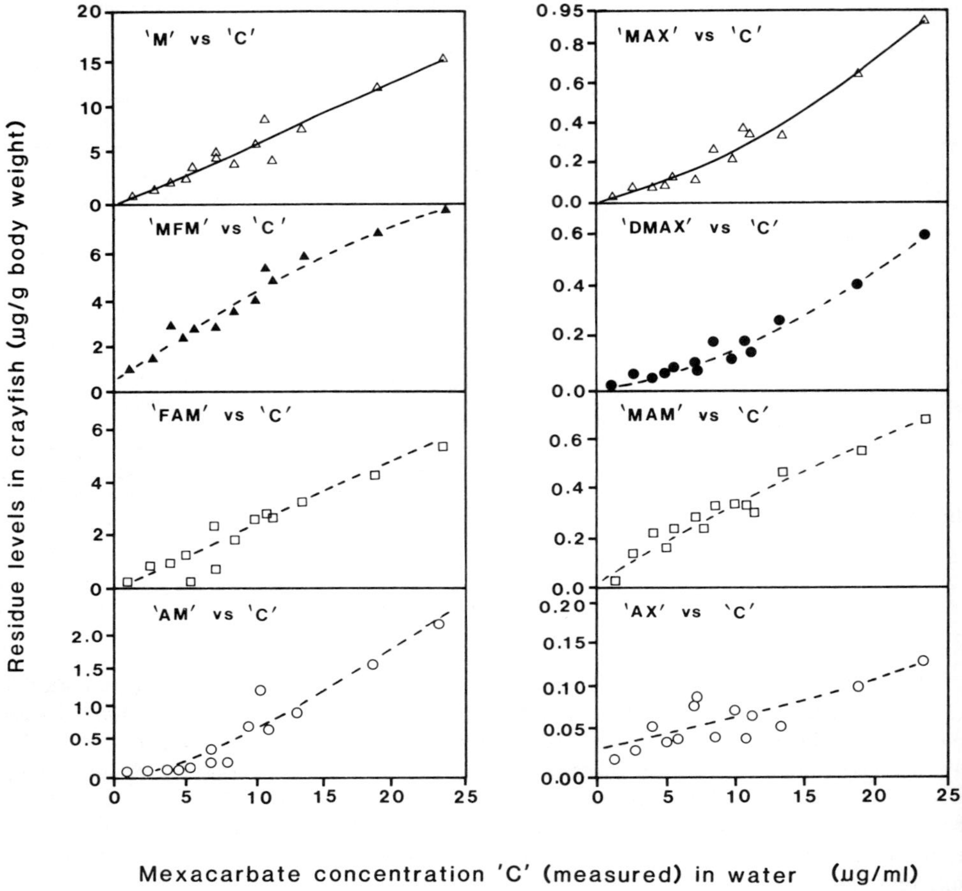

FIG. 6—*Relationships between mexacarbate levels in water (measured) and residues in crayfish after 144 h of exposure. (See Fig. 2 for key to metabolite abbreviations.)*

Acknowledgment

The author thanks Reginald Nott, Terry Giddings, and John W. Leung for their technical assistance in this investigation.

References

[1] Worthing, C. R., *The Pesticide Manual: A World Compendium,* 7th ed., British Crop Protection Council, Croyden, England, 1974, p. 590.

[2] Flake, H. W. and Lyon, R. L., "Insecticide Tests Against Larvae of *Hemerocampa,* New Species, a Tussock Moth That Defoliates Boxelder in New Mexico," *Journal of Economic Entomology,* Vol. 60, 1967, pp. 607–608.

[3] Blais, J. R., Benoit, P., and Martineau, R., "Aerial Control Operations Against the Spruce Budworm in Quebec," in *Aerial Control of Forest Insects in Canada,* Catalogue No. Fo 23/19/1975, M. L. Prebble, Ed., Information Canada, Ottawa, Ontario, K1A OS9, 1975, pp. 113–125.

[4] Wolfenbarger, D. A., "Toxicity of Methylcarbamates and Their Acetylated and Chloroacetylated Analogues Against the Bollworm and the Tobacco Budworm," *Journal of Economic Entomology*, Vol. 66, 1973, pp. 272–274.
[5] Robertson, J. L., Lyon, R. L., Shon, F. L., and Page, M., "Western Blackheaded Budworm: Toxicity of Five Insecticides to Two Populations," *Journal of Economic Entomology*, Vol. 66, 1973, pp. 274–275.
[6] Miskus, R. P., Look, M., Andrews, T. L., and Lyon, R. L., "Biological Activity as an Effect of Structural Changes in Aryl *N*-methylcarbamates," *Journal of Agricultural and Food Chemistry*, Vol. 16, 1968, pp. 605–607.
[7] Maksymiuk, B., Neisess, J., Waite, R. A., and Orchard, R. D., "Distribution of Aerially Applied Mexacarbate in a Coniferous Forest and Correlation with Mortality of *Choristoneura occidentalis* (Lepid., Tortricidae)," *Zeitschrift fur Angewandte Entomologie*, Vol. 79, 1975, pp. 194–204.
[8] Williams, C. B. and Walton, G. S., "Effects of Naled and Zectran on the Budworm *Choristoneura occidentalis* and Associated Insects in Montana," *Journal of Economic Entomology*, Vol. 61, 1968, pp. 784–787.
[9] Lyon, R. L., Flake, H. W., and Ball, L., "Laboratory Tests of 55 Insecticides on Douglas-Fir Tussock Moth Larvae," *Journal of Economic Entomology*, Vol. 63, 1970, pp. 513–518.
[10] Tucker, R. K. and Crabtree, D. G., "Toxicological Studies of a New Spruce Budworm Control Agent (Zectran) with Selected Wildlife Species," *Journal of Economic Entomology*, Vol. 62, 1969, pp. 1307–1310.
[11] Schmiege, D. C., Crisp, C. E., Lyon, R. L., Miskus, R. P., Roberts, R. B., and Shea, P. J., *Evaluation Report on Zectran^R, A Substitute for DDT in the Control of Western Spruce Budworm*. Insecticide Evaluation Project, Research Units 2203, Pacific Southwest Forest and Range Experiment Station, USDA Forest Service, California 94701, 1970, pp. 9–14.
[12] Pieper, G. R. and Miskus, R. P., "Determination of Zectran Residues in Aerial Forest Spraying," *Journal of Agricultural and Food Chemistry*, Vol. 15, 1967, pp. 915–916.
[13] Crosby, D. G., Leitis, E., and Winterlin, W. L., "Photodecomposition of Carbamate Insecticides," *Journal of Agricultural and Food Chemistry*, Vol. 13, 1965, pp. 204–207.
[14] Abdel-Wahab, A. M. and Casida, J. E., "Photooxidation of Two 4-Methylaminoaryl Methylcarbamate Insecticides (Zectran and Matacil) on Bean Foliage and of Alkyl-Aminophenyl Methylcarbamates on Silica Gel Cromatoplates," *Journal of Agricultural and Food Chemistry*, Vol. 15, 1967, pp. 479–487.
[15] Poirier, D. G. and Surgeoner, G. A., "Laboratory Flow-Through Bioassays of Four Forestry Insecticides Against Stream Invertebrates," *The Canadian Entomologist*, Vol. 19, 1987, pp. 755–763.
[16] Kovacicova, J., Batora, V., and Truchlik, S., "Hydrolysis Rate *in vitro* Anticholinesterase Activity of Fenitrothion and S-Methyl Fenitrothion," *Pesticide Science*, Vol. 4, 1973, pp. 759–763.
[17] Kovacicova, J., Masud, Z., Batora, V., Kovac, J., and Truchlik, S., "Laboratory Purification of Fenitrothion," *Pesticide Science*, Vol. 2, 1971, pp. 101–102.
[18] Murty, A. S., "Toxicity Tests and Test Methodology," in *Toxicity of Pesticides to Fish*, Vol. I, CRC Press, Inc., Florida, 1986, Chap. 4, pp. 117–154.
[19] Abbott, W. S., "A Method of Computing the Effectiveness of an Insecticide," *Journal of Economic Entomology*, Vol. 18, 1925, pp. 265–267.
[20] Finney, D. J., *Probit Analysis*, 3rd ed., Cambridge University Press, Cambridge, England, 1971.
[21] *SAS User's Guide: Statistics, Version 5 Edition*, SAS Institute Inc., Box 800, Cary, North Carolina 27511-8000, 1985, pp. 639–645.
[22] Sundaram, A., Villeneuve, D. C., Chu, I., Secours, V., and Becking, G. C., "Subchronic Toxicity of Photomirex in the Female Rat: Results of the 28- and 90-Day Feeding Studies," *Drug and Chemical Toxicology*, Vol. 3, 1980, pp. 105–134.
[23] Sundaram, K. M. S. and Szeto, S. Y., "A Study of the Lethal Toxicity of Aminocarb to Freshwater Crayfish and its *in vivo* Metabolism," *Journal of Environmental Science and Health*, Vol. B14, 1979, pp. 589–602.
[24] McLeese, D. W., Metcalfe, C. D., and Zitko, V., "Lethality of Permethrin, Cypermethrin and Fenvalerate to Salmon, Lobster and Shrimp," *Bulletin of Environmental Contamination and Toxicology*, Vol. 25, 1980, pp. 950–955.
[25] Ralston, M., "Derivative-Free Non-Linear Regression," in *BMDP Statistical Software*, W. J. Dixon, Ed., University of California Press, Berkeley, 1983, pp. 305–314.
[26] Bradbury, S. P., Coats, J. R., and McKim, J. M., "Differential Toxicity and Uptake of Two Fenvalerate Formulations in Fathead Minnows *(Pimephales promelas)*," *Environment Toxicology and Chemistry*, Vol. 44, 1985, pp. 553-541.
[27] Schimmel, S. C., Garnas, R. L., Patrick, J. M., and Moore, J. C., "Acute Toxicity, Bioconcentration, and Persistence of AC222, 705, Benthiocarb, Chlorpyrifos, Fenvalerate, Methyl Parathion and Permethrin in the Estuarine Environment," *Journal of Agricultural and Food Chemistry*, Vol. 31, 1983, pp. 104–113.

[28] *Aminocarb: The Effects of Its Use on The Forest and The Human Environment,* NRCC No. 18979, National Research Council Canada, Environmental Secretariat, Ottawa, Ontario, Canada, 1982, pp. 111–118.

[29] Sundaram, K. M. S., Nott, R., Holmes, S. B., and Boyonoski, N., *The Distribution and Fate of Mexacarbate in a Forest Aquatic Ecosystem,* Government of Canada, Canadian Forestry Service, Forest Pest Management Institute Information Report FPM-X-73, Sault Ste. Marie, Ontario, Canada, 1986.

Sediment Toxicity Testing

K. John Scott[1] *and Michele S. Redmond*[1]

The Effects of a Contaminated Dredged Material on Laboratory Populations of the Tubicolous Amphipod *Ampelisca abdita*

REFERENCE: Scott, K. J. and Redmond, M. S., **"The Effects of a Contaminated Dredged Material on Laboratory Populations of the Tubicolous Amphipod *Ampelisca abdita*,"** *Aquatic Toxicology and Hazard Assessment: 12th Volume, ASTM STP 1027,* U. M. Cowgill and L. R. Williams, Eds., American Society for Testing and Materials, Philadelphia, 1989, pp. 289–303.

ABSTRACT: Short-term and full life cycle toxicity tests have been conducted with *Ampelisca abdita* as part of the joint COE-EPA field Verification Program (FVP). These studies comprised a portion of the effects assessment component of a risk analysis describing the hazards associated with aquatic disposal of dredged materials from Black Rock Harbor (BRH), Connecticut. The specific objective of these toxicity tests was to document the sensitivity of *Ampelisca*'s chronic endpoints to a range of contaminated suspended particulate concentrations. In long-term exposures of 56 days, amphipod growth and intrinsic rate of population growth, *r*, were impaired at all BRH sediment exposures. In all cases, the reduction in population growth rate was a function of slower growth of females, causing a longer time to maturity, which was coupled with a reduced egg production at maturity because the females were smaller.

KEY WORDS: amphipods, *Ampelisca abdita,* chronic toxicity, contaminated sediments, survival, growth, reproduction, demography

It is well recognized that acute toxicity data generated from 96-h exposure periods, while useful, are insufficient to identify acceptable toxicant concentrations that do not adversely affect growth and reproduction [*1–3*]. This limitation has been addressed by the development of chronic toxicity tests designed to assess pollutant effects on survival, growth, and reproduction over long periods of exposure, often an entire life cycle. Unfortunately, these chronic effects are measured at the individual organism level of biological organization and are not coupled in a predictive manner to population responses [*4,5*].

The National Research Council's report, *Testing for the Effects of Chemicals on Ecosystems* [*6*], recommends that appropriate and relevant decisions regarding the release of potentially toxic chemicals into the environment be based upon a hierarchy of biological tests, with those at the level of populations and communities being particularly crucial to such an assessment. The intrinsic rate of population growth, *r*, integrates age-specific survivorship and all facets of reproduction, including age at reproductive maturity, number of offspring per brood, frequency of reproduction, and reproductive longevity. Using this demographic parameter, the growth rate and reproductive potential of a population under both contaminant and environmental stress can be determined. The life history of the benthic amphipod *Ampelisca abdita* lends itself to this type of analysis in that all aspects of the reproductive process can be readily discerned

[1]Senior scientist and associate biologist, respectively, Science Applications International Corp., c/o EPA, Environmental Research Laboratory, Narragansett, RI 02882.

and quantified, providing a biological and ecologically defensible basis for regulatory decisions.

Ampelisca abdita is a tube-dwelling amphipod which constructs a soft, upright, membranous tube, 3 to 4 cm long, in surface sediments. It ranges from Maine to Louisiana from the intertidal zone to depths of 60 m [*7*]. This amphipod is a particle feeder, ingesting either surface-deposited particles or particles in suspension. Fertilization occurs in the water column, and they brood their young after mating. The egg-bearing females return to the bottom, and, after a brood period of approximately two weeks, young are released into the surrounding sediment. They are reproductively active at about 5 mm and grow to a maximum length of 7 to 9 mm [*8*]. Available information indicates that each female reproduces only once.

The present study was part of the joint Army Corps of Engineers—U.S. Environmental Protection Agency Field Verification Program (FVP), whose objective was to field verify existing test methodologies for predicting the environmental consequences of dredged material disposal under aquatic, intertidal, and upland conditions. The Environmental Research Laboratory—Narragansett was responsible for conducting the aquatic portion of the FVP with a primary goal to evaluate the risk of the open water disposal of contaminated dredged materials from Black Rock Harbor, (BRH), Connecticut. The aquatic portion of the FVP assessed the relative sensitivity and utility of various biological endpoints in both the laboratory and in the field following the disposal of BRH sediments [*9*]. The chronic experiments with *Ampelisca* were designed to simulate the fringe areas of a dredged material mound where suspended sediments are the primary route of exposure. Acute studies were conducted to simulate the short-term nature of a disposal operation and near-field plume transport. Long-term, chronic exposures examined the biological responses to mixtures of contaminated and noncontaminated sediments in the suspended phase with the sediments reduced or fully oxidized. The latter condition represented the potential effects of the resuspension and transport of aged dredged material. The objectives of the experiments described in this paper were (1) to assess the feasibility of conducting full life cycle tests using *A. abdita* and (2) to determine the relative sensitivity of the various biological responses.

Methods

Sediment Collection, Dosing, and Exposure

Two sediment types were utilized to conduct these laboratory tests. A reference sediment (REF) was collected from the South Reference site in Long Island Sound (40°7.95′N and 72°52.7′W) by Smith-MacIntyre grap (0.1 m^2), press sieved through a 2-mm sieve, and stored at 4°C until used. Prior to dredging, a contaminated sediment was collected from Black Rock Harbor, Connecticut (41°9′N and 73°13′W) with a gravity box corer (0.1 m^2) to a depth of 1.21 m, thoroughly mixed, press sieved through a 2-mm sieve, and refrigerated (4°C) until used. Details of sediment collection and storage procedures may be found in Rogerson et al. [*10*]. These authors also describe in detail the chemical properties of these two sediments. In all experiments, sediments were allowed to reach test temperature and homogenized prior to use.

Two identical sediment-dosing systems were constructed to provide either BRH or REF material as suspended sediment to the exposure system [*9*]. Each dosing system (Fig. 1) consisted of a conical-shaped slurry reservoir placed in a chilled fiberglass chamber, a diaphragm pump, a 4-L separatory funnel, and several return loops that directed the particulate slurry through microprocessor-controlled dosing valves. The slurry reservoirs (40 cm diameter by 55 cm high) contained 40 L of slurry composed of 37.7 L of filtered seawater and 2.3 L of either BRH or REF sediment. The fiberglass chamber (94 by 61 by 79 cm high) was maintained between 4° and 10°C using an externally chilled water source to minimize microbial degradation during the

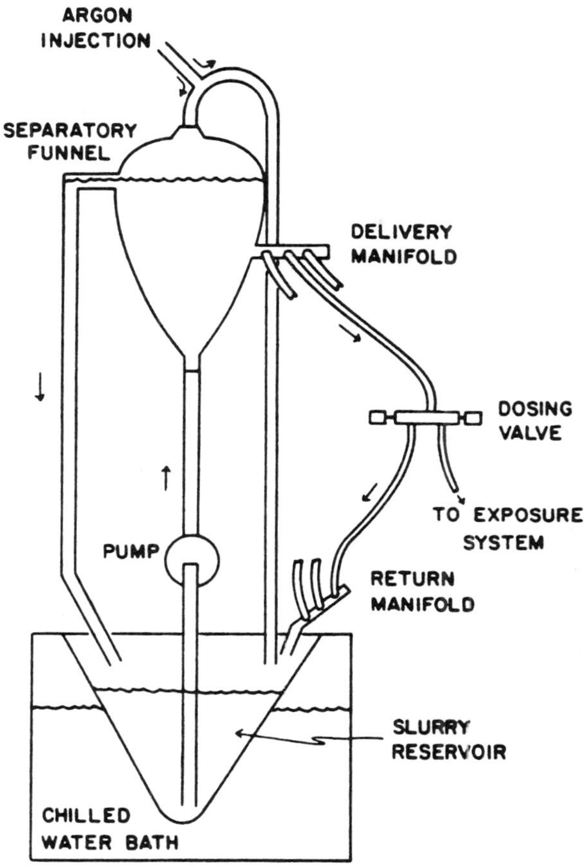

FIG. 1—*Suspended sediment dosing system.*

test. Upon dosing, the slurries were mixed with Narragansett Bay seawater filtered (to 15 μm) through sand filters to provide the desired concentrations for the toxicity tests.

In the first set of experiments, sediments were maintained under reduced (anaerobic) conditions by providing argon gas at the rate of 200 mL/min to the reservoir and separatory funnel. In the second experimental series, the sediments were fully oxidized prior to use in the dosing and exposure systems by transferring 2 L of sediment to an inverted polycarbonate carboy, diluting it to 19 L with filtered natural seawater at room temperature, and aerating for three to four days. Chemical oxygen demand measurements indicated that this time period was sufficient to satisfy the immediate oxygen demand of the sediments.

For the acute range-finding tests, the slurry was delivered to a mixing chamber (glass 4 L reagent bottle) in the amphipod exposure system, where it was initially diluted with seawater at a preset flow rate (right side of Fig. 2). The diluted suspension then passed from the bottom of the bottle to a distribution chamber (17 cm diameter by 9 cm high) fitted with a standpipe to maintain a constant water level. The suspension flowed via a siphon, at a flow rate controlled by head pressure, to a collection funnel which distributed the material to the exposure system.

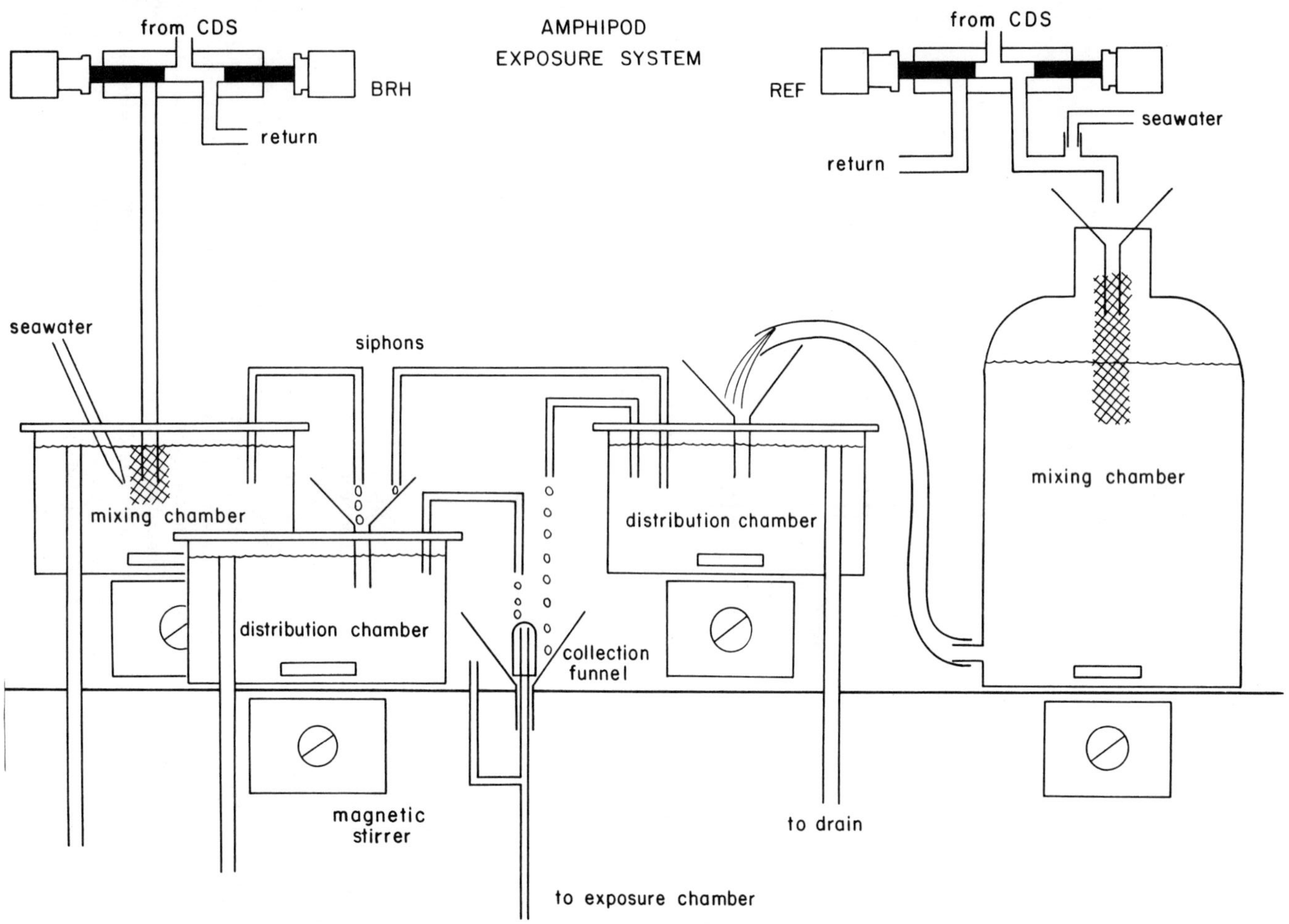

FIG. 2—*Suspended sediment proportional diluter system for short- and long-term tests with* A. abdita.

Each collection funnel was fitted with an umbrella siphon acting as a flow accelerator; these fed the exposure chambers through a polypropylene tube and tee to split the flow to each of two replicates. The mixing and distribution chambers were set up in duplicate, where one mixing chamber was dosed with the BRH sediment while the other was fed filtered seawater only.

For the longer-term tests, a constant suspended particulate density of 50 mg/L was desired, where BRH sediments were diluted with REF sediment instead of with seawater. To achieve this exposure, REF sediment was dosed to one mixing chamber and BRH sediment to the other so that there was an equal particle concentration in each distribution chamber. The limiting factor of this dosing design is the minimum flow rate, 10 mL/min, that can readily be attained from the siphons. Since the chronic tests required much lower BRH sediment concentrations, a modified design was implemented (left side of Fig. 2). The BRH concentration in the distribution chamber was set to 50% by feeding that chamber 50 mL/min of REF and 50 mL/min of BRH at equal particle densities.

The exposure chambers for all suspended particulate tests were glass gallon jars (Fig. 3) filled with 0.75 L of REF sediment (5 cm deep). The suspension was maintained in the gallon jar by aeration and was removed from the jar by a siphon which collected material from just above the sediment surface. The effluent entered a water trap with a screened standpipe which permitted the monitoring of swimming activity and the presence of mating adults in the water column.

The flow rate to each gallon jar was 40 mL/min for the acute tests and was reduced to 20 mL/min for the long-term tests to reduce the turnover rate and, hence, maximize the concentration of food supply. In the latter tests, the diatom *Phaeodactylum tricornutum* was cultured [*11*] and delivered to each gallon jar at 1 mL/min using a Harvard peristaltic pump. The total suspended solids concentration in the exposure system was monitored as total dry weight in mg/L. One-hundred-millilitre samples were taken twice weekly from each gallon jar using a free-standing siphon to minimize disturbance. Temperature and salinity were measured daily. In all experiments, temperature was maintained at 20 ± 1°C. Salinity varied between 28 and 32 parts per thousand. Exposure system flow rates were checked biweekly and adjusted when necessary. The photoperiod for all suspended phase tests was 14-h light and 10-h dark.

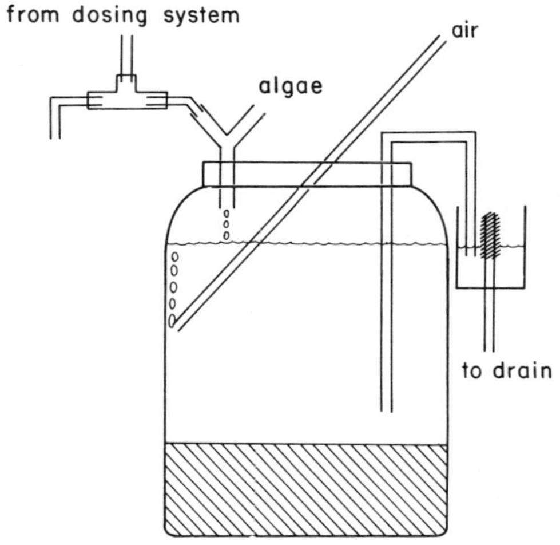

FIG. 3—*Chronic exposure chamber for suspended sediment toxicity tests with* A. abdita.

Experimental Procedure

Ampelisca were collected from tidal flats in the Pettaquamscutt River, a small estuary flowing into Narragansett Bay, Rhode Island. Bulk sediment containing the amphipods was immediately transported to the laboratory and was sieved through a 0.5-mm screen. The *Ampelisca* were collected by flotation from the air-water interface. When collection temperatures were lower than the test temperature, amphipods were acclimated at 1°C/day up to 20°C. During this acclimation, these animals were fed, ad libidum, laboratory-cultured *Phaeodactylum tricornutum*. At the initiation of a test, animals were sieved from the holding sediment and randomly distributed to the appropriate exposure chamber. The solid phase treatment for all suspended phase tests was REF sediment. Two 96-h tests were conducted with 200, 100, 50, and 25 mg/L of BRH suspended sediment and a seawater control. In the first range-find test with reduced sediments, 50 subadult amphipods were exposed in one chamber per treatment; in the second test, there were 25 subadult amphipods in each of two replicates per treatment. These 96-h range-find tests provided data on the acute responses to the raw dredged material.

Since the chronic tests were designed to use a mixture of REF and BRH sediments, a second set of acute tests was also conducted in order to set the BRH concentrations for the long-term exposures. These tests were of intermediate duration and lasted 18 days using reduced sediments and 10 days with oxidized BRH sediments. There were two replicates per treatment with 100 and 50 subadult *A. abdita* per replicate in the reduced and oxidized tests, respectively. Each of these tests, as well as each of the chronic tests, was conducted at a nominal constant suspended solids concentration of 50 mg/L.

The mortality data from the two acute test series were used to set concentrations for the two chronic tests. The concentrations of BRH sediment were 10 and 5% in the first chronic test with reduced sediments and 4 and 2% in the second test with oxidized sediments. Fifteen egg-bearing females were added to each of four replicates per treatment. The size and number of eggs of 30 females were determined prior to the start of the test. Two replicates were sieved after 32 and 28 days of exposure, and the remaining two replicates were harvested after 58 and 56 days of exposure for reduced and oxidized sediments, respectively.

The overflow water trap was checked daily, and the amphipods were enumerated and sexed when possible. Live *A. abdita* were returned to the exposure chambers. At the end of each test, all containers were sieved and the amphipods were enumerated. Any animals not accounted for were considered dead.

In addition, animals recovered from the chronic test were sexed and then measured using a computerized digitizer and camera lucida. The data collected at the 32 and 28-day sampling of the chronic experiments included survival, mean size, and proportion of amphipods that were mature. In addition, the number of eggs held by each ovigerous female was enumerated and the developmental stage determined. These same data were collected at the termination of the test, as was the total number of young amphipods produced per chamber. When the number of young per chamber exceeded 100, these young were subsampled for size measurements using a Folsom plankton splitter, which randomly splits the sample in halves. A minimum sample size of 30 was arrived at by examining the size variability found in previous tests and by determining the minimum sample size necessary to observe a treatment effect at the 5% probability level.

The intrinsic rate of growth, r, is typically calculated by constructing life tables [*12*] to determine age specific survivorship and fecundity. The Euler equation [*13*] is then used to calculate r by successive approximation. This technique has been previously applied [*9*] to these data to demonstrate the utility of this population statistic, but because of the tenuous nature of the underlying assumptions, a similar and more direct approach is presented here.

The intrinsic rate of growth, r, is a direct function of the difference between per capita birth and death rates. *Ampelisca* birth rates were calculated as the number of young produced by the F_1 generation, those young found at the end of the experiment, divided by the predicted number of adults in the F_1 generation. This value is then divided by the time interval in days over which their young were produced. The death rates were calculated as the number of surviving adults in the F_1 generation, those remaining at the end of the experiment, by the predicted number of young released by the initially introduced ovigerous females. This value was divided by the time interval in days over which these deaths occurred. For both of these determinations, it was assumed that the F_1 generation was released from the brood sac between Days 0 and 14 for the birth rate calculation and that no deaths occurred subsequent to the release and establishment of juvenile *Ampleisca* in their tubes. The final and most basic assumption in this analysis is that these populations are experiencing exponential growth.

Acute toxicity data were analyzed using the probit method [*14*]. Analysis of variance was used to analyze survival, growth, and reproductive data from all tests. Significant treatment differences were identified from Dunnett's or Tukey-Kramer's pairwise comparison tests [*15*]. Reproductive data for *A. abdita* were evaluated by analysis of covariance to account for differences in female size.

Results and Discussion

Short-Term Exposures

Exposure concentrations and resultant *Ampelisca* mortality data for the 96-h range-find tests with reduced and oxidized BRH sediments are shown in Table 1. The measured suspended sediment concentrations were close to the nominal concentrations for the lower treatments, while the measured concentrations for the two highest treatments (100 and 200 mg/L) had greater variability. *Ampelisca* mortalities exhibited a consistent dose response in both tests,

TABLE 1—Ampelisca abdita *mortalities and suspended sediment concentrations ($\bar{x}$ + SD) for BRH sediment exposures of four days.*

Nominal Concentration, mg/L	Measured Concentration BRH, mg/L, $\bar{x}$ + SD (n)	Mortality, %
	REDUCED EXPOSURES[a]	
0	...	4
25	34 ± 11 (2)	20
50	55 ± 9 (2)	20
100	92 ± 45 (2)	50
200	239 ± 30 (2)	88
	OXIDIZED EXPOSURES[b]	
0	...	6
25	31 ± 9 (3)	26
50	51 ± 13 (3)	30
100	76 ± 4 (2)	48
200	167 ± 38 (3)	74

NOTE: $\bar{x}$ = mean; s = standard deviation; n = number of replicates.

[a] LC_{50} = 91 mg/L (95% confidence interval = 76 to 111). Initial size of *A. abdita* ($\bar{x} \pm s$), 3.39 ± 0.89 mm (n = 25).

[b] LC_{50} = 80 mg/L (95% confidence interval = 64 to 106 mg/L). Initial size of *A. abdita* ($\bar{x} \pm s$), 6.40 ± 0.89 mm (n = 50).

yielding LC_{50}s of 91 and 80 mg/L BRH sediment for the reduced and oxidized exposures, respectively.

Based on the 96-h mortality data, exposures of a longer duration were designed to examine the acute responses when BRH sediments were diluted with REF sediments. Since moderate toxicities occurred at approximately 30 mg/L (Table 1), dilution concentrations were chosen to provide BRH levels at or below this concentration. In the reduced sediment test, amphipods were exposed for 18 days to 75, 50, 25, and 0% (control) BRH sediment at a constant suspended solids concentration of 50 mg/L. Significant mortalities were observed at all concentrations (Table 2), even in the 25% BRH treatment (12 to 13 mg/L BRH sediment). Another experiment was conducted with oxidized sediments exposing *Ampelisca* for ten days at 50, 25, 12.5, and 0% (control) BRH sediment. The toxicities observed in this test were slightly lower than those seen in the reduced exposures and were likely the result of a shorter exposure period.

Long-Term Exposures

Mortality—Suspended sediment concentrations for the reduced sediment exposures had a range of mean concentrations of 40 to 50 mg/L across all replicates, and there were no significant treatment or replicate differences. The total suspended solids concentrations in the oxidized exposures were slightly higher, ranging from 50 to 57 mg/L, and similarly showed no significant treatment differences. The calculated BRH concentration for each of the treatments in these two tests and their respective chronic mortalities are shown in Table 3.

TABLE 2—*Suspended sediment concentrations and* Ampelisca *mortality for an 18-day exposure to reduced and a 10-day exposure to oxidized BRH sediments.*

Treatment	Replicate Number	Total Suspended Sediment Concentration, $\bar{x} \pm s$ (n)	Calculated BRH Concentration, mg/L	Mortality, %
		REDUCED—18 DAY[a]		
REF control	1	56 ± 28 (6)	0	14
	2	53 ± 36 (6)	0	9
25% BRH	1	50 ± 22 (6)	13	90
	2	51 ± 23 (6)	13	98
50% BRH	1	51 ± 12 (6)	25	100
	2	56 ± 9 (6)	28	100
75% BRH	1	56 ± 6 (6)	42	100
	2	54 ± 7 (6)	40	100
		OXIDIZED—10 DAY[b]		
REF control	1	56 ± 4 (5)	0	0
	2	53 ± 3 (5)	0	2
12.5% BRH	1	61 ± 8 (5)	8	12
	2	59 ± 6 (5)	7	14
25% BRH	1	51 ± 5 (5)	13	70
	2	51 ± 8 (5)	13	74
50% BRH	1	41 ± 9 (5)	21	88
	2	44 ± 11 (5)	22	72

[a]The number of amphipods per replicate is 100. Initial size of *A. abdita*, ($\bar{x} \pm s$), 3.11 ± 0.48 mm (n = 91).

[b]The number of amphipods per replicate is 50. Initial size of *A. abdita*, ($\bar{x} \pm s$), 4.74 ± 0.43 mm (n = 50).

TABLE 3—*Estimated* Ampelisca *adult mortality in long-term exposures to reduced and oxidized BRH suspended sediments.*

Treatment	Calculated BRH Sediment Concentration, mg/L	Number of Ovigerous Females Introduced	Harvested/ Number Predicted	Estimated % Mortality
		DAY 32—REDUCED		
REF control		30	495/532	7.0
5% BRH	2.0	30	585/532	0.0
10% BRH	4.5	30	417/532	21.6
		DAY 56—REDUCED		
REF		30	286/532	46.2
5% BRH	2.3	30	191/532	64.1
10% BRH	5.0	30	384/532	27.8
		DAY 28—OXIDIZED		
REF control		28	304/320	5.0
2% BRH	1.0	28	332/320	0.0
4% BRH	2.0	29	285/331	13.9
		DAY 56—OXIDIZED		
REF control		30	123/342	64.0
2% BRH	1.0	29	172/331	48.0
4% BRH	2.0	30	115/342	66.4

Survival in each exposure was estimated by comparing the number of *Ampelisca* harvested with the estimated initial number of young, assuming 100% survival of all eggs of the ovigerous females that were introduced into the containers at Day 0. For reduced exposures, these females had a mean egg number of 17.7 eggs ($s = 6.73$) to yield 532 initial young. As shown in Table 3, survival in the control and 5% exposures was better than in 10% BRH at Day 32. The 95% confidence interval around the mean egg number per female was 15.2 to 20.2 eggs yielding 454 and 606 initial young; the 10% BRH falls significantly outside this range. The mortality threshold appears to occur at 4.5 mg/L BRH sediment based upon these 32-day data.

For the oxidized exposures the initial sample of ovigerous females had an egg number of 11.4 ($s = 4.98$) yielding 342 young per 30 females. The number of amphipods harvested on Days 28 and 56 are also shown in Table 3. The predicted number of juveniles was adjusted to 160 for some containers because one ovigerous female died in each of those replicates on the first day. The 95% confidence limits are 9.5 to 13.3 eggs/female resulting in 266 to 372 offspring for 28 females, 276 to 386 offspring for 29 females, and 285 to 399 for 30 females. For the 28-day exposure period there were no significant mortalities.

Mortalities in the adult population at the 56-day harvest ranged from 27.8 to 64.1% for the reduced exposures and from 48.0 to 66.4% for the oxidized exposures. These mortalities are not considered to be due to BRH sediment exposure. They represent a six to eight-week life span in laboratory populations with a significant contribution to the mortalities being males, who die subsequent to mating. Of the 138 mortalities observed in the overflow cups across all treatments, 76% were adult males.

The mortalities observed in the 10% BRH treatment of the reduced exposures are noteworthy. The presence of significant mortalities at Day 32 was the reason for lowering the nominal concentrations in the second test with oxidized sediments. One would expect that this effect would be compounded after a 56-day exposure, however the converse was true; this treatment period exhibited the lowest mortality. As will be shown below, this result is not inconsistent and actually reflects an effect of BRH sediment on *Ampelisca*'s growth and the eventual structure of the population.

Population Structure and Individual Growth—Individuals can be sexed and classified into stages of sexual maturity based on morphological characteristics. Morphologically, adult males (M) are the most distinctive with a carinate urosome for swimming, which facilitates mating in the water column. They do not molt again because they die after mating. Whether they mate with more than one female is unknown.

Female *A. abdita* can be divided into four progressively distinct groups. The earliest stage, after sexual differentiation, is here termed developing female (FDV). This stage is distinguished by the presence of a brood plate (oostegite) on the interior of each of the first five coxal appendages. As the female grows, the brood plate increases in size and eggs begin to develop in the oviduct, which is dorsal to the digestive tract and can be seen through the exoskeleton. This stage is termed developing egg female (FE). After a series of molts the brood plates develop long setae and the eggs are deposited through a gonopore into a fully developed brood pouch. This stage is termed ovigerous female (FOV). Presumably, mating occurs at this time. Since all eggs in a brood pouch are in the same stage of development, all eggs in the oviduct are assumed to be deposited at once. After releasing their young, the females do not immediately die but remain in the population for some undetermined time before death. These females may have eggs in the oviduct, but do not have eggs in the brood pouch, at the same time retaining the setose oostegites. This stage is termed spent females (FS).

There is one other group, termed undifferentiated (UD). These include juveniles and undifferentiated subadult females and males. In most cases where this group was encountered, the animals were larger than 4.0 mm. These animals are probably subadult males since females can be distinguished at a smaller size when the brood plates first develop.

The percentage of amphipods in each population class is shown in Table 4. In the first experiment with reduced sediments, the 32-day harvest revealed that the majority of the population was as yet immature (undifferentiated), constituting 45% or greater of the population in all these treatments. The control and 5% BRH exposures did have higher proportions of adult males and females. This trend became more obvious in the 58-day samples where these two treatments contained much higher proportions of ovigerous and spent females and females with eggs in the oviduct. Conversely, the 10% BRH treatment was dominated by developing females and undifferentiated amphipods. These results signify a retardation in the development and maturation at the highest BRH sediment exposures. Because the number of mature males present in the 10% BRH treatment was reduced at 58 days, and because natural mortality of mature males was a major contribution to total mortality at this exposure period, mortality in the 10% BRH treatment was less than that in the other treatments, as noted previously.

This same pattern was repeated in the second experiment using oxidized sediments, although the differences were not as dramatic. At 28 days, the undifferentiated group dominated the population structure. There were no apparent treatment differences in the rate of maturity. After 56 days the population had matured, as expected, with most of the females being ovigerous. There was a trend for there to be fewer immature animals (UD, FDV) in the control treatment, which also produced many more spent females. These data indicate that the control exposure was producing a slightly more reproductively active population.

Examination of the mean lengths of each of these population classes helps to explain the differences observed in the population structure. These data are presented in Table 5. At Day 32, in the reduced sediment test, there were significant treatment differences in the mean lengths of the undifferentiated group, with the control amphipods largest and the 10% BRH exposed animals the smallest. The developing females and those with eggs in the oviduct were also smaller as a result of this exposure. At 58 days, the 10% BRH exposed amphipods were

TABLE 4—*Percent of* A. abdita *in each population category for long-term exposures to reduced and oxidized BRH suspended sediments.*

Treatment (BRH Sediment)	Category[a]					
	UD	FDV	FE	FOV	FS	M
		DAY 32—REDUCED				
REF control	50.5	22.3	14.2	5.2	...	7.8
5% BRH (2.0 mg/L)	45.3	23.1	16.6	4.8	...	10.3
10% BRH (4.5 mg/L)	63.8	25.2	10.1	0.7	...	0.2
		DAY 58—REDUCED				
REF control	17.5	14.7	14.0	29.7	9.4	14.7
5% BRH (2.3 mg/L)	10.5	6.3	13.6	43.5	16.8	9.4
10% BRH (5.0 mg/L)	43.8	32.8	7.8	6.5	0.8	8.3
		DAY 28—OXIDIZED				
REF control	69.5	21.2	6.4	2.9	0	0
2% BRH (1.0 mg/L)	68.3	24.6	5.3	1.9	0	0
4% BRH (2.0 mg/L)	80.7	16.1	0.3	2.6	0	0.4
		DAY 56—OXIDIZED				
REF control	8.1	4.1	1.6	38.3	31.0	17.0
2% BRH (1.0 mg/L)	16.7	17.0	12.4	26.8	14.5	12.8
4% BRH (2.0 mg/L)	14.6	12.2	6.7	40.2	11.6	14.9

[a] See text for designation of categories.

smaller than controls in all population classes. The developing egg females and ovigerous females in the 5% BRH treatment were also smaller than those in the control treatment. These data clearly show that slower growth of exposed amphipods accounts for the changes in the population structure and overall maturity.

In the oxidized sediment experiment, the growth pattern described above was repeated. At Day 28, the undifferentiated group and the developing females were smallest at the 4% BRH sediment exposure. At Day 56 the ovigerous and spent females were also smallest at this exposure level. The adult males in the two BRH exposures were larger than those in the controls. Again, there were not as many significant size differences in this second experiment. This lack of differences was probably due to the lower exposure concentrations used in this test. When comparing the common concentration in the reduced (5% BRH) and oxidized (4% BRH) experiments, where exposures were approximately 2.0 mg/L BRH sediment, the size data and significant differences for ovigerous females are surprisingly consistent. The lengths of these females were significantly different from those in controls in both exposures, resulting from the slower growth that would have been predicted from the 28-day data.

The population proportion and size data presented here are consistent with estimates of growth and time to first reproduction described by Mills [*8*] for field populations of *A. abdita* from Barnstable Harbor, Massachusetts. He found that for summer populations, at approximately 20°C, growth was essentially linear and that newly released juveniles grew to breeding adults in 40 to 80 days. Adult females grew to a mean size of 5.5 to 6.0 mm. The *Ampelisca* in these two experiments grew from hatching at approximately 1.5 mm to 6 to 7 mm in a minimum of about 50 days. This would be a minimum time span since all juveniles were not released on Day 1 but rather over a two-week time span.

To summarize the population structure and growth data, there appears to be an adverse

TABLE 5—*Mean length (mm) of* A. abdita *in long-term exposures to reduced and oxidized BRH suspended sediments.*

Treatment, (BRH Sediment)	Sex Category UD	FDV	FE	FOV	FS	M
			DAY 32—REDUCED			
REF control	5.74	5.29[a]	6.37	6.14	...	5.50
5% BRH (2.0 mg/L)	5.47	5.22	6.39	6.34	...	5.59
10% BRH (4.5 mg/L)	4.42	4.67	6.04	6.75	...	5.25
			DAY 58—REDUCED			
REF control	7.33	7.18	7.58	6.89	NM	6.40
5% BRH (2.3 mg/L)	7.23	6.88	6.99	6.25	NM	6.24
10% BRH (5.0 mg/L)	6.15	5.80	6.69	6.12	NM	5.83
			DAY 28—OXIDIZED			
REF control	3.77	4.07	4.62	6.59	...	...
2% BRH (1.0 mg/L)	3.77	4.11	5.03	6.98	...	...
4% BRH (2.0 mg/L)	3.21	3.78	4.41	6.44	...	...
			DAY 56—OXIDIZED			
REF control	6.69	5.69	6.13	6.77	6.74	5.83
2% BRH (1.0 mg/L)	6.60	6.28	6.95	6.88	6.94	6.17
4% BRH (2.0 mg/L)	6.32	5.95	7.13	6.21	6.09	6.16

NOTE: NM = not measured.
[a] Means connected by the same line are not significantly different from each other at $P < 0.05$.

impact on growth, which, although small, does have an effect on maturation. This effect on maturation was more evident at 28 or 32 days than it was at the 56 or 58-day sampling. As will be shown below, the impact on growth may be accompanied by more subtle effects of exposure to dredged material on egg production and juvenile survivability.

Reproduction—Fecundity, the number of eggs/female, in *A. abdita* is a function of female size, with larger females producing larger broods [*8*]. Treatment effects were assessed using analysis of covariance (ANCOVA), with egg number as the dependent variable and female size as the covariate. Once the quantitative relationship between egg number and female size is determined, the differences in egg number/female can be analyzed by statistically adjusting the fecundity estimates based on the regression relationship.

The raw and size-adjusted fecundity estimates for the 58- and 56-day harvests in the two sediment exposures are shown in Table 6. The number of eggs/female in the reduced sediment exposures showed no significant treatment effects when adjusted for female size. The number of young produced, however, were dramatically different among treatments. There were no young observed at the 10% BRH sediment exposure even though there were spent females present (Table 6). This would indicate some problem with egg viability or hatching and survival success. Young were observed in the 5% BRH treatment, but there densities were much lower than those in the control treatment.

TABLE 6—*Reproductive characteristics of* Ampelisca *exposed to reduced and oxidized Black Rock Harbor suspended sediments for 58 and 56 days, respectively.*

Treatment	Female Size, mm	No. Eggs/Female: Raw	No. Eggs/Female: Size Adjusted	Young Produced
		REDUCED		
REF	6.89	13.6	12.4	1212
5% BRH (2.3 mg/L)	6.25[a]	12.3	13.5	318
10% BRH (5.0 mg/L)	6.12	9.2	10.9	0
		OXIDIZED		
REF	6.77	15.8	15.4	1012
2% BRH (1.0 mg/L)	6.88	15.8	15.0	117
4% BRH (2.0 mg/L)	6.21	10.8	12.0	122

[a]Means connected by the same line are not significantly different.

Brood size of *A. abdita* was also reported by Mills and, more recently, by Nelson [*16*] for North Carolina populations. Mills found 26.0 eggs per female for a mean female size of 5.4 mm, while Nelson observed 13.7 eggs per female for females 5.3 mm in length. The egg numbers reported here are within the range of those found previously, as are the sizes of ovigerous females.

In the oxidized sediment exposures, there were significant treatment effects on fecundity. In the 4% BRH treatment, the ovigerous females were not only smaller (Table 5), and, thus, produced fewer eggs, but also had lower fecundities when compared with similar-sized animals in the other two treatments.

The effect of oxidized dredged material exposure on fecundity and growth is further borne out by examining the number of young produced (Table 6). There were over eight times the total number of young produced in the controls as compared with the 4% BRH treatment. Unexpectedly, however, the productivity of the 2% BRH treatment was also lowered, even though fecundity was equal to that of the control. This again indicates possible effects on either hatching success or juvenile survival, neither of which were directly measured in this experiment.

The mean sizes of these juveniles from the oxidized sediment exposures were analyzed by ANOVA, which shows the controls to be significantly larger ($P < 0.05$) than the other two treatments (Control = 2.29 mm, 2% BRH = 1.65 mm, 4% BRH = 1.72 mm). The control juveniles were either released sooner or were growing faster. In either case, the assumption is that, with continued exposure, they would have matured and reproduced sooner than juveniles in the two BRH treatments. These size data and the productivity results are the first evidence in these experiments of detrimental impacts at the lowest BRH exposure levels.

These results, in conjunction with the previous information on population structure and individual growth, support the conclusion that the impact of dredged material on reproductive output tends to be compounded with each generation. In this particular case, the exposed parental populations grew slower and exhibited delayed maturation. The mature females were not only smaller but also produced fewer eggs, resulting in fewer young in the next generation. Because these filial young hatch at a later date, time to first reproduction, as compared with control populations, is already adversely affected. The implications of effects on growth and time to first reproduction are, that in temperate (seasonal) environments, survival of late brooders may be severely impacted by being out of synchrony with the normal water temperatures and food supplies that are necessary for naturally growing populations.

Population Growth—The utility of measuring the various population parameters described above is that they can be used in a predictive nature. A means to accomplish this is to make projections or estimates of population growth, stability, or both, using population response parameters such as intrinsic rate of growth.

The population responses parameters, birth rate, death rate, and intrinsic rate of growth (r) are shown in Table 7 for reduced and oxidized exposures. Example calculations of birth and death rates for the control treatment in the reduced exposures are as follows. The number of young produced by the F_1 generation was 1212 (Table 6). This value was divided by the predicted number of F_1 adults in this treatment, 532 (Table 3). The quotient was then divided by the number of days over which these young were produced, 49, assuming Day 7 as the release date, to arrive at a birth rate of 0.045 individuals per capita per day. The death rate is the number of *Ampelisca* dying between Days 7 and 56 divided by the predicted number of organisms in that treatment (Table 3), or $[(532 - 286)/532]/49$. The difference between birth and death rates, r, was less than zero in all BRH sediment exposures, indicating that these populations will eventually become extinct under continued exposure to these sediments. The death rates for all exposures, including the controls, are relatively similar and reflect the fact that these suspended sediment concentrations were neither acutely nor chronically toxic. In fact, the lowest death rate was observed in the highest BRH sediment exposure (10% BRH). As described above, this low death rate is related to slower individual maturation, which delayed reproduction, mating, and senescent death. As can be seen in Table 7, the major factor controlling the intrinsic rate of growth, using this analysis, was the birth rate. This factor was severely depressed in all BRH sediment treatments.

The primary reason for the differences in population responses noted is the much lower reproductive output of the two BRH treatments (Table 6), which is a function of delayed maturation, in combination with lower fecundities in those treatments. As discussed above, the smaller mean size of the progeny at the 56-day harvest would cause even later maturation of those individuals, leading to the potential for continued negative r values.

Conclusions

The biological responses of acute mortality and growth did not differ greatly between the reduced and oxidized sediment exposures. However, amphipod fecundity was affected more dramatically by exposure to oxidized sediments. In both exposures, there were survival problems in the second generation juveniles that were not apparent in the first generation. The

TABLE 7—*Population responses for* A. abdita *exposed to reduced and oxidized BRH suspended sediments for 58 and 56 days, respectively.*

Treatment	Death Rate, individuals/capita/day	Birth Rate, individuals/capita/day	Intrinsic Rate of Growth
	REDUCED		
REF Control	0.009	0.045	0.036
5% BRH, 2.3 mg/L	0.013	0.012	−0.001
10% BRH, 5.0 mg/L	0.005	0.000	−0.005
	OXIDIZED		
REF Control	0.013	0.060	0.047
2% BRH (1.0 mg/L)	0.010	0.007	−0.003
4% BRH (2.0 mg/L)	0.014	0.007	−0.007

results suggest that the long-term exposure of the maturing females to contaminated sediments may affect the viability of developing embryos.

It is clear that the primary biological response to these contaminated sediments is slower growth. Slower growth delayed maturation, increasing the time to first reproduction and decreasing the size of breeding females. The implication to the population is lower numbers of young being produced over a longer time period, therefore forcing the population out of synchrony with natural breeding cycles which are normally temperature and food-source dependent.

Additionally, this research demonstrates the ability to culture *Ampelisca abdita* under laboratory conditions through at least two generations. The population size increased 2.5 to 3.0 times between the interim, 28 and 32-day sampling, to the final harvest at 56 and 58 days. If all ovigerous females were allowed to brood their young to hatching, much larger numbers would be expected after 56 days.

Acknowledgements

The authors wish to thank John Sewall, SAIC, who was instrumental in the design and maintenance of the composite dosing system. We also acknowledge Jack Gentile, EPA—Narragansett, for his unfailing cooperation and stimulation in all aspects of these studies. Wayne Munns, SAIC—Narragansett, and Thomas Wright and Thomas Dillon, COE—Waterways Experiment Station, provided critical reviews of this work throughout all of its phases and their efforts are greatly appreciated. This research was supported by EPA contract No. 68-03-3236 to Science Applications International Corporation. The contents of the manuscript do not necessarily reflect views or policies of the U.S. Environmental Protection Agency. Mention of trade names or commercial products does not constitute endorsement or recommendation for use by the U.S. Environmental Protection Agency. This is ERL-N contribution number X129.

References

[1] Mount, D. I., *Water Research*, Vol. 2, 1968, pp. 215–233.
[2] Sprague, J. B., *Water Research*, Vol. 5, 1971, pp. 245–266.
[3] Sprague, J. B., *Journal of the Fisheries Research Board of Canada*, Vol. 33, 1976, pp. 1988–2022.
[4] Rand, G. M. in *Introduction to Environmental Toxicology*, F. E. Guthrie and J. J. Perry, Eds., 1980, pp. 390–403.
[5] Hansen, S. R. and Garton, R. R., *Canadian Journal of Fisheries and Aquatic Sciences*, Vol. 39, 1982, pp. 1273–1288.
[6] National Research Council, *Testing for Effects of Chemicals on Ecosystems*, National Academy Press, Washington, DC, 1981.
[7] Bousfield, E. L., *Shallow Water Gammaridean Amphipoda of New England*, Cornell University Press, Ithaca, NY, 1973.
[8] Mills, E. L., *Journal of the Fisheries Research Board of Canada*, Vol. 24, 1967, pp. 305–355.
[9] Gentile, J. H., Scott, K. J., Lussier, S. M., and Redmond, M. S., "The Assessment of Black Rock Harbor Dredged Material Impacts on Laboratory Population Responses," Technical Report D-87-3, U.S. Army Engineer Waterways Experiment Station, Vicksburg, MS, 1987.
[10] Rogerson, P. F., Schimmel, S. C., and Hoffman, G. L., "Chemical and Biological Characterization of Black Rock Harbor Dredged Material," Technical Report D-85-9, U.S. Army Engineer Waterways Experiment Station, Vicksburg, MS, 1985.
[11] Guillard, R. R. and Ryther, J. H., *Canadian Journal of Microbiology*, Vol. 8, 1962, pp. 229–239.
[12] Birch, L. C., *Journal of Animal Ecology*, Vol. 17, 1948, pp. 15–26.
[13] Euler, L., *Theoretical Population Biology*, Vol. 1, 1970, pp. 307–314.
[14] Stephan, C. E. in *Aquatic Toxicology and Hazard Evaluation (First Conference), ASTM STP 634*, American Society for Testing and Materials, Philadelphia, 1977, pp. 65–84.
[15] Snedecor, G. W. and Cochran, W. G., *Statistical Methods*, 7th ed., The Iowa State University Press, Ames, IA, 1980.
[16] Nelson, W. G., *Sarsia*, Vol. 65, 1980, pp. 61–71.

Ronald J. Breteler,[1] K. John Scott,[2] and Susan P. Shepherd[1]

Application of a New Sediment Toxicity Test Using the Marine Amphipod *Ampelisca abdita* to San Francisco Bay Sediments

REFERENCE: Breteler, R. J., Scott, K. J., and Shepherd, S. P., **"Application of a New Sediment Toxicity Test Using the Marine Amphipod *Ampelisca abdita* to San Francisco Bay Sediments,"** *Aquatic Toxicology and Hazard Assessment: 12th Volume, ASTM STP 1027,* U. M. Cowgill and L. R. Williams, Eds., American Society for Testing and Materials, Philadelphia, 1989, pp. 304–314.

ABSTRACT: Sediment bioassays are becoming widely recognized as effective tools to determine the biological significance of the contamination found in coastal sediments. This paper describes the application of a new sediment toxicity test with *Ampelisca abdita,* a marine amphipod, to sediments from three stations at each of four sites in San Francisco Bay and one reference site in Tomales Bay. The tests were conducted under static and flow-through conditions, and all treatments and controls were maintained in quintuplicate (20 organisms per replicate). The exposure chambers were monitored daily for mortalities and emerged amphipods. Results indicated that the range of mortality within each station was generally less than 10%. Statistical analyses indicated highly significant site differences ($p < 0.001$) for both mortality and emergence. San Pablo Bay and Tomales Bay sediments were, as expected, the least toxic, with mean ten-day mortalities of 8.7 and 11.0%, respectively. Yerba Buena and Vallejo sediments were slightly, but significantly, more toxic at 15.0 and 14.7%, respectively; and the Oakland Harbor sediments were the most toxic, with 25.7% mortality. The data indicate that this sediment toxicity test can differentiate small differences in sediment quality. The emergence data were surprisingly consistent and showed the same response pattern, indicating that this sublethal response may be equal in sensitivity to survival. Static tests appeared slightly more sensitive than tests conducted under flow-through conditions.

KEY WORDS: sediment bioassays, aquatic toxicology, amphipod, *Ampelisca abdita,* San Francisco Bay

Because organism responses integrate physical and chemical characteristics of the water column and benthic environments, sediment bioassays are increasingly being considered as effective tools to determine the biological significance of the contamination found in coastal sediments.

The tube building infaunal amphipod, *A. abdita,* is found along the North American coast from Maine to northern Florida and the eastern Gulf of Mexico, from the intertidal zone to depths of 60 m [*1*]. It was introduced into the mud flats of San Francisco Bay when seed oysters were imported from the East Coast. The species is thriving there, although it is not known whether it has spread to other parts of the West Coast [*2,3*]. It is most abundant in protected embayments and estuaries in infralittoral zones with fine sand, silty sand, or silt substrates [*1*]. In their natural habitats, the animals often form dense aggregations (commonly 70 000+ and

[1]Springborn Life Sciences, Inc., Wareham, MA 02571.
[2]SAIC, Admiral's Gate, Newport, RI 02840-6669.

up to 110 000/m^2), and there is evidence of stiff intraspecific competition for space [*4–6*]. The thickly aggregated tube mats often cover areas hundreds of square meters in size [*7*]. This amphipod species occurs in waters ranging from fully marine to brackish conditions, tolerating considerable lowering of salinities for extended periods as often occurs in estuaries, particularly following heavy rainstorms [*1,3,8*]. *A. abdita* can survive the relatively wide seasonal (0°C in winter to >25°C in summer) and daily fluctuations in temperature which accompany life in shallow waters [*1*].

This amphipod constructs soft, membranous, solid, upright tubes 3 to 4 cm long in surface sediments. The tube is flattened and composed of grey, nonchitinous, parchment-like material and adhered sediment particles. The tube is sealed at the bottom when completed. The upper end is left open but closes tightly when the amphipod withdraws. Each animal builds a new tube as growth makes the old one too small. *A. abdita* is both a deposit and a suspension feeder. It feeds while propped at the mouth of its tube and by whirling the second antennae in a circular motion [*9*]. The water currents thus produced bring particulates to the mouthparts, where they are processed. The second antennae also scrape the surface deposits and carry that material to the mouthparts. Mineral particles are scraped clean and organic detritus is ingested. Because of its feeding activities, *A. abdita* is in constant contact with particles and particulate-bound contaminants at the sediment-water interface.

Although *A. abdita* is not itself of commercial or recreational value, it is a significant component of food chains leading directly to humans. Benthic communities dominated by detritus and filter-feeding infauna are the ecological zones of value for bottom-feeding fish [*10*]. Ampeliscids are a common food for these bottom fish, particularly flounder [*11,12*]. The fish consume the amphipods by ingesting the sediment and tubes while they are living in the substrate as well as by preying on them when they leave their tubes to mate.

Laboratory studies with *A. abdita* have focussed on the development of acute testing methodologies with water-borne and sediment-associated contaminants. These tests have been applied in the development of water quality criteria for four metals [*13–16*] and the assessment of sediment toxicity in two East Coast harbors, Black Rock Harbor, Connecticut [*17*], and New Bedford Harbor, Massachusetts [*18*]. Chronic test methods using this species have also been developed and applied for the evaluation of suspended dredged materials [*19*]. This report describes the results of the ten-day, flow-through *Ampelisca abdita* sediment toxicity test, which was evaluated for its sensitivity and applicability to routinely monitor sediments of coastal waters. In addition to flow-through tests, static tests were conducted with three different sediments to determine if these exposures would elicit a more sensitive biological response in *A. abdita* as compared to the flow-through tests.

Methods

Test Organism Collection and Holding

A. abdita were collected from tidal flats in Bourne Cove, a small inlet in Buzzards Bay, Massachusetts. The sediment containing the amphipods was immediately transported to the laboratory. The sediment was sieved through a 0.5-mm screen using unfiltered seawater at ambient temperature of 3°C. *A. abdita* were collected by floatation from the air/water interface [*19*]. The animals were transferred to 3-L jars (200 to 300 animals/jar) containing approximately 5 to 7 centimeters of collection site sediment in gently aerated ambient seawater. The amphipods were gradually acclimated under static seawater conditions at 1 to 3°C/day up to the test temperature of 20°C. During acclimation, these animals were fed laboratory cultured *Skeletonema* sp. (1.0×10^7 cells/mL) at the rate of 200 mL of algae per 3-L jar daily. Prior to feeding, approximately 1 L of seawater was removed from each 3-L jar and replaced with 800 mL of unfiltered ambient seawater and 200 mL of algal suspension. This served to maintain constant salinity and pH in each jar. No mortality of the amphipods was observed throughout the acclimation period (based on absence of amphipods on the sediment surface). Temperature, dissolved oxygen, salinity, and pH were measured in one jar daily with each jar measured at least once during the course of the acclimation period.

Test Water and Sediment Collection

Natural filtered (5-μm polypropylene core filter) seawater from Cape Cod Canal was used as dilution water for the test. Test sediments were collected from various sites in San Francisco Bay and Tomales Bay (Fig. 1) using an 18-ft Boston whaler. The sites were selected based on their known or suspected level of contamination. Data from previous studies, including those conducted as part of the National Status and Trends Programs administered by the National Oceanic and Atmospheric Administration, indicated that concentrations of chemicals in the sediments of the five selected sites would differ by an order of magnitude. A manuscript describing the results of detailed physical and chemical analyses of each sediment tested in this program, which included grain size analysis, and analysis for 17 metals, 19 polynuclear aromatic hydrocarbons, 6 isomers of DDT/DDD/DDE, 9 pesticides other than DDT, and 9 chlorinated levels of polychlorinated biphenyls, is currently in preparation [*20*]. The results of these analyses generally confirmed the expected range of concentrations. Oakland Inner Harbor was chosen as the "hot" site, characterized by high concentrations of organic and inorganic sediment contaminants. The remaining three sites in San Francisco Bay were near Vallejo (off Semple Point), expected to be lightly contaminated; the southern part of San Pablo Bay, also lightly contaminated; and off Yerba Buena Island, due west of the Emeryville Marina, with evidence of moderate contamination. Tomales Bay was selected as a noncontaminated inlet removed from industrial and urban influences. Three stations were sampled at each site, and five grab samples were obtained from each sampling station. The sediment collection device consisted of a Teflon-coated, modified Van Veen grab sampler. The five grab samples were dropped into a Teflon-coated stainless steel pan, thoroughly stirred with a Teflon-coated stainless steel spoon, and added to half-gallon polypropylene bottles. Thus, a single, uniform sample was available for chemical analyses and toxicity testing from each station sampled. The bottles were closed tightly, placed on ice inside a cooler, and shipped within 72 h to the laboratory where they were used immediately for the toxicity tests. The maximum time between sediment collection and usage for the toxicity tests was five days.

Test Conditions

For each station (that is, treatment group), sediments were rehomogenized thoroughly and press-sieved wet through 2-mm mesh to remove any native *A. abdita*, as well as large shell material and other debris. Approximately 12 h prior to the initiation of the test, 200 mL (about 4 cm) of the test sediment was placed into the exposure chambers with flowing, gently aerated seawater. This allowed oxygenation of the surface sediments and minimized release of potentially toxic sulfides. Control sediments, obtained from the *Ampelisca* collection site, were tested in the same manner as the test sediments. Test results with control sediment were used only as a check on animal performance. The exposure chamber (Fig. 2) consisted of a 1-L mason jar with a 2-cm diameter hole covered with a 0.5-mm mesh Nitex screen near the top to serve as an overflow. An inverted 9-cm finger bowl with a 2-cm diameter hole functioned as a lid. Air delivery and seawater delivery tubes were fixed through a hole in the lid. The air delivery tube was directed toward the side of the jar, approximately 5 cm from the sediment surface to minimize sediment disturbance. Inflowing filtered seawater was delivered via an intermittent flow system at a rate of approximately 18 turnovers/vessel/day. The 90% volume replacement time was estimated to be approximately 3 h.

Sediments collected from one station at each of three sites (Oakland, Yerba Buena, and Tomales Bay), as well as control sediment, were used to conduct static exposures in addition to

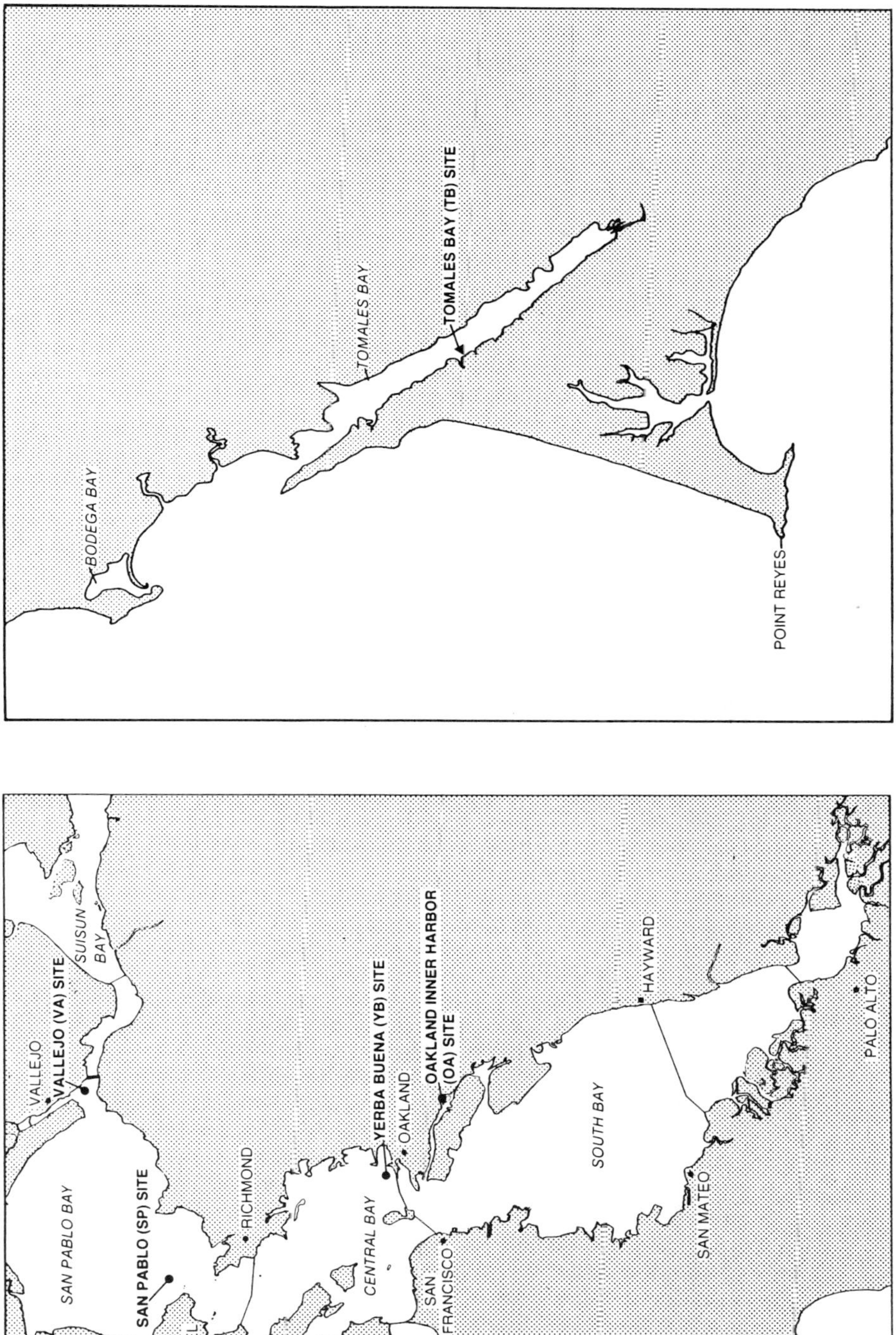

FIG. 1—*Sediment collection sites within San Francisco Bay and Tomales Bay.*

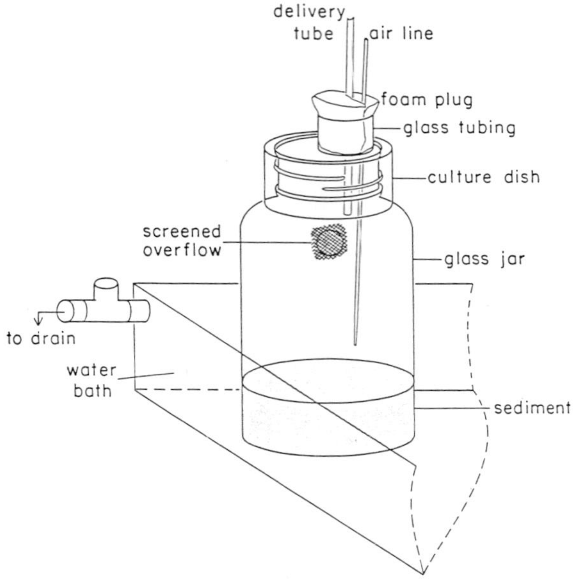

FIG. 2—*Schematic presentation of exposure chambers.*

flow-through exposures. For static exposures, only a capillary tube was inserted through each lid to function in air delivery. Salinity was closely monitored in these static exposure chambers, and deionized water was added when salinity exceeded 34 ppt. Exposure chambers were held in a temperature-controlled water bath at 20 ± 1°C and under 24-h illumination. Constant light conditions were selected to avoid possible emergence of the amphipods during the dark cycle. In addition to the treatment groups, static and flow-through seawater controls were maintained during the study. All treatments and controls were maintained in quintuplicate, that is, the sediment sample was divided into five portions which were each tested separately. This procedure allows determination of the experimental error, but does not allow comparison of sediments collected from different stations.

At the initiation of the test, the *A. abdita* were sieved from the holding sediments (that is, control sediments taken from the *A. abdita* collection site) and sorted into 100-mL plastic beakers containing filtered seawater. Prior to adding *A. abdita* to the test vessels, sediments were carefully examined for the presence of *A. abdita* tubes indicating the presence of a living native ampeliscid which had survived the sediment screening process. If any such tubes were found, the tubes and sediments immediately surrounding it were removed.

Twenty amphipods of uniform size were placed into each beaker, and moribund or outsized animals were replaced. Additionally, mature males were replaced with either subadults or females [5] to avoid the natural mortality of males associated with reproductive activity. The beakers, each representing a replicate, were randomized and the animals then transferred to exposure chambers containing experimental sediments. One hour after introduction of the animals to the exposure chambers, the number of individuals which had not burrowed were recorded and were then removed and replaced with others from the sieved population. The test organisms were not fed during the ten-day exposure.

The exposure chambers were monitored daily for mortalities and emerged amphipods. *A. abdita* were considered dead if, on gentle prodding with a glass probe, they did not curl into their natural position or if there was no movement of the appendages. These animals were removed and recorded. Moribund animals exhibiting twitching of appendages were considered alive and returned to the test chamber. Additionally, the number of emerged animals (on the sediment and water surface) and cast exoskeletons were recorded daily for each test chamber and the molts removed. At the end of the ten-day exposure, all sediments were sieved through a 0.5-mm mesh screen. All survivors were enumerated and those unaccounted for were considered dead.

Temperature and salinity were measured daily in one replicate vessel of each control solution throughout the ten-day exposure. The dissolved oxygen concentration and pH were measured daily in one replicate vessel of each treatment level and control solutions.

Statistical Procedures

Data analyses initially consisted of a two-way analyses of variance (ANOVA) with station variability and site variability as independent factors. All percentage data were arcsine square-root transformed. To compare the toxic response resulting from exposure to sediments from each site, the mean of five response values available for each of the three stations per site were used in the statistical analysis using a one-way analysis of variance (ANOVA) followed by a Student-Newman-Keuls (SNK) multiple-range test. All survival counts in each individual beaker were expressed as percentage of the initial number of *A. abdita* added. In some cases, in excess of 20 *A. abdita* could be accounted for at the conclusion of the test, indicating that occasionally a small number (generally 1 or 2) of *A. abdita* had passed through the sieves used initially to screen out the native amphipods. To account for this experimental error, the percentage mortality and number emerged was based on the total number of test organisms present at test initiation, that is, the number added to the test vessels plus the number of residual native *A. abdita* determined *a posteriori.* Emergence data were treated in a similar fashion as the survival data. No transformations were applied to this data set, in contrast to the survival data.

The cumulative percent mortalities were graphically displayed based on the sum of the cumulative percent mortalities determined from daily mortality counts, and the number of "unaccounted for" organisms were determined as the difference of the live animals at test termination and the cumulative mortality. Unaccounted for organisms were expressed as a percentage of the total number of test organisms present at test initiation and proportioned over each test day.

Results and Discussion

Flow-Through Exposure

The calculated mean percent mortality of *A. abdita* at each station within each site, based on five replicate exposures, was generally consistent (Table 1). The cumulative percent mortalities for each station have been graphically displayed in Fig. 3, while Fig. 4 illustrates the mean cumulative percent mortality by site. A two-way factorial analysis of variance indicated that the station-to-station differences were not statistically significant at the 95% level of confidence, but that there were highly significant differences among sites ($p < 0.001$). Further analyses indicated that San Pablo Bay was the least, and Oakland Harbor the most, toxic with mean ten-day mortalities of 8.7 and 25.9%, respectively. Yerba Buena, Tomales Bay, and Vallejo sediments were not statistically significant from one another (SNK test, $p < 0.05$). Similarly, statistical analyses conducted with the emergence data (Table 1) showed that the amphipods emerged in the following increasing order: San Pablo Bay = Tomales Bay < Yerba Buena < Vallejo < Oakland Harbor. Both toxic endpoints indicated Oakland Harbor as the most contaminated sediment and San Pablo Bay as the least contaminated site within San Francisco Bay. These results were in agreement with the detailed analyses for sediment pollutants, to be discussed elsewhere [*21*].

Sediment texture (grain size analyses) indicated that, with the exception of the Vallejo site, sediments at most stations consisted predominantly of clay (43 to 63%), followed by silt (28 to

TABLE 1—*Mean mortality (%) and mean number of* A. abdita *emerged (*n = *5) following ten-day toxicity test under flowing seawater conditions with sediments from each of three stations at five sites in San Francisco Bay and Tomales Bay.*

Site	Station	Mean % Mortality	Mean Number Emerged[a]
San Pablo Bay	1	9.1	2.9
	2	9.0	1.4
	3	7.9	0.8
	Mean ± SD	8.7 ± 0.7	1.7 ± 1.1
Tomales Bay	1	14.0	2.0
	2	4.9	1.2
	3	15.8	3.2
	Mean ± SD	11.6 ± 5.8	2.1 ± 1.0
Yerba Buena	1	15.5	3.7
	2	14.7	3.3
	3	14.9	5.3
	Mean ± SD	15.0 ± 0.4	4.1 ± 1.1
Vallejo	1	18.6	6.5
	2	15.0	4.8
	3	11.0	3.2
	Mean ± SD	14.9 ± 3.8	4.8 ± 1.7
Oakland Harbor	1	34.0	6.8
	2	24.0	6.8
	3	19.6	4.5
	Mean ± SD	25.9 ± 7.4	6.0 ± 1.3

[a]Normalized for the number of organisms present, for example, if the number of organisms exposed was 22, and the number of emerged organisms was 2, the normalized number emerged was 20/22 × 2 = 1.8.

48%), and sand and silt (generally <5%). The Vallejo site contained significantly more sand (23 to 59%) and less clay (10 to 32%). Detailed grain size analyses are described elsewhere [*20*].

Water quality variables were consistent throughout the flow-through ten-day tests. Mean dissolved oxygen concentrations ranged between 7.4 and 7.6 mg/L (99 to 101% of saturation at 32 ‰ salinity and 20°C, respectively). The pH values ranged between 7.5 and 8.2, temperatures were 20°C (coefficient of variation <2%), and salinity was 32 ‰.

Static Exposure

Station 1 sediments from Oakland Harbor, Yerba Buena, and Tomales Bay were chosen for the static tests. Table 2 indicates that, in all three cases, test sensitivity was slightly increased, that is, more test animals died during the ten-day exposure under static conditions than under flow-through conditions. The largest difference in response was found with the Oakland Harbor sediments. However, none of these differences were statistically significant (pairwise t-tests of the arcsine square root transformed mortality data). Similarly, none of the static versus flow-through emergence responses were significantly different. It is noteworthy that the variability in both the survival and emergence response tended to be consistently higher under static exposures. The only exception was survival in the Yerba Buena sediment. It appears that, although sensitivity may be slightly increased, the decrease in precision may outweigh the advantage gained using static exposures.

In the static exposures, dissolved oxygen concentrations, pH, and temperatures were within

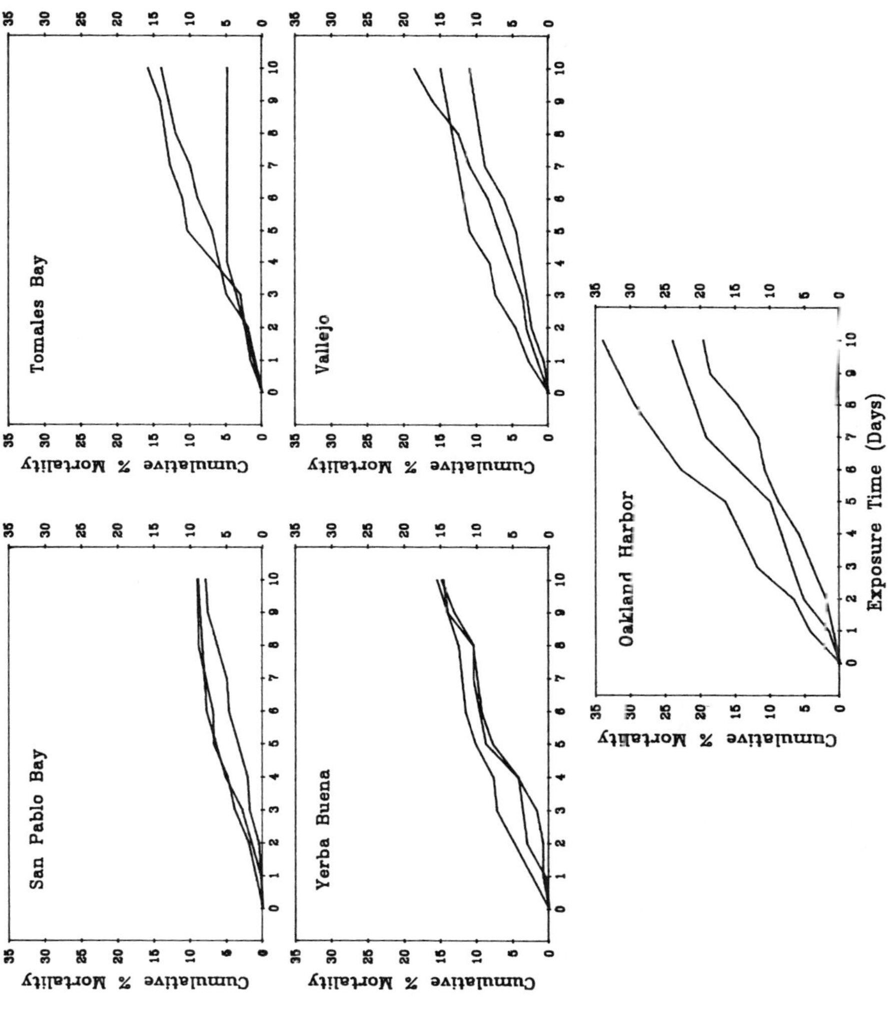

FIG. 3—*Cumulative mortality of* A. abdita *from various stations within San Francisco Bay and Tomales Bay.*

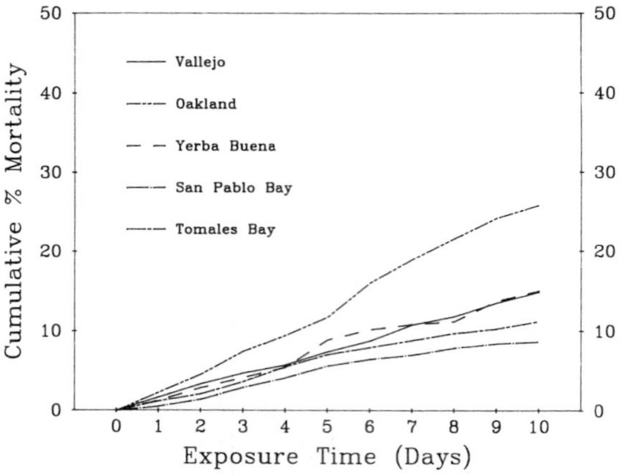

FIG. 4—*Comparative cumulative survival of* A. abdita *from various sites within San Francisco Bay and Tomales Bay.*

TABLE 2—*Survival and number of emerged* A. abdita *following completion of ten-day static exposure to various sediments from Station 1 at each of three sites in San Francisco Bay and Tomales Bay.*

Site	Station	Replicate	Initial Number Exposed[a]	Number Alive	Cumulative Number Emerged[b]
OA	01	A	20	9	3
	01	B	20	6	3
	01	C	20	12	2
	01	D	20	14	9
	01	E	20	10	8
		Mean ± SD	20 ± 0	10.2 ± 3.0	5.0 ± 3.2
			(20 ± 0)[c]	(13.2 ± 1.6)	(6.8 ± 1.9)
YB	01	A	20	16	2
	01	B	20	20	6
	01	C	20	17	0
	01	D	20	16	5
	01	E	20	17	7
		Mean ± SD	20.0 ± 0.0	17.0 ± 1.7	4.0 ± 2.9
			(20.6 ± 0.9)	(17.4 ± 2.5)	(3.8 ± 2.2)
TB	01	A	20	18	0
	01	B	21	20	3
	01	C	20	12	1
	01	D	20	15	1
	01	E	20	20	0
		Mean ± SD	20.2 ± 0.4	17.0 ± 3.5	1.0 ± 1.2
			(20.0 ± 0.0)	(17.2 ± 1.8)	(2.0 ± 0.7)

NOTE: OA = Oakland Harbor; YB = Yerba Buena; TB = Tomales Bay.

[a] Estimated based on amphipods present at test termination and number of mortalities recorded during the exposure.

[b] Based on the combined, cumulative total number of organisms found daily at the sediment surface and those found on the water surface.

[c] Mean and SD in parentheses are corresponding data from the flow-through exposures for each station.

the range of those measured in the flow-through exposures. Because of evaporation, salinity values ranges were wider, although salinity was adjusted periodically with deionized water to maintain it at natural seawater level. Based on the water quality variables, test organisms were not subject to stress which might have explained subtle nuances in the toxic responses between replicate stations or between the San Pablo Bay and Tomales Bay sites.

Conclusions

From the results of this study it is clear that the ten-day *A. abdita* test has the ability to discriminate between highly or moderately contaminated sites and uncontaminated sites. We feel that the generally low mortalities caused by the San Francisco Bay sediments is real and not an artifact of the presence of native amphipods or *A. abdita*'s insensitivity.

The choice of test performance under flow-through versus static conditions is an important one since static tests do not require elaborate seawater facilities. Furthermore, because contaminants will partition into overlying waters, static tests should elicit a more sensitive response. On the other hand, the release of contaminants into the water column, thereby eliciting a response mediated through the water column, does not conceptually appear to suit the objectives of "sediment" toxicity tests. This is especially true for test organisms such as *A. abdita*, which are exposed to water column contaminants. Although, as expected, *A. abdita*'s sensitivity under static conditions was slightly increased, variability in static exposures was also higher, causing the precision of the test to be lower. We recommend that *A. abdita* tests not be conducted under static conditions.

The presence of "native" test organisms may pose a serious problem. There are several potential methods for removing these animals from test sediments, all having varying degrees of success and subsequent effects on sediment characteristics. Sieving the sediments prior to use has been the most effective method where *A. abdita* densities are low. When high densities are encountered (San Pablo Bay), this method is not as attractive. Killing resident organisms by freezing or heat stress would cause unacceptable changes in the sediment properties. Purging the sediment of oxygen is a potential means of removing *A. abdita*, since this species is very sensitive to oxygen stress and will leave the sediments as an escape response. The duration of this low dissolved oxygen exposure, however, would have to be examined very closely to evaluate effects on contaminant bioavailability in the exposed sediments. This method may be the least disruptive of all, and, thus, it deserves further work.

The test duration of ten days was chosen to be consistent with the more widely used *Rhepoxynius* test [*21*]. Previous unpublished studies conducted by one of the authors (Scott) have shown that increasing the test duration from four to ten days nearly doubled the acute response. Similarly, studies that assessed the mortality of *A. abdita* exposed to suspended sediments yielded an acute (4-day test) to chronic (28-day test) ratio of 12. These studies have shown that increasing the exposure time certainly increases sensitivity, but the longer exposures would require that a food source be provided.

A broader application of the *A. abdita* sediment bioassay requires that a greater range of sediments with varying contamination and grain sizes be tested to provide a more comprehensive data base. As with all sediment test organisms, grain size sensitivity must also be determined. There is an additional need for the development of a test for low salinity sediments. Research on *A. abdita* should be undertaken to address both of these issues.

Acknowledgments

The enthusiasm and organizational skills of NOAA's project administrator, Ed Long, are greatly appreciated. Michele Redmond assisted with the test animal collections and performance of the tests. This study was funded by the National Oceanic and Atmospheric Administration, Ocean Assessments Division under Contract No. 50ABNC700086 to Springborn Life Sciences, Inc.

References

[1] Bousfield, E. L., *Shallow-Water Gammaridean Amphipods of New England*, Cornell University Press, Ithaca, NY, 1973.
[2] Nichols, F. H., *Estuaries*, Vol. 8, No. 2a, 1985, pp. 136–144.
[3] Nichols, F. H. and Thompson, J. K., *Marine Ecology Progress Series*, Vol. 24, 1985, pp. 83–97.
[4] Stickney, A. P. and Stringer, L. D., *Ecology*, Vol. 38, 1957, pp. 11–122.
[5] Mills, E. L., *Journal of Fisheries Research Board of Canada*, Vol. 24, 1967, pp. 305–355.
[6] Santos, S. L. and Simon, J. L., *Marine Ecology Progress Series*, Vol. 3, No. 4, 1980, pp. 347–356.
[7] Mills, E. L., *Journal of the Fisheries Research Board of Canada*, Vol. 26, 1969, pp. 2425–2428.
[8] Hyland, J. L., "Comparative Structure and Response to (Petroleum) Disturbance in Two Nearshore Infaunal Communities," Ph.D. thesis, University of Rhode Island, Kingston, 1981.
[9] Enequist, P., *Zoologische Bidragen Fran Uppsala*, Vol. 28, 1980, pp. 297–492.
[10] Zatsepin, V. I., "On the Significance of Various Ecological Groups of Animals in the Bottom Communities of the Greenland, Norwegian, and the Barents Seas," *Marine Food Chains*, J. H. Steele, Ed., University of California Press, Berkeley, CA, 1970, pp. 207-221.
[11] Nichards, S. W., *Bulletin of the Bingham Oceanography College*, Vol. 18, 1983, pp. 32–101.
[12] Wakabara, Y., Kawakani de Regende, E., and Tararam, A. S., *Marine Biology*, Vol. 68, 1982, pp. 67–70.
[13] "Ambient Water Quality Criteria for Lead—1984," EPA 440/5-84-027, USEPA, Washington, DC, 1985.
[14] "Ambient Water Quality Criteria for Copper—1984," EPA 440/5-84-031, USEPA, Washington, DC, 1985.
[15] "Ambient Water Quality Criteria for Cadmium—1984," EPA 440/5-84-032, USEPA, Washington, DC, 1985.
[16] "Ambient Water Quality Criteria for Arsenic—1984," EPA 440/5-84-033, USEPA, Washington, DC, 1985.
[17] Rogerson, P., Schimmel, S., and Hoffman, G., "Chemical and Biological Characterization of Black Rock Harbor Dredged Material," Technical Report D-85-9, USEPA, Environmental Research Laboratory, Narragansett, RI, and U.S. Army Engineers Waterways Experiment Station, Corps of Engineers, Vicksburg, MI, 1985.
[18] Scott, K. J., Redmond, M. S., and Pruell, R. J., "The Acute Response of the Amphipod *Ampelisca abdita* to Contaminated Sediments from New Bedford Harbor, MA," in preparation.
[19] Gentile, J. H., Scott, K. J., Lussier, S. M., and Redmond, M. S., "The Application of Population Responses to the Risk Assessment of the Aquatic Disposal of Black Rock Harbor Dredged Material," Technical Report D-86-xx, USEPA, Narragansett, RI, and U.S. Army Engineer Waterways Experiment Station, Vicksburg, MI, 1987.
[20] Long, E. R., Buchman, M. F., Bay, S. M., Breteler, R. J., Carr, R. S., Chapman, P. M., Hose, J. E., Lissner, A. L., Scott, K. J., and Wolfe, D. A., "Comparative Evaluation of Five Toxicity Tests of Sediments from San Francisco Bay and Tomales Bay, California" (marine environmental research), in press.
[21] Swartz, R. C., DeBen, W. A., Jones, J. K. P., Lamberson, J. O., and Cole, F. A., "Phoxocephalid Amphipod Bioassay for Marine Sediment Toxicity," *Aquatic Toxicology and Hazard Assessment: Seventh Symposium, ASTM STP 854*, R. D. Cardwell, R. Purdy, and R. C. Bahner, Eds., American Society for Testing and Materials, Philadelphia, 1984, pp. 284–307.

Peter F. Landrum,[1] *Warren R. Faust,*[1] *and Brian J. Eadie*[1]

Bioavailability and Toxicity of a Mixture of Sediment-Associated Chlorinated Hydrocarbons to the Amphipod *Pontoporeia hoyi*

REFERENCE: Landrum, P. F., Faust, W. R., and Eadie, B. J., **"Bioavailability and Toxicity of a Mixture of Sediment-Associated Chlorinated Hydrocarbons to the Amphipod *Pontoporeia hoyi*,"** *Aquatic Toxicology and Hazard Assessment: 12th Volume, ASTM STP 1027,* U. M. Cowgill and L. R. Williams, Eds., American Society for Testing and Materials, Philadelphia, 1989, pp. 315-329.

ABSTRACT: The toxicity and bioavailability of a mixture of 13 sediment-associated chlorinated hydrocarbons were determined for the amphipod *Pontoporeia hoyi.* The processes were traced with two radiolabeled compounds, ^{14}C-2,2′,4,4′-tetrachlorobiphenyl (2244TCB) and ^{3}H-benzo(a)pyrene (BaP). The toxicity of the chlorinated hydrocarbon mixture resulted primarily from the lindane and dieldrin components. The 72-h LC_{50} was 14.5 nmol g^{-1} as the sum of the added chlorinated hydrocarbons and 4.7 nmol g^{-1} as the sum of the lindane and dieldrin. At 144 h the LC_{50}s had dropped to 1.51 and 0.53 nmol g^{-1}, respectively. At the lowest dose, 2.4 nmol g^{-1} sediment, the time to yield 50% mortality (LT_{50}) was 119 h. The toxicokinetics for the radiolabeled compounds yielded sediment uptake clearance constants of 0.0018 and 0.018 g dry sediment g^{-1} animal h^{-1} for BaP and 2244TCB in the absence of added chlorinated hydrocarbons. At the lowest chlorinated hydrocarbons dose, the clearance constants of both BaP and 2244TCB were approximately doubled. The clearance constants were not proportional to the freely dissolved concentration of the radiolabeled compounds in the interstitial water but were presumed to be proportional to the compound desorption rates. For the chlorinated hydrocarbons, the uptake clearances were inversely proportional to the log of the octanol-water partition coefficient. The bioaccumulation factor (BAF, concentration in the organism divided by the concentration in the sediment) appeared to peak at about a log octanol-water partition coefficient of 6. However, the BAF ranged from less than 1 to greater than 30 for compounds with a log K_{ow} near 6.

KEY WORDS: bioavailability, toxicokinetics, sediments, partitioning, toxicity, LC_{50}, amphipod, mixture, chlorinated hydrocarbons, bioaccumulation, *Pontoporeia hoyi*

Contaminated sediments, considered a major problem in many areas throughout the Great Lakes Basin, are associated with biological effects such as tumors in bullheads [*1*] or changes in benthic species composition [*2*]. Further, sediment-associated contaminants are an important source of contaminants to benthic organisms [*2,3*] for potential food chain bioaccumulation. Understanding the degree of biological availability is important since sediment-associated contaminants can remain present in the biologically inhabited portions of the sediments for decades. However, the significance of these contaminated sediments as contaminant sources to the benthic food chain or to the pelagic system via resuspension is not known [*4*]. This is particularly true for mixtures of xenobiotics in sediments. Previous laboratory studies exposing *Pontoporeia hoyi* to sediment-associated polycyclic aromatic hydrocarbons (PAH) indicate that the

[1]Great Lakes Environmental Research Laboratory, NOAA, Ann Arbor, MI 48105, Contribution No. 596.

accumulation is dependent on the specific compound desorption rate from the sediment matrix with subsequent accumulation from the interstitial water. Additional accumulation occurs through ingestion of contaminated particles by *P. hoyi* [*5*]. A similar conclusion was reached for the accumulation of hexachlorobenzene by the clam *Macoma nasuta* [*6*] and oligochaetes [*7–9*].

The objectives were: (*a*) to seek a no effect level for the chlorinated hydrocarbon mixture at which to measure the toxicokinetics of these sediment-associated chlorinated hydrocarbons; (*b*) to compare the toxicokinetics for selected compounds in the mixture with previous kinetics determined on single compounds; and (*c*) to compare the uptake kinetics with partitioning between sediments and interstitial water. Measures of the partitioning between the sediment interstitial water and particules for the radioactive tracers, in combination with relationships between the bioavailability and the physical chemical properties of the different chlorinated hydrocarbons, are used to examine the equilibrium partitioning approach, one of the primary means for the development of sediment quality criteria by EPA [*10–13*].

Materials and Methods

Collection of Environmental Samples

P. hoyi were collected from Lake Michigan off Grand Haven, Michigan at a depth of 24 m with a PONAR grab sampler. The *P. hoyi* were gently removed from the sediment with a screen, placed in clean lake water, and kept cold with ice during transport to the laboratory. *P. hoyi* were kept in aquaria containing about 1 in. of sediment and 4 in. of lake water at 4°C [*14*].

Water used throughout the work was Lake Michigan surface water (collected about 1 m below the surface) and stored at 4°C.

Lake Michigan sediment used in the experiment was obtained by PONAR grab approximately 5 miles off Grand Haven, Michigan at a 45-m depth. The sediment was sieved at 1 mm to remove animals and large pieces of debris and kept at 4°C. The sediment was characterized by wet sieving a subsample to determine the amount of material, on a dry weight basis, passing an 88-μm sieve. The organic carbon content of the sediment and the sieved fractions were determined [*15*].

Dosing of Sediments

The sediment was dosed with two radioactive tracers, ^{14}C-2,2′,4,4′-tetrachlorobiphenyl (2244TCB, universally labeled, specific activity 15.33 mCi mmol^{-1}) and ^{3}H-benzo(a)pyrene (BaP, specific activity 33.1 Ci mmol^{-1}), and an additional 13 cold chlorinated hydrocarbons: hexachlorobenzene (HCB), lindane) (LIN), 2,2′,5-trichlorobiphenyl (225TCB), 2,4,4′-trichlorobiphenyl (244TCB), 2,2′,6,6′-tetrachlorobiphenyl (2266TCB), 2,2′,3,4-tetrachlorobiphenyl (2234TCB), 3,3′,4,4′-tetrachlorobiphenyl (3344TCB), 2,2′,4,4′,6,6′-hexachlorobiphenyl (224466HCB), 2,2′,4,4′,5,5′-hexachlorobiphenyl (224455HCB), dieldrin (DIEL), p,p′-DDE, p,p′-DDT, and octachlorodibenzo-p-dioxin.

A solution of chlorinated hydrocarbons was prepared by dissolving 5 to 6 mg of each the nonlabeled compounds in acetone (10.5-mL total). The mixture was filtered through glass fiber filters to remove undissolved crystals of some of the compounds. The dioxin proved to be so insoluble that it was effectively removed by the filtering; no dioxin was subsequently detected by gas chromatography in either sediments or the test organisms.

The 1-mm sieved sediment was mixed until it appeared to be homogeneous. An amount of wet sediment corresponding to 700 g dry weight was placed in a 4-L beaker, and 2500-mL lake water was added. The following solutions were added dropwise while the mixture was stirred vigorously at room temperature: 700 μL methanol containing ^{14}C-2244TCB (3.15 μCi/700 g dry sediment); 800 μL methanol containing ^{3}H-BaP (15.8 μCi/700 g dry sediment); and 7.11 mL acetone containing the appropriate amount of the above chlorinated hydrocarbon solution for the contaminant level being prepared. The amount of acetone and methanol carriers added to each solution was held constant. The mixture was stirred at room temperature for 4 h and kept at 4°C thereafter.

The nonlabeled compounds were tested at five different dose levels (Table 1) with the radiotracers added at the same concentration for each dose level, 0.31 ± 0.02 nmol g^{-1} for 2244TCB and 0.75 ± 0.04 pmol g^{-1} for BaP. The "control" level contained the same amount of radiotracers as did all the other levels but only the natural background concentrations of the nonlabeled chlorinated hydrocarbons (Table 1).

Experimental Procedure

Two days after dosing the sediment, the overlaying water was decanted, removing most of the acetone and methanol carriers, the sediment was stirred for homogeneity, and 40-g wet sediment transferred to 400-mL beakers. The amount of acetone (<325 $\mu L\ L^{-1}$) and methanol (<32 $\mu L\ L^{-1}$) remaining were not expected to affect the toxicokinetics [*16*]. The estimate of the amount of carrier assumes 80% of the water was decanted, and there was no loss to volatilization. The following samples were taken in triplicate from each dose level at the start, middle, and end of the distribution of sediment to beakers: 2-g sediment for analysis of nonlabeled chlorinated hydrocarbons by gas chromatography (GC); 2-g sediment for wet to dry weight determination; and 5-g sediment dessicated with sodium sulfate, extracted with ethyl acetate, and counted for radioisotope activity. Lake water (250 mL) was then carefully added to each beaker, minimizing the disturbance of the sediment; on the following day (three days after dosing the sediment) an initial water sample was taken and 20 test organisms added to each beaker.

Groups of 20 *Pontoporeia hoyi* were exposed to each sediment in triplicate for each sample point, 3, 6, 12, 19, and 31 days. Thus, the total number of beakers was 15 per dose level. Two additional beakers at each concentration containing no organisms were sampled on Days 15 and 31 for radiotracer concentration in sediment interstitial water.

TABLE 1—*Concentration[a] of contaminants in dosed sediments measured by gas chromatography.*

Compound Recovery[b]	Molecular Weight	Level 0	Level 1	Level 2	Level 3	Level 4	Level 5
HCB 82.4(1.9)	284.8	0.05	0.14	1.34	1.80	6.77	17.97
LIND 82.4(2.2)	290.9	0.32	0.71	3.95	5.23	11.45	16.05
225TCB 76.6(1.4)	256	0.13	0.12	0.28	1.42	8.11	19.36
2266TCB 77.4(1.5)	290	0.02	0.17	0.27	1.70	7.77	18.19
244TCB 80.3(1.8)	256	0.02	0.22	0.88	1.30	5.72	11.74
2234TCB 81.7(1.5)	290	0.22	0.25	1.34	1.60	9.09	19.95
224466HCB 77.9(1.4)	358	ND	0.18	0.94	1.36	6.40	13.49
DIEL 82.3(1.4)	381	ND	0.13	0.60	0.81	5.11	9.92
p,p′-DDE 98.8(1.2)	318.5	ND	0.22	0.99	1.31	7.76	14.00
3344TCB 95.3(3.6)	290	ND	0.07	0.92	1.51	6.48	7.31
224455HCB 95.1(1.5)	358	0.03	0.08	0.59	0.77	6.45	10.13
p,p′DDT 101.2(1.3)	354.5	0.15	0.06	0.35	0.63	4.56	8.38
Total nmoles g^{-1}		0.94	2.4	12.5	19.9	87.9	171.6
LIND Plus DIEL Total nmoles g^{-1}		0.32	0.84	4.6	6.0	16.6	26.0

[a] Concentrations in nmoles g^{-1}. Level 0 was the natural sediment concentration.
[b] Percent recovery for spiked specimens with coefficient of variation (%) in parentheses. All concentrations are corrected for percent recovery.

Distribution of the radiolabeled compounds in the overlaying water was determined from an initial water sample of 25 mL, in duplicate, from each dose level; 2 mL were counted for total activity, 20 mL were filtered for particle bound contaminant, with a back-up filter to correct for sorption and 2 mL of filtrate counted, 15 mL was run through a C-18 Sep Pak (Waters Associates) and subsequently eluted with 3 mL of methanol that was counted to determine the amount of freely dissolved compound in the water column [*17,18*].

Three beakers from each dose level were sampled at random at each time point (3, 6, 12, 19, and 31 days) and the following information obtained:

1. *Overlying water*—total, free, bound radioisotope activity as described above and oxygen concentration [*19*].
2. *Test organisms*—the number in overlying water was recorded to check for sediment avoidance behavior, number surviving was recorded for LC_{50} determinations, and the radioisotope concentration of two individuals (wet weight basis) was measured for toxicokinetic determinations; another two individuals were measured for lipid content [*20*], and the remaining live organisms were combined weighed and extracted by homogenizing in a tissue grinder with acetone and cyclohexane for GC determination of their chlorinated hydrocarbon content.
3. *Sediment*—a sample for wet to dry weight ratio, a sample for later extraction and determination of chlorinated hydrocarbon content via GC, and a sample was dessicated, extracted, and counted for radioisotope activity as indicated above.
4. *Pore water*—on Days 15 and 31 the overlaying water was removed by aspiration from beakers having no organisms, the sediment was transferred to centrifuge tubes and centrifuged at approximately 1000 $\times$ g for 10 min, and the total, free, and bound concentration of the two radioisotopes was determined for the resulting pore water as described above for the overlying water.

Pesticide Toxicity Experiment

In a separate experiment, ten test organisms were exposed to sediment (45 g) containing various combinations of the nonlabeled chlorinated hydrocarbons to determine which were largely responsible for the observed mortality. Sediment was prepared similarly to the above procedure at the nominal concentration equivalent to Contaminant Level 3, 0.6 $\mu g\ g^{-1}$ for each constituent. The following combinations were tested: acetone carrier only; each of the PCB congeners listed above combined; and each of the following in the presence of the combined PCB congeners—HCB, DDE, DDT, LIN, and DIEL. Mortality data were collected after six and nine days exposure to the sediments. All LC_{50} values were calculated using a probit analysis routine [*21*].

Analyses

Approximately 2 g (dry) of sediments were soxhlet extracted for 24 h in a 2:1 mixture of hexane and methanol. The extracted specimens, sediments and *P. hoyi* (extraction described above), were washed twice with an equal amount of hexane-extracted distilled water. The washed extract was rotary evaporated to less than 1 mL. The sample was transferred to a 2-mL glass bottle, brought to final volume of 1 mL with cyclohexane, and stored at 4°C until analysis. Four microlitres were splitless auto-injected into a Hewlett-Packard 5880A capillary gas chromatograph fitted with a 25-m DB-1 fused silica capillary column and electron capture detector. The temperature program of the analysis ran from 100 to 270°C at 4°C min^{-1}.

Scintillation counting was performed on an LKB 1217 liquid scintillation counter using RPI scintillation cocktail 3a70B. The data were corrected for quench using the external standards ratio method after correcting for background.

Results

Sediment Characteristics

The bulk dry sediments for these studies had an organic carbon content of 1.0 ± 0.2%. The ashfree dry weight was 1.9 ± 0.2% of the dry weight. The sediment particle size was distributed between 43.6% for >88 μm and 56.4% for <88 μm on a dry weight basis. The organic carbon content of the fractions was 0.26 ± 0.06% for >88 μm and 2.1 ± 0.3% for <88 μm fraction.

The homogenity of the mixed sediments was determined through the analysis of the radiotracers for samples taken at the beginning, middle, and end of the distribution of sediment to the exposure beakers. The coefficients of variation were low 3.5 ± 2.4% (mean ± 1 SD) for the BaP and 3.1 ± 0.5% for the 2244TCB at the various dose levels. The recoveries for these radiolabeled compounds were high and averaged 103 ± 9% for BaP and 98.4 ± 7.1% for 2244TCB. Thus, because of the high recoveries and the low coefficients of variation at time zero, the sediment was assumed to be homogenously dosed.

The overlying water contained low amounts of radiolabel with higher quantities in the samples containing organisms. The overlying water in the samples with the organisms was somewhat cloudy. Greater than 75% of the activity for the ^{3}H-BaP and nearly 90% of the ^{14}C-2244TCB were either on particles or associated with dissolved organic matter. Previous studies have shown that *P. hoyi* do not obtain a substantial amount of their body burden from the overlaying water when the sediment is the major source [*5,22*].

Toxicity

No overt avoidance of the sediment was found with the chlorinated hydrocarbons. Only one or two amphipods maximum were found per beaker swimming above the sediments at any of the dose levels, and there was no pattern to the potential avoidance as a result of dose or duration. Throughout the experiment the oxygen concentration remained above 6 mg L^{-1} except for the last sample at Day 31 where the oxygen content was 5.0 ± 0.4 mg L^{-1} for animals exposed to the radiolabeled compounds only. The oxygen concentration was higher in the water where the concentration of toxicant was greater. The lipid content of the *P. hoyi* was 27.9 ± 2.2% on a dry weight basis after three days exposure, and no significant difference was found on Day 31 for the controls or on Day 6 for Level 1 exposed organisms. The one organism from Level 1, Day 12 had a higher lipid content of 46.3%. From this data, it was concluded that the organisms were not suffering from nutrient limitations.

Independent toxicity experiments with either the PCB mixture or the PCB mixture with one of the pesticides indicated that the toxicity was likely due primarily to the presence of LIN and DIEL in the chlorinated hydrocarbon mixture (Table 2). All of the other combinations of mixtures produced no mortality except one acetone control. However, the second acetone control did not replicate the high mortality.

Toxicity of the total mixture of chlorinated hydrocarbons was determined by examining the mortality data in two ways (Table 3). First, the LC_{50} was determined across the range of concentrations tested. The control, Level 0, mortality was 5% or less for all the samples except at Day 31 when the control mortality ranged from 0 to 16%. The probit calculations accounted for the percent mortality in the controls. The log concentrations used in the probit analysis were for the total number of nmol g^{-1} dry sediment for all the nonlabeled chlorinated hydrocarbons added and the total of the number of nmol g^{-1} dry sediment of LIN and DIEL only. The mixture yielded 72-h LC_{50}s of 14.45 and 4.68 nmol g^{-1} for the total for the organic contaminants and the sum of LIN and DIEL, respectively. The data had considerable variability at 72 h, and the confidence limits were not estimated by the probit analysis program [*21*]. At 144 h the respective LC_{50}s were 1.51 [0.73 to 2.21, 95% confidence interval (CI)] and 0.53 (0.26 to 0.78, 95% CI) nmol g^{-1} dry sediment from the two methods of identifying the dose.

TABLE 2—*Toxicity screen for portions of the mixture.*

	Exposure Duration, percent mortality	
Toxicants[a]	6 Days	9 Days
Control sediment only	0	0
Acetone carrier control	60	0
PCBs and HCB	0	Lost
PCBs and Lindane	90	90
PCBs and Dieldrin	20	60
PCBs and DDE	0	0
PCBs and DDT	0	0
PCBs and OCS	0	0
PCBs only	0	0

[a]The concentrations used were nominally equivalent to the amount of the compounds found at Level 3 in Table 1. The amount of carrier used to dose the sediments was the same proportion as for the kinetics experiments.

TABLE 3—*Mortality of* Pontoporeia hoyi *Exposed to a Mixture of Chlorinated Hydrocarbons.*

Dose Level[a]	Replicate	Duration of Exposure, h: 72	144	288	456	744
0	A	20[b]	19	18	20	16
	B	20	19	19	20	16
	C	17	20	19	20	20
1	A	17	5	10	0	0
	B	19	5	1	0	0
	C	18	8	0	0	0
2	A	18	1	0	0	0
	B	20	1	0	0	0
	C	18	0	0	0	0
3	A	1	0	0	0	0
	B	1	0	0	0	0
	C	1	0	0	0	0
4		All dead at all exposure durations				
5		All dead at all exposure durations				

[a]Concentrations for each dose level can be found in Table 1 plus the concentration of the radiotracers, 0.31 nmol g^{-1} dry sediment for 2244TCB and 0.075 pmol g^{-1} dry sediment for BaP.
[b]Number alive out of 20 per group.

The second method of analyzing the data was to determine the time required for 50% mortality at the lowest dose, 2.4 nmoles g^{-1} nonlabeled chlorinated hydrocarbons (0.84 nmoles g^{-1} LIN and DIEL). The LT_{50} was 118.97 h with no confidence limits calculated by the probit analysis program [*21*]. Therefore, the toxicokinetics, discussed below, determined for the nonlabeled compounds were estimated for organisms that were responding to the toxic effects of the mixture.

Since the concentration of the compounds in the live organisms was determined for toxicokinetic purposes, the body burden that produced the resultant LC_{50} could be determined for comparison with the LC_{50}s calculated from sediment concentrations. The 72-h body burden

LC_{50} was 13.3 (12.0 to 15.1 95% CI) and 6.3 (5.7 to 7.3 95% CI) nmol g^{-1} wet weight organism for the nonlabeled chlorinated hydrocarbons and the sum of LIN and DIEL, respectively. Variabilities in the experiment were better defined on a body burden basis for the short-term response than was obtained using the sediment concentrations as a basis for calculating the LC_{50}s. For the LT_{50}, the body burden required to produce the effect was 11.8 and 4.0 nmol g^{-1} wet weight organism for the total mixture and the sum of LIN and DIEL, respectively. Variability in the data was sufficiently great that the program could not estimate the confidence intervals around the LT_{50} estimates. The amount of compound required to produce the two 50% mortality estimates are quite similar for the 72-h LC_{50} and the 118.97-h LT_{50}.

Bioaccumulation

With the extent of toxicity encountered for this mixture of chlorinated hydrocarbons, the bioaccumulation was estimated only for the radiolabeled compounds at Levels 0 and 1 and for the nonlabeled compounds at Level 1. Estimations of the uptake clearances were generally performed using a simplification of a general sediment uptake model [*5*].

$$dCa/dt = K_s Cse^{-\lambda t} - K_d Ca$$

where

Ca = the concentration in the animal (nmol g^{-1}),
K_s = the uptake clearance constant (g dry sediment g^{-1} wet weight organism h^{-1}),
λ = the rate constant at which compounds become biologically unavailable (h^{-1}),
K_d = the elimination rate constant (h^{-1}),
Cs = the concentration in the sediment (nmol g^{-1}) and
t = the duration of the experiment (h).

For most cases, the model could be simplified to a one- or two-compartment donor dependent model with λ equal to zero [*5*].

Uptake clearance from sediments for the radiolabeled compounds, determined in the presence and absence of the nonlabeled chlorinated hydrocarbons, were 0.018 $\pm$ 0.001 g dry sediment g^{-1} organism h^{-1} (note that all clearances used in this manuscript will use these units of g dry sediment g^{-1} organism h^{-1}) and 0.029 $\pm$ 0.003 for 2244TCB and 0.0018 $\pm$ 0.0002 and 0.0024 $\pm$ 0.0005 for BaP at Levels 0 and 1, respectively. Clearances for both compounds increased significantly, approximately doubled, between the two exposure conditions, with no added chlorinated hydrocarbons and the lowest level of added chlorinated hydrocarbons. Uptake clearance constants increased proportionately, implying a general change in the compound bioavailability or physiology under the differing exposure conditions.

The potential of the carrier to effect the toxicokinetics and mortality data was expected to be minimal. The "control" sediment was dosed with the same total amount of carrier as the other dose levels but exhibited minimal or no toxicity except possibly at 744 h (Table 3). The presence of low levels of water-missible solvents had previously been found not to affect kinetics data [*16*]. Further, the rate constant for BaP accumulation was not statistically different from previous determinations [*23*] that had estimated carrier concentrations of <4 μL L^{-1}.

Radiotracer concentration in interstitial water was measured for the controls, no animals, only because the samples containing animals had to be screened to insure that all organisms were found for accurate mortality data. The amount of the two radiolabeled compounds, added at constant concentration to all study levels, yielded nearly constant fractions of freely dissolved, sorbed to suspended particles and bound to dissolved organic carbon (DOC) across the concentration range of nonlabeled chlorinated hydrocarbons (Table 4). Radiotracer partitioning between the freely dissolved interstitial water concentration was not different at the various exposure levels (Table 4). Further, the partitioning was not at equilibrium when the experiment

TABLE 4—*Fractionation of 2,2′,4,4′-tetrachlorobiphenyl and benzo(a)pyrene in interstitial water and partitioning to sediment.*

Category, Sample Day	Benzo(a)pyrene	2,2′4,4′-tetrachlorobiphenyl
	15-DAY SAMPLE	
Percent freely dissolved	43.0 ± 10.3	30.2 ± 7.0
Percent dissolved organic carbon bound	44.7 ± 4.2	57.5 ± 17.5
Percent particle bound[a]	13.0 ± 5.2	14.1 ± 10.8
K_p[b]	378 ± 18.4	884 ± 147
Koc[c]	37 800 ± 1840	88 400 ± 14 700
	31-DAY SAMPLE	
Percent freely dissolved	32.7 ± 5.0	17.8 ± 10.0
Percent DOC bound	56.0 ± 10.9	51.6 ± 10.2
Percent particle bound[a]	6.4 ± 3.0	8.6 ± 2.7
K_p[b]	578 ± 42	1673 ± 107
Koc	57 800 ± 4200	167 300 ± 10 700

[a] Percent of compound remaining on unsedimented particles.
[b] Partition coefficient between the freely dissolved compound in the interstitial water and the sedimentary solid material.
[c] Partition coefficient normalized to the organic carbon content of the sediment. Organic carbon content of the sediment was 1.0 ± 0.2%.

started and increased by a factor of approximately two between the two sample times. For both 15 days and 31 days, the calculated K_p (sediment concentration/interstitial water concentration) of 2244TCB was greater than BaP by a factor of about 2.5. In spite of the higher partitioning, the 2244TCB had the higher uptake clearance by approximately an order of magnitude compared to BaP.

Average radiotracer concentrations in the organisms were not statistically different at Day 3 at exposure Levels 0, 1, and 2 but declined sharply to between 20 and 30% of the control value (Level 0) for the animals in Level 3. At this exposure the mortality was at 95%.

Uptake clearance for the nonlabeled chlorinated hydrocarbons (Table 5) ranged from the low of 0.006 for 224455HCB to a high of 0.15 for 3344TCB. Most of the data were fit to the one-compartment donor dependent simplification of the general model, while LIN appeared to reach a steady state and was fit to a two-compartment model with a resultant K_d of 0.029 ± 0.007 h^{-1}. The variability in the data for p,p′-DDE, 3344TCB and 224455HCB was so great that only an estimate of the uptake clearance could be made. The estimate employed initial rate assumptions and the slope of the uptake between time equal zero and the first sample at three days to find an estimated uptake clearance. In addition to estimating the uptake clearance, a bioaccumulation factor (BAF) was determined for each compound from Level 1 after twelve days exposure (Table 5). The uptake clearances and BAF data were determined only from live organisms, and the twelve days was the longest exposure time used for calculations. At twelve days the mortality even at Level 1 was on average greater than 50%, LT_{50} 5.5 days.

The log of the uptake clearance (K_s) for the chlorinated hydrocarbons was inversely proportional to the log K_{ow} for the chlorinated hydrocarbons (Fig. 1). There are a few outliers in the data set. These outliers occur, in part, for two reasons. First, some of the uptake clearances were only rough estimates; this is particularly true for 3344TCB. Second, BaP is from a different class of compounds than the chlorinated hydrocarbons. If just these two compounds are removed from the data set, then the regression of log K_s versus log K_{ow} is significant (log $K_s = -0.38 \pm 0.57 - 0.22 \pm 0.10$ log K_{ow}, $n = 12$, $r^2 = 0.33$, $p < 0.05$). Leaving these two

TABLE 5—*Uptake clearance and twelve-day bioaccumulation factor.*

Compound	Log K_{ow}	K_s[a]	BAF
Lindane	3.24[b]	0.085 ± 0.017	4.2
Dieldrin	4.56[b]	0.032 ± 0.015	12.2
225TCB	5.44[c]	0.012 ± 0.005	7.9
HCB	5.47[d]	0.043 ± 0.004	15.8
244TCB	5.51[c]	0.008 ± 0.004	3.3
p,p′-DDE	5.69[e]	0.022 ± 0.013	2.4
2234TCB	5.84[c]	0.060 ± 0.004	31.7
2266TCB	5.92[c]	0.057 ± 0.014	16.0
2244TCB[f]	5.92[c]	0.030 ± 0.003	8.8
3344TCB	5.95[c]	0.15 ± 0.07	11.1
p,p′-DDT	6.19[e]	0.014 ± 0.008	17.9
BaP[f]	6.2[d]	0.0024 ± 0.0005	0.8
224455HCB	6.77[c]	0.006 ± 0.004	3.5
224466HCB	6.88[c]	0.012 ± 0.006	3.5

[a] Uptake clearance has units of g dry sediment g^{-1} organism h^{-1}.
[b] Ref *32.*
[c] Ref *33.*
[d] Ref *34.*
[e] Ref *35.*
[f] Radiolabeled compounds.

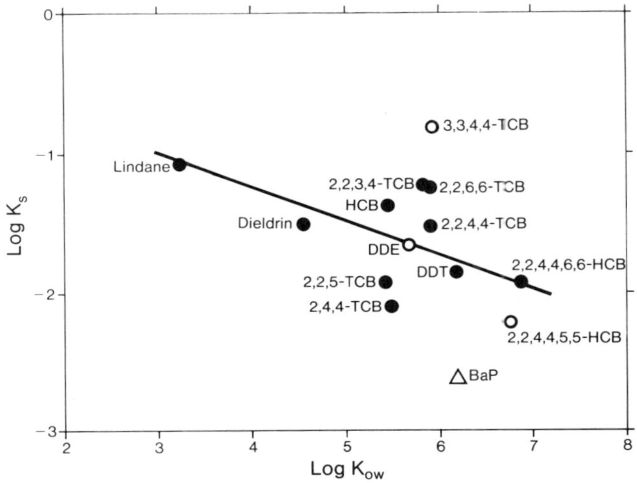

FIG. 1—*Scatter plot of log* K_s *against log* K_{ow} *shows the inverse relationship between the uptake clearance and hydrophobicity of test compounds (log* $K_s = -0.38 - 0.22$ *log* K_{ow}, $r^2 = 0.33$ *when 3344TCB and BaP are omitted from the regression calculation). The uptake clearance from sediment was estimated from a one- or two-compartment donor dependent model for compounds with dark circles. The uptake clearance for the compounds with the open circles was estimated from the first uptake sample time. The open triangle represents benzo(a)pyrene, the only compound that is not a chlorinated hydrocarbon. The uptake clearance for BaP was from the one-compartment model.*

compounds as part of the data set does not statistically change the slope or the intercept but reduces the significance of the regression to $p < 0.15$.

For the BAF, an apparent "peak" was observed at about a log K_{ow} of 6. The data at the "peak" BAF ranged from 2 to 31 or about one order of magnitude (Fig. 2) for the chlorinated hydrocarbons. The 12-day BAF for BaP was 0.75 and much lower than any of the chlorinated hydrocarbons.

Discussion

At least two mechanisms of action for acute toxicity, narcosis [*24,25*] and neurotoxicity [*26*], have been attributed to various chlorinated hydrocarbons used in this study. The toxicity of the mixture was assessed by evaluating the sediment molar concentration of the nonlabeled chlorinated hydrocarbons and the molar concentration of LIN and DIEL as a pair on the assumption that the toxicity of mixtures of organic contaminants likely occur from additive molar concentrations of compounds acting with the same mechanism of action. Based on the independent experiments, the toxicity of the mixture to the amphipod, *P. hoyi,* apparently results primarily from the toxicity of the LIN and DIEL. When *P. hoyi* were exposed independently to LIN and DIEL, the ten-day LC_{50}s were 0.38 and 0.09 nmol g^{-1} dry sediment for a sediment with a mean organic carbon content of 0.3 ± 0.07% [*27*]. However, the other compounds in the mixture may contribute significantly to the toxicity. Concentrations in the organism required to produce 50% mortality in 72 h yielded a better effects estimate than the concentrations in sediment for which no confidence limits could be generated by the probit analysis program due to the variability in the data. Further, the concentration of the LIN and DIEL in the organism relative to the rest of the compounds in the mixture was greater than the LIN and DIEL concentrations in the sediment at 72 h and 50% mortality. Therefore, part of the reason LIN and DIEL exhibit greater toxicity than the other chlorinated pesticides in the mixture may occur because of enhanced bioavailability relative to the other mixture components. The concentrations in the organism required to produce toxicity were two and a half orders of magnitude lower than that

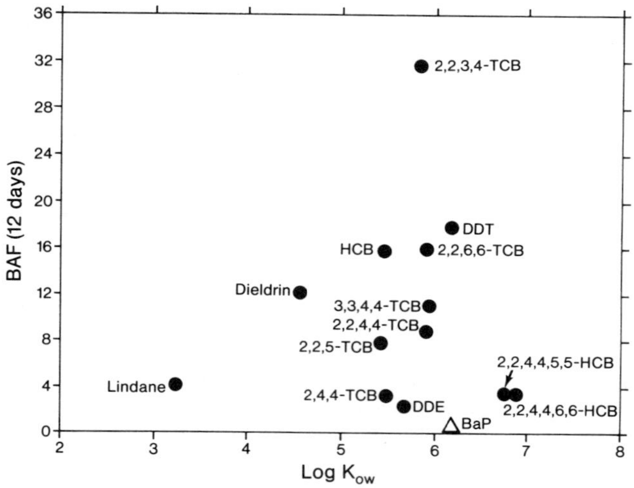

FIG. 2—*The scatter plot of the bioaccumulation factor (BAF) concentration in the organism divided by the concentration in the sediment against log* K_{ow} *tends to exhibit a peak in BAF. However, the data are quite scattered. The open triangle represents benzo(a)pyrene, the only compound that is not a chlorinated hydrocarbon.*

estimated to produce acute toxicity by a narcotic mechanism of action [28]. This helps confirm that the mechanism of toxic action is likely a specific mechanism of action and likely the neurotoxic action of the pesticides LIN and DIEL.

The time required to yield 50% mortality at even the lowest dose level was short, 118.97 h. The chlorinated hydrocarbon concentrations in the organisms required to yield 50% mortality at 118 h were only slightly less than that required to yield 50% mortality at 72 h with a similar result for the sum of the LIN and DIEL. Slopes of the acute mortality curves were steep and the effect of the compounds dropped off rapidly with reduction in concentration. These data further support the hypothesis of a specific mechanism of action producing the mortality.

In spite of all the data available on the toxicity of the mixture, determination of the exact compounds responsible for the mortality was not possible. Although it seems clear that inclusion of LIN and DIEL in the mixture accounts for most of the mortality, it was also not possible to determine what fraction of the toxicity was attributable to these compounds. The apparent enhanced biological availability compared to the rest of the compounds does suggest that these compounds will be ones of environmental concern.

Since the freely dissolved interstitial water radiotracer concentration was essentially constant across the various exposure levels, the freely dissolved chlorinated hydrocarbon concentrations were expected to depend on their individual partitioning to sediments and would reflect the total amount of materials at each dose. The implication of this partitioning, in this single sediment system, is that toxicity is proportional to the freely dissolved interstitial water concentration. However, the relative contribution of LIN and DIEL at the LC_{50} is about 50% lower in the sediment than in the organisms. Thus, a differential accumulation of LIN and DIEL is important for eliciting the toxic response. This differential accumulation does not support the hypothesis of equilibrium partitioning with subsequent uptake from interstitial water but supports differential desorption from the sorbed phase. Therefore, reliance on the interstitial water concentrations for prediction of a toxic response would be an inappropriate assumption.

The basic purpose of the toxicity data was to find a no effect level for the added compounds and to determine the toxicokinetics at the no effect level. Since even the lowest level of added chlorinated hydrocarbons produced effects, the toxicokinetics data for the nonlabeled compounds was that of organisms under stress. Uptake clearance for BaP, determined in the absence of added chlorinated hydrocarbons, was not different from that previously determined for *P. hoyi* from sediment [23]. The uptake clearance for both BaP and 2244TCB in the absence of added chlorinated hydrocarbons was about half that found in the presence of the lowest level of added chlorinated hydrocarbons. The rate for the 224455HCB at level one was not different from previous determinations in the absence of added chlorinated hydrocarbons [5]. However, the 224455HCB is not a very good compound for comparison of effects because the uptake estimate had to be made from the first sample point due to the extreme variability in the analytical data. Therefore, the uptake clearances for these stressed organisms are apparently an estimated factor of two greater than would be observed if no toxic effects occurred. Increased clearances imply either some stimulation of the organism or increased bioavailability. Determining which potential mechanism is responsible for the increases is not possible from the data. However, stimulation may be a reasonable mechanism because chlorinated hydrocarbons, particularly the chlorinated pesticides, are known to enhance the firing rate of neurons, and animals exhibiting toxic symptoms often show elevated activity levels [29]. Elevations in activity may result in increased contact with fresh interstitial water and/or increased ingestion of contaminated particles that would yield increased clearances. The potential of enhanced bioavailability as a mechanism for the increased clearances was not testable with the current data because of the mortality at higher dose levels.

The absence of proportionality between the freely dissolved compound in the interstitial water and the uptake clearance indicates that the accumulation of the compounds cannot occur strictly as a result of accumulation from sediment interstitial water. These data support our previous model of the accumulation of sediment-associated compounds by benthic organisms

(Fig. 3) [5]. Accumulation of sediment-associated compounds is thought to occur from ingestion of contaminated particles and uptake of freely dissolved compound from the sediment interstitial water. The sediment interstitial water pool of compound is perceived to be of limited size and not well mixed; as a result the concentration of compounds in the interstitial water is easily depleted by the organism. To the extent that the organism moves from one area of the sediments to another, it should encounter fresh sediment interstitial water. However, the main mechanism for maintaining the sediment interstitial water concentration is compound desorption from the particles and the dissolved organic matter pools. The particle pool is the largest, on a mass basis, and should provide the greatest mass of compound. The particle pool is, however, thought to be divided into two subpools, a readily reversible pool and a slowly reversible pool [*30,31*]. Previous data suggest that the material in the slowly reversible pool is biologically unavailable [*5*] over the short time scales of laboratory experiments. Thus, the rate of desorption from the rapidly reversible pool will dictate the rate of supply to the interstitial water and subsequently the rate of accumulation from the interstitial water by the biota. Accumulation through ingestion will depend on the rate of particle ingestion, gut throughput, and assimilation efficiency. Estimates for *P. hoyi* suggest that, for the more nonpolar compounds, the ingestion route could supply 50 to 100% of the accumulated dose, while the interstitial water is more important for the more water-soluble compounds with approximately 10% from the ingested sediments for compounds such as phenanthrene and anthracene [*5*].

If desorption rate is an important feature of sediment-associated compound accumulation by benthos, the rate of accumulation (uptake clearance) should be proportional to sediment partitioning within a compound class. This proportionality assumes that compounds within a class associate with the same sites and bind with the same mechanism. Different classes of compounds may have different binding mechanisms or binding sites for particles, and a relationship with partitioning between classes may not hold. This difference between classes is suggested from the clearance and partition data of BaP and 2244TCB in this work. Since partitioning is strongly related to the nonpolar molecular character, there should be a relationship between uptake clearance and the octanol-water partition coefficient (K_{ow}). The regression line for log K_s versus log K_{ow} for the chlorinated hydrocarbons confirms that the partitioning to sediments for a single compound class should be proportional to the hydrophobicity of the molecules while the desorption rate is inversely proportional to the hydrophobicity of the contaminants and is driving the accumulation rate. The range of the data is still large and may reflect, in part,

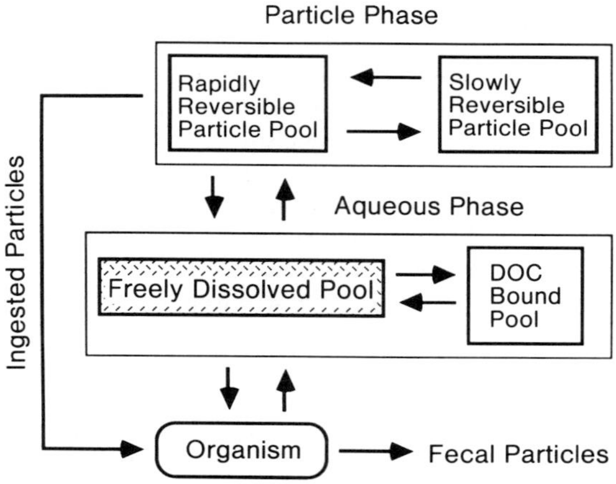

FIG. 3—*Conceptual model of the sources of sediment-associated contaminants to benthic organisms.*

differences in binding among the different chlorinated hydrocarbons, some of which are alicyclic and some aromatic.

The result of these complex interactions between accumulation and the toxicokinetics of the organisms results in a complex picture of the relative accumulation of compounds by *P. hoyi* as represented by the BAF. The BAF was quite variable for compounds at any specific log K_{ow}. This variability reflects the variability in the accumulation processes including desorption from particles as well as the variability in the ability of the organism to eliminate compounds. The end result of these interactions is a suggested peak in the BAF at approximately log K_{ow} of 6. However, even at log K_{ow} of 6 the range in BAF for the chlorinated compounds was greater than an order of magnitude. Some of this variability may have resulted from not attaining steady state since the BAFs were determined after only 12-day exposure. Thus, compound concentrations in the organism may have been at different fractions of steady state. Variability may also have been enhanced because of the effects of the compounds on the organisms. This suggested peak in BAF is similar to that found for oligochaetes feeding on both laboratory-dosed and natural sediments [*7,8*].

In summary, toxicity of the chlorinated hydrocarbon mixture was high even at the lowest concentration employed and resulted from the presence of the chlorinated pesticides, LIN and DIEL. Identification of the pesticides as the most probable toxicants was based on the body burden required to produce the toxicity and the independent experiments. In spite of the effects at the lowest level of exposure, estimates of the chlorinated hydrocarbon uptake clearances were obtained. These clearances were estimated to be about two times greater than would have occurred at a no effect level based on the BaP and 2244TCB kinetics in the presence and absence of the other added chlorinated hydrocarbons. The accumulation appeared to be driven by the desorption rate of the compounds from sediment. For a single class of compounds within a single sediment, the clearance is approximately inversely proportional to K_{ow}. The K_{ow} represents the strength of binding to sediments and therefore the inverse of the desorption rate. These data, in addition to that previously reported [*5*], indicate that the equilibrium partitioning approach to bioavailability of sediment-associated pollutants is inappropriate and the bioavailability is likely driven by the desorption rate. Further, the BAF is a complex measure of both sediment desorption, organism accumulation, and elimination and appears to result in an apparent BAF maximum for chlorinated hydrocarbons at approximately a log K_{ow} of 6.

References

[*1*] Baumann, P. C., "Cancer in Wild Freshwater Fish Populations with Emphasis on the Great Lakes," *Journal of Great Lakes Research*, Vol. 10, 1984, pp. 251–253.

[*2*] Nalepa, T. F. and Landrum, P. F., "Benthic Invertebrates and Contaminant Levels in the Great Lakes: Effects, Fates and Role of Cycling," *Toxic Contaminants and Ecosystem Health: A Great Lakes Focus*, M. S. Evans, Ed., John Wiley & Sons, New York, NY, 1988, pp. 77–102.

[*3*] Eadie, B. J., Nalepa, T. F., and Landrum, P. F., "Toxic Contaminants and Benthic Organisms in the Great Lakes: Cycling, Fate and Effects," *Toxic Contamination in Large Lakes, Volume I: Chronic Effects of Toxic Contaminants in Large Lakes*, N. W. Schmidthe, Ed., Lewis Publishers, Chelsea, MI, Chap. 8, pp. 161–178.

[*4*] Knezovich, J. P. and Harrison, F. L., "The Bioavailability of Sediment-Sorbed Organic Chemicals: A Review," *Water, Air and Soil Pollution*, Vol. 32, 1987, pp. 233–245.

[*5*] Landrum, P. F., "Bioavailability and Toxicokinetics of Polycyclic Aromatic Hydrocarbons Sorbed to Sediments for the Amphipod, *Pontoporeia hoyi*," *Environmental Science and Technology*, Vol. 23, 1989, in press.

[*6*] Lee, H., Winson, M., Randall, R. C., Specht, D. T., and Boese, B. L., "Sediment and Water Pollutant Uptake Vector Determinations in the Deposit Feeding Clam, *Macoma nasuta*," Eighth annual meeting of the Society of Environmental Toxicology and Chemistry, Pensacola, FL, 9–12 Nov. 1987, Abstract No. 182, 1987.

[*7*] Oliver, B. G., "Uptake of Chlorinated Organics from Anthropogenically Contaminated Sediments by Oligochaete Worms," *Canadian Journal of Fisheries and Aquatic Sciences*, Vol. 41, 1984, pp. 878–883.

[*8*] Oliver, B. G., "Biouptake of Chlorinated Hydrocarbons from Laboratory-Spiked and Field Sediments by Oligochaete Worms," *Environmental Science and Technology*, Vol. 21, 1987, pp. 785–790.

[9] Frank, A. P., Landrum, P. F., and Eadie, B. J., "Polycyclic Aromatic Hydrocarbons Rates of Uptake, Depuration, and Biotransformation by Lake Michigan *Sytlodrilius heringianus,*" *Chemosphere,* Vol. 15, 1986, pp. 317-330.
[10] Kadeg, R. D., Pavlou, S. P., and Duxbury, A. S., "Sediment Criteria Methodology Validation: Elaboration of Sediment Normalization Theory for Nonpolar Hydrophobic Organic Chemicals," prepared for U.S. Environmental Protection Agency, Criteria and Standards Division, Office of Water Regulation and Standards, Washington, DC, 1986.
[11] Neff, J. M., Bean, D. J., Cornaby, B. W., Vaga, R. M., Gubransen, T. C., and Scanlon, J. A., "Sediment Quality Criteria Methodology Validation: Calculation of Screening Level Concentrations from Field Data," prepared for U.S. Environmental Protection Agency, Criteria and Standards Division, Office of Water Regulation and Standards, Washington, DC, 1986.
[12] Pavlou, S. P. and Weston, D. P., Eds., "Initial Evaluation of Alternatives for Development of Sediment related Criteria for Toxic Contaminants in Marine Waters: Phase II: Development and Testing of the Sediment-Water Equilibrium Partitioning Approach," EPA Report 910/9-83-117, U.S. Environmental Protection Agency, Washington, DC, 1984.
[13] Poston, T. M. and Prohammer, L. A., "Protocol for Sediment Toxicity Testing of Nonpolar Organic Compounds," prepared for U.S. Environmental Protection Agency, Criteria and Standards Division, Office of Water Regulation and Standards, Washington, DC, 1986.
[14] Landrum, P. F., "Uptake, Depuration and Biotransformation of Anthracene by the Scud, *Pontoporeia hoyi,*" *Chemosphere,* Vol. 11, 1982, pp. 1049-1057.
[15] Menzel, D. W. and Vaccaro, R. F., "The Measurement of Dissolved Organic and Particulate Carbon in Sea Water," *Limnology and Oceanography,* Vol. 9, 1964, pp. 138-142.
[16] Landrum, P. F., "The Effect of Co-contaminants on the Bioavailability of Polycyclic Aromatic Hydrocarbons to *Pontoporeia hoyi,*" *Polynuclear Aromatic Hydrocarbons: Seventh International Symposium on Formation Metabolism and Measurement,* M. W. Cooke and A. J. Dennis, Eds., Battelle Press, Columbus, OH, 1983, pp. 731-743.
[17] Landrum, P. F., Nihart, S. R., Eadie, B. J., and Gardner, W. S., "Reverse-Phase Separation for Determining Pollutant Binding to Aldrich Humic Acid and Dissolved Organic Carbon of Natural Waters," *Environmental Science and Technology,* Vol. 18, 1984, pp. 187-192.
[18] Landrum, P. F., Nihart, S. R., Eadie, B. J., and Herche, L. R., "Reduction in Bioavailability of Organic Contaminants to the Amphipod, *Pontoporeia hoyi* by Dissolved Organic Matter of Sediment Interstitial Waters," *Environmental Toxicology and Chemistry,* Vol. 6, 1987, pp. 11-20.
[19] Grasshoff, K., Ehrhardt, M., and Kremling, K., *Methods of Seawater Analysis,* 2nd revised and extended ed., Verlag Chemie, Weinheim, FRG, 1983, pp. 61-72.
[20] Gardner, W. S., Frez, W. A., Chichocki, E. A., and Parish, C. C., "Micromethod for Lipid Analysis in Aquatic Invertebrates," *Limnology and Oceanography,* Vol. 30, 1985, pp. 1099-1105.
[21] *SAS Users' Guide: Statistics,* Version 5, SAS Institute, Inc., Cary, NC, 1985.
[22] Eadie, B. J., Landrum, P. F., and Faust, W., "Polycyclic Aromatic Hydrocarbons in Sediments, Porewater and the Amphipod, *Pontoporeia hoyi* from Lake Michigan," *Chemosphere,* Vol. 11, 1982, pp. 847-858.
[23] Landrum, P. F., Eadie, B. J., Faust, W. R., Morehead, N. R., and McCormick, M. J., "Role of Sediment in the Bioaccumulation of Benzo(a)pyrene by the Amphipod, *Pontoporeia hoyi,*" *Polynuclear Aromatic Hydrocarbons: Eighth International Symposium on Mechanisms, Methods and Metabolism,* M. W. Cooke and A. J. Dennis, Eds., Battelle Press, Columbus, OH, 1985, pp. 799-812.
[24] Veith, G. D., Call, D. J., and Brooke, L. T., "Estimating the Acute Toxicity of Narcotic Industrial Chemicals to Fathead Minnows," *Aquatic Toxicology and Hazard Assessment: Sixth Symposium, ASTM STP 802,* W. E. Bishop, R. D. Cardwell, and B. B. Heidolph, Eds., American Society for Testing and Materials, Philadelphia, 1983, pp. 90-97.
[25] Call, D. J., Brooke, L. T., Knuth, M. L., Poirier, S. H., and Hoglund, M. D., "Fish Subchronic Toxicity Prediction Model for Industrial Organic Chemicals That Produce Narcosis," *Environmental Toxicology and Chemistry,* Vol. 4, 1985, pp. 335-341.
[26] O'Brian, R. D., *Insecticides Action and Metabolism,* Academic Press, New York, NY, 1967.
[27] Stehly, G. R., personal communication, Great Lakes Environmental Research Laboratory, Ann Arbor, MI 48105.
[28] McCarty, L. S., "The Relationship Between Aquatic Toxicity QSARS and Bioconcentration for Some Organic Chemicals, *Environmental Toxicology and Chemistry,* Vol. 5, 1986, pp. 1071-1080.
[29] Gilman, A. G., Goodman, L. S., and Gilman, A., *Goodman and Gilman's The Pharmacological Basis of Therapeutics,* 6th ed., Macmillan Publishing Co., New York, NY, 1980, pp. 1647-1655.
[30] Ditoro, D. M., Horzempa, L. M., Casy, M. M., and Richardson, W., "Reversible and Resistant Components of PCB Adsorption-Desorption: Adsorbant Concentration Effects," *Journal of Great Lakes Research,* Vol. 8, 1982, pp. 336-349.
[31] Karickhoff, S. W. and Morris, K. R., "Sorption Dynamics of Hydrophobic Pollutants in Sediment Suspensions," *Environmental Toxicology and Chemistry,* Vol. 4, 1985, pp. 469-479.

[32] Leo, A., Hansch, C., and Elkins, D., "Partition Coefficients and Their Uses," *Chemical Reviews,* Vol. 71, 1971, pp. 525–616.
[33] Burkhard, L. P., "Physical-Chemical Properties of the Polychlorinated Biphenyls: Measurement, Estimation and Application to Environmental Systems," doctoral dissertation, University of Wisconsin at Madison, 1984.
[34] Miller, M. M., Wasik, S. P., Huang, G., Shiu, W., and Mackay, D., "Relationships Between Octanol-Water Partition Coefficient and Aqueous Solubility," *Environmental Science and Technology,* Vol. 19, 1985, pp. 522–529.
[35] Verschueren, K., *Handbook of Environmental Data on Organic Chemicals,* 2nd ed. Van Nostrand Reinhold Co., New York, NY, 1983.

Culturing and Testing of New Organisms

William L. Goodfellow, Jr.,[1] *and William J. Rue*[1]

Evaluation of a Chronic Estimation Toxicity Test Using *Mysidopsis bahia*

REFERENCE: Goodfellow, W. L., Jr., and Rue, W. J., **"Evaluation of a Chronic Estimation Toxicity Test Using *Mysidopsis bahia*,"** Aquatic Toxicology and Hazard Assessment: 12th Volume, ASTM STP 1027, U. M. Cowgill and L. R. Williams, Eds., American Society for Testing and Materials, Philadelphia, 1989, pp. 333–344.

ABSTRACT: Traditionally, chronic toxicity tests have utilized life-cycle tests or early-life-stage tests. The U.S. Environmental Protection Agency's national "Policy for the Development of Water Quality-Based Permit Limitations for Toxic Pollutants" and the associated "Technical Support Document for Water Quality-Based Toxics Control" emphasized development of less-expensive chronic estimation testing alternatives for National Pollution Discharge Elimination System (NPDES) effluent testing requirements.

Mysid seven-day short-term chronic estimation tests, developed in accordance with EPA water quality-based effluent testing, were conducted with the reference toxicants potassium dichromate (hexavalent chromium) and sodium pentachlorophenate (pentachlorophenol). The seven-day estimated chronic values (MATCs) of the three tests ranged between 0.49 and 0.69 mg/L Cr^{+6} for hexavalent chromium and from 298 to 702 μg/L PCP for pentachlorophenol tests. The mean ($\pm$SD) MATC values for hexavalent chromium was 0.62 ($\pm$0.12) mg/L with a coefficient of variation of 19.4%, while the mean MATC ($\pm$SD) for pentachlorophenol was 498 ($\pm$202) μg/L with a coefficient of variation of 40.5%.

KEY WORDS: short-term chronic, reference toxicant, *Mysidopsis bahia*, mysid, potassium dichromate, sodium pentachlorophenate, chronic value, MATC

The U.S. Environmental Protection Agency (EPA) has emphasized toxicity testing for evaluating toxic pollutants in effluents. Tests estimating chronic toxicity are being included as biomonitoring requirements in an increasing number of NPDES permits. Chronic toxicity tests have been life-cycle tests or early-life-stage tests; however, the long exposure duration of these tests increases the cost and ability of laboratories to successfully conduct the tests. The EPA published short-term chronic estimation toxicity test methodologies for freshwater species [*1*] and developed short-term chronic estimation toxicity test methodologies for marine and estuarine species [*2,3*]. These saltwater short-term chronic estimation methods were developed by EPA researchers in Narragansett, Rhode Island [*2*] and were developed or adapted from existing short-term chronic methods for two species of invertebrates, two fish species and a red macroalgae.

One of the tests to estimate chronic toxicity developed for saltwater use was an opossum shrimp or "mysid" (*Mysidopsis bahia* Molenock 1969) survival, growth, and reproduction test [*4*]. This short-term chronic estimation test evaluates the effect effluent has on mysids during a seven-day exposure. The EPA-developed short-term chronic test methodologies were intended for the evaluation of effluent and receiving waters only [*1,3*]; however, they can be used for estimating the chronic toxicity of single chemicals. The Narragansett test method protocol [*4*]

[1]Scientist and senior scientist, respectively, EA Engineering, Science, and Technology, Inc., Hunt Valley/Loveton Center, 15 Loveton Circle, Sparks, MD 21152.

has since been modified for inclusion in the draft short-term chronic methods manual for marine and estuarine organisms [3].

Limited data are available on the use of *M. bahia* with regard to the short-term chronic estimation tests. The purpose of this study was to evaluate the sensitivity and reproducibility of the *M. bahia* short-term chronic test methodology using two reference toxicants, hexavalent chromium and sodium pentachlorophenate, and determine the case of test performance in a (contract) laboratory.

Life History and Ecology

The estuarine opossum shrimp, *Mysidopsis bahia* Molenock, has been used since the mid-1970s for acute toxicity tests. Nimmo, Mirenda, and Carlson [5] state that a large acute data base has been developed for this species compared to other saltwater invertebrates. Very few saltwater invertebrate species are commonly used for toxicity testing because of difficulties in culturing or testing techniques. Mysids have been one of the saltwater species of choice for toxicity testing because they have a short life cycle, marsupial brood pouch reproduction without a larval stage, and are relatively sensitive to a wide variety of pollutants [6]. The mysid juveniles can be sexually differentiated within ten days after release from the broodsac, develop eggs in the oviducts by Day 12, have eggs in the broodsac by Day 14, and release young at 18 to 20 days [4]. This short life-cycle is a major factor used in selection of a test species for chronic toxicity testing [4, 7].

The general ecology of the genus *Mysidopsis* is largely unknown [6]. The published ranges of *M. bahia* are Laguna de Tamihua, Mexico eastward along the coast of the Gulf of Mexico to Cottonwood Channel, Cape Sable, Florida [6]. Specimens of *M. bahia* have also been collected on the northern coast of Virginia [6]. *M. bahia* have been collected at salinities as high as 40 ppt [6], but have not been found at salinities below 9 ppt [8]. Two additional species, *M. almyra* and *M. bigelowi*, have also been used for toxicity testing.

Mysids are largely predatory on other small crustaceans and can be cannibalistic in culture. They are also omnivorous or multivorous to some extent and will disturb bottom sediments to resuspend food particles [6]. Mysids are an important food source for many fishes and other larger invertebrates and represent a vital link in aquatic food chains [7].

Materials and Methods

Test Materials and Dilution Water

Reagent grade (99.96% purity) potassium dichromate (Lot No. 6772 KXAS, Mallinckrodt, Inc., Paris, Kentucky) was used to prepare stock solutions. Hexavalent chromium (1 g/L) solutions were prepared by dissolving the dichromate salt in deionized water. Concentrations of chromium reported in this study are based on nominal concentrations and expressed as mg/L hexavalent chromium determined in volume to volume measurements. Hexavalent chromium has been proposed as a reference toxicant in effluent toxicity tests [9, 10]. Advantages of its use include stability in solution and similarity of nominal to measured test concentrations [9–11].

Technical grade (90% purity) sodium pentachlorophenate (Lot No. 229964-1084, Dowicide G, Fluka Chemical Corp., Ronkonkoma, New York) was used for the PCP studies. Fresh stock solutions were prepared daily by dissolving the organic salt in deionized water. All values reported are based on nominal concentrations, corrected for purity, and are expressed as μg/L of PCP determined in volume to volume measurements.

Water for culture and testing was artificial seawater [formulated from a dry sea salts mixture (Forty Fathoms Marine Mix, Marine Enterprises, Towson, Maryland) and activated carbon-filtered, dechlorinated, municipal water]. Artificial seawater was prepared in large-capacity polyethylene tanks equipped with recirculating pumps to facilitate dissolution and mixing. It was then adjusted to a final salinity of 25 ppt. This dilution water has been acceptable at EA for the testing of aquatic organisms, as evidenced by maintenance of the multigenerational *M. bahia* cultures, with no evident loss of fecundity.

Test Organisms

Mysidopsis bahia used in the three hexavalent chromium assays were obtained from the multigenerational cultures of EA Engineering, Science, and Technology, Inc., Aquatic Toxicology Laboratory (Sparks, Maryland). The *M. bahia* used in the three pentachlorophenol assays were purchased from Mote Marine Laboratory (Sarasota, Florida). It was necessary to use mysids from a commercial vendor for the pentachlorophenol assays because of a system malfunction in the culturing system that resulted in a "crash" of the laboratory cultures.

The mysid cultures at EA were in 18.9-L and 37.8-L, all-glass aquaria, outfitted with undergravel filtration and a substrate of crushed coral. The temperature of the system was maintained at 25 ± 2°C, a salinity of 25 ppt, and a 16-h light/8-h dark photoperiod. When juvenile mysids were needed, gravid females were isolated into a brood chamber/trap. The females were isolated in a screened chamber of the chamber/trap, and the juveniles (after release) were captured in a siphon trap. The juveniles were harvested daily, and only juveniles released during a 24-h period were used for testing.

Mysids purchased from Mote Marine Laboratory were cultured similarly as described in the EA procedure above. Each time the mysids were shipped in a 3.7-L Cubitainer with approximately 1 cm of head space remaining in the container. The Cubitainer was packed in a corrugated container insulated with styrofoam packing chips and shipped via overnight carrier. Upon receipt at EA, the temperature and salinity were measured and ranged from 20.3 to 21.1°C and 20.0 to 23.0 ppt for the three shipment lots used for testing, respectively. Two lots of mysids were stressed during shipment due to lower temperatures and were immediately discarded. The organisms were slowly acclimated to the test conditions of 25 ± 2°C and 25 ± 2 ppt. The mysids were four days old at the time of receipt from Mote Marine Laboratory. The mysids were held for all tests in a constant flow-recirculating holding system at 25 ± 2°C until the test organisms were seven (hexavalent chromium study) or eight (pentachlorophenol study) days old. The additional holding day was used for the organisms purchased from Mote Marine Laboratory, allowing the organisms more time to mature due to the one day of reduced temperature as a result of shipment. Horning and Weber [*3*] suggest if the temperature cannot be maintained at the test conditions, it is advisable to hold the juveniles an additional day before testing. Mysids were fed ad libidum <24-h-old *Artemia salina* (Columbia strain, Aquarium Products, Glen Burnie, Maryland) a minimum of twice daily during the acclimation/holding period.

Testing

General toxicity test methods followed Lussier et al. [*4*] and Horning and Weber [*3*] and are summarized in Table 1. The mysids were exposed in 10-cm, round glass culture bowls containing 100 mL of test solution. Each treatment group consisted of 8 to 10 replicate test chambers with five organisms randomly assigned individually to each chamber. Test solutions were measured with Class A volumetric glassware. Solutions were dispensed equally to all chambers and renewed daily. All testing was performed in an environmentally controlled chamber at a temperature of 25 ± 2°C and a salinity of 25 ± 2 ppt. The photoperiod was 16-h light/8-h dark with a light intensity of 75 fc using wide-spectrum fluorescent lighting.

During the solution renewal, uneaten and dead *Artemia* and other debris were removed from the bottom of the test chambers using a siphon tube. The water level in each chamber was lowered, leaving approximately 25% of test solution. New test solution was prepared in volumetric flasks on the day of the renewal and was dispensed equally (100 mL) into all replicate test

TABLE 1—*Description of test conditions.*

Test organism	*Mysidopsis bahia*
Age of organism	Cr^{+6}: 7 days PCP: 8 days
Temperature	25 ± 2°C
Photoperiod	16-h light/8-h dark
Light intensity	75 ft-candles (15 $\mu E/M^2/s$)
Dilution water	25 ± 2 ppt Forty Fathoms Marinemix
Food type	*Artemia salina* (<24 h old)
Feeding rate	Twice daily, 0.25 mL of 0.05 g/mL suspension
Number of concentrations	5 + 1 control
Number of replicates	8 to 10
Number of animals per replicate	5
Test chambers	10-cm culture bowls
Test volume	100 mL
Toxicant renewal frequency	Daily
Test duration	7 days
Endpoints	Survival; average dry weight/animal for males and females; females with eggs

chambers. Twice daily, the juveniles were fed 0.25 mL of a concentrated (0.05 g/mL) suspension of newly hatched (<24 h old) *Artemia*.

Mortality observations were made at 24-h intervals (during solution renewal). Dead organisms were removed as they were observed. Temperature, pH, dissolved oxygen, and salinity were measured daily from all daily prepared solutions. Temperature was measured in all individual test chambers at test initiation, before solution renewal, and at the test termination. The pH, dissolved oxygen, and salinity were measured, from a minimum of the control, low, medium, and highest concentrations, by pooling the test solutions after solution renewal and test termination. Pooling of test solutions was performed by syphoning the solution from the test chambers into a 1000-mL beaker (with the syphon tube below the water level in beaker) in order to minimize the aeration of test solutions. Routine water quality determinations were performed in accordance with *Standard Methods* [*12*].

At test termination the mysids were sexed using the methodology outlined in Lussier et al. [*4*]. Mysids were placed individually in a ten depression spot plate slide and examined under a dissecting microscope (×250). Mysids were classified into groups of females with eggs in oviducts or broodsacs, females without eggs in oviducts or broodsacs, males, or immature mysids. The mysids from each individual replicate chamber were then pooled into two groups—females and males/immatures—placed in pretared aluminum weigh pans, and dried in an oven at 105°C. After a minimum of 12 h, the weigh pans containing test organisms were removed from the oven, cooled in a desiccator, and measured to the nearest 0.01 mg using a Cahn Model 4700 Automatic Electrobalance (Cerritos, California).

Statistical Analysis

Chronic endpoints were determined for survival, pooled dry weight of mysid females per replicate, pooled dry weight of mysid males/immatures per replicate, and the percent of females with eggs in the oviducts or broodsacs using the statistical program TOXSTAT [*13*]. Arc sine square root transformation was applied to the replicate proportions of surviving mysids and females with eggs. The transformed survival and females with eggs were then analyzed by Dunnett's mean comparison test. Pooled dry weight per replicate for females and males/immature mysids were also analyzed by Dunnett's mean comparison test. If the data (even after transformation) were not normally distributed (failed chi-square test) or the variances did not pass homogeneity (failed Bartlett's test), the data were analyzed using the nonparametric Steel's many-one rank test. The definition of chronic endpoint points used in this study were: no observable effect concentration (NOEC), the highest treatment concentration that is not statistically significant ($\alpha = 0.05$) from the controls; lowest observable effect concentration (LOEC), the lowest treatment concentration that is statistically significant from the controls; and the chronic value (MATC), the geometric mean of the NOEC and LOEC.

Results and Discussion

Chronic Toxicity

The toxicity data for *Mysidopsis bahia* exposed to chromium and pentachlorophenate are presented in Tables 2 and 3. Control survival in all of the seven-day tests was greater than 80% survival. The pooled dry weights in all tests for the control females and males/immature were greater than 0.20 mg/mysid and 0.19 mg/mysid, respectively. The percent females with eggs in the oviducts or broodsac was greater than 90.3% for all of the controls. The seven-day MATCs ranged from 0.49 to 0.69 mg/L Cr^{+6} for the three short-term chronic tests using hexavalent chromium and ranged from 298 to 702 μg/L PCP for the three pentachlorophenol tests. The mean ($\pm$SD) MATCs for the two reference toxicants were 0.62 ($\pm$0.12) mg/L and 498 ($\pm$202) μg/L (Table 4).

The percent females having eggs in the oviducts or broodsac was the most sensitive indicator of *M. bahia* chronic toxicity resulting from exposure to hexavalent chromium (Fig. 1). This chronic endpoint was always at least one (hexavalent chromium) concentration treatment more sensitive than survival or growth (Table 2). In all three of the tests, no statistical significant ($\alpha = 0.05$) reduction in growth compared to the controls was observed for either females or males/immature mysids. The pooled female dry weight for the 3.2 mg/L hexavalent chromium treatment group in Test 3 was significantly larger ($\alpha = 0.05$) than the controls; however, this is probably a data anomaly as a result of only one female mysid surviving after the seven-day exposure period in this treatment group.

Growth of female mysids, expressed as dry weight, was consistently the most sensitive chronic endpoint for the three PCP short-term chronic tests (Fig. 2). In pentachlorophenol Tests 1 and 2, survival and growth of males/immature mysids were as sensitive a chronic indicator as the growth of female mysids (Table 3). Unlike the hexavalent chromium tests, percent females with eggs in the oviduct or broodsac was not the most sensitive chronic endpoint for the pentachlorophenol tests. The percent females with eggs in the 365 μg/L PCP treatment group in Test 2 was significant ($\alpha = 0.05$) from the controls; however, the next two higher test concentrations were not significant from the controls (Table 3). It should be noted that only one mysid (a female) remained in the 870 μg/L PCP treatment. This female was observed with eggs in the oviducts, which is why a significant difference when compared to the controls was not observed while survival in this treatment concentration was only 2%.

Very little information exists specifically on the acute and chronic toxicity of *Mysidopsis bahia* to hexavalent chromium, and no studies could be found in the literature on *M. bahia* toxicity with regards to pentachlorophenol. Even less information exists on the chronic toxicity of marine organisms to these two common reference toxicants. Lussier et al. [*14*] reported a 96-h LC_{50} for *M. bahia* of 2.03 mg/L Cr^{+6} and a chronic life-cycle MATC of 0.132 mg/L Cr^{+6}. They reported that the number of young produced was the only life-cycle parameter effected by hexavalent chromium. Their results and observations on the most sensitive chronic endpoint are consistent with the results of this study. Jop et al. [*9*] and Dorn et al. [*10*] reported 48-h EC_{50} values from eight acute tests ranged from 5.1 to 7.2 mg/L Cr^{+6} with a mean EC_{50} value of 6.0

TABLE 2—*Results of* Mysidopsis bahia *seven-day survival, growth, and reproduction test conducted with hexavalent chromium.*

Test No.	Chromium (Cr^{+6}) Concentration, mg/L	Survival, %	Pooled Dry Weight, mg/mysid ±SD: Females	Pooled Dry Weight, mg/mysid ±SD: Males/Immatures	Pooled Percent Females With Eggs, ±SD	
1	Control	83	0.24 (0.01)	0.26 (0.06)	90.3 (15.3)	
	0.06	67	0.21 (0.04)	0.25 (0.02)	77.8 (34.4)	
	0.14	82	0.28 (0.03)	0.23 (0.04)	43.8 (42.7)	
	0.32	60	0.25 (0.04)	0.24 (0.02)	47.2 (36.0)	MATC = 0.49 mg/L[a]
	0.75	69	0.23 (0.02)	0.20 (0.03)	39.6 (41.7)*	
	1.80	67	0.23 (0.02)	0.20 (0.03)	11.9 (20.9)*	
2	Control	80	0.31 (0.06)	0.27 (0.03)	100.0 (0.0)	
	0.06	88	0.33 (0.02)	0.27 (0.03)	82.0 (17.0)	
	0.16	75	0.31 (0.04)	0.29 (0.03)	71.9 (19.4)	MATC = 0.69 mg/L[a]
	0.42	75	0.30 (0.02)	0.32 (0.06)	61.5 (42.0)	
	1.15	68	0.30 (0.03)	0.28 (0.01)	33.3 (31.8)*	
	3.20	42*	0.28 (0.06)	0.23 (0.03)	0.0 (0.0)*	
3	Control	86	0.20 (0.04)	0.19 (0.04)	96.8 (8.8)	
	0.06	90	0.22 (0.03)	0.22 (0.03)	80.2 (24.4)	
	0.16	72	0.22 (0.02)	0.21 (0.05)	65.0 (38.8)	MATC = 0.69 mg/L[a]
	0.42	88	0.21 (0.03)	0.22 (0.03)	46.3 (30.9)	
	1.15	68	0.20 (0.04)	0.21 (0.02)	11.9 (20.9)*	
	3.20	22*	0.28 (0.0)*	0.21 (0.03)	0.0 (0.0)*	

[a]Most sensitive chronic endpoint used for calculation of the chronic value was percent females with eggs.
*Statistically significant difference ($\alpha = 0.05$) from controls using Dunnett's mean comparison test.

TABLE 3—*Results of* Mysidopsis bahia *seven-day survival, growth, and reproduction test conducted with pentachlorophenol.*

Test No.	PCP Concentration, μg/L	Survival, %	Pooled Dry Weight, mg/mysid ±SD		Pooled Percent Females With Eggs, ±SD	
			Females	Males/Immatures		
1	Control	90	0.34 (0.04)	0.34 (0.02)	91.7 (15.4)	
	135	91	0.37 (0.05)	0.31 (0.04)	89.3 (19.7)	
	322	70	0.32 (0.03)	0.32 (0.04)	79.2 (35.4)	MATC = 494 μg/L[a]
	757	48*	0.23 (0.03)*	0.25 (0.07)*	33.3 (31.2)**	
	1827	0*	...*	...*	...**	
	4263	0*	...*	...*	...**	
2	Control	85	0.44 (0.06)	0.40 (0.05)	95.8 (11.8)	
	157	65	0.42 (0.14)	0.40 (0.08)	92.8 (18.9)	
	244	78	0.41 (0.06)	0.38 (0.05)	77.1 (36.7)	MATC = 702 μg/L[b]
	365	100	0.40 (0.07)	0.40 (0.06)	69.8 (33.9)*	
	566	92	0.39 (0.06)	0.35 (0.06)	82.3 (23.4)	
	870	2*	0.35 (0.0)**	...*	100.0 (0.0)	
3	Control	90	0.45 (0.09)	0.39 (0.05)	92.5 (16.9)	
	157	84	0.44 (0.12)	0.40 (0.14)	85.0 (25.5)	
	244	88	0.44 (0.06)	0.39 (0.11)	94.2 (12.4)	MATC = 298 μg/L[c]
	365	84	0.33 (0.06)*	0.31 (0.07)	78.3 (20.5)	
	566	66*	0.31 (0.06)*	0.35 (0.04)	65.7 (23.0)*	
	870	0*	...*	...*	...*	

[a]All of the chronic endpoints were equally sensitive and were used for the calculation of the chronic value.
[b]The chronic endpoints of survival and dry weight of males and females were used for the calculation of the chronic value.
[c]Dry weight of females was the most sensitive endpoint and was used for the calculation of the chronic value.
*Statistically significant difference ($\alpha = 0.05$) from controls using Dunnett's mean comparison test.
**Statistically significant difference ($\alpha = 0.05$) from controls using Steel's many-one rank test.

TABLE 4—*Summary of the most sensitive chronic endpoints of* Mysidopsis bahia *seven-day survival, growth, and reproduction tests conducted with hexavalent chromium and pentachlorophenol.*

Test No.	Most Sensitive Endpoint Parameter	Chronic Endpoints		
		NOEC	LOEC	MATC
	HEXAVALENT CHROMIUM (EXPRESSED AS MG/L)			
1	Percent females with eggs	0.32	0.75	0.49
2	Percent females with eggs	0.42	1.15	0.69
3	Percent females with eggs	0.42	1.15	0.69
	Mean MATC (±SD)			0.62 (±0.12)
	Coefficient of Variation (CV)			19.4
	PENTACHLOROPHENOL (EXPRESSED AS μG/L)			
1	All (survival, dry weight, and percent females with eggs)	322	757	494
2	Survival and dry weight	566	870	702
3	Dry weight (females)	244	365	298
	Mean MATC (±SD)			498 (±202)
	Coefficient of Variation (CV)			40.5

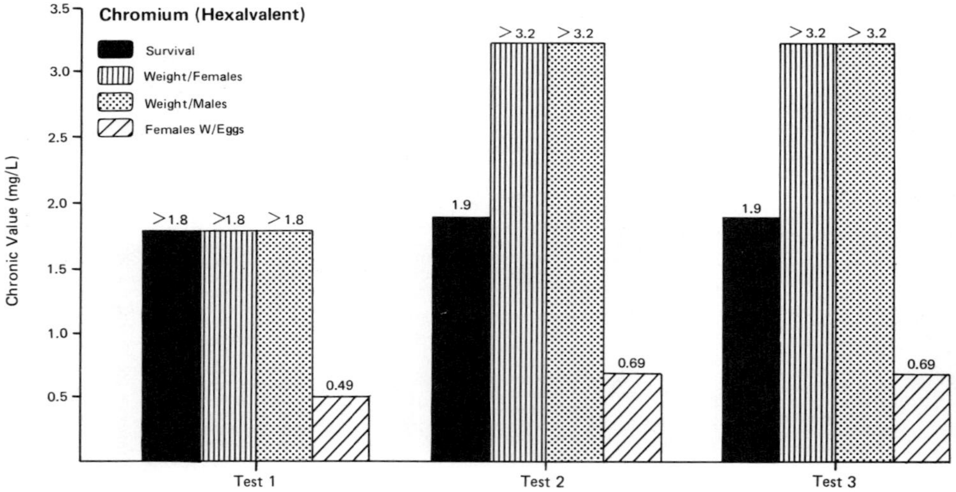

FIG. 1—*Calculated chronic values for* Mysidopsis bahia *seven-day survival, growth, and reproduction test.*

mg/L Cr^{+6}. The acute toxicity of two other species of mysid, *M. almyra* and *M. bigelowi*, were reported having a 48-h EC_{50} of 5.1 mg/L Cr^{+6} [*10*] and a 96-h LC_{50} of 4.4 mg/L Cr^{+6} [*15*], respectively.

Other investigators have evaluated the acute toxicity of hexavalent chromium to marine crustaceans. The copepods *Pseudodiaptomus coronatus* had a 96-h LC_{50} of 3.65 mg/L Cr^{+6} and *Acartia clausi* and 6.6 mg/L Cr^{+6}, respectively [*15*]. The 96-h LC_{50}s for the hermit crab, *Pagurus longicarpus*, was determined by Eisler and Hennekey [*16*] as 10 mg/L Cr^{+6} and for blue crabs, *Callinectes sapidus*, as 3.4 mg/L Cr^{+6} [*17*]. Generally, marine crustaceans are the most

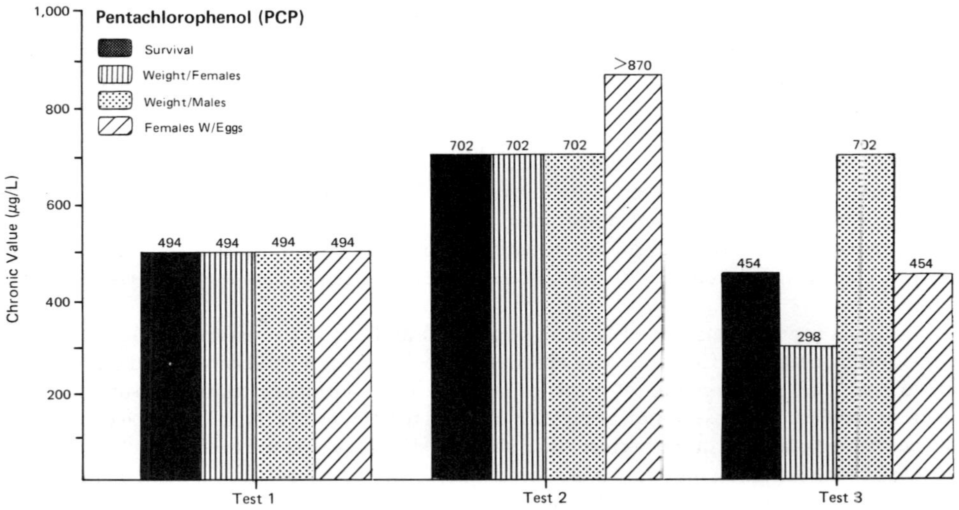

FIG. 2—*Calculated chronic values for* Mysidopsis bahia *seven-day survival, growth, and reproduction test.*

sensitive marine species to hexavalent chromium. Of the five most sensitive genus in the U.S. EPA *Ambient Water Quality Criteria for Chromium—1984* [*18*], three are marine crustaceans, with the genus *Mysidopsis* the second most sensitive species tested. *M. bahia* do not appear to be as sensitive to hexavalent chromium as other heavy metals. Lussier et al. [*14*] evaluated the acute toxicity of *M. bahia* to ten heavy metals and observed that hexavalent chromium was the second least toxic heavy metal to mysids of the heavy metals evaluated.

Even fewer acute toxicity tests have been performed on pentachlorophenol using marine invertebrates as the test organism. Hauch et al. [*19*] evaluated the acute toxicity of PCP to the copepod *P. coronatus* and reported a 96-h LC_{50} value of 62.8 μg/L, while Adema and Vink [*20*] found the copepod *Temora longicornus* to be less sensitive to PCP with a 96-h LC_{50} of 170 μg/L. The acute toxicity of grass shrimp *Palaemonetes pugio,* reported as 96-h LC_{50} value, range from approximately 400 to 2400 μg/L [*21,22*], while the pink shrimp was even less sensitive to PCP with a 96-h LC_{50} of 5600 μg/L [*23*]. Marine crustaceans do not appear to be the most acutely sensitive group to pentachlorophenol [*24*]. Only the copepod *P. coronatus,* which is ranked fourth, is among the five most sensitive species [*24*].

No chronic values for pentachlorophenol could be found for marine crustaceans. The only chronic value for a marine species that could be found in the literature was for the sheepshead minnow *Cyprinodon variegatus*. Parrish et al. [*25*] determined the chronic toxicity of sheepshead minnows after 168 days of exposure as 64.3 μg/L PCP. Hughes et al. [*26*] compared the results of 28-day and 7-day chronic tests with sheepshead minnows and reported values of 53 μg/L PCP and 106 μg/L, respectively. These chronic values are lower than that found in this study (Table 3); however, many marine fishes were determined to be more sensitive to PCP than marine crustaceans [*24*].

Rue et al. [*27*], in a review of existing inter- and intralaboratory variability of effluent toxicity test methodologies, found that 89.2% of the intralaboratory studies yielded coefficients of variation (CV) of ≤40%. The coefficients of variation determined from this study (hexavalent chromium CV = 19.4% and pentachlorophenol CV = 40.5%) agree with the intralaboratory values determined from the other studies evaluated by Rue et al. [*27*].

Comments on Short-Term Chronic Estimation Protocol

The reproductive endpoint of this test is not measured as viable young or the production of young, but rather as fecundity. No evaluation is made on the survival or general health of the embryos during exposure, nor is any evaluation made on the numbers of young produced. This test evaluates only whether a mysid female has at least one egg in her oviduct. Because the more traditional life-cycle test for *M. bahia* does evaluate embryo viability and numbers of young produced [*28*], it is difficult to compare the two tests. If a toxicant such as hexavalent chromium does appear to effect reproductive potential or the number of young produced, the estimated chronic toxicity will most likely underestimate the chronic effects of the toxicant. This is probably reflected in the differences of the chronic values reported in Lussier et al. [*14*] and this study.

This test's procedure is relatively straight forward and has been performed successfully by numerous laboratories. However, only experienced toxicologist/technicians should evaluate the mysid's sex and determination of presence of eggs. This step of the test is very difficult and requires considerable observation of mysid males and females under the microscope before confidence can be obtained in the performance of this task. Because of the need for detailed observations of each surviving mysid, this test is more expensive and requires more training than the normal short-term chronic test. However, this test still meets the major objectives of performing a short-term chronic test [*29*] because it can provide chronic toxicity information less expensively and considerably faster than the 28-day mysid life-cycle test.

The original short-term chronic test required ten replicates of five organisms each. This was later reduced to eight replicates (in a later EPA draft methods). Two of the tests (one hexavalent chromium, one PCP) performed in this study had ten replicates of five organisms each, while the other four tests had only eight replicates (in later EPA draft methods). Test Number 3 of both Tables 2 and 3 were the two studies with ten replicates. These data support the reduction of replicate numbers to eight since it does not appear that the additional replicates were necessary. In fact, the experimental design of this short-term chronic test does not agree with the experimental design of other established seven-day survival/growth tests (for example, fathead minnow [*1*]). Four replicates of ten organisms each would be easier to perform, require less space, and substantially reduce the chance of replicates without females or males. Such missing data are not useful and may in fact reduce the power of the test. The use of four replicates would also standardize the techniques utilized in the seven-day short-term chronic tests [*1*], which may reduce the chance of experimental error. In addition, similar statistics and computer programs already established could then be used for the mysid test.

Finally, we (like many other laboratories) have found it difficult to obtain from our cultures and commercial vendors the necessary number of mysids released during a single day required to perform several of these tests concurrently. Approximately 300 mysids per test (a minimum of 240 mysids assuming no mortality during the holding period) are required to initiate the seven-day holding (grow-up) period of the test. Because of the need for at least 300 mysids per test, this test needs more prior notice of the intent to perform the test. Also, the number of tests that can be performed in a given time frame may be reduced due to the problems in obtaining enough $<$24-h-old mysids from the cultures. Because of the test organism logistic problems, the reduced numbers of tests that can be conducted over short time frames, and increased costs, the seven-day mysid chronic test may not be an appropriate test for large testing efforts as compared to other short-term chronic tests.

Conclusions

Based on the reference toxicant tests performed in this study, the seven-day mysid survival, growth, and reproduction test appears to have similar sensitivity to the longer 28-day life-cycle test (for hexavalent chromium). In addition, the coefficients of variation determined for this study are within the expected variability of other toxicity tests for intralaboratory variability [27]. It should be cautioned, however, that the reproductive endpoint of this test is not measured as viable young or the production of young, but rather as fecundity. If a toxicant does appear to effect reproduction, this test may underestimate the chronic toxicity of the toxicant. Finally, as a result of the test organism logistic problems, the reduced numbers of tests that can be conducted over short time frames, and the increased costs, this chronic estimation test may not be as useful a test for large testing programs.

Acknowledgments

We would like to thank Phil Dorn, Ursula Cowgill, and Wayne McCulloch for their valuable suggestions and comments concerning this manuscript, as well as the anonymous peer reviewers. We acknowledge D. Lemarie, J. Boraczek, V. Sohn, M. Fisher, M. Williams, and S. Leigh for their technical assistance in this study. And finally, we wish to thank EA's Communications Department for the preparation of this project. This study was funded by The American Petroleum Institute.

References

[1] Horning, W. B. II and Weber, C. I., "Short-Term Methods for Estimating the Chronic Toxicity of Effluents and Receiving Waters to Freshwater Organisms," EPA-600/4-85-014, U.S. Environmental Protection Agency, Environmental Monitoring and Support Laboratory, Cincinnati, OH, December 1985.

[2] *Users Guide to the Conduct and Interpretation of Complex Effluent Toxicity Tests at Estuarine/Marine Sites*, Schimmel, S. C. Ed., Contribution No. 796, Environmental Research Laboratory, U.S. Environmental Protection Agency, Narragansett, RI, 1986.

[3] "Short-Term Methods for Estimating the Chronic Toxicity of Effluents and Receiving Waters to Marine and Estuarine Organisms," Horning, W. B. and Weber, C. I. Eds., Draft, EPA-600/4-87-000, U.S. Environmental Protection Agency, Environmental Monitoring and Support Laboratory, Cincinnati, OH, 1987.

[4] Lussier, S. M., Kuhn, A., and Sewall, J., "Guidance Manual for Conducting Seven Day Mysid Survival/Growth/Reproduction Study Using the Estuarine Mysid, *Mysidopsis bahia*," Contribution No. X106, *Users Guide to the Conduct and Interpretation of Complex Effluent Toxicity Tests at Estuarine/Marine Sites*, S. C. Schimmel, Ed., Environmental Research Laboratory, U.S. Environmental Protection Agency, Narragansett, RI, 1986, pp. 86–124.

[5] Nimmo, D. R., Mirenda, R. J., and Carlson, C. A., "Feasibility of Culturing and Testing the Mysid, *Mysidopsis bahia*, under Artificial Conditions," Cooperative Agreement CR-811691-01-0, report for U.S. Environmental Protection Agency, Environmental Monitoring and Support Laboratory, Cincinnati, OH, 1987.

[6] Rodgers, J. H., Jr., Dorn, P. B., Duke, T., Parrish, R., and Venables, B., "*Mysidopsis* sp.: Life History and Culture," report from a workshop held in Gulf Breeze, Florida, 15–16 Oct. 1986, American Petroleum Institute, U.S. Environmental Protection Association, and North Texas State University, 1986.

[7] Nimmo, D. R., Bahner, L. H., Rigby, R. A., Sheppard, J. M., and Wilson, A. J., Jr., "*Mysidopsis bahia:* An Estuarine Species Suitable for Life-Cycle Toxicity Test to Determine the Effects of a Pollutant," in *Aquatic Toxicology and Hazard Evaluation (First Conference), ASTM STP 634*, F. L. Mayer and J. L. Hamlink, Eds., American Society for Testing and Materials, Philadelphia, 1977, pp. 109–116.

[8] Odum, W. E. and Heald, E. J., *Bulletin of Marine Science*, Vol. 22, No. 10, 1972, pp. 671–738.

[9] Jop, K. M., Rodgers, J. H., Dorn, P. B., and Dickson, K. L. in *Aquatic Toxicology and Environmental Fate: Ninth Volume, ASTM STP 921*, T. M. Postan and R. Purdy, Eds., American Society for Testing and Materials, Philadelphia, 1986, pp. 390–403.

[10] Dorn, P. B., Rodgers, J. H. Jr., Jop, K. M., Raia, J. C., and Dickson, K. L., *Environmental Toxicology and Chemistry*, Vol. 6, 1987, pp. 435–444.

[11] Jop, K. M., Parkerton, T. F., Rodgers, J. H. Jr., Dickson, K. L., and Dorn, P. B., *Environmental Toxicology and Chemistry*, Vol. 6, 1987, pp. 697–703.

[12] American Public Health Association (APHA), American Water Works Association, and Water Pollution Control Federation, *Standard Methods for the Examination of Water and Wastewater*, 16th ed., APHA, Washington DC, 1985.

[13] Gulley, D. D., Boelter, A. M., and Bergman, H. L., "TOXSTAT/DATASYS Release 1.0," University of Wyoming, Laramie, WY, 1987.
[14] Lussier, S. M., Gentile, J. H., and Walker, J., *Aquatic Toxicology*, Vol. 7, 1985, pp. 25–35.
[15] Gentile, S. M., "Memorandum to John H. Gentile, U.S. Environmental Protection Agency, Narragansett, Rhode Island," in *U.S. Environmental Protection Agency, Ambient Water Criteria for Chromium—1984*, Office of Water and Standards, Criteria and Standards Division, Washington, DC, 1985.
[16] Eisler, R. and Hennekey, R. J., *Archives of Environmental Contamination and Toxicology*, Vol. 6, 1977.
[17] Frank, P. M. and Robertson, P. B., *Bulletin of Environmental Contamination and Toxicology*, Vol. 21, 1979.
[18] U.S. Environmental Protection Agency, *Ambient Water Criteria for Chromium—1984*, Office of Water and Standards, Criteria and Standards Division, Washington, DC, 1985.
[19] Hauch, R. G., Norris, D. R., and Pierce, R. H., Jr., *Bulletin of Environmental Contamination and Toxicology*, Vol. 25, 1980, pp. 562–568.
[20] Adema, D. M. M. and Vink, G. J., *Chemosphere*, Vol. 10, 1981, pp. 533–554.
[21] Conklin, P. J. and Rao, K. R., *Bulletin of Environmental Contamination and Toxicology*, Vol. 20, 1978, pp. 275–279.
[22] Borthwick, P. W. and Schimmel, S. C. in *Pentachlorophenol, Chemistry, Pharmacology and Environmental Toxicology*, K. R. Rao, Ed., Plenum Publishing Company, New York, NY, 1978, pp. 141–146.
[23] Bentley, R. E., Heitmuller, T., Sleight, B. H., and Parrish, R. R., "Acute Toxicity of Pentachlorophenol to Bluegill *(Lepomis macrochirus)*, Rainbow Trout *(Salmo gairdneri)*, and Pink Salmon *(Penaeus duorarum)*," WA-6-99-1414-B, U.S. Environmental Protection Agency, Criterial Branch, Washington, DC, 1975.
[24] U.S. Environmental Protection Agency, "Ambient Aquatic Life Water Quality Criteria for Pentachlorophenol," Office of Research and Development, Environmental Research Laboratories, Duluth, MN and Narragansett, RI, 1986.
[25] Parrish, P. R., Dyar, E. E., Enos, J. M., and Wilson, W. G., "Chronic Toxicity of Chlordane, Trifluration, and Pentachlorophenol to Sheepshead Minnows *(Cyprinodon variegatus)*," EPA-600/3-78-010, National Technical Information Service, Springfield, VA, 1978.
[26] Hughes, M. M., Heber, M. A., Schimmel, S. C., and Berry, W. J., "Guidance Manual for Conducting Complex Effluent and Receiving Water Larval Fish Growth-Survival Studies with the Sheepshead Minnow (Cyprinodon variegatus)," *Users Guide to the Conduct and Interpretation of Complex Effluent Toxicity Tests at Estuarine/Marine Sites*, S. C. Schimmel, Ed., Environmental Research Laboratory, U.S. Environmental Protection Agency, Narragansett, RI, 1986.
[27] Rue, W. J., Fava, J. A., and Grothe, D. R., "A Review of Inter- and Intralaboratory Effluent Toxicity Test Method Variability," *Aquatic Toxicology and Hazard Assessment, 10th Volume, ASTM STP 971*, W. J. Adams, G. A. Chapman, and W. G. Landis, Eds., American Society for Testing and Materials, Philadelphia, 1988, pp. 190–203.
[28] ASTM Guide for Conducting Life-Cycle Toxicity Tests with Saltwater Mysids (E 1191-87), American Society for Testing and Materials, Philadelphia, PA, 1987.
[28] DeGraeve, G. M., Cooney, J. D., Carr, R. S., and Neff, J. M., "Identification, Comparison and Evaluation of Existing Freshwater, Estuarine, and Marine Rapid Chronic Bioassays," American Petroleum Institute, Washington, DC, 1985.

Carroll L. Missimer,[1] *David P. Lemarie,*[2] *and William J. Rue*[1]

Evaluation of a Chronic Estimation Toxicity Test Using *Skeletonema costatum*

REFERENCE: Missimer, C. L., Lemarie, D. P., and Rue, W. J., "**Evaluation of a Chronic Estimation Toxicity Test Using *Skeletonema costatum*,**" *Aquatic Toxicology and Hazard Assessment: 12th Volume, ASTM STP 1027,* U. M. Cowgill and L. R. Williams, Eds., American Society for Testing and Materials, Philadelphia, 1989, pp. 345–354.

ABSTRACT: The marine diatom *Skeletonema costatum* was cultured and tested as part of a comprehensive study to evaluate chronic toxicity estimation testing methods for five marine and estuarine species. This species was cultured successfully in a medium prepared with Atlantic Ocean water. A medium prepared from an artificial seawater formulation produced inconsistent culturing results. Triplicate 96-h tests were conducted using hexavalent chromium and sodium pentachlorophenate (PCP) as reference toxicants to evaluate intralaboratory variability. Three biological endpoints, number of cells, biomass, and *in vivo* fluorescence, were quantified. Mean 96-h chronic values were calculated as 1.2 mg/L Cr^{+6} and 61.5 μg/L PCP, which were similar to other reported estimates of toxicity for these chemicals. Estimated chronic values increased during the 96-h test period for both toxicants. Standardization of a clonal variety and further studies to determine the intra- and interlaboratory variability are recommended before this test is adopted as a routine effluent monitoring tool.

KEY WORDS: chronic estimation toxicity test, *Skeletonema costatum*, hexavalent chromium, pentachlorophenol

The use of algal bioassays to estimate the eutrophication potential and/or toxicity of wastewaters dates back to the development and adoption of the *Selenastrum capricornutum* Printz bottle test in the late 1960s and early 1970s [*1*]. Soon, thereafter, similar bottle tests were developed for marine and estuarine algal species [*2*], with the same intent of evaluating both eutrophication potential and toxicity. One of the earliest references to the use of marine algae as test organisms involved their use as a screening technique for estimating metal toxicity to estuarine phytoplankton [*3*]. Currently, standard procedures for conducting freshwater and/or marine algal toxicity tests are available in *Standard Methods* [*4*], EPA [*5–7*], and in draft procedure from ASTM [*8*]. All of these methods recommend the use of microalgae. The marine microalgae most frequently recommended [*2,4,5,6,8*] are *Skeletonema costatum* (Grev.) Cleve. and *Thalassiosira pseudonana* (Hasle and Heimdal).

While the use of microalgae bottle tests as estimators of eutrophication potential is reasonably well defined, the use of these same tests as estimators of toxicity is less clearly defined. Specifically, it is not clear whether the results of algal bottle tests are to be interpreted as being indicative of acute toxicity, chronic toxicity, or both. While several of the procedures listed

[1]Scientist and senior scientist, respectively, EA Engineering, Science, and Technology, Inc., Hunt Valley/Loveton Center, 15 Loveton Circle, Sparks, MD 21152.

[2]Fisheries biologist, U.S. Fish and Wildlife Service, Fish Culture Research Lab, Kearneysville, WV 25430.

above explicitly label these tests as acute tests, at least one [8] considers the results of such tests as being interpretable of either acute or chronic toxicity. EPA guidelines for developing water quality criteria [9] also consider such tests as reflecting chronic toxicity. However, EPA's latest draft of marine chronic estimation methods (rapid chronic tests) [10] does not include a microalgae test but instead includes the *Champia parvula* macroalgae reproduction test. This test is clearly a chronic estimation test, very similar to the other marine chronic estimation tests, because it quantifies a reproductive endpoint, fertilization, that is typically considered a sensitive estimator of chronic toxicity.

This paper evaluates the use of *Skeletonema costatum* as an estimator of chronic toxicity to estuarine and marine organisms. The goals of this research project were to evaluate culturing methodologies, testing procedures, and the suitability of hexavalent chromium and pentachlorophenol as reference toxicants.

Life History

Skeletonema costatum (Grev.) Cleve. is a marine diatom (Class: Bacillariophyceae, Family: Thalassiosiraceae) common to coastal waters around the world, including those of the United States and Canada. It is found in a wide range of salinities, including those as low as 3 parts per thousand (ppt) [11]. At salinities of 5.31 ppt, the growth rate of *S. costatum* is well below maximum [11]. Present evidence suggests that many different clones of *S. costatum* have adapted to a variety of ecological conditions [11].

Culturing Procedures

Environmental conditions for the culturing of *S. costatum* were taken from the proposed ASTM procedure [8] and are listed in Table 1. The nutrient medium used for routine culturing and as dilution water during testing was also adopted from ASTM [8] and is listed in Table 2.

The clone SKEL of *S. costatum* was obtained from the Culture Collection of Marine Phytoplankton at the Bigelow Laboratory of Ocean Sciences, West Boothbay Harbor, Maine. This variety of *S. costatum* was collected from Milford Harbour, Connecticut by Guillard. Culturing conditions accompanying the starter culture list a salinity range of 12 to 32 ppt and a temperature range of 12 to 30°C.

Original attempts to culture *S. costatum* using an artificial sea salt formulation as the basis of the nutrient medium were not successful; growth was slow and cultures seldom reached densi-

TABLE 1—*Culturing conditions for* Skeletonema costatum.

Temperature	20° ± 2°C
Lighting	
Type	Cool white fluorescent
Intensity	400 foot candles
Photoperiod	14-h light/10-h dark
pH	8.2 ± 0.1
Salinity	30 ± 1 ppt
Culture volume	50 mL
Culture flask	250-mL borosilicate glass Erlenmeyer
Nutrient medium	See Table 2
Agitation	By hand, twice daily

TABLE 2—*Saltwater medium.*

Stock Solution	Reagent	Amount
Metal stock solution[a]	$FeCl_3 \cdot 6H_2O$	0.045 g
	$MnCl_2 \cdot 4H_2O$	0.144 g
	$ZnSO_4 \cdot 7H_2O$	0.045 g
	$CuSO_4 \cdot 5H_2O$	0.157 mg
	$CoCl_2 \cdot 6H_2O$	0.404 mg
	H_3BO_3	1.140 g
Minor salt stock solution[a]	K_3PO_4	3.0 g
	$NaNO_3$	50.0 g
	$Na_2SiO_3 \cdot 9H_2O$	20.0 g
Vitamin stock solution[b]	Thiamine hydrochloride	50.0 mg
	Biotin	0.01 mg
	B_{12}	0.10 mg

15 mL of metal stock solution, 1 mL of minor stock solution, and 0.5 mL of vitamin stock solution are added to 1 L of salt water.

[a] Reagents added in given order to 900 mL deionized water, then diluted to 1 L.

[b] Reagents added in given order to 90 mL deionized water, then diluted to 1 L.

ties needed to inoculate test solutions. An attempt was also made to culture *S. costatum* using the f/2 medium and the artificial seawater formulation. While cultures did well in this medium for short periods of time, sustained growth of successive inoculations lasted for less than one month. These difficulties suggested that further attempts to use artificial seawater formulations would be useless, and Atlantic Ocean water was therefore adopted as the basis of the nutrient medium. Atlantic Ocean water collected at a depth of 2 m off the coast of New Jersey was polished using an activated carbon filter and purified by filtering through a 0.22-μm membrane filter. The 35-ppt ocean water was diluted to 30 ppt using 0.22-μm filtered, deionized water.

Trial and error experimentation indicated that culture renewal should occur at a minimum of every three or four days. A standard procedure of renewing cultures every Monday and Thursday has produced successive generations of healthy, actively reproducing cells. During each culture renewal, a small (3-4 mL) sample of the culture was removed from each culture flask, examined, and the number of cells counted using a Bright-Line hemocytometer (Reichert Scientific, Buffalo). The flask having the healthiest, most abundant, and least contaminated culture was used to inoculate the next round of cultures. The maintenance of a purely axenic culture proved impractical. The most meticulous technique inevitably resulted in the presence of an unidentified protozoan of unknown origin. However, using the above culture renewal technique the densities of this organism were kept very low. Bacterial contamination was never an apparent problem.

Materials and Methods

Materials and methods were taken from the draft ASTM microalgae protocol [*8*] and are briefly outlined below.

Dilution Water

All waters used as dilution waters were prepared in a manner identical to the preparation of the culturing medium as described above. Salinity was adjusted to 30 ppt by dilution with deionized water; pH was adjusted, if necessary, to 8.2 with dilute HCl or NaOH.

Algal Cultures

Skeletonema costatum was cultured, using the materials and methods described above, for a period of over three weeks prior to use as an inoculum in toxicity tests.

Experimental Conditions

All tests were conducted in 250-mL borosilicate glass tissue culture flasks equipped with metal covers. All flasks and covers were identical and were prepared as specified in ASTM [*8*]. Environmental conditions during the 96-h exposure period were identical to the culturing conditions listed in Table 1. In each test there were three replicates of the control and each of the five exposure concentrations of each of the toxicants. Exposure concentrations likely to elicit chronic responses were estimated from available literature values. Serial dilutions (1:1) of the highest concentration yielded a series of exposure concentrations, each of which was half the concentration of the next highest concentration. Exposure concentrations of toxicants were not analytically verified.

Each exposure chamber received 50-mL of test solution and was inoculated with a quantity of cells to yield an initial density of 20 000 cells/mL of test solution [*8*].

Test Materials

Three 96-h assays were conducted with each of two standard reference toxicants: hexavalent chromium as potassium dichromate, and pentachlorophenol as sodium pentachlorophenate. Chemical characterization of the test materials is given in Table 3.

Biological Data

Three biological endpoints: number of cells, biomass, and *in vivo* fluorescence were determined for each exposure chamber at the end of each 96-h exposure period. Cell number was determined using a hemocytometer, and *in vivo* fluorescence was determined using a Turner Designs Model No. 111 fluorometer. Biomass were determined by filtering the entire sample

TABLE 3—*Chemical characterization of test materials.*

Characteristic	Potassium Dichromate	Sodium Pentachlorophenate
Molecular formula	$K_2Cr_2O_7$	Cl_5C_6ONa
Molecular weight	294.2	288.32
Purity, %	99.96	>90
Photo oxidation	0	50% (66 h at 90 to 95°C) [*12*]
Source	Mallinckrodt	Fluka AG
Lot number	6772 KXAS	229964 1084

(including the small volume used for fluorescence measurements) through predried, preweighed, prerinsed 0.45-μm membrane filters. Filters were dried at 70°C for 2 h, cooled in a desiccator, and weighed to the nearest 0.1 mg. The final weight was subtracted from the filter weight to yield the total biomass. The total biomass was divided by the final volume of each exposure chamber to determine the biomass per millilitre of solution.

During four tests (Cr 2 and PCP 1, 2, and 3) *in vivo* fluorescence was measured at the end of 24, 48, and 72 h in addition to being measured at the termination of the test. This was accomplished by pouring 3 to 4 mL of test solution directly into a fluorometer cuvette, measuring the fluorescence, and returning the solution to the original exposure chamber. The cuvette was rinsed thoroughly with distilled water between measurements which were conducted from low to high toxicant concentrations.

Data Analysis

The TOXSTAT computer program supplied by the University of Wyoming was used to analyze each data set. Each test was examined for normality and homogeneity of variance using the chi-square test and Bartlett's test, respectively. Raw data were transformed, if necessary, to $\log_{10}$ or square root values in order to meet the tests for normality and homogeneity of variance. Each data set was then analyzed using Williams' test to determine the no observed effect concentration (NOEC) and lowest observed effect concentration (LOEC). Williams test, like the Dunnett's procedure, compares each treatment to the control, but Williams test, unlike Dunnett's, is designed to be sensitive to increasing toxicant concentrations [*13*]. The maximum allowable toxicant concentration (MATC) value was calculated as the geometric mean of the NOEC and LOEC.

Results and Discussion

The test results using potassium dichromate and sodium pentachlorophenate are shown in Tables 4 and 5, respectively. Results from the tests using potassium dichromate are reported as mg/L Cr. Results from the tests using sodium pentachlorophenate are reported as μg/L PCP with a correction for the purity of the stock compound.

Hexavalent Chromium

As indicated in Table 6, 96-h MATCs for hexavalent chromium ranged from <0.1 mg/L Cr to 2.0 mg/L Cr depending on the particular test and the endpoint used. The mean MATC of the three assays was 1.2 mg/L Cr using MATCs that were calculated as the geometric mean of the NOEC and LOEC. Cell counts and *in vivo* fluorescence apparently gave the most reliable estimate of the MATC. MATCs calculated from biomass (Tests Cr 1 and Cr 3) were lower than MATCs calculated from cell counts or *in vivo* fluorescence. In test Cr 3 the biomass endpoint yielded a MATC of <0.1 mg/L Cr, the lowest exposure concentration, while the cell counts and *in vivo* fluorescence endpoints yielded MATCs of 2.0 and 1.1 mg/L Cr, respectively.

While values are not directly comparable, because different effects were measured, an estimated MATC of 1.2 mg/L Cr for *Skeletonema costatum* appears to agree with other estimates of the toxicity of chromium to aquatic plants as reported in the chromium water quality criteria document [*14*]. Estimates of various effective concentrations of hexavalent chromium range from 0.002 mg/L Cr for the blue alga *Microcystis aervginosa* to 7.8 mg/L Cr for the diatom *Nitzschia linearis*. No results for comparable marine microalgae are presented. Concentrations of hexavalent chromium that have inhibiting effects on the rate of photosynthesis of the macroalgae (kelp) *Macrocystis pyrifera* range from 1.0 to 5.0 mg/L Cr.

TABLE 4—*Results of* Skeletonema costatum *chronic estimation tests conducted with hexavalent chromium.*

Test No.	Chromium (CR^{+6}) Concentration, mg/L	Cell Count (±SD)[a]	Biomass (mg/mL ± SD)	In Vivo Fluorescence (±SD)[b]
1	Control	85.7 (32.1)	0.0866 (0.0094)	68.2 (5.3)
	0.3	71.0 (13.5)	0.0657 (0.0042)	59.8 (4.8)
	0.6	67.7 (9.0)	0.0561 (0.0054)*	54.3 (1.5)
	1.2	59.3 (7.0)*	0.0587 (0.0105)*	54.0 (1.0)
	2.4	1.0 (1.0)*	0.0402 (0.0121)*	0.7 (0.5)*
	4.9	0.2 (0.3)*	0.0375 (0.0094)*	0.0 (0.0)*
2	Control	81.7 (19.6)	...[c]	66.2 (6.4)
	0.2	77.0 (24.8)	...	58.5 (7.1)
	0.4	50.3 (3.2)	...	64.2 (1.4)
	0.8	63.0 (2.0)	...	62.8 (1.8)
	1.6	40.0 (7.5)*	...	36.0 (12.2)*
	3.2	1.2 (0.8)*	...	0.7 (0.3)*
3	Control	64.3 (14.3)	0.0977 (0.0466)	61.0 (4.3)
	0.1	56.7 (5.7)	0.0458 (0.0024)*	60.8 (2.6)
	0.3	62.7 (18.9)	0.0673 (0.0186)*	57.7 (0.6)
	0.7	62.7 (4.9)	0.0391 (0.0078)*	58.2 (1.3)
	1.4	47.3 (27.7)	0.0575 (0.0238)*	52.5 (16.5)*
	2.8	2.0 (3.5)*	0.0666 (0.0016)*	1.5 (0.9)*

[a] Numbered cells/0.1 μL.
[b] Fluorescence units (relative).
[c] Not calculated because of error in drying samples.
*Statistically significant difference ($\alpha = 0.05$) from controls using Williams' test.

Pentachlorophenol

Calculated MATCs for PCP ranged from <2.7 μg/L PCP to >87.0 μg/L PCP (Table 6). Test PCP 2 gave inconsistent results (see Table 4) between the three measured endpoints even though the same MATC was calculated using all three endpoints. Test PCP 1 failed to yield a definitive MATC. Cell counts and *in vivo* fluorescence indicated the MATC to be in excess of the highest exposure concentration, 43.5 μg/L PCP. This was consistent with tests PCP 2 and PCP 3 where MATCs were calculated five out of six times as 61.5 μg/L PCP.

These results agree reasonably well with other estimates of the effective concentration of PCP on *S. costatum* and *Thalassiosira pseudonana,* another member of the family Thalassiosiraceae. The geometric mean of three 96-h EC_{50} assays with *S. costatum* reported from [*15*] is 18.5 μg/L PCP, while the geometric mean of three 96-h EC_{50} assays with *T. pseudonana,* reported from the same reference, is 190.7 μg/L PCP.

Temporal Changes In MATCs

Skeletonema costatum is estimated to undergo cell division two to three times per day under the optimal light and nutrient conditions of the experimental design [*16*]. Any cell therefore that goes through one or more mitotic division under exposure conditions has effectively gone through a complete life cycle. With this taken into consideration, it does not appear unreasonable to consider that 24, 48, and 72-h tests are of sufficient duration to also be indicative of chronic effects. It is interesting to note therefore that chronic values change remarkably over a 96-h test period.

TABLE 5—*Results of* Skeletonema costatum *chronic estimation tests conducted with pentachlorophenol.*

Test No.	PCP Concentration, μg/L	Cell Count (±SD)[a]	Biomass (mg/ml ± SD)	In Vivo Fluorescence (±SD)[b]
1	Control	49.3 (4.2)	0.1846 (0.1304)	50.3 (6.3)
	2.7	59.7 (20.4)	0.0548 (0.0099)*	45.3 (2.8)
	5.4	47.0 (15.1)	0.0538 (0.0122)*	53.8 (12.7)
	10.8	41.0 (6.2)	0.0662 (0.0090)*	43.5 (2.5)
	21.5	50.3 (2.9)	0.0559 (0.0020)*	41.0 (1.0)
	43.0	55.3 (2.5)	0.0496 (0.0041)*	46.2 (1.3)
2	Control	88.1 (14.2)	0.0464 (0.0268)	74.3 (11.4)
	5.4	90.3 (25.4)	0.0488 (0.0036)	78.3 (6.8)
	10.8	121.3 (21.7)	0.0502 (0.0113)	77.2 (2.0)
	21.5	107.3 (19.3)	0.0601 (0.0159)	78.5 (8.8)
	43.0	98.8 (12.0)	0.0721 (0.0284)	84.7 (3.5)
	87.0	47.5 (5.4)*	0.0802 (0.0117)*	86.0 (4.4)*
3	Control	74.3 (13.8)	0.0604 (0.0295)	85.0 (9.5)
	5.4	78.9 (36.6)	0.0436 (0.0057)	77.3 (1.2)
	10.8	91.0 (36.4)	0.0485 (0.0088)	84.7 (3.2)
	21.5	91.3 (0.6)	0.0530 (0.0185)	86.0 (2.3)
	43.0	104.8 (33.8)	0.0521 (0.0166)	88.2 (3.9)
	87.0	44.0 (23.6)*	0.0364 (0.0139)	29.5 (14.3)*

[a]Numbered cells /0.1 μL.
[b]Fluorescence units (relative).
*Statistically significant different ($\alpha = 0.05$) from controls using Williams' test.

TABLE 6—*Calculated MATC values for 96-h hexavalent chromium and PCP reference toxicant tests.*

Test	96-h MATC Values by Endpoint: Cell Counts	Biomass	In Vivo Fluorescence	Arithmetic Mean of Endpoints
Cr 1 (mg/L)	0.8	0.4	1.7	1.0
Cr 2 (mg/L)	1.1	. . .[a,b]	1.1	1.1
Cr 3 (mg/L)	2.0	<0.1[b]	1.00	1.5
			Arithmetic Mean of 3 Tests =	1.2 mg/L
PCP 1 (μg/L)	>43.5[b]	<2.7[b]	>43.5[b]	. . .[c]
PCP 2 (μg/L)	61.5	61.5	61.5	61.5
PCP 3 (μg/L)	61.5	>87.0[b]	61.5	61.5
			Arithmetic Mean of 2 Tests =	61.5 μg/L

[a]Not calculated because of error in drying samples.
[b]Value not used in further calculations.
[c]Not calculated because all MATC values were < or >.

Figure 1 indicates (see also Table 7) that the MATC for hexavalent chromium increased from less than 0.18 mg/L Cr at 24 h to 0.26 mg/L at 48 and 72 h, to 1.08 mg/L at the end of the 96-h test. Likewise, as indicated in Fig. 2, the MATCs for PCP increased during the 96-h exposure period in each of the three PCP tests.

While possible explanations for these observations are advanced below, more definitive hy-

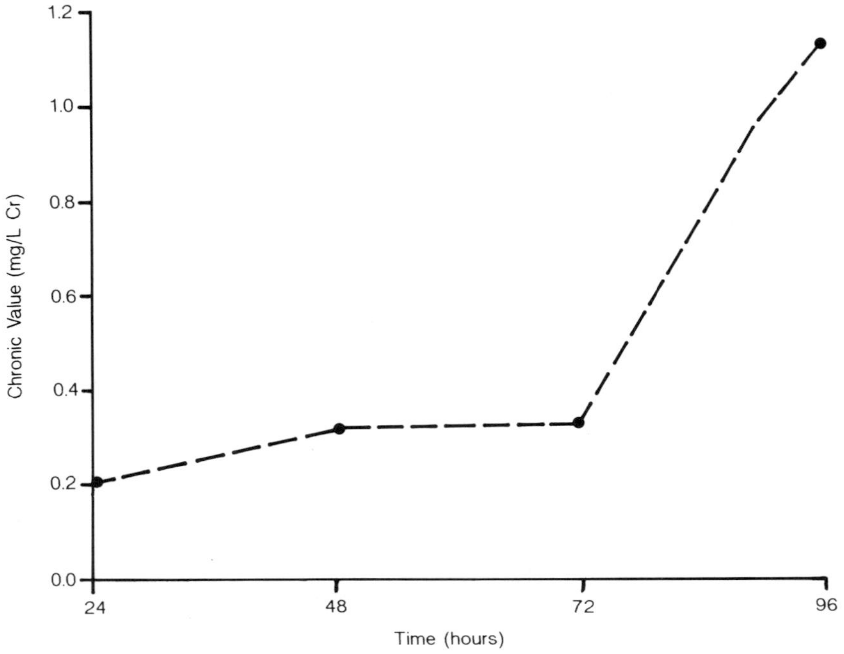

FIG. 1—*Chronic values during 96-h algal bioassay for test Cr No. 2.*

TABLE 7—*Time series MATC values for one chromium test and three PCP tests based on* in vivo *fluorometric measurements.*

		Time, h			
Test	Endpoint	24	48	72	96
Chrome 2	NOEC (mg/L)	...	0.2	0.2	0.8
	LOEC (mg/L)	<0.2	0.4	0.4	1.6
	MATC (mg/L)	<0.2	0.3	0.3	1.1
PCP 1	NOEC (μg/L)	21.8	21.8	21.8	>43.5
	LOEC (μg/L)	43.5	43.5	43.5	
	MATC (μg/L)	30.8	30.8	30.8	>43.5
PCP 2	NOEC (μg/L)	...	...	...	43.5
	LOEC (μg/L)	<5.5	<5.5	<5.5	87.0
	MATC (μg/L)	<5.5	<5.5	<5.5	61.5
PCP 3	NOEC (μg/L)	10.9	21.8	43.5	43.5
	LOEC (μg/L)	21.8	43.5	87.0	87.0
	MATC (μg/L)	15.4	30.8	61.5	61.5

pothesis testing will, of course, require further experimentation. PCP, as indicated in Table 3, is subject to photodegradation which may explain the apparent increase in MATC during the 96-h exposure period. Algae exposed to relatively low initial concentration of PCP may be temporarily inhibited (a chronic response) from undergoing normal logarithmic growth. However, once photodegradation reduced PCP to less than inhibiting concentrations, cell division may return to a normal rate [*17*]. At the same time, the higher initial exposure concentrations may

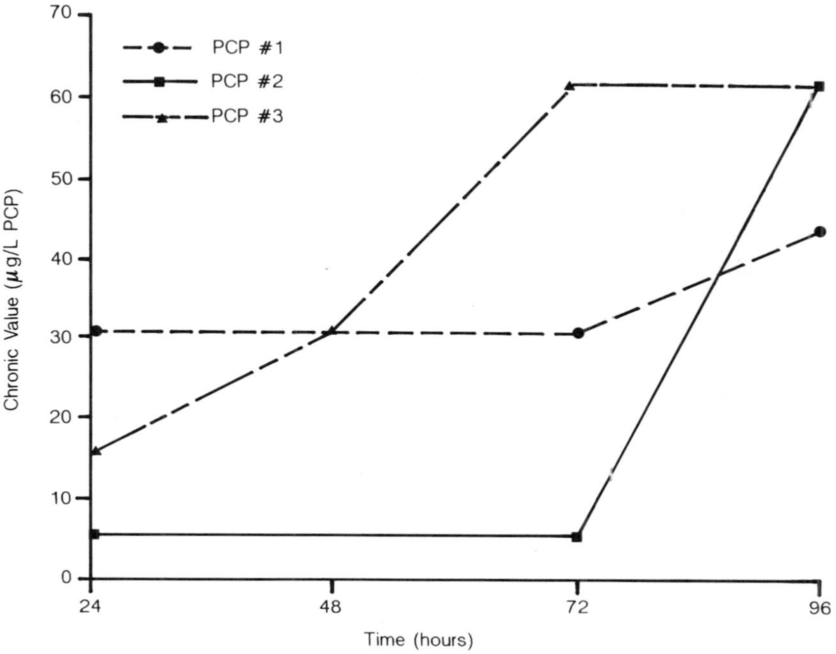

FIG. 2—*Chronic values during three 96-h algal bioassays with PCP.*

be sufficient to immediately kill (an acute response) cells and therefore reduce the algal population to levels that cannot recover even if PCP concentrations decrease to less than inhibiting levels.

Similar explanations for the same observed change in MATC over time with hexavalent chromium are not satisfactory since the chromate ion, $CrO_4^=$, is not photodegraded. However, it has been demonstrated that the toxicity of hexavalent chromium to marine and estuarine diatoms is closely linked to the sulfate concentration in the medium [*18*]. The similarity of the molecular structure between the chromate and sulfate ions has suggested that the toxicity of chromium is a result of competition between chromate and sulfate for active transport across the cell membrane [*18*]. Once inside the cell, hexavalent chromium is converted to trivalent chromium. Chromate uptake has been demonstrated to continue in *Skeletonema costatum* cultures where growth and sulfate uptake were totally inhibited [*19*]. At relatively low concentrations of chromium, cell growth may be temporarily inhibited until extracellular concentrations are reduced to a level where chromium is no longer able to effectively compete with sulfate for uptake into the cell.

Conclusions and Recommendations

The following conclusions can be drawn with respect to the stated goals of this research project. The basic culturing and testing procedures outlined in the draft ASTM microalgae protocol [*8*] are appropriate for the marine diatom *Skeletonema costatum*. However, the results indicate that healthy, actively reproducing stock cultures of *S. costatum* can be maintained in the laboratory only when natural seawater is used as the basis of the culturing media and cultures are renewed twice a week. PCP and hexavalent chromium appear to be suitable reference toxicants; however, further experiments should involve the analytical determination of exposure concentrations during the course of the experiment in order to confirm the preliminary indications given here that exposure concentrations of these chemicals may decrease during the 96-h assay.

Additionally, this research has highlighted several questions that are worthy of further investigation before this toxicity test is used for strict regulatory compliance. The accuracy of different biological endpoints needs to be further quantified. Biomass may not be an appropriate endpoint for bioassays involving diatoms where between 20 and 98% of the cell biomass [*20*] is refractory silica. It would also seem appropriate that future investigations should seek to quantify any differences that may exist in the relative sensitivity of different clonal varieties of *Skeletonema costatum.* It may be desirable and necessary to standardize the clonal variety in order to make the results of assays performed by different laboratories comparable.

References

[*1*] "Algal Assay Procedure: Bottle Test," U.S. Environmental Protection Agency, Corvalis, OR, 1971.

[*2*] "Marine Algal Assay Procedure: Bottle Test," U.S. Environmental Protection Agency, Corvalis, OR, 1974.

[*3*] Erickson, S. J., Lackie, N., and Maloney, T. E., *Journal of the Water Pollution Control Federation,* Vol. 42, No. 8, August 1970, pp. R270–R278.

[*4*] *Standard Methods for the Examination of Water and Waste Water,* 16th ed., American Public Health Association (APHA), Washington, DC, 1985, pp. 726–737.

[*5*] "Freshwater and Marine Algae Acute Toxicity Tests," EG-22, Office of Toxic Substances, U.S. Environmental Protection Agency, Washington, DC, 1983.

[*6*] "Bioassay Procedures for the Ocean Disposal Permit Program," EPA-600/9-76-010, U.S. Environmental Protection Agency, Gulf Breeze, FL, 1976, pp. 2–25.

[*7*] "The *Selenastrum capricornutum* Printz Assay Bottle Test," EPA-600/9-78-018, U.S. Environmental Agency, Corvalis, OR, 1978, p. 19.

[*8*] American Society for Testing and Materials, "Proposed New Standard Guide for Conducting Static 96-hour Toxicity Tests with Microalgae," Draft 12.

[*9*] "Guidelines for Deriving Numerical National Water Quality Criteria for the Protection of Aquatic Organisms and Their Uses," U.S. Environmental Protection Agency, Washington, DC, 1985.

[*10*] "Draft Short-term Methods for Estimating the Chronic Toxicity of Effluents and Receiving Waters to Marine and Estuarine Organisms," EPA-600/4-87-000, U.S. Environmental Protection Agency, Cincinnati, OH, 1987.

[*11*] Hasle, G. R. and D. L. Evensen, *Phycologia,* Vol. 14, No. 4, 1975, pp. 283–297.

[*12*] Verschueren, K., *Handbook of Environmental Data on Organic Chemicals,* 2nd. ed., Van Nostrand Reinhold Company, New York, NY, 1983, p. 954.

[*13*] Gelber, R. D., Lewin, P. T., Mehta, C. R., and Schoenfeld, D. A. in *Fundamentals of Aquatic Toxicology,* G. M. Rand and S. R. Petrocelli, Eds., Hemisphere Publishing Corp., Washington, DC, 1985, pp. 110–123.

[*14*] "Ambient Water Quality Criteria for Chromium—1984," EPA 440/5-84-029, U.S. Environmental Protection Agency, Washington, DC, 1985.

[*15*] "Ambient Aquatic Life Water Quality Criteria for Pentachlorophenol," draft, U.S. Environmental Protection Agency, Duluth, MN, 1986.

[*16*] "Technical Support Document for Algae Acute Toxicity Test," ES-5, Office of Toxic Substances, U.S. Environmental Protection Agency, Washington, DC, 1982.

[*17*] Giddings, J. M., Stewart, A. J., O'Neill, R. V., and Gardner, R. H. in *Aquatic Toxicology and Hazard Assessment: Sixth Symposium, ASTM STP 802,* American Society for Testing and Materials, Philadelphia, PA, 1983, pp. 445–459.

[*18*] Reidel, G. F., *Journal of Phycology,* Vol. 20, 1984, pp. 496–500.

[*19*] Reidel, G. F., *Aquatic Toxicology,* Vol. 7, 1985, pp. 191–204.

[*20*] Reynolds, C. S., *The Ecology of Freshwater Phytoplankton,* Cambridge University Press, New York, NY, 1984, p. 33.

Wayne L. McCulloch[1] *and William J. Rue*[1]

Evaluation of a Seven-Day Chronic Toxicity Estimation Test Using *Cyprinodon variegatus*

REFERENCE: McCulloch, W. L. and Rue, W. J., "**Evaluation of a Seven-Day Chronic Toxicity Estimation Test Using *Cyprinodon variegatus*,**" *Aquatic Toxicology and Hazard Assessment: 12th Volume, ASTM STP 1027,* U. M. Cowgill and L. R. Williams, Eds., American Society for Testing and Materials, Philadelphia, 1989, pp. 355–364.

ABSTRACT: The seven-day larval survival and growth test with the sheepshead minnow *(Cyprinodon variegatus)* is a modification of the previously published freshwater test with the fathead minnow *(Pimephales promelas).* Three sheepshead minnow larval survival and growth tests were conducted with hexavalent chromium and four tests were conducted with sodium pentachlorophenate to evaluate the intralaboratory precision and sensitivity of the draft protocol and determine the readiness of the test for NPDES permit biomonitoring as would be used by industrial dischargers. The mean MATC for the three hexavalent chromium tests was 6.8 mg/L, with a coefficient of variation of 51.5%. The mean MATC and coefficient of variation for the four sodium pentachlorophenate tests was 430 μg/L and 43.5%, respectively.

KEY WORDS: chronic estimation test, reference toxicant, *Cyprinodon variegatus,* sheepshead minnow, hexavalent chromium, sodium pentachlorophenate, MATC

Increasing emphasis is being placed on the evaluation of industrial and municipal effluents permitted through the National Pollutant Discharge Elimination System (NPDES) for acute and chronic toxicity to aquatic organisms. U.S. EPA's national "Policy for the Development of Water Quality-Based Permit Limitations for Toxic Pollutants" [*1*] and associated "Technical Support Document for Water Quality-Based Toxics Control" [*2*] have resulted in increased NPDES permit biomonitoring.

Methods [*3–5*] for measuring the acute toxicity of chemicals or effluents to freshwater, estuarine, and marine organisms have been available for a number of years, and acute studies are historically the type of tests most commonly required in water quality-based discharge permits. Full life-cycle or partial life-cycle chronic tests were generally viewed as too expensive or too time consuming for routine monitoring purposes.

Rapid, relatively economical procedures for estimating the chronic toxicity of effluents to freshwater organisms became available with the development of shorter duration chronic test methods [*6–8*] and the release of an EPA guidance manual for estimating chronic effects with freshwater organisms [*9*]. Until recently, however, EPA-approved short-term methods for estimating the chronic toxicity of effluents to saltwater organisms have not been available. As a result, EPA researchers in Narragansett, Rhode Island have developed toxicity test methods for estimating the chronic effects of effluents entering estuarine or marine waters [*10*]. Methods were developed or adapted for two species of invertebrates, two fish species, and a red macroalgae.

[1]Senior scientists, EA Engineering, Science, and Technology, Inc., Hunt Valley/Loveton Center, 15 Loveton Circle, Sparks, MD 21152.

One of the tests which was adapted for saltwater use was a sheepshead minnow (*Cyprinodon variegatus* Lacépède) larval fish growth/survival study [*11*]. This procedure was a modification of the freshwater protocol developed for fathead minnows [*7*]. The Narragansett protocol has since been modified for inclusion into a draft methods manual for marine and estuarine organisms [*12*]. Two potential shortcomings to the test method are: (1) preliminary work by EPA researchers suggests that the test should not be conducted at salinities less than 20 ppt due to inadequate control of growth, thus limiting the method's usefulness in brackish water situations; and (2) from a acute toxicity standpoint, the test species is not normally considered to be among the most sensitive of the saltwater species routinely tested.

One of the potential advantages of the test is that the protocol states that the test can be successfully conducted with uncontaminated seawater, diluted brine, or artificial seawater. The purpose of this study was to evaluate sensitivity and reproducibility of the *C. variegatus* protocol using two reference toxicants, hexavalent chromium and sodium pentachlorophenate (NaPCP).

Materials and Methods

Test Water

An artificial saltwater formulation (Forty Fathoms Marine Mix, Marine Enterprises, Towson, Maryland) was used for all testing, holding, and acclimation. Artificial saltwater was prepared by dissolving the dry salts in dechlorinated tap water from the City of Baltimore municipal water systems. Before entering the laboratory, the water is passed through a high-capacity activated carbon dechlorination system to remove chlorine and hexane soluble organics. Each newly prepared batch of seawater was aerated and allowed to age for approximately three days before use. This water source has been used for the past six years for the culture and testing of *Mysidopsis bahia*. Test salinity for the *Cyprinodon* studies was 25 ppt (± 2 ppt).

Test Organisms

During this study, sheepshead minnow embryos were obtained from three commercial suppliers. Ideally, in a study of this type it is preferable to obtain all test organisms from a single population of fish to minimize the variability in response from different lots of organisms. Due to a limited demand, none of the commercial firms maintained a constant supply of *C. variegatus* embryos. Selection of the source was based on availability and previously observed quality (that is, percent fertilization) of the embryos. The commercial suppliers used included: Aquatic Research Organisms, Hampton, New Hampshire (five lots), Aquatic Biosystems, Inc., Fort Collins, Colorado (one lot), and Quality Pet Supply, Canton, Massachusetts (one lot).

Developing embryos were incubated at the temperature (25°C) and salinity (25 ppt) that would be used in the chronic tests. The embryos were held in a 2-L glass separatory funnel, and gentle aeration was provided to maintain the embryos in the water column. Incubation water was renewed daily and unfertilized/dead embryos were removed when observed. Tests were initiated with newly hatched (<24 h old) larvae. Larvae that appeared to hatch prematurely or were malformed in any way were discarded.

Reference Toxicants

Reagent grade potassium dichromate $K_2Cr_2O_7$ (Mallinckrodt, Inc., Paris, Kentucky, Lot No. 6772, purity 99.96%) was used for the hexavalent chromium studies. Stock solutions of hexavalent chromium (1 g/L) were prepared by dissolving potassium dichromate in deionized water. Concentrations of chromium reported in this study are based on nominal concentrations. Hexavalent chromium has been proposed as a reference toxicant in effluent toxicity tests [*13–14*]. Advantages of its use include stability in solution and similarity of nominal to measured test concentrations [*13–15*].

Toxicant source for the PCP studies was technical grade (90% purity) sodium pentachlorophenate (Dowicide G, Fluka Chemicals, Lot No. 229964-1084). Fresh stock solutions were prepared daily by dissolving PCP in deionized water. All values reported are based on nominal concentrations. Sodium pentachlorophenate concentrations expressed as PCP were corrected for purity.

Test Procedures

General test methodology and test conditions (Table 1) followed that described in the Narragansett guidelines [*10*], with the following minor modifications. The tests were conducted with four replicates of 10 organisms for each test concentration rather than with three replicates of 15 organisms each. This modification was made to allow the use of the nonparametric Steel's many-one rank test (minimum of four replicates necessary) when the data did not meet the assumptions of Dunnett's test. The second modification was the substitution of a 16-h-light:8-h-dark photoperiod for the 14:10 photoperiod specified in the protocol. These modifications were made in an attempt to standardize, where feasible, the sheepshead test specifications with the existing freshwater larval fish survival and growth protocol [*9*] without altering the study design.

The tests were performed in 1-L glass beakers containing 500 mL of test solution. Test concentration series were based on preliminary range-finding tests. Ten newly hatched *C. variegatus* larvae (<24 h old) were transferred randomly to each beaker and, following loading, each beaker received a similar random placement for the test. During testing, the larvae were fed newly hatched brine shrimp nauplii *(Artemia)* twice daily (Aquarium Systems, Glen Burnie, Maryland). Each feeding consisted of the additions of 0.5 to 1 mL (ration increased with larval

TABLE 1—*Test conditions for sheepshead minnow* (Cyprinodon variegatus) *larval survival and growth test.*

Test duration	7 days
Test type	Static renewal (daily)
Effects measured	Survival and growth
Temperature	25 ± 2°C
Light quality	Wide spectrum fluorescent light
Light intensity	50 to 100 ft-candles
Photoperiod	16-h light/8-h dark
Dissolved oxygen	≥40% saturation
Salinity	20 to 30 ± 2 ppt
Test chamber	1-L beaker
Test volume	500 mL
Replicate chambers per concentration	4
Number of animals per chamber	10
Age of test animals	≤24 h
Food source	Newly hatched *Artemia* nauplii
Ration	5 to 10 mg per larvae per day
Dilution water	Uncontaminated seawater, diluted brine, or artificial sea salts

age) of a 0.05 g/mL stock concentration of brine shrimp nauplii to each test vessel. Larvae were not fed on Day 7 of the test.

Test solutions were renewed daily, at which time dead and uneaten brine shrimp were siphoned from the test chambers and the number of live sheepshead minnow larvae in each replicate recorded. Larvae that did not respond to gentle prodding with a pasteur pipet were considered dead. At the termination of each test, surviving larvae from each replicate were rinsed with deionized water and were placed in 4% buffered formalin for subsequent weight determination.

Data from a hexavalent chromium range-finding test were used to evaluate the effect of drying time on final larval dry weights. Pooled larvae from each replicate were placed in pretared aluminum pans and were dried at 60°C in a Blue M (Blue Island, Illinois) drying oven for 2 h per protocol specifications. The pans were placed in a desiccator to cool and were then weighed to the nearest 0.01 mg using a Cahn model 4700 automatic electrobalance (Cerritos, California). After weighing, the pans were dried at 60°C for an additional 16 h, and the weights were then determined.

The test conditions for the last two NaPCP tests (Tests No. 3 and 4) were nearly identical. The two tests were conducted with a single lot of sheepshead embryos from the same brood stock; the two tests were initiated one day apart, and with the exception of Day 0 of Test No. 3 and Day 7 of Test No. 4, the two tests shared the same dilution water, NaPCP stock solutions, and daily *Artemia* nauplii ration.

Statistical Analyses

The TOXSTAT statistical program supplied by the University of Wyoming was utilized for the data analysis. Survival data from each test concentration were first converted to a proportion and were then transformed using the arcsine square-root transformation. Data were tested for normality and for heterogeneity of variance using chi^2 analysis and Bartlett's test, respectively. Statistically significant differences ($p = 0.05$) in mean survival between control and treatment groups were determined by performing one-way analysis of variance (ANOVA) and Dunnett's test. Mean individual dry weight data from each test were first evaluated for normality and homogeneity of variance as described above to determine if transformation was necessary. Significant differences in growth were determined by ANOVA and Dunnett's test. Survival and growth endpoints determined for each test were the lowest observed effect concentration (LOEC), no observed effect concentration (NOEC), and the MATC—defined as the geometric mean of the LOEC and the NOEC.

Results

Drying Time

A preliminary study was performed on a range-finding test to determine if the recommended drying procedure (2 h at 60°C) specified in the protocol was adequate for sheepshead minnow larvae, or whether a longer drying time was necessary. Extending the drying time at 60°C from 2 to 18 h (Table 1) decreased the pooled larval dry weight per replicate by an average of only 2.0% (range 1.0 to 2.5%) and the differences in weight (expressed as percent) were relatively uniform across treatment groups. Although the decrease in dry weight observed in the 18-h drying period may not significantly affect the test results, there is no additional labor effort in overnight drying, and therefore it was felt that drying the larvae for at least 18 h in subsequent studies was warranted.

Hexavalent Chromium

Results of three replicate sheepshead minnow seven-day survival and growth tests using hexavalent chromium are presented in Table 2. Mean control survival for the three tests was 94.3%. In Test 1, both survival and growth (as individual mean dry weight) were significantly lower than the control mean at a concentration of 15.5 mg/L Cr^{+6}. The decrease in growth at 15.5 mg/L was substantial and represented an 80% reduction compared to the controls. The 7.5 mg/L test concentration did not significantly affect growth or survival; thus the MATC for this test was 10.8 mg/L, based on survival and growth.

Growth was the most sensitive chronic endpoint in the second chromium test. The NOEC for growth was 3.2 mg/L as compared to an NOEC of 18.0 mg/L for survival. Growth was significantly reduced at all concentrations $\geq$5.6 mg/L, yet there was 100% survival at 10.0 and 18.0 mg/L. Mean dry weight of the controls in Test 2 (0.85 mg $\pm$ 0.06) was above the 0.6 mg per larvae minimum criteria specified in the protocol, but was considerably less than the 1.37 mg $\pm$ 0.09 observed in Test 1. The hexavalent chromium MATC based on growth for Test 2 was 4.2 mg/L.

In the third chromium test, growth was significantly lower than the control mean at all concentrations $\geq$7.5 mg/L. Survival was significantly lower than control at 7.5 mg/L, but was not significantly different at 15.5 mg/L. The majority of the mortality at 7.5 mg/L occurred in one replicate (50% mortality), while mean survival in the remaining replicates was 80%. The larval growth MATC for this test was 5.3 mg/L. The MATCs for the three tests were 10.8, 4.2, and 5.3 mg/L Cr^{+6}, with a mean MATC of 6.8 mg/L Cr^{+6}. The coefficient of variation of 51.4% indicates that the test results were not highly reproducible. The observed variability could be due, in part, to different organism stock sensitivities since different populations of test organisms were used.

Sodium Pentachlorophenate

Survival and growth data for the four replicate tests conducted with sodium pentachlorophenate are presented in Table 3. For all NaPCP tests, growth was the most sensitive chronic end-

TABLE 2—*Effect of drying time (2 versus 18 h) at 60°C on pooled larval dry weight.*

			Pooled Larval Dry Weight Per Replicate, mg				
			Drying Time				
Group	Replicate	*n*	2 h	18 h	Difference	Percent Difference	Final Mean Dry Weight/Larvae
Control	A	10	18.94	18.56	0.38	2.0	1.86
	B	10	16.48	16.14	0.34	2.1	1.61
0.1 mg/L Cr^{+6}	A	10	18.24	17.93	0.31	1.7	1.79
	B	10	17.34	16.96	0.38	2.2	1.70
1.0 mg/L Cr^{+6}	A	10	18.62	18.18	0.44	2.4	1.82
	B	10	12.88	11.99	0.29	2.3	1.20
10.0 mg/L Cr^{+6}	A	10	6.13	6.04	0.06	1.0	0.60
	B	9	3.56	3.47	0.09	2.5	0.39
						Mean = 2.0 $\pm$ 0.5	

TABLE 3—*Results of* Cyprinodon variegatus *seven-day survival and growth tests conducted with hexavalent chromium.*

Test No.	Chromium Concentration, mg/L	Mean Percent Survival	Mean Dry Weight (mg) ± SD	MATC[a] in mg/L	
				By Weight	Survival
1	Control	95	1.37 (0.09)		
	1.8	95	1.43 (0.20)		
	3.7	98	1.27 (0.16)	10.8	10.8
	7.5	88	1.32 (0.20)		
	15.5	68*	0.27 (0.04)*		
	32.0	23*	0.11 (0.06)*		
2	Control	93	0.85 (0.06)		
	3.2	98	0.70 (0.14)		
	5.6	98	0.64 (0.05)*	4.23	24.0
	10.0	100	0.56 (0.02)*		
	18.0	100	0.32 (0.03)*		
	32.0	20*	0.16 (0.01)*		
3	Control	95	0.83 (0.06)		
	1.8	93	0.83 (0.08)		
	3.7	95	0.77 (0.03)	5.3	22.3
	7.5	73*	0.50 (0.16)*		
	15.5	88	0.46 (0.05)*		
	32.0	58*	0.23 (0.19)*		

[a]MATC—defined as the geometric mean of the lowest observed effect concentration (LOEC) and the no observed effect concentration (NOEC).

*Significantly different ($P = 0.05$) from the control based on Dunnett's test.

point. In Tests 1 and 4, survival was not significantly affected at any of the concentrations tested. Mean control survival for the four tests was 98.3%.

The MATC for the NaPCP tests were 492, 209, 368, and 651 μg/L as PCP, with a mean of 430 μg/L for the four tests, as determined by significant differences in dry weight. The coefficient of variation for the four PCP tests was 43.5%. The range of mean dry weights of the controls in the NaPCP tests was similar to that observed in the hexavalent chromium tests. The variation in replicate mean dry weight within a given control group was small, especially in Tests 2, 3, and 4, as evidenced by the control weight standard deviations.

There was a somewhat surprising difference in results obtained in the two NaPCP tests conducted concurrently (Tests 3 and 4) with embryos from a single supplier shipment. As mentioned before, however, the eggs may have been from different parentage spawns on successive days. The final mean dry weights of the two control groups (Table 4) differed by a factor of 1.4, with the larvae in Test 4, which hatched one day later than those from Test 3, being the smaller of the two groups. The MATC for these two tests differed by a factor of 1.8, and it was the smaller fish (based on final dry weights) in Test 4 that appeared to be less sensitive to NaPCP.

Discussion

Previous Studies with Hexavalent Chromium and PCP

Available literature on the toxicity of hexavalent chromium and sodium pentachlorophenate to sheepshead minnows was reviewed to compare the results of this study to previously reported values. Dorn et al. [*14*] reported the results of four replicate 96-h sheepshead minnow tests

TABLE 4—*Results of* Cyprinodon variegatus *7-day survival and growth tests conducted with sodium pentachlorophenate.*

Test No.	PCP Concentration, μg/L	Mean Percent Survival	Mean Individual Dry Weight (mg) ± S.D.	MATC[a] in μg/L	
				Dry Weight	Survival
1	Control	100	0.98 (0.14)		
	43.5	98	0.91 (0.06)		
	87	98	0.91 (0.12)	492	>696
	174	100	0.91 (0.03)		
	348	95	0.84 (0.07)		
	696	98	0.66 (0.06)*		
2	Control	93	1.26 (0.07)		
	87	95	1.31 (0.02)		
	157	98	1.21 (0.08)	209	368
	278	85	1.03 (0.09)*		
	487	68*	0.75 (0.08)*		
	870	15*	0.40 (0.17)*		
3	Control	100	1.23 (0.05)		
	87	100	1.23 (0.07)		
	157	100	1.16 (0.03)	368	651
	278	100	1.07 (0.06)		
	487	98	1.00 (0.18)*		
	870	78*	0.62 (0.07)*		
4	Control	100	0.88 (0.04)		
	87	100	0.92 (0.08)		
	157	100	0.88 (0.04)	651	>870
	278	98	0.89 (0.05)		
	487	98	0.81 (0.04)		
	870	95	0.59 (0.05)*		

[a] MATC—defined as the geometric mean of the Lowest Observed Effect Concentration (LOEC) and the No Observed Effect Concentration (NOEC).

*Significantly different ($P = 0.05$) from the control based on Dunnett's test.

conducted with hexavalent chromium. The 96-h LC_{50} values ranged from 16.3 to 26.8 mg/L Cr^{+6}, with a mean LC_{50} of 21.4 mg/L. Similarly, Jop et al. [*13*] reported the results of five 96-h sheepshead minnow acute tests with hexavalent chromium and found a mean LC_{50} of 23.2 mg Cr^{+6}/L and a coefficient of variation (CV) of 25%. No studies were found reporting the chronic toxicity of hexavalent chromium to sheepshead minnows or other marine fish species. The mean MATC of 6.8 mg Cr^{+6}/L derived in this study for sheepshead minnow is a factor of slightly greater than 3 lower than the acute values discussed above.

Borthwick and Schimmel [*16*] performed static acute tests with pentachlorophenol using sheepshead minnows ranging in age from one day to six weeks. The LC_{50} values ranged from 223 to 392 μg/L in five tests conducted at a salinity of 10 ppt. Parrish et al. [*17*] reported the results of a flow-through pentachlorophenol test with juvenile sheepshead minnows at a salinity of 24 ppt. The 96-h LC_{50} for the test was 442 μg/L PCP.

Parrish [*17*] also conducted a chronic (166-day) PCP study with sheepshead minnows. Significant mortality of parental fish was observed at PCP concentrations ≥88 μg/L. The reported MATC for the study was >44<88 μg/L and the application factor limits (that is, chronic toxicity ÷ acute toxicity) were 0.11 to 0.20. In this chronic study, growth was not a sensitive indicator of PCP toxicity. Exposure to PCP concentrations up to 389 μg/L for a 34-day period did not affect sheepshead growth.

Hughes [11] compared the results of the 7-day sheepshead minnow larval growth/survival test to the results of a 28-day early life stage test (ELS) using pentachlorophenol. For both tests, survival was the most sensitive chronic endpoint. The MATC for the ELS study was 53 μg/L, whereas the MATC for the seven-day growth/survival test was 106 μg/L. Based on growth, the MATC for the two tests were 106 and 212 μg/L, respectively. MATCs in this study (range is 209 to 651 μg/L; $\bar{x}$ (mean) is 494 μg/L) were comparable, but consistently higher than MATCs previously reported for PCP.

Sensitivity of the Test Endpoints

Ward and Parrish [18] evaluated the results of early life stage tests with sheepshead minnows and found that growth was not a statistically sensitive endpoint in 16 of the 18 (89%) tests which they considered. Based on those results, they believed that growth could be eliminated from ELS studies and that an accurate estimate of the MATC and application factor would still be obtained. This position was supported by Mayer et al. [19], who reviewed the results of 34 chronic toxicity studies with fish.

A similar conclusion was reached by Woltering [20] after reviewing the results of 173 partial or full life cycle fish chronic and early life stage tests. Fry survival was the single most sensitive endpoint in 32% of the studies, whereas fry growth was the single most sensitive endpoint in only 14% of the 173 studies evaluated.

Comparison of survival and growth endpoints are presented in Fig. 1 for hexavalent chromium and in Fig. 2 for sodium pentachlorophenate. Results of the present study indicate that for the two toxicants tested, larval growth was the more sensitive of the two test endpoints. Growth was the single most sensitive endpoint in five out of seven (71%) of the tests and equally as sensitive as survival in the remaining two tests. The estimation of chronic toxicity for the two

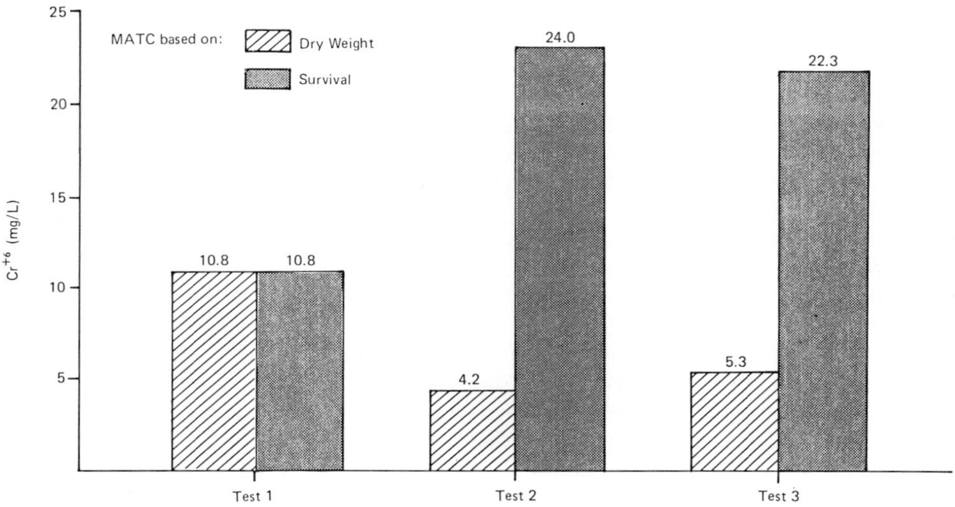

FIG. 1—*Hexavalent chromium MATCs for seven-day larval survival and growth test with* Cyprinodon variegatus.

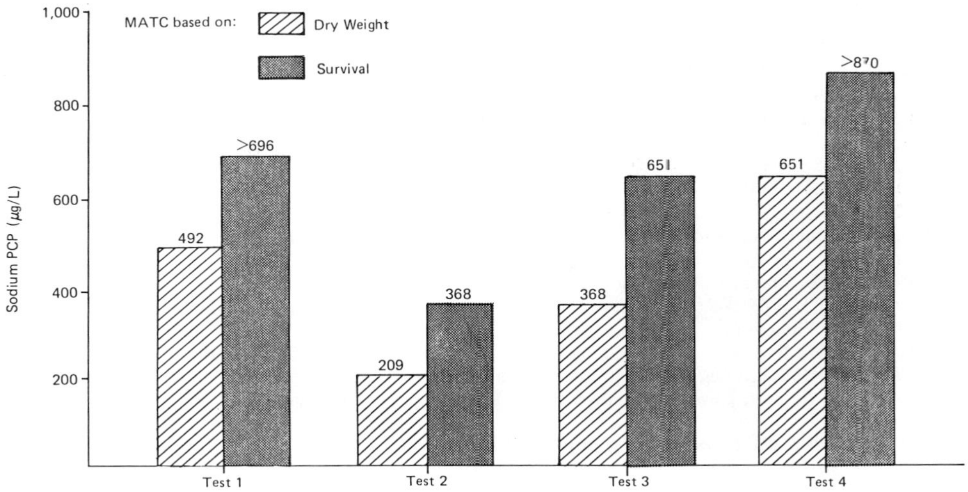

FIG. 2—*Sodium pentachlorophenate MATCs for seven-day larval survival and growth test with* Cyprinodon variegatus.

compounds tested would have been significantly higher if only survival had been considered. These results confirm that growth should be included as a test endpoint in early life-stage studies with *C. variegatus.*

Based on observations of uniform brine shrimp densities in the test vessels, it is our opinion that the closely controlled feeding rate specified in the seven-day sheepshead larval growth/survival protocol decreases the variation in the growth endpoint within each treatment group. The within-test variability of the mean dry weights in the control replicates was very low (indicated by small SD). The coefficient of variability (CV) of the within-test control group dry weights ranged from 4.1 to 14.3% for the seven tests reported, with a mean C.V. of 7.1%.

Conclusions

The advantages to the *C. variegatus* test include: (1) the test organism is easily cultured and spawned in the laboratory; (2) the test procedure is straightforward and, because of its comparability to the freshwater fish growth test, minimal equipment purchase and training is necessary; (3) the test can be conducted in uncontaminated natural seawater, diluted brine, or artificial seawater; (4) the test endpoints are easy to measure; and (5) the within-test variability in control survival and weight are low. Disadvantages noted during the present study were that the between test variability was relatively high (possibly due to different populations of test organisms), and the MATC values for both reference toxicants were higher than existing literature values.

References

[1] "Development of Water Quality-Based Permit Limitations for Toxic Pollutants; National Policy," *Federal Register,* Vol. 49, No. 48, March 1984, pp. 9016–9019.

[2] "Technical Support Document for the Development of Water Quality-Based Toxics Control," Office of Water, U.S. Environmental Protection Agency, Washington, DC, 1985.

[3] American Public Health Association (APHA), American Water Works Association, and Water Pollution Control Federation, *Standard Methods for the Examination of Water and Wastewater,* 16th ed., APHA, Washington, DC, 1985.
[4] ASTM Practice for Conducting Acute Toxicity Tests with Fishes, Macroinvertebrates, and Amphibians (E 729), American Society for Testing and Materials, Philadelphia.
[5] Peltier, W. H. and Weber, C. I., "Methods for Measuring the Acute Toxicity of Effluents to Freshwater and Marine Organisms," 3rd ed., EPA-600/4-85-013, U.S. Environmental Protection Agency, Environmental Monitoring and Support Laboratory—Cincinnati, March 1985.
[6] Mount, D. I. and Norberg, T. J., *Environmental Toxicology and Chemistry,* Vol. 3, 1984, pp. 425–434.
[7] Norberg-King, T. J. and Mount, D. I., *Environmental Toxicology and Chemistry,* Vol. 4, 1985, pp. 711–718.
[8] Birge, W. J., Black, J. A., Hudson, J. E., and Bruser, D. M. in *Aquatic Toxicology (Second Conference), ASTM STP 667,* L. L. Marking and R. A. Kimerle, Eds., American Society for Testing and Materials, Philadelphia, 1979, pp. 131–147.
[9] Horning, W. B. II and Weber, C. I., "Short-Term Methods for Estimating the Chronic Toxicity of Effluents and Receiving Waters to Freshwater Organisms," EPA-600/4-85-014, U.S. Environmental Protection Agency, Environmental Monitoring and Support Laboratory, Cincinnati, OH, December 1985.
[10] *Users Guide to the Conduct and Interpretation of Complex Effluent Toxicity Tests at Estuarine/Marine Sites,* Contribution No. 796, S. C. Schimmel, Ed., Environmental Research Laboratory, U.S. Environmental Protection Agency, Narragansett, RI, 1986.
[11] Hughes, M. M., Heber, M. A., Schimmel, S. C., and Berry, W. J., "Guidance Manual for Conducting Complex Effluent and Receiving Water Larval Fish Growth-Survival Studies with the Sheepshead Minnow (Cyprinodon variegatus)," *Users Guide to the Conduct and Interpretation of Complex Effluent Toxicity Tests at Estuarine/Marine Sites,* S. C. Schimmel, Ed., Contribution No. X104, 1986 Contribution No. 796, Environmental Research Laboratory, U.S. Environmental Protection Agency, Narragansett, RI.
[12] "Short-Term Methods for Estimating the Chronic Toxicity of Effluents and Receiving Waters to Marine and Estuarine Organisms," draft, EPA-600/4-87-000, W. B. Horning and C. I. Weber, Eds., U.S. Environmental Protection Agency, Environmental Monitoring and Support Laboratory, Cincinnati, OH, 1987.
[13] Job, K. M., Rodgers, J. H., Dorn, P. B., and Dickson, K. L. in *Aquatic Toxicology and Environmental Fate: Ninth Volume, ASTM STP 921,* Vol. 9, T. M. Postan and R. Purdy, Eds., American Society for Testing and Materials, Philadelphia, 1986, pp. 390–403.
[14] Dorn, P. B., Rodgers, J. H. Jr., Jop, K. M., Raia, J. C., and Dickson, K. L., *Environmental Toxicology and Chemistry,* Vol. 6, 1987, pp. 435–444.
[15] Jop, K. M., Parkerton, T. F., Rodgers, J. H. Jr., Dickson, K. L., and Dorn, P. B., *Environmental Toxicology and Chemistry,* Vol. 6, 1987, pp. 697–703.
[16] Borthwick, P. W. and Schimmel, S. C. in *Pentachlorophenol, Chemistry, Pharmacology and Environmental Toxicology,* K. R. Rao, Ed., Plenum Publishing Co., New York, NY, 1978, pp. 141–146.
[17] Parrish, P. R., Dyar, E. E., Enos, J. M., and Wilson, W. G., "Chronic Toxicity of Chlordane, Trifluration, and Pentachlorophenol to Sheepshead Minnows *(Cyprinodon variegatus),*" EPA-600/3-78-010, National Technical Information Service, Springfield, VA, 1978.
[18] Ward, G. S. and Parrish, P. R. in *Aquatic Toxicology (Third Conference), ASTM STP 707,* J. G. Eaton, P. R. Parrish and A. C. Hendricks, Eds., American Society for Testing and Materials, Philadelphia, 1980, pp. 243-247.
[19] Mayer, F. L., Mayer, K. S., and Ellersieck, M. R., *Environmental Toxicology and Chemistry,* Vol. 5, 1986, pp. 737-748.
[20] Woltering, D. M., *Aquatic Toxicology,* Vol. 5, 1984, pp. 1-21.

Nutrition and Testing

Kathleen I. Keating,[1] *Paula B. Caffrey,*[1] *and Kathryn A. Schultz*[1]

Inherent Problems in Reconstituted Water

REFERENCE: Keating, K. I., Caffrey, P. B., and Schultz, K. A., "**Inherent Problems in Reconstituted Water,**" *Aquatic Toxicology and Hazard Assessment: 12th Volume, ASTM STP 1027,* U. M. Cowgill and L. R. Williams, Eds., American Society for Testing and Materials, Philadelphia, 1989, pp. 367-378.

ABSTRACT: The assumption that widely reported reconstituted water [*1*] associated problems in cladoceran bioassays are due to unrecorded, unidentified, and unisolated materials is unwarranted. It is possible that distillation/deionization systems could add materials to the waters they produce, but the production of substances of sufficient toxicity in sufficient quantity to exert measurable interference in the culture and testing of Cladocera is rare. The generally reported difficulties with reconstituted water (RCW) [*1*] more likely reflect a problem inevitably associated with RCW or with any simplified medium. By fiat it contains only four salts. The full complement of trace elements demonstrated essential to Eukaryotes is missing. Assuming that this omission will be corrected by unknown additions of trace elements from contaminants either in food materials or in the few salts intentionally added is initially appealing but imprudent.

To satisfy the animal's needs and to assure test validity, trace elements must be both present in sufficient (but not excessive) quantity and consistently available. Evidence, based on work with zinc and selenium, suggests that, at least for these elements, the latter requirement is not readily met by solid foods because the preferred route of uptake is via the liquid in which the animals dwell. With this in mind indications that the U.S. EPA is considering requiring RCW both for maintenance and testing of invertebrates are disturbing. It is likely that other essential trace nutrients will behave in a fashion similar to that of Zn and Se. We have found that employing A-MS algal medium with carefully timed harvest of food algae and planned feeding regimes can minimize the damage introduced by general use of RCW. While there is considerable room for improvement, the nutritionally more complete A-MS offers clearcut advantages over U.S. EPA and O.E.C.D. (Organization for Economic Cooperation and Development) media.

KEY WORDS: daphnids, nutrition, reconstituted water, feeding regime, *Daphnia pulex, Daphnia magna,* algal food, zinc, selenium

Marking and Dawson [*1*] published a medium which they called "reconstituted water" (RCW). It was intended to mimic, in fish bioassays, the various levels of water hardness encountered in nature. It is comprised of only four salts (nine elements: O, H, C, Cl, S, K, Mg, Na, Ca) and could in no sense be considered a substitute for natural water, but it has proven capable of fulfilling its original intent, that of mimicking various water hardnesses in fish bioassays. It appears that fish maintained in RCW obtain the various trace elements essential to their survival from the fish chows and other solid foods provided, and from random, opportune, contamination resulting either from exposure of test systems to ambient air or from leaching of trace elements from the materials employed in holding tanks, vessels, and cement ponds. There is, however, no basis in the literature for assuming that daphnids are as adept as fish at gleaning needed trace nutrients from their solid foods, and the commonplace use of this medium for

[1] Associate professor, research associate, and graduate student, respectively. Department of Environmental Science, Cook Campus, Rutgers, New Brunswick, NJ 08903.

rearing and bioassaying daphnids is not supportable as either best practical technology (BPT) or best available technology (BAT). Thus, in the absence of improvements in common practice (inevitable if that practice becomes a required procedure), the often stated U.S. EPA ultimate goal for analytical procedures, BAT, can not be achieved in bioassay.

Currently, ours happens to be the only lab with the capacity to partition the trace elements entering a daphnid-rearing system via the solid food from those entering via the water. We have observed consistent distinctions in availability between these two presentation modes. Thus, when serious consideration was given at the federal level to the requirement for use of RCW for daphnid rearing and bioassay [*2–4*], we decided to explore daphnid growth in RCW to determine if, in fact, the dichotomy of availability which we observe for certain trace elements when tested in our MS system [*5*] could be demonstrated in RCW.

The recently expressed interest of Donald Mount and his coworkers [*6*] in the use of Perrier additions to daphnid cultures, while often discussed with humor, is encouraging in scientific terms. This is because even this rather unusual natural water would be an improvement over the use of RCW in that it would provide some trace nutrients. However, problems with precipitation when trapped CO_2 is driven off force use of a 10% Perrier/90% distilled water combination for daphnid rearing and testing. Thus, the concentrations of natural trace elements are significantly diminished. Also, Perrier quality assurance would require that the product be consistently suitable for human consumption, but would not require that all dissolved substances be identical in different production runs. Ultimately, a genuinely functional, wholly defined, chemical medium for invertebrate maintenance and testing must be employed if laboratories in every state are to employ similar testing circumstances.

Methods

Animal and algal cultures were maintained microbe-free in all experiments. Excepting in experiments which required that animals be reared in RCW, animals were reared in MS media. All animals were fed algae, *Ankistrodesmus convolutus* Corda and *Nitzschia frustulum* Kützing produced in A-MS medium [*5*]. Algae were harvested after approximately 14 days of culture. These procedures assure that all material additions to the animal's culture circumstances, whether directly (in the animal medium) or indirectly (from the algal medium or via the algae to the animal medium), are certified inorganic salts. The one exception is that vitamin B_{12} is added as Sigma crystalline B_{12}. This practice limits the level of possible trace element contamination to that of the certified salts. Our cleaning procedures and the use of disposable borosilicate class culture vessels introduce considerably less contamination than certified salts. Details of the MS system can be found elsewhere [*5*]. U.S. EPA algal medium [*2*] and OECD medium [*7*] employed in algal growth tests were provided by B. Tepper of U.S. Testing, Hoboken, New Jersey. In acute bioassays animals from the eighth broods of four RCW mothers, or fourth broods of two RCW mothers, were bioassayed, blocked on brood; that is, equal numbers of animals from each brood were placed in each concentration and were identified as to source brood. This permits comparison of the performances of different broods, as well as performances of the entire group of animals, for their capacity to withstand high concentrations of a toxicant.

Methods Specific to Food Versus Water Experiments

All experiments were run in the MS system [*5*]. *Daphnia magna* broodmates were subjected to one of three treatments: some were reared in the standard MS medium (25 ppb Zn) and fed 14-day-old algae harvested from the standard A-MS algal medium (25 ppb Zn); a second group was reared in zinc-free MS and fed 14-day-old algae from zinc-deprived algal cultures; the third group was reared in zinc-free MS, but fed the 14-day-old algae harvested from standard A-MS (25 ppb Zn). Critical zinc concentrations were determined with an atomic absorption spectrometer. Selenium studies were carried out in the MS system; for purposes of these experiments selenium was included in media at 1 ppb; the same array of partial deprivation as employed in the zinc experiments was employed. Additional details of selenium studies can be found [*8*].

Methods Specific to RCW+ Experiments

It was postulated that the feeding of algae (with the medium in which they were produced) from very young algal cultures might represent a method to improve materially the health of animals held in RCW. To assist with the decision as to what food would be provided to animals in these experiments, an algal growth comparison test was run with three media: that recommended by U.S. EPA, that recommended by O.E.C.D., and A-MS. Algae reared in A-MS were selected.

Initially *Daphnia pulex* broodmates were placed in either RCW or MS and fed algae produced in A-MS. This experiment was repeated once. Second generation RCW animals were taken from a third brood of RCW animals. Broodmates were returned to MS as comparison-controls. Based on the observations that a distinction between the two treatments was clearly visible after the fifth brood, it was decided that an imposed improvement in the sixth brood of RCW animals would be a useful means of demonstrating that the health of RCW animals could be improved—to the betterment of any bioassay employing such animals.

At the fourth brood of these second generation animals, the feeding regime was changed. Food algae were harvested from 4-day-old, instead of 14-day-old, cultures. Introduction of the new feeding regime at this time was necessary if the sixth brood were to be affected by the changed feeding regime because it is after the fourth brood that the eggs which will become the sixth brood are selected for maturation in the ovary. These matured eggs are dropped from the ovary into the brood chamber of the animal immediately after the fifth brood is delivered. The resulting embryos mature inside the brood chamber, essentially independent of the mother's condition, into the neonates which comprise the sixth brood. For the remainder of the experiment, all animals (RCW and MS, alike) were fed according to the new feeding regime.

Acute bioassays were run in MS system conditions [*5*] excepting that acute bioassays with animals produced in MS or in RCW media were tested in RCW when this is stated in the text.

Results

Food Versus Water Experiments: I. Selenium

Early in the progress of what proved to be a long-term study of selenium deficiency [*8*], we observed that selenium is more readily taken from the water than from the solid food. When algae carrying selenium are fed to animals, the animals consume the algae and make good use of the calories provided. Yet the animals do not adequately extract the selenium from the algae as the algae pass through their guts and are digested. Additionally, if algae are reared in selenium-deprived circumstances, those deficient algae compete directly with the animal (when the algae are added to the animal culture as foods) for any stray selenium which might enter the "selenium-free" cultures as inadvertent, but inevitable, contamination in the certified salts of other elements. Again, once incorporated by the algae, the selenium is no longer available to the animals. This results in greater evidence of selenium deprivation in animals fed selenium-deprived algae.

This is not a reflection of a loss of selenium which might be provided by the algae were they nondeprived for this element. An increase in the number of selenium-deprived algae induces a proportional increase in the overt manifestations of selenium deprivation in the animals. This proportionality is so consistent that it was at first interpreted [*9*] as an indication of an allelochemical [*10–11*] effect. Allelochemistry was ruled out when no biologically active molecule could be partitioned from the algal materials which introduced the problem. This was a selenium deficiency exaggerated by the algal consumption of the last remnants of selenium in the animal culture. The algae were more efficient than the daphnids at gleaning the last selenium ions from these circumstances of severe deficit.

Food Versus Water Experiments: II. Zinc

When *D. magna* is deprived of zinc, reproduction is obviously diminished (compare Lines A and B, Fig. 1). The loss of reproductive capacity shown by *Daphnia magna*, when zinc is provided solely via the solid foods (algae) (Line C, is quite similar to that of a zinc-deprived animal (Line B). This is true even though the algal food portion (comprised of both algae and the spent algal medium in which those algae were produced) added to the animal culture carries 25 ppb of zinc. Zinc appears to be taken only from solution, not from the solid food, that is, not from the algae. The animals consume the algae and make good use of the calories they provide. Yet, the animals do not adequately extract the zinc from the algal material as it passes through their guts and is digested.

Algal Experiments

Since (1) when compared to the other algal media under consideration, the A-MS formulation contains more of the trace elements known to be required by Eukaryotes generally, and (2) algal growth in A-MS was considerably greater than that in the other media (Fig. 2), all food algae for experiments were produced in A-MS.

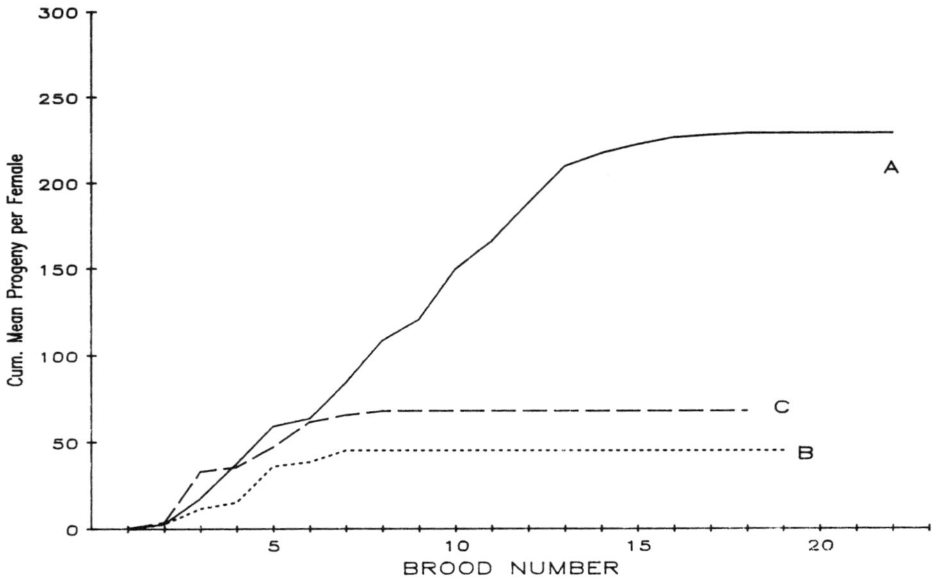

FIG. 1—Daphnia magna *reproduction compared to availability of zinc: food versus water. A: Control: MS medium with 25 ppb zinc. B: Zinc-free animal medium with 0 measurable zinc (MS-Z), fed algae from standard algal medium (A-MS) with 25 ppb zinc; detection limit 0.5 ppb. C: Zinc-deprived: 0 measurable zinc in animal medium and 0 measurable zinc in algal medium.*

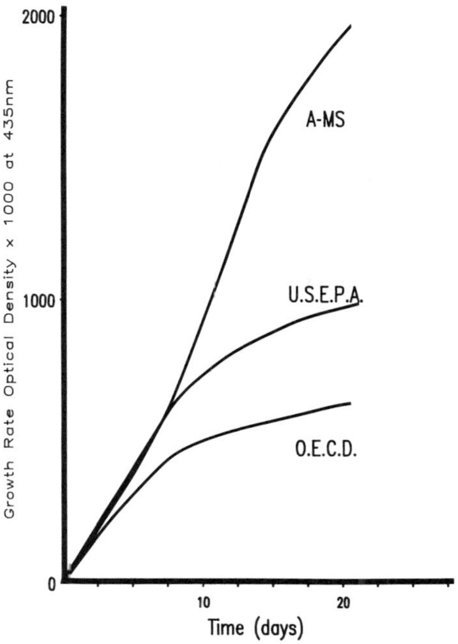

FIG. 2—*Comparison of algal growth:* Ankistrodesmus convolutus *in three media. A-MS* [5]; *U.S. EPA* [2]; *O.E.C.D.* [7].

RCW Chronic Tests

Animals reared in RCW and fed the same diet as broodmates reared in MS were less fecund in the first generation (compare Lines A and B, Fig. 3). Initially, we had difficulty establishing a second generation in RCW because we followed our practice of taking older animals (usually the tenth or later broods from the 15 to 17 broods of our *D. pulex*) to initiate the next generation. These RCW animals were grossly malnourished and exhibited a Lansing [*12–14*] effect. Ultimately, a functional second generation was established with third brood animals (Fig. 4).

RCW+ Chronic Tests

Reproduction of second generation animals were quite similar to that in the first generation (compare the first five broods of Fig. 3 with the first five broods of Fig. 4 for either MS or RCW). In the second generation the feeding regime was changed in a manner which would be reflected in the animal's reproductive capacity at the sixth brood. A difference between reproduction in the two feeding regimes at the sixth brood is evident (compare reproduction after Brood 5 in Figs. 3 and 4).

The coefficient of variation (CV) of MS animals is consistently lower than that of RCW animals. This can be seen in both Figs. 3 and 4. The CV of RCW animals improves greatly after the fifth brood in Fig. 4. This followed the introduction of the change in feeding regime.

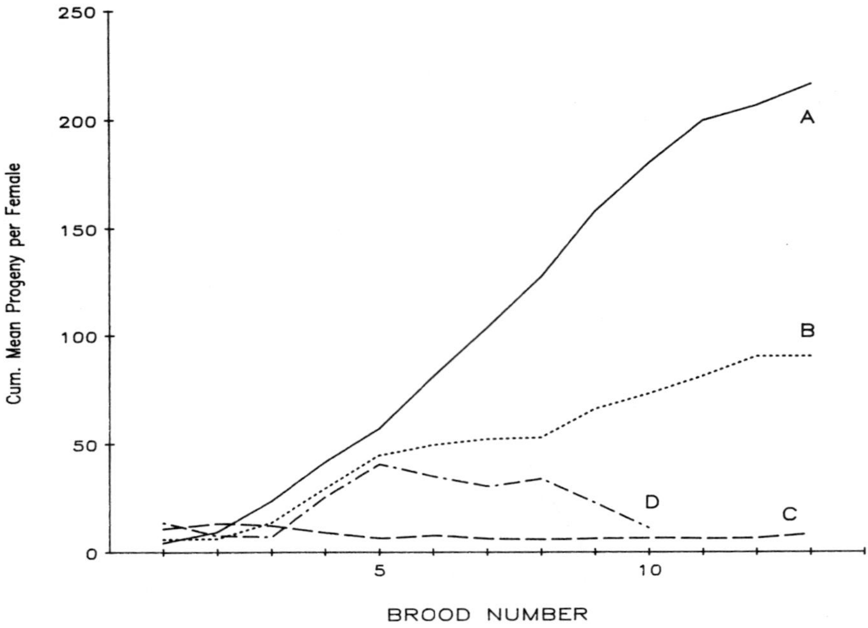

FIG. 3—*Comparison of* Daphnia pulex *reproduction in MS versus RCW: first generation in RCW; broodmates born in MS. A. Cumulative reproduction in MS. B. Cumulative reproduction in RCW (parental group for second generation). C. CV × 100 for A. D. CV × 100 for B.*

Acute Tests

While we do not run conventional bioassays, it is our practice to measure the vigor of animals by cumulative reproduction, by CV level and stability, and by stressing them in acute, 48-h, unfed bioassays. In such a test, animals produced and tested in RCW appeared to respond to excessive zinc concentrations (presented as chloride) with an LC_{50} similar to that of the control (MS-reared) animals whether tested in RCW or MS. For control animals the threshold of death due to zinc concentration is approximately 2500 ppb—no deaths at or below 2000 ppb; 50% deaths at 2500 ppb; 100% deaths at 3000 ppb and above. Since animals from RCW mothers at all concentrations 2500 ppb and below, including no-add controls tested in RCW, and an extra set of no-add controls in MS, died irregularly during the 48-h test, results from RCW animals would not be valid.

It was clear that some of the RCW animals were active at 2500 ppb after 48 h, and that none above that concentration survived for more than a few hours. Animals from the eighth brood of an RCW mother were more irregular than those from a fourth brood. Since fourth brood controls survived, the test could technically be considered valid if limited to results produced by fourth brood animals. It would, however, have given a "calculated" LC_{50} below the 2500-ppb figure because many deaths were dispersed among the lower concentrations. We consider 2500 ppb more accurate than some calculated lower number. The irregular performance of RCW animals reflected something other than zinc concentration.

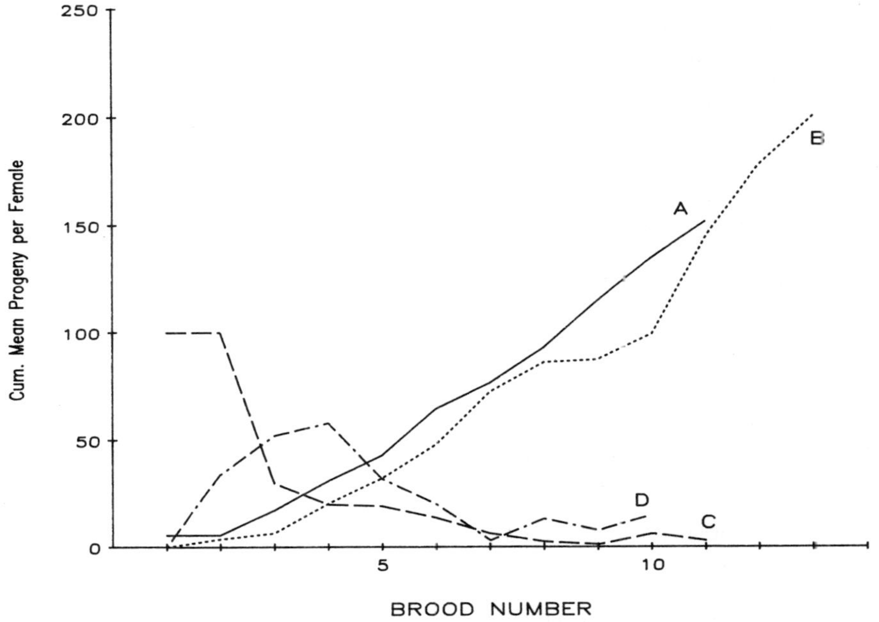

FIG. 4—*Comparison of* Daphnia pulex *reproduction in MS versus RCW/RCW+: before and after introduction of new feeding regime. Second generation in RCW; broodmates born in RCW. A. Cumulative reproduction in MS; animals born in RCW and returned to MS. B. Cumulative reproduction in RCW, second generation in RCW. C. CV × 100 for A. D. CV × 100 for B.*

Discussion

That so many trace elements are essential relates to the complex set of metabolic or biochemical pathways which we are only now beginning to understand [*16*]. A few differences among the multitudes of living organisms depend on unique pathways—the presence or absence of chlorophyll or the capacity for anaerobic respiration or nitrogen fixation—but only a few. The existence of every pathway need not be proven for each species or even for each phylum. Daphnid trace element requirements will ultimately prove similar to those of other Eukaryotes. Recent reviews [*17–18*] include lists of essential trace elements (Table 1). Daphnids evolved in habitats filled with natural water and even the most imbalanced natural waters, for example, naturally carbonated from a well, are likely to carry all of the essential trace elements. Sometimes a water

TABLE 1—*Required inorganic nutrients.*

BULK: H, O, C, N, Ca, P, Na, K, S, Cl, Mg
TRACE: Fe, I, Cu, Zn, Mn, Co, Cr, Mo, Ni, V, Se, As, F, Si, Li
LIKELY: Cd, Pb, Sn, Rb

NOTE: Table adapted from Ref *17* and *18*.

has a poor mix for one creature or another; a natural population of those creatures will probably not be found in that water.

For any essential nutrient there can be too much or too little present. When too much of some inorganic micronutrient is added to a culture, it is referred to as a toxic and eliminated. In contrast, when too little is added, no one realizes it is missing. A most helpful description of the concept of the total dose/response of an organism to any substance is presented by Walter Mertz [*19*]. Essentially, there is a deficiency response at low concentrations, an optimum response (for example, maximum sustainable reproduction) in the central range of concentrations, a negative response at high concentrations, and a final response, death, at toxic concentrations. Environmental toxicology has been devoted to this final response range and has been feigning blindness regarding all of the others.

Food Versus Water Experiments

Just as sufficient calories will not assure a balanced diet, the mere presence of trace nutrients will not assure their availability. There is an assumption that RCW will serve because all unidentified trace requirements will come into cultures as contaminants in food. Many would argue that this works—and "we shouldn't fix what ain't broke!" Sadly, this one is broken. The evidence offered in Figs. 1 and 4 from our earlier studies of selenium requirements and from our current studies of other trace elements strongly supports this conclusion. It appears that, although present in adequate amounts in the solid portion of the diet, certain trace inorganics are somewhat unavailable to the animal. This could provide explanation for much of the unpredictability of bioassay cultures because attention is rarely paid to the trace elements dissolved in the water. Marking and Dawson designed RCW to provide macroinorganic requirements for fish, which they assumed would obtain all other essential nutrients from standard commercial fish chows. In terms of incorporating trace elements (availability), daphnids may differ significantly from fish. Our studies of selenium and zinc indicate that at least these essential microinorganics, possibly several others, should be offered from the solution in which the animal dwells rather than from solid foods.

Common wisdom has been that aquatic animals, like terrestrial animals, obtain their trace element needs from solid food materials. This may simply be one more example of the egocentricity of we humans since the assumption was that all creatures follow our pattern. If the evolutionary history of the daphnids is considered, it is much simpler to assume that they never developed the complicated metabolic pathways required to wrest trace element ions from their incorporated positions in complex organic molecules (for example, phytates, oxalates, and or mucopolysaccharides). The underlying basis for the lessened availability of selenium, zinc, or any other metal or metalloid is probably somewhat idiosyncratic to each plant-animal pair, and no studies concerning the daphnids have yet been published. However, considerable work concerning the production of metal-complexing, intracellular metabolites (especially phytates) which are recalcitrant to terrestrial animal digestive processes has been published [*20*], and daphnids appear to have even greater digestive limitations.

Comparison of RCW and RCW+ Chronic Results

The procedures employed in commercial bioassay labs easily mask the obvious problems presented by deprived animals. Mothers which produce test animals are rarely monitored as individuals; none are monitored for even half of their lives. There are no requirements, nor are there voluntary efforts, to compare performances of animals reared in RCW with broodmates reared in more adequate circumstances. Because RCW animals are routinely held for only five broods, the problems indicated in Fig. 3 are avoided. Five broods represent 20 to 30% of a normal lifespan for *Daphnia pulex* or *D. magna* (less for *Ceriodaphnia dubia*) [*5,21*]. The limitation of "chronic" daphnid tests to five broods, or 21 days, is rationalized both by a claimed absence of additional useful information after that time and by the assumption that wholesale deaths after the fifth brood will often occur. This is strong evidence that animals are routinely sick.

Increased numbers of progeny may indicate increased health, but the number of progeny is often controlled by such factors as the number of available calories which can be invested in progeny production or by such physical factors as vessel dimension or temperature, rather than by the health status of the producing mother. CV is a more reliable indicator of basic differences in animal health, thus, in animal test performance.

The CV of MS animals is consistently low because the standard deviation of MS animal reproductive performance increases only in proportion to the increase in the number of progeny not in proportion to the mother's age. The CV of RCW reproduction, in contrast, increases as the animal's exposure to RCW increases, that is, as the mother matures. Thus, the standard deviation and CV of reproduction from animals reared in RCW will be higher in general and will increase with age. The increased variability associated with RCW is sufficiently important after the fifth brood (mothers often die without apparent cause) to be considered as justification for ending "chronic" tests at that early age.

In Figs. 3 and 4 CV is expressed as percent to allow it to be drawn to scale on the graph representing mean cumulative reproduction. This facilitates both examination of the change in CV with each passing brood and comparison between reproductive quality (CV) and quantity (mean progeny). By comparing the CVs of reproduction in the two animal media, it can be seen that (Fig. 3) the CV of RCW animals increases after Brood 3. They give evidence of a stress not experienced by the MS animals. At Brood 6 (which is comprised of animals which began to mature in the ovary just after the production of Brood 3) the reproduction of RCW animals drops by comparison to that of MS animals. Note that the CV of the RCW animals is also more irregular. These distinctions become more pronounced as the animals age. There are additional, more subjective, differences between broodmates reared in these two sets of circumstances; for example, animals in RCW are sluggish, eat poorly, and die younger on the average.

After the fifth brood clear-cut distinctions in both CV and reproduction can be readily identified in Fig. 3. This fifth brood "break point" is eliminated in Fig. 4; second generation animals (from the third brood of a first generation animal) in RCW versus broodmates *returned* to MS. The original (their mother's) pattern of cumulative reproduction for RCW occurs only in the first five broods. The CV for MS animals returns to its original usual low level as the animals (born to RCW mothers in an RCW culture) acclimate to the improved circumstances. In contrast to Fig. 3, however, after the fifth brood, instead of dropping away from the MS curve, the cumulative progeny curve for RCW+ accompanies it.

The CV for the RCW+ animals in Fig. 4 is even more telling. It shows an immediate positive response to the improved conditions resulting from the new feeding regime which offered the animal trace elements in solution. This new diet was comprised of portions of four-day-old algal cultures including both the cells and the incompletely spent algal medium (which carries trace elements in solution). Thus, the change in feeding regime offered immediate, quantifiable improvement in the general health of the daphnids in RCW. The resultant RCW+ animals were not actually as healthy as MS animals and most died quite a bit earlier. One, however, did live to its thirteenth brood. Nevertheless, their reproduction did greatly improve. The change had no visible effect on MS animals.

There is no secure proof of a specific threshold concentration for daphnids for trace nutrients other than B_{12} [*22*], selenium [*8*], and zinc (we have identified 10 to 15 ppb as a desirable minimum, unpublished). Thus, it is unlikely that any commercial lab would be willing to establish minimum required concentrations for other trace nutrients. This means that, although the need for many trace elements is certain, these trace elements will not be included in many test/rearing systems. The only solution to this problem at this time is to foster the use of a reasonable natural water (not an imbalanced, precipitation-prone one) because this will assure the presence of minimal concentrations of all trace elements. The worst-case scenario would be the mandated general use of Marking and Dawson's RCW, because it is exceptionally inadequate in terms of trace elements.

Recently, U.S. EPA and Battelle [*15*] did a survey and concluded that the problems associated with *Ceriodaphnia* culture seemed, rather consistently, to show up in labs employing RCW. It was suggested that the widespread problems were associated with some toxic substance coming out of many stills in many labs. No identification or isolation has been generally attempted. Unless Perrier is an antitoxin (or a detergent) this explanation is odd because a 10% addition of Perrier detoxifies the distilled waters from all those "dirty" stills [*6*]. Since Perrier is a natural well water, with all the trace element inclusions of any natural water, it is more reasonable that the addition of Perrier is helpful because it adds small quantities of missing essential trace elements to RCW.

Diluting Perrier to one tenth its normal TDS (total dissolved solids) cuts the major elements which are present in relatively high concentrations. Unfortunately, when TDS is diluted not only are macroinorganics diluted (a positive), but also all microinorganics are diluted (a negative). Thus, some other natural water which could be employed at concentrations closer to normal strength would be preferable. The natural water employed need not generate the highest reproduction possible. A desirable water which happened to be oligotrophic would support fewer animals and algae than that which would be supported by many eutrophic waters. This does not reflect the health of the animal. Daphnids often respond in nature by producing numbers of progeny appropriate to the circumstances in their environment.

Acute Tests

In acute circumstances RCW neonates from a fourth brood responded to added zinc chloride more consistently than those from an eighth brood. Thus, even for acute testing purposes, the general practice in commercial bioassay labs to discard animals after the fifth brood is desirable. The increased variability of performance is the inevitable result of the damage to RCW animals which increases with age, and of the same, previously discussed, irregular and high CV in chronic circumstances.

When deaths in acute bioassays are reported as dispersed around the LC_{50}, statistical manipulations are used to incorporate them into the LC_{50}. In fact a myth that "partial kills" are desirable has evolved. This is highly questionable. In acute tests our experience has been that, providing test concentrations are very close to each other, healthy animals die at a single concentration, showing very little variation in vigor. Since the selected concentrations in acute bioassays are quite distinct, each concentration being a doubling of the last, the dispersal of deaths among several concentrations in acute tests is a sign of undesirable variability as much as a response to the toxicant.

It could be argued that since our zinc-deprived animals offer LC_{50}s similar to those offered by healthy animals, it does not matter if test animals are deprived. In fact, there may be a recurrent pattern of this sort because very rapid death by poisoning (acute bioassay) is likely to occur at some unreasonable concentration whether an animal is healthy or sick. However, similar results with one (zinc chloride) or a dozen toxics cannot assure similar results for the next toxic [*23*]. It should also be noted that the use of animals recognized as sick could never be characterized as BAT, nor BPT.

While RCW should not be used for rearing test daphnids, it might serve as a functional acute test medium for some materials. We have observed, however, at least one situation in which the ready precipitation of RCW interfered with acute testing. Very high concentrations of copper (as chloride) were being studied. In RCW 100% of MS-reared test animals were active after 48 h of unfed exposure to 1500 ppb of copper. None were alive above that concentration. Those animals dying during the test died as the RCW produced increasing amounts of precipitate. Precipitate production was proportional to the amount of copper chloride added to each test replicate. Deaths appeared to be due to the precipitate since no animals exhibited the split open carapace which we have learned to anticipate when a daphnid dies as the result of a toxic exposure.

Summary

Comparison of broodmates reared in MS and RCW indicates that for RCW animals overt signs of reduced fecundity appear after the fifth brood and increased variability appears after the third. In contrast, the same parameters improve in MS animals at these points. The feeding and activity of RCW daphnids is less vigorous than that of MS animals. Consecutive generations in RCW can only be maintained by selecting the next generation from an early brood. Responses of RCW daphnids in acute tests are also variable. In both acute and chronic circumstances, variability increases with the ordinal position of the brood from which test animals are taken.

Daphnids are considerably better equipped to take essential trace quantities of (at least) selenium, zinc, and B_{12} from the solution in which they dwell than from solid foods.

Compensating for the Inadequacies of RCW

Avoidance procedures are suggested. The amount of time during which animals must be kept in RCW prior to testing (a required acclimation period) must be kept at a minimum. There is a "carryover" of trace elements which can be advantageous. An animal which (a) itself has spent the beginning of its life in reasonable trace element circumstances, or (b) (less desirable) is born to a mother who has spent her life in reasonable trace element circumstances, will suffer in proportion to the length of time the animal has been *out* of the suitable circumstances. It is helpful that tests are only required for a few broods (three for *Ceriodaphnia dubia;* five for the larger daphnids). The feeding regime tested in our RCW+ experiments will permit some trace elements to be incorporated into an adequate solution in which the animal must be held for extended periods.

The Value of Inherent Inadequacy

The greatest selling point of RCW is that the pH, or hardness, can be set by manipulating the relative concentrations of the four salts of which it is comprised. This makes the medium of chemical, not biological, value. RCW is also favored because every lab in the country could use the same water for tests if they all used it. This offers more bureaucratic, than biological, legitimacy. Additionally, maintaining animals in this medium is supposed to be convenient because RCW is easy to assemble. To the extent that it helps to eliminate stocks, it would cut back on the work required to take care of stocks.

The only evidence currently extent suggests that solid foods are not a good source of trace elements but that water is. There are many open questions concerning minimum requirements for most elements. Wholly ignoring the daphnid's need for trace elements is questionable science. Daphnids, as Eukaryotes, require an extensive array of trace nutrients. Considering the current state of knowledge, compensating for the inadequacy of RCW by providing trace nutrients via the solid food materials rather than as solutes in the animal's culture medium presents a serious procedural flow.

Acknowledgements

We wish to acknowledge the assistance of George Doza with the determination of zinc concentrations. This work was supported in part by the New Jersey Department of Environmental Protection Spill Fund and in part by the New Jersey Agricultural Research Station, NJAES Publication Number D-07496388.

References

[1] Marking, L. L. and Dawson, V. K., "Toxicity of Quinaldine Sulfate to Fish," *Investigations Fish Control,* No. 48, U.S. Fish and Wildlife Service, Dept. of the Interior, Washington, DC, 1973.
[2] Beisinger, K. E., Williams, L. R., and van der Schalie, W. H., "Procedures for Conducting *Daphnia Magna* Toxicity Bioassays," EPA/600/8-87/011, Government Printing Office, U.S. EPA, 1987, pp. viii and 57.
[3] Peltier, W. H. and Weber, C. I., "Methods for Measuring the Acute Toxicity of Effluents to Freshwater and Marine Organisms," 3rd ed., EPA/600/4-85/-13, Government Printing Office, U.S. EPA, National Technical Information Service, 1985, pp. xv and 216.
[4] Hornig, W. B. and Weber, C. I., "Short-Term Methods for Estimating the Chronic Toxicity of Effluents and Receiving Waters to Freshwater Organisms," EPA/600/4-85/014, Government Printing Office, U.S. EPA, National Technical Information Service, 1985, pp. x and 162.
[5] Keating, K. I., "A System of Defined (Sensu Stricto) Media for Daphnid (Cladocera) Culture," *Water Research,* Vol. 19, 1985, pp. 73–78.
[6] Mount, D. I., Keen, R. E., and Taraldsen, J., "A Reference Test Water for Cladocerans," 11th Symposium on Aquatic Toxicology and Hazard Assessment, Abstracts, p. 34, 1987.
[7] "*Section 2: Effects on biotic systems No. 201, Guidelines for Testing of Chemicals,*" Organization for Economic Cooperation and Development, Paris, France, 1981.
[8] Keating, K. I. and Dagbusan, B. C., "The Effect of Selenium Deficiency on Cuticle Integrity in the Cladocera (Crustacea)," *Proceedings of the National Academy of Sciences,* Vol. 81, 1984, pp. 3433–3437.
[9] Vedagiri, U., "The Effect of the Presence in Culture of *Nitschia frustulum* on the Growth and Reproduction of *Daphnia pulex,* Master's thesis, Rutgers University, New Brunswick, NJ, 1982.
[10] Keating, K. I., "Allelopathic Influence on Blue-Green Bloom Sequence in a Eutrophic Lake," *Science,* Vol. 196, 1977, pp. 885–887.
[11] Keating, K. I., "Blue-Green Algal Inhibition of Diatom Growth: Transition from Mesotrophic to Eutrophic Community Structure," *Science,* Vol. 199, 1978, pp. 971–973.
[12] Lansing, A., "Calcium and Growth in Aging and Cancer," *Science,* Vol. 106, 1947, pp. 187–188.
[13] Lansing, A., "A Transmissible, Cumulative, and Reversible Factor in Aging," *Journal of Gerontology,* Vol. 2, 1947, pp. 228–239.
[14] Murphy, J. and Davidoff, M., "The Results of Improved Nutrition on the Lansing Effect in *Moina macrocopa,*" *Biological Bulletin,* Vol. 142, 1972, pp. 302–309.
[15] Degreave, M., *New England Interstate Water Pollution Control Commission,* Andover, MA, 1986.
[16] Martin, D. W., Jr., Mayes, P. A., Rodwell, V. W., and Granner, D. K., *Harper's Review of Biochemistry,* 20th ed., Lange Medical Publications, Los Altos, CA, 1985, pp. xiv and 718.
[17] Frieden, E. in *Biochemistry of the Essential Ultra Trace Elements,* E. Frieden, Ed., 1984, Chap. 1, pp. 1–15.
[18] Schrauzer, G. in *Bichemistry of the Essential Ultra Trace Elements,* E. Frieden, Ed., 1984, Chap. 2, pp. 17–31.
[19] Mertz, W., "The Essential Trace Elements," *Science,* Vol. 213, 1981, pp. 1332–1338.
[20] Prasad, A., "Clinical and Biochemical Spectrum of Zinc Deficiency in Human Subjects," *Current Topics in Nutrition and Disease,* Vol. 6, A. Prasad, Ed., 1982, Chap. 1, pp. 3–62.
[21] Cowgill, U., Keating, K., and I. Takahashi, "Fecundity and Longevity of *Ceriodaphnia dubia/affinis* at Two Different Temperatures," *Journal of Crustacean Biology,* Vol. 5, 1985, pp. 420–429.
[22] Keating, K. I., "The Influence of B_{12} Deficiency on *Daphnia pulex* de Geer (Cladocera) Reproduction," *Journal of Crustacean Biology,* Vol. 5, 1985, pp. 130–136.
[23] Keating, K. I. and Griffith, J., "The pH Tolerance Range of Daphnids," *Society Environmental Toxicology and Chemistry Symposium,* Hudson Delaware Chapter, Abstracts Volume, 1986.

U. M. Cowgill[1] *and D. P. Milazzo*[1]

The Culturing and Testing of Two Species of Duckweed

REFERENCE: Cowgill, U. M. and Milazzo, D. P., **"The Culturing and Testing of Two Species of Duckweek,"** *Aquatic Toxicology and Hazard Assessment: 12th Volume, ASTM STP 1027*, U. M. Cowgill and L. R. Williams, Eds., American Society for Testing and Materials, Philadelphia, 1989, pp. 379–391.

ABSTRACT: The objective of this study was to develop rearing conditions and a successful long-term culture medium for the maintenance of *Lemna gibba* L. G-3, *Lemna minor* L. clone 6591 (California), 7102 (= LMS) (Kansas), 7101 (LMY) (Connecticut), and 7136 (46) (Illinois). A variation of Hoagland's medium was employed. Growth of *L. gibba* G-3 was found to become enhanced by additions of selenium, (4.2 μgL^{-1}, Na_2SeO_4 = 100 μgL^{-1}), tin (456.6 μgL^{-1}, $SnCl_4$ = 10 mgL^{-1}), cobalt [20.3 μgL^{-1}, $Co(No_3)_2$ 6 H_2O = 1 mgL^{-1}], and vanadium (25.6 μgL^{-1}, $VOSO_4$ 2 H_2O = 1 mgL^{-1}). Vanadium failed to elicit any increase in growth on addition to Hoagland's medium from *L. minor* 7102, 7101, or 7136. There was no response shown by *L. minor* 6591 7136, and 7101 to Co nor by *L. minor* 6591 and 7136 to Sn. The 7136 clone failed to respond to addition of Se. Root elongation, biomass (dry weight at 60°C), number of fronds, number of plants, Kjeldahl nitrogen, and percent chlorophyll a and b are among the endpoints examined. A comparison of these endpoints is presented. All biomass endpoints (growth) are highly related, and dry weight should be used as a final endpoint.

KEY WORDS: duckweed, *Lemna gibba*, *Lemna minor*, nutrition, growth medium, selenium, cobalt, tin, vanadium

Lemna spp are the organisms of choice [*1,2*] for the detection of the phytotoxic effects of chemicals introduced into aquatic environments. The genus is ubiquitous and is an integral portion of the food chain of waterfowl, fishes, and aquatic mammals. As a result, plants belonging to this genus have gained prominence as test organisms in biomonitoring. The purpose of the present study is to establish and describe a successful rearing system and culture medium.

Materials and Methods

Axenic samples of *Lemna gibba* G-3, duckweed, were procured from Dr. C. F. Cleland of the Smithsonian Institution of Washington, D.C. on 23 Feb. 1987. Stock cultures of this organism have been maintained axenically by weekly transfer of 15 fronds into fresh sterile medium.

On 30 Mar. 1987, axenic cultures of *Lemna minor* L. were received from Professor E. Landolt of the Geobotanisches Institut in Zürich, Switzerland. Four clones were received: 6591 (California), 7102 (= LMS) (Kansas), 7101 (LMY) (Connecticut), and 7136 (46) (Illinois) [*3*]. These clones have been maintained axenically by weekly transfer of 30 fronds into fresh sterile medium.

Parenthetically, it is interesting to note that *L. gibba* grows only in the southwestern portion of the United States, while *L. minor* is ubiquitous in the United States.

[1] Associate environmental consultant and research chemist, respectively, The Dow Chemical Co., Mammalian and Environmental Toxicology, 1702 Building, Midland, MI 48674.

The culture conditions are listed in Table 1. The initial growth medium is Hoagland's, which was further developed by Hillman [4] and revised by Cleland and Briggs [5] (Table 2).

There are four trace elements thought to be necessary for plant growth [6], the optimum quantities of which are yet to be determined for *Lemna* spp. These are selenium (Se), cobalt (Co), tin (Sn) and vanadium (V). Cobalt, Se, Sn, and V are involved in a variety of biochemical processes in lower plants, and generally deficiency symptoms have been produced in some higher plants. To estimate the optimum amounts of these elements, laboratory-acclimated plants were exposed to a concentration series for seven days. The compounds used were sodium selenate (Na_2SeO_4), cobalt nitrate $Co(NO_3)_2$ 6 H_2O, stannic chloride ($SnCl_4$), and vanadyl sulfate ($VOSO_4 \cdot 2\,H_2O$). The initial series used concentrations an order of magnitude apart; that is, 0.1, 1.0, 10.0, 100, 1000 μgL^{-1}, etc. Control solutions were examined for the presence of Se, Co, Sn, and V. None of these elements were detected at a level of 0.10 μgL^{-1}. It was therefore presumed that the growth solution would not compromise this nutrition study. The concentration series were developed employing a six-place balance and are to be considered nominal.

Seven endpoints were selected. These were root elongation, biomass (dry weight to constant weight at 60°C), number of fronds, number of plants, percent chlorophyll a, percent chlorophyll b, and percent Kjeldahl nitrogen.

Root elongation was measured with the aid of a metric ruler and a dissecting microscope.

A method was developed for the extraction and analysis of chlorophyll a and b from duckweed and will be reported in detail elsewhere [7]. Suffice to say, this method utilizes high performance liquid chromatography (HPLC); the usual (spectrophotometric) procedure tends to overestimate chlorophyll a and underestimate chlorophyll b due to the overlapping of their respective absorbance spectra [8].

TABLE 1—*Culturing conditions of two species of* Lemna.

Temperature, °C	25 ± 2
Light, lux	
L. gibba	6461 ± 323
L. minor	5385 ± 323
Medium	Hoagland's E+ (cf. Table 2)
pH	4.8 to 5.2
Vessel, mL	250, glass Erlenmeyer Shimadzu closure
Vessel content, mL	100
Habitat maintenance	
Septic conditions	Axenic
Frequency of renewal	Weekly
Number of fronds/vessel	*L. gibba* 15 *L. minor* 30
Number of plants/vessel	*L. gibba* 5 *L. minor* 10
Acclimatization, weeks	8
Health criteria	
Plants	Increase of threefold in 7 days
Fronds	Increase of threefold in 7 days
Endpoints	Number of fronds Number of plants Root length Biomass (dry wt at 60°C) Chlorophyll a Chlorophyll b Kjeldahl N

TABLE 2—*Chemical composition of Hoagland's E+ medium* [4,5].

Substance	Stock Solution, gL^{-1}	Growth Medium, $mL\ L^{-1}$
A. $Ca(NO_3)_2\ 4\ H_2O$	59.00	20
KNO_3	75.76	
KH_2PO_4	34.00	
6 mL 6 NHCl		
B. Tartaric acid	3.00	1
C. $FeCl_3\ 6\ H_2O$	5.40	1
D. EDTA	9.00	1
8 mL 6 N KOH		
E. $MgSO_4\ 7\ H_2O$	50.00	10
F. H_3BO_3	2.86	1
$ZnSO_4\ 7\ H_2O$	0.22	
$Na_2MoO_4\ 2\ H_2O$	0.12	
$CuSO_4\ 5\ H_2O$	0.08	
$MnCl_2\ 4\ H_2O$	3.62	
G. Sucrose		10.00 g
H. Yeast extract		0.10 g
I. Bactotryptone		0.60 g

Make growth medium up to 1 L with glass distilled water. Adjust the pH to 4.60 with KOH/HCl. Autoclave 20 min at 121°C and 124.1 kPa.

NOTE: EDTA = ethylenediamine tetraacetic acid.

Statistical procedures involved the use of linear correlation coefficients (*r*) and the use of chi-square when means were compared. Usually, the highest growth peak in a series was selected and then verified by exposing plants to concentrations below the highest growth peak as well as above. Once the best concentration for each element was established, mixtures were tested that utilized both the peak obtained with single element testing as well as quantities above and below these peaks. *Lemna gibba* is used as a specific example in this paper, while only final results, that is, those that provided consistently the best growth response for the four clones of *Lemna minor,* are shown.

Results

Tables 3 to 11 show the data for *L. gibba* G-3, and Tables 12 to 15 contain the *L. minor* information. Tables 3 (Se), 5 (V), 7 (Co), and 9 (Sn) show *L. gibba*'s response to varying concentrations of Se, V, Co, and Sn separately, and the intervening tables (4, 6, 8, and 10) show the linear correlation coefficients calculated between various pairs of endpoints. The response to Se, as measured by the number of fronds, the number of plants, the number of roots, dry weight, and total root length is greatest at 4.2 μgL^{-1}. Percent N and chlorophyll a and b fail to show any enhancement. The correlation coefficients calculated among the number of plants, the number of fronds, the number of roots, the dry weight, and the total root length are all highly statistically significant at the 0.1% level or better. This observation suggests that this collection of variables measures the same parameter, namely growth. The percent N fails to show any statistically significant relationship with any growth variable (cf. Tables 4, 6, 8) except for Sn (cf. Table 10). Chlorophyll a exhibits a slightly significant response with the number of fronds, the number of roots, and the total root length, but chlorophyll b lacks concordance with these observations. These results essentially imply the presence of three types of endpoints: growth as measured by the number of fronds, the number of plants, the number of roots, dry

TABLE 3—*Endpoint comparison for* L. gibba *exposed to varying concentrations of selenium.*

Quantity Se, μgL^{-1}	Fronds	Number Plants	Roots	Dry Wt, g	Total Root Length, mm	N, %	Chlorophyll, μgg^{-1} a	b
ND(0.1)[a]	306	81	221	0.09345	2919	5.52	3872	1085
0.42	407	113	295	0.13436	2674	5.12	4035	1256
4.2	487	126	375	0.16183	5145	5.36	5057	1410
41.8	360	85	251	0.10476	3293	5.44	4052	1214
417.9	307	69	223	0.09534	2952	5.48	5676	1526
4 179	279	56	202	0.09430	2608	5.00	4955	1452
41 790	42	27	24	0.02445	290	4.74	1752	1683

[a]Not detected at level noted in parenthesis.

TABLE 4—*Linear correlation coefficients* (r) *relating the various endpoints for selenium.*

	Quantity	Fronds	Plants	Roots	Dry Wt	Root Length	N	Chlorophyll a	b
Quantity	1.000	−0.901	−0.744	−0.879	−0.849	−0.819	−0.802	−0.875	+0.756
No. fronds	−0.901	1.000	+0.951	+0.997	+0.987	+0.990	+0.653	+0.717	−0.574
No. plants	−0.744	+0.951	1.000	+0.954	+0.960	+0.945	+0.552	+0.504	−0.568
No. roots	−0.879	+0.997	+0.954	1.000	+0.993	+0.997	+0.634	+0.719	−0.535
Dry wt	−0.849	+0.987	+0.960	+0.993	1.000	+0.987	+0.544	+0.689	−0.486
Rt length	−0.819	+0.990	+0.945	+0.997	+0.987	1.000	+0.648	+0.723	−0.508
N	−0.802	+0.653	+0.552	+0.634	+0.544	+0.648	1.000	+0.629	−0.647
Chl a	−0.875	+0.717	+0.504	+0.719	+0.689	+0.723	+0.629	1.000	−0.197
Chl b	−0.756	−0.574	−0.568	−0.535	−0.486	−0.508	−0.649	−0.197	1.000
$P =$	0.1	0.05	0.02	0.01	0.001				
$r =$	0.6694	0.7545	0.8329	0.8745	0.95074				

P = probability.

TABLE 5—*Endpoint comparison for* L. gibba *exposed to varying concentrations of vanadium.*

Quantity V, μgL^{-1}	Fronds	Number Plants	Roots	Dry Wt, g	Total Root Length, mm	N, %	Chlorophyll, μgg^{-1}	
							a	b
ND(0.1)[a]	306	81	221	0.09354	2 919	5.52	3872	1085
0.26	782	314	385	0.16855	5 512	6.10	5139	1385
2.6	873	466	441	0.23475	5 924	5.23	4864	1439
25.6	2022	678	706	0.32190	11 381	5.00	3663	1203
256	692	327	445	0.17770	5 976	5.45	4192	1148
2 560	599	245	385	0.16262	3 567	5.37	3217	1048
25 600	352	172	199	0.07953	934	5.24	1800	663

[a]Not detected at level noted in parenthesis.

TABLE 6—*Linear correlation coefficients* (r) *relating the various endpoints for vanadium.*

	Quantity	Fronds	Number Plants	Roots	Dry Wt	Root Length	N	Chlorophyll a	Chlorophyll b
Quantity	1.000	−0.482	−0.571	−0.706	−0.701	−0.694	−0.213	−0.840	−0.884
No. fronds	−0.482	1.000	+0.947	+0.950	+0.940	+0.952	−0.418	+0.213	+0.399
No. plants	−0.571	+0.947	1.000	+0.951	+0.977	+0.925	−0.427	+0.320	+0.513
No. roots	−0.706	+0.950	+0.951	1.000	+0.976	+0.971	−0.360	+0.353	+0.524
Dry wt	−0.701	+0.940	+0.977	+0.976	1.000	+0.954	−0.393	+0.412	+0.604
Root length	−0.694	+0.952	+0.925	+0.971	−0.954	1.000	−0.273	+0.500	+0.592
N	−0.213	−0.418	−0.427	−0.360	−0.393	−0.273	1.000	+0.507	+0.317
Chl a	−0.846	+0.213	+0.320	+0.353	+0,412	−0.500	+0.507	1.000	+0.962
Chl b	−0.884	+0.399	+0.513	+0.524	+0.604	+0.592	+0.317	+0.962	1.000
$P =$	0.1	0.05	0.02	0.01					
$r =$	0.6694	0.7545	0.8329	0.8745	0.95074				

TABLE 7—*Endpoint comparison for* L. gibba *exposed to varying concentrations of cobalt.*

Quantity Co, μgL^{-1}	Fronds	Number Plants	Roots	Dry Wt, g	Total Root Length, mm	N, %	Chlorophyll, μgg^{-1} a	b
ND(0.1)[a]	306	81	221	0.09354	2919	5.52	3872	1085
0.20	1325	530	599	0.24572	9281	5.42	2951	1039
2.0	584	248	81	0.16076	4484	7.17	4204	1239
20.3	659	307	356	0.16644	5127	5.80	4754	1211
202.5	425	171	270	0.10845	3285	5.70	5319	1412
2 025	383	149	230	0.10224	2845	5.46	4646	1328
20 250	265	82	141	0.07157	1125	5.23	2485	854

[a]Not detected at level noted in parenthesis.

TABLE 8—*Linear correlation coefficients* (r) *relating the various endpoints for cobalt.*

	Quantity	Fronds	Number Plants	Roots	Dry Wt	Root Length	N	Chlorophyll a	b
Quantity	1.000	−0.481	−0.554	−0.574	−0.603	−0.608	−0.426	−0.676	−0.759
No. fronds	−0.481	1.000	+0.988	+0.985	+0.972	+0.983	+0.040	−0.242	−0.097
No. plants	−0.554	+0.988	1.000	+0.975	+0.986	+0.977	+0.111	−0.144	−0.015
No. roots	−0.574	−0.985	+0.975	1.000	+0.962	+0.992	−0.0003	−0.122	+0.003
Dry wt	−0.603	+0.972	+0.986	+0.962	1.000	+0.983	+0.186	−0.116	+0.008
Root length	−0.608	+0.983	+0.977	+0.992	+0.983	1.000	+0.109	−0.115	−0.077
%N	−0.426	+0.040	+0.111	−0.0003	+0.186	+0.109	1.000	+0.312	+0.365
Chl a	−0.676	−0.242	−0.144	−0.122	−0.116	−0.115	+0.312	1.000	+0.952
Chl b	−0.759	−0.097	−0.015	+0.003	+0.008	−0.077	+0.365	+0.952	1.000
$P =$	0.1	0.05	0.02	0.01	0.001				
$r =$	0.6694	0.7545	0.8329	0.8745	0.95074				

TABLE 9—*Endpoint comparison for* L. gibba *exposed to varying concentrations of tin.*

Quantity Sn, μgL⁻¹	Number						Chlorophyll, μgg⁻¹	
	Fronds	Plants	Roots	Dry Wt, g	Total Root Length, mm	N, %	a	b
ND(0.1)[a]	81	306	221	0.09354	2919	5.52	3872	1085
0.46	621	250	322	0.14573	4402	5.89	5224	1409
4.6	647	323	245	0.1096	3538	6.48	3386	1011
45.7	825	356	340	0.17895	4887	5.64	4377	1222
456.6	921	489	484	0.20171	7359	5.45	4227	1294
4 566	805	380	367	0.19687	5853	5.49	4191	1205
45 660	184	60	112	0.05158	1030	6.60	4030	1216

[a]Not detected at level noted in parenthesis.

TABLE 10—*Linear correlation coefficients* (r) *relating the various endpoints for tin.*

	Quantity	Number						Chlorophyll	
		Fronds	Plants	Roots	Dry Wt	Root Length	N, %	a	b
Quantity	1.000	−0.887	−0.827	−0.763	−0.772	−0.769	+0.622	−0.180	+0.677
No. fronds	−0.887	1.000	+0.723	+0.853	+0.894	+0.864	−0.378	+0.246	+0.335
No. plants	−0.827	+0.723	1.000	+0.895	+0.847	+0.913	−0.746	−0.036	−0.003
No. roots	−0.763	+0.853	+0.895	1.000	+0.953	+0.994	−0.738	+0.342	+0.426
Dry wt	−0.722	+0.894	+0.847	+0.953	1.000	+0.964	−0.745	+0.360	+0.400
Root length	−0.769	+0.864	+0.913	+0.994	+0.964	1.000	−0.735	+0.276	+0.361
N	+0.622	−0.378	−0.746	−0.738	−0.745	−0.735	1.000	−0.329	−0.279
Chl a	−0.180	+0.246	−0.036	+0.360	+0.360	+0.276	−0.329	1.000	+0.936
Chl b	+0.677	+0.335	−0.003	+0.426	+0.400	+0.361	−0.279	+0.936	1.000
P =	0.1	0.05	0.02	0.01	0.001				
r =	0.6694	0.7545	0.8329	0.8745	0.95074				

TABLE 11—*Final comparison of mixed concentrations (Se, Co, V, Sn) in Hoagland's E+ medium for* L. gibba *G-3.*

		Chlorophyll, μgg^{-1}	
Quantity,* μgL^{-1}	Dry Wt, g	a	b
0	0.13980	9150	2435
0	0.14073	7459	2144
0	0.14139	9457	2574
Mean ± standard deviation	0.14064 ± 0.0008	8689 ± 1076	2384 ± 219
1	0.11406	3598	1129
1	0.11056	3242	1370
1	0.10948	1950	938
Mean ± standard deviation	0.11137 ± 0.0024	2930 ± 867	1146 ± 216
2	0.20551	8253	2389
2	0.19924	7742	2749
2	0.19785	8377	2319
Mean ± standard deviation	0.20087 ± 0.0041	8124 ± 337	2486 ± 231
3	0.12451	7030	2360
3	0.13357	8628	2492
3	0.12989	8899	2544
Mean ± standard deviation	0.12932 ± 0.0046	8186 ± 1010	2465 ± 95
4	0.17965	6896	1962
4	0.17534	7078	1918
4	0.17355	8630	2365
Mean ± standard deviation	0.17618 ± 0.0031	7535 ± 953	2082 ± 246
5	0.21001	9202	2638
5	0.20723	7520	2856
5	0.20185	10106	2901
Mean ± standard deviation	0.20636 ± 0.0041	8943 ± 1312	2798 ± 141
6	0.04007	2419	740
6	0.04012	2473	849
6	0.04345	2542	760
Mean ± standard deviation	0.04121 ± 0.0019	2478 ± 62	783 ± 58

*Quantity	Se	Sn	Co	V
1	4.2	457	41	51
2	2.1	228	10	13
3	8.4	228	41	51
4	2.1	913	61	77
5	4.2	685	20	26
6	13 215	14 439	6 456	8 095

$r_{\text{dry wt versus chl a}} =$	0.778	$df = 19$	$P < 0.001$
$r_{\text{dry wt versus chl b}} =$	0.833	$df = 19$	$P < 0.001$

NOTE: df = degrees of freedom.

TABLE 12—*Final data for* L. minor *6591, which showed no response to additions of Sn and Co to Hoagland's medium but discernable response in growth to Se and V.*

		Chlorophyll, μgg^{-1}	
Quantity, μgL^{-1}	Dry Wt, g	a	b
0	0.07515	1640	533
0	0.07351	663	235
0	0.08594	1302	496
Mean ± standard deviation	0.07820 ± 0.007	1202 ± 496	421 ± 162
4.2 Se, 25.6 V	0.06966	1476	661
4.2 Se, 25.6 V	0.09728	1232	724
4.2 Se, 25.6 V	0.08567	160	188
Mean ± standard deviation	0.08420 ± 0.0139	956 ± 700	524 ± 293
2.1 Se, 51.2 V	0.10201	6664	2137
2.1 Se, 51.2 V	0.08045	6510	1385
2.1 Se, 51.2 V	0.10267	6625	2172
Mean ± standard deviation	0.09504 ± 0.0126	6600 ± 80	1898 ± 445
8.4 Se, 12.8 V	0.11889	5118	1202
8.4 Se, 12.8 V	0.09729	5367	1123
8.4 Se, 12.8 V	0.13096	3137	923
Mean ± standard deviation	0.11571 ± 0.0171	4541 ± 1222	1083 ± 144
13 215 Se, 8095 V	0.03793	616	390
13 215 Se, 8095 V	0.03569	490	192
13 215 Se, 8095 V	0.03208	784	270
Mean ± standard deviation	0.03523 ± 0.0030	630 ± 147	284 ± 100
$r_{\text{dry wt versus chl a}}$ =	−0.252	*df* = 13	not significant
$r_{\text{dry wt versus chl b}}$ =	−0.298	*df* = 13	not significant

weight, and total root elongation; % Kjeldahl N as a second endpoint; and chlorophyll a and b as the third endpoint. Test concentrations of Se are significantly related to all variables examined. The level of significance varies from about 7 to 0.1%.

The growth peak for V (Table 5) occurs at 25.6 μgL^{-1}. Again, the various measurements that signify growth are highly correlated (Table 6) with each other, while Kjeldahl N and chlorophyll a and b show no statistical significance except that chlorophyll a and b are highly related as would be expected. The concentration of V is statistically related at various levels of significance to the number of roots, dry weight, and chlorophyll a and b.

The growth peak for Co (Table 7) occurs at 20.3 μgL^{-1} and at 0.2 μgL^{-1}. The growth peak at 20.3 μgL^{-1} has been confirmed in a sequence of concentrations that also included 0.05 μgL^{-1} and 0.1 μgL^{-1}. The peak in chlorophyll a and b occurs at 202.5 μgL^{-1} and for Kjeldahl N at 2 μgL^{-1}. Despite the double peak found among the growth variables, the results are highly significant (cf. Table 8). However, Kjeldahl N and chlorophyll a and b fail to show any significant relationship with any of the growth variables. Concentrations of Co are only significantly related to chlorophyll a and b.

Tin (Table 9) shows only one growth peak at a concentration of 456.6 μgL^{-1}. The highest levels noted for chlorophyll a and b appear at 0.46 μgL^{-1}, while those for Kjeldahl N were noted at 4.6 μgL^{-1} and at the highest concentration tested. Significance levels of linear correlation coefficients between any of the possible pairs of growth variables range between about 7 and

TABLE 13—*Final data for* L. minor *7102 which showed no response to additions of V to Hoagland's medium but discernible response in growth to Se, Sn, Co.*

		Chlorophyll, $\mu g g^{-1}$	
Quantity, $\mu g L^{-1}$	Dry Weight, g	a	b
0	0.08178	1505	507
0	0.09951	1614	900
0	0.07565	1335	454
Mean ± standard deviation	0.08565 ± 0.0124	1485 ± 141	620 ± 244
4.2 Se, 228 Sn, 20.3 Co	0.06717	1313	684
4.2 Se, 228 Sn, 20.3 Co	0.05077	993	331
4.2 Se, 228 Sn, 20.3 Co	0.06656	1039	420
Mean ± standard deviation	0.06150 ± 0.0093	1115 ± 173	478 ± 184
2.1 Se, 913 Sn, 61 Co	0.11164	1872	605
2.1 Se, 913 Sn, 61 Co	0.16080	6905	1483
2.1 Se, 913 Sn, 61 Co	0.13298	4567	1862
Mean ± standard deviation	0.13514 ± 0.0247	4448 ± 2519	1317 ± 645
8.4 Se, 685 Sn, 10.2 Co	0.18340	975	989
8.4 Se, 685 Sn, 10.2 Co	0.18716	5248	1928
8.4 Se, 685 Sn, 10.2 Co	0.16911	6532	2087
Mean ± standard deviation	0.17989 ± 0.0095	4252 ± 2909	1668 ± 593
13 215 Se, 14 439 Sn, 6456 Co	0.06263	163	383
13 215 Se, 14 439 Sn, 6456 Co	0.05530	666	358
13 215 Se, 14 439 Sn, 6456 Co	0.07117	350	136
Mean ± standard deviation	0.06303 ± 0.0079	393 ± 254	292 ± 136
$r_{\text{dry wt versus chl a}}$ =	0.752	$df = 13$	$p < 0.002$
$r_{\text{dry wt versus chl b}}$ =	0.851	$df = 13$	$p < 0.001$

0.1%. Kjeldahl N is significantly related to the number of plants, the number of roots, dry weight, and root length. Chlorophyll a and b are related only to each other. Concentrations of Sn are significantly related to all the growth variables and chlorophyll b.

To summarize, the optimum growth is achieved for *L. gibba* with 4.2 $\mu g L^{-1}$ of Se, 25.6 $\mu g L^{-1}$ of V, 20.3 $\mu g L^{-1}$ of Co, and 457 $\mu g L^{-1}$ of Sn. Even though this may be true for individual elements, these peaks need to be confirmed by testing concentrations on either side of the peak and the peak itself. These numbers have been confirmed eight times during this study.

Now, it is important to discover what happens to growth when these concentrations are all in the same flask. Table 11 shows the final definitive test employing the various concentrations noted in the footnote. These concentrations include the optimum quantities derived from the single element studies as well as higher and lower amounts to bracket the peaks. Number 2 and Number 5 are very similar. Since the chlorophyll a and b content are higher in 5 than in 2 it was decided to select Number 5, which is concordant with the results of the single element experiments with the exception of Sn. The Sn content was reduced to 457 $\mu g L^{-1}$ since it was observed that over time it gave a better growth response. It is interesting to note that dry weight and both chlorophyll a and b are highly correlated at a significance level beyond the 0.1% level.

Lemna minor clones behave rather differently than *Lemna gibba*. Since the growth variables gave essentially the same results, only dry weight was used in the *L. minor* clone study.

TABLE 14—*Final data for* L. minor *7101, which showed no response to additions of V and Co to Hoagland's medium but discernible response in growth to Se and Sn.*

		Chlorophyll, μgg^{-1}	
Quantity, μgL^{-1}	Dry Weight, g	a	b
0	0.18585	5924	1348
0	0.18210	6661	2328
0	0.18434	6485	2274
Mean ± standard deviation	0.18410 ± 0.0019	6357 ± 385	1983 ± 551
4.1 Se, 228 Sn	0.05790	1539	670
4.1 Se, 228 Sn	0.06487	1877	931
4.1 Se, 228 Sn	0.05092	1201	407
Mean ± standard deviation	0.05790 ± 0.0070	1539 ± 338	669 ± 262
2.1 Se, 913 Sn	0.15745	6274	2069
2.1 Se, 913 Sn	0.16182	6718	2207
2.1 Se, 913 Sn	0.21695	6565	1515
Mean ± standard deviation	0.17874 ± 0.0332	6519 ± 226	1930 ± 366
8.4 Se, 685 Sn	0.19958	4975	1656
8.4 Se, 685 Sn	0.25450	5593	1980
8.4 Se, 685 Sn	0.21126	6483	2265
Mean ± standard deviation	0.22178 ± 0.0289	5684 ± 758	1967 ± 305
13 215 Se, 14 439 Sn	0.07731	714	254
13 215 Se, 14 439 Sn	0.06108	561	206
13 215 Se, 14 439 Sn	0.07759	738	244
Mean ± standard deviation	0.07199 ± 0.0095	671 ± 96	235 ± 25
$r_{\text{dry wt versus chl a}}$ =	0.895	$df = 13$	$p < 0.001$
$r_{\text{dry wt versus chl b}}$ =	0.836	$df = 13$	$p < 0.001$

The clone 6591 (Table 12) shows no increase in growth on additions of Sn and Co to Hoagland's medium. Its peak in biomass occurs with a combination of 8.4 μgL^{-1} Se and 12.8 μgL^{-1} V. Dry weight and chlorophyll a and b are not related, and the chlorophyll peak occurs with 2.1 μgL^{-1} Se and 51.2 μgL^{-1} V.

Clone 7102 (Table 13) shows no response to additions of V but exhibits a maximum biomass at 8.4 μgL^{-1} Se, 685 μgL^{-1} Sn and 10.2 μgL^{-1} Co. This is also shared by chlorophyll b, but chlorophyll a is highest at 2.1 μgL^{-1} Se, 913 μgL^{-1} Sn, and 61 μgL^{-1} Co. In this clone there is a highly statistically significant relationship between dry weight and chlorophyll a and b.

Clone 7101 (Table 14) again shows no increase in growth on additions of V and Co to the medium but responds favorably to small quantities of Se and Sn. The highest biomass was achieved with additions of 8.4 μgL^{-1} Se and 685 μgL^{-1} Sn. Chlorophyll b was also at a maximum with these additions but the peaks in chlorophyll a occurred at 2.1 μgL^{-1} Se and 913 μgL^{-1} Sn. The dry weight and chlorophyll a and b are significantly related beyond the 0.1% level.

Clone 7136 (Table 15) results are rather unexpected. An examination of the growth data would suggest that it really does best with no added Sn, V, Se, or Co. There is no measurable relationship between dry weight and chlorophyll a and b.

To summarize, a small table below shows the results (μgL^{-1}):

TABLE 15—*Final data for* L. minor *7136 which showed no response to additions of Sn and V to Hoagland's medium but discernible response in growth to Se and Co, though not in excess of the controls.*

		Chlorophyll, μgg^{-1}	
Quantity, μgL^{-1}	Dry Weight, g	a	b
0	0.15881	4917	1253
0	0.24178	4482	1661
0	0.29453	4324	1859
Mean ± standard deviation	0.23171 ± 0.0684	4574 ± 307	1591 ± 309
4.2 Se, 20.3 Co	0.15841	2179	801
4.2 Se, 20.3 Co	0.20169	2686	898
4.2 Se, 20.3 Co	0.19410	1180	792
Mean ± standard deviation	0.18473 ± 0.0231	2015 ± 766	830 ± 59
2.1 Se, 61 Co	0.16979	6118	2076
2.1 Se, 61 Co	0.19121	5908	2033
2.1 Se, 61 Co	0.14627	6207	1416
Mean ± standard deviation	0.16909 ± 0.0225	6078 ± 154	1842 ± 369
8.4 Se, 10.2 Co	0.17004	6325	2126
8.4 Se, 10.2 Co	0.16615	5454	2111
8.4 Se, 10.2 Co	0.12299	5800	1256
Mean ± standard deviation	0.15306 ± 0.0261	5860 ± 439	1831 ± 498
13 215 Se, 6456 Co	0.12595	396	145
13 215 Se, 6456 Co	0.11140	523	265
13 215 Se, 6456 Co	0.09749	305	156
Mean ± standard deviation	0.11161 ± 0.0142	408 ± 109	189 ± 66
$r_{\text{dry wt versus chl a}} = 0.303$	$df = 13$	not significant	
$r_{\text{dry wt versus chl b}} = 0.549$	$df = 13$	not significant	

		Se	Co	Sn	V
L. minor	6591	8.4	0	0	12.8
L. minor	7102	8.4	10.2	685	0
L. minor	7101	8.4	0	685	0
L. minor	7136	0	0	0	0
L. gibba	G-3	4.2	20.3	457	25.6

It is clear that *L. gibba* has a higher requirement of V and Co and a lesser need for Se and probably Sn than the first three clones of *L. minor.* The Se requirement is quite uniform, but only one clone requires Co and only one requires V. The Sn requirements appear high.

Discussion

The first point to be emphasized is that species and clone identification is of utmost importance as even the clones have been noted to exhibit different requirements.

It should be noted that the addition of these extra elements to the growth medium have increased vegetative growth and essentially prevented premature flowering. Thus, a sufficient number of plants are produced within seven days (see tables) to provide adequate results. Carrying out tests for 14 days creates some interesting problems. For example, *Lemna* in 14-day tests tends to adjust to the presence of the toxicant. Toxic response was noted to a concentration of 45 660 μgL^{-1} in 7 days, but *L. gibba* had recovered completely from the toxic response by 14 days. This has not been observed to occur with any toxicant tested within 7 days. Second, the requirement for counting plants and fronds [*1,2*] resulting from 14-day tests cannot be done by one person. The plants continue to grow during the counting process, and thus it must be completed within a day. A 14-day test of *L. gibba* requires 61.5 h to count the number of plants and the number of fronds, while any of the clones of *L. minor* require 108 h to count. This problem makes the cost of carrying out a 14-day *Lemna* test prohibitive. A 7-day *Lemna* test, on the other hand, requires fewer counters. A 7-day *L. gibba* test can be completed by one person in 5 h, while the *L. minor* clones need 10 h to amass the necessary data.

This study indicates that there are only three real endpoints, and that those associated with growth measurement are so highly correlated that any one would suffice. It will be shown in another paper to be published shortly that the EC_{50} is the same (statistically as well as toxicologically) for the number of plants, the number of fronds, and the dry weight in a 7-day *Lemna* test. Dry weight is actually a much better endpoint as it is less subject to human error when measured properly, is reproducible, and can be carried out in routine manner.

Conclusions and Recommendations

1. *Lemna gibba* G-3 and clones of *Lemna minor* respond rather differently to additions of Se, Sn, V, and Co to Hoagland's medium.
2. The growth indicators, namely the number of fronds, the number of plants, the number of roots, dry weight, and total root length are all the same endpoint and are highly related to each other.
3. It is recommended that dry weight be used as the final growth endpoint as it is more reproducible, less subject to error, and cheaper than counting the number of plants and fronds.
4. It is also recommended that the present 14-day test be reduced to 7 days as *Lemna* tends to adjust to toxicants after 7 days.
5. Kjeldahl N and chlorophyll a and b are three endpoints that measure the response to a toxicant that is generally unrelated to growth. Kjeldahl N and chlorophyll a and b are sometimes correlated with some growth measurements, but the response is erratic.

References

[*1*] Holst, R. W. and Ellwanjer, T. C., "Pesticide Assessment Guidelines, Subdivision J, Hazard Evaluation: Non-target Plants," EPA 540/9-82-020, 1982, Washington, DC.
[*2*] Holst, R. W., "Standard Evaluation Procedure Non-target Plants: Growth and Reproduction of Aquatic Plants-Tiers 1 & 2," EPA 540/9-86-134, 1986, Washington, DC.
[*3*] Landolt, E., *Biosystematic Investigations in the Family of Duckweeds (Lemnaceae)*, Vol. 2, Veröffentlichungen des Geobotanischen Institutes ETH, Stiftung Rubel, Zürich, Switzerland, 1986.
[*4*] Hillman, W. S., *American Journal of Botany*, Vol. 48, 1961, pp. 413–419.
[*5*] Cleland, C. F. and Briggs, W. R., *Plant Physiology*, Vol. 42, 1967, pp. 1538–1561.
[*6*] Kabata-Pendias, A. and Pendias, H., *Trace elements in soils and plants*, CRC Press, Inc., Boca Raton, FL, 1984.
[*7*] Milazzo, D. P., Cowgill, U. M., and Murphy, P. G., *Bulletin Environmental Contamination and Toxicology*, in press.
[*8*] Sartory, D. P., *Water Research*, Vol. 19, 1985, pp. 605–610.

Francis M. Gersich,[1] *Dennis P. Milazzo,*[1] *and Bryce D. Landenberger*[1]

A Comparison of Seven Diets Used to Culture *Tanytarsus Dissimilis*

REFERENCE: Gersich, F. M., Milazzo, D. P., and Landenberger, B. D., **"A Comparison of Seven Diets Used to Culture *Tanytarsus Dissimilis*,"** *Aquatic Toxicology and Hazard Assessment: 12th Volume, ASTM STP 1027,* U. M. Cowgill and L. R. Williams, Eds., American Society for Testing and Materials, Philadelphia, 1989, pp. 392–401.

ABSTRACT: The objective of this study was to identify an appropriate diet for maintaining a healthy culture of the parthenogenetic chironomid, *Tanytarsus dissimilis* Johannsen. The effects of seven different diets were examined, including an alfalfa-based diet, an algal diet (*Selenastrum capricornutum* Printz), and various combinations of the two. A full life-cycle study was conducted for each diet. The studies lasted approximately 13 to 15 days. The critical endpoints examined were percent hatch, survival, and reproduction. Egg hatching in all cases was completed within the first 48 to 72 h of the test. Pupation and adult emergence occurred primarily on Days 11 through 12 and 13 through 14, respectively. The results of these studies indicated that the algae/alfalfa mix diets were the most adequate. Additionally, the 0.13 mg algae/0.2 mL alfalfa diet provided the best survival (92%) and reproduction (129.3 ± 14.1 eggs/adult). Generally, a successful diet would yield >80% survival consistently and approximately 120 eggs/adult.

KEY WORDS: *Tanytarsus dissimilis,* midge, culturing, toxicity test, diet, nutrition

The advantages of incorporating a representative species of the dipteran family, Chironomidae, into the battery of test organisms used for environmental toxicity screening have been discussed by numerous researchers [*1–4*]. Ecologically, the chironomids are important because they are the most diverse and widely distributed family of aquatic insects. Their presence and or absence may indicate various water quality conditions (for example, hardness, alkalinity, pH, etc.), temperature and oxygen regimes, and particle size distribution of sediment. These organisms play a major role in aquatic food chains. Chironomids are often the dominant aquatic organisms in terms of density and biomass in lentic and/or lotic ecosystems.

The advantages associated with chironomids from a testing standpoint would include that many species have been successfully adapted to laboratory culturing (for example, *Chironomus tentans, C. riparius, C. plumosus, C. attenuatus, C. thummi,* and *Tanytarsus dissimilis*). Additionally, these species are good candidates for both acute and chronic testing because of their relatively short life cycles (2 to 4 weeks). For many of the species, test methodologies for short (acute) and life cycle exposures have been developed [*1*].

The species chosen for this study was *Tanytarsus dissimilis* Johannsen 1905 [*5*]. Before a testing program can begin with a newly introduced laboratory species, the health of the culture must be determined. The endpoints associated with many test methods are concerned with survival, growth, and reproduction. These endpoints are among those used to measure culture health. The relationship between these endpoints and nutrition have often been overlooked [*6,7*]. The objective of this study was to determine which of seven diets would produce healthy

[1]The Dow Chemical Co., Mammalian and Environmental Toxicology, Midland, MI 48674.

cultures of *T. dissimilis*. To determine the effectiveness of these diets, endpoints associated with egg hatching success, survival (larval, pupal, and adult), and reproduction (number of eggs/egg mass) were examined.

Methods and Materials

Culture Water

The water used in this study was pumped from the upper Saginaw Bay of Lake Huron, off Whitestone Point. The water was limed and flocculated with ferric chloride by the Midland Water Treatment Plant (Midland, Michigan). The water, upon arrival at the laboratory, was carbon filtered, ultraviolet (UV) irradiated, and pH adjusted with CO_2 prior to use. Additionally, before use for culturing, the water was autoclaved at 121°C and 124.1 kPa for 35 min. Typical analyses of the dilution water can be seen in Table 1.

TABLE 1—*Typical chironomid culture water analyses, μg/L.*

	1/19/84	4/4/84
Aluminum	ND(50)[a]	ND(50)
Ammonia	ND(10)	ND(5)
Arsenic	0.6	. . .[b]
Barium	15	. . .
Beryllium	ND(1)	. . .
Bismuth	ND(1)	. . .
Boron	230	240
Bromine	26	. . .
Cadmium	ND(1)	. . .
Calcium	45 800	36 800
Chlorides	57 000	. . .
Chromium	ND(5)	ND(5)
Cobalt	ND(5)	. . .
Copper	ND(5)	ND(5)
Fluoride	70	80
Iron	ND(10)	20
Lead	ND(5)	ND(5)
Lithium	ND(5)	. . .
Magnesium	8300	9760
Manganese	ND(1)	ND(5)
Mercury	ND(1)	. . .
Molybdenum	ND(5)	. . .
Nickel	ND(5)	. . .
Phosphorus	30	. . .
Potassium	1050	1170
Silicon	2800	3520
Silver	ND(5)	. . .
Sodium	4300	5840
Sulfur	5300	6300
Tin	ND(5)	. . .
Zinc	ND(10)	20
Total dissolved solids	282 000	194 000
Total organic carbon	7780	6130

[a]Not present at or below the level in parenthesis.
[b]Not determined.

Test Species

The organism used in this study was the chironomid, *Tanytarsus dissimilis*. The life cycle of *T. dissimilis* has been described by Johannsen [*4*], Nebeker [*8*], and Anderson [*1*].

The egg mass of *T. dissimilis* is cylindrical, up to 5.0 mm in length, and contains a single row of diagonally placed eggs. The egg mass is very sticky and difficult to handle. The eggs are often deposited in two separate masses. The first egg mass contains 120 to 190 eggs, while the second mass is much smaller (30 to 50 eggs). The eggs will hatch in about 48 h at 22°C. The entire egg mass usually hatches within 4 h.

The development of *T. dissimilis* progresses through four larval instars. The first instar is approximately 0.6 mm in length and is planktonic for about 24 h. After 24 h the larva settles to the bottom and constructs a tube of silk into which bottom debris is woven. All larval development is completed in 12 to 14 days. The maximum length of the fourth instar larva approaches 5.0 mm.

The pupal stages lasts approximately 48 h. A spindle-shaped pupation tube is formed by the fourth instar larva. The pupa is about 4.0 mm in length. At the end of the pupal stage, the pupa immediately swims to the surface, where the adult emerges. The adult is parthenogenic, and egg laying occurs 4 to 5 h after adult emergence.

Test Methodology

Many diets have been recommended for culturing chironomids [*1, 9*]. These diets suggest the use of Cerophyl, trout-chow granules, leaves, and many types of green algae. In this study seven different diets were examined (Table 2). The alga used in this study was *Selenastrum capricornutum* Printz [*10*]. The amount of algal suspension fed to the chironomids was determined with the aid of a Coulter counter. Feeding rates ranged between 0.06 and 0.25 mg dry wt of algae/0.1 L of water (Table 2).

The second part of the diet consisted of an alfalfa solution. The solution was prepared by adding 10 g of alfalfa powder to 250 mL of culture water and mixing vigorously in a blender for 5 min. The alfalfa solution was fed at a rate of 0.2 mL/0.1 L of water (Table 2). Alfalfa is an extremely nutritious leaf, the chemical composition of which is relatively constant [*11*].

The test chambers were Pyrex jars approximately 9.5 cm tall by 8.5 cm wide. Each jar contained 100 mL of water and the appropriate amounts and type of diet. Each diet had five replicates with each replicate containing ten individual eggs. Egg masses were collected in the following manner for both setting the test and monitoring the reproduction endpoint of the study. When adult emergence occurred the adults were captured and individually placed in a 4-dram vial containing 3 mL of water. The adults laid egg masses within 1 to 2 h. To begin the study, egg masses were removed from the vials, and with the use of a dissecting microscope, groups of ten individual eggs were dissected from the mass. At the end of the study the reproduction end

TABLE 2—*The seven diets used to culture* Tanytarsus dissimilis.

Diet	Alfalfa Solution, mL/0.1 L of Water	Algae, mg dry wt/0.1 L of Water
1	0.2	0
2	0.2	0.06
3	0.2	0.13
4	0.2	0.25
5	0	0.06
6	0	0.13
7	0	0.25

point (number of eggs/egg mass) was examined by collecting the eggs and counting the number of individual eggs per egg mass. The jars were kept in a temperature-controlled environmental chamber set at 20 ± 1°C. The photoperiod was 16 h daylight/8 h darkness.

Each Monday, Wednesday, and Friday the organisms were fed and the test monitored for dissolved oxygen, pH, and temperature. The endpoints of this study were percent hatch, survival (larval, pupal and adult), and reproduction (number of eggs per egg mass).

Statistical Analyses

Statistical analysis was conducted using a one-way ANOVA [*12*], with diet as the factor. This procedure was used for both the reproduction and survival variables. Observations on reproduction were taken from each surviving larva to maturity; thus, while there was a potential for 350 observations, the actual number was 215. For the survival endpoint, each replicate was treated as an observation; thus, each diet group had a total of five observations.

Linear contrast statements were used with the ANOVAs to examine effects of the following diets on the survival and reproduction endpoints: alfalfa suspension, algae, and alfalfa suspension and algae. In the event of significant interactions ($\alpha = 0.01$), separate ANOVAs for the alfalfa and no alfalfa diets were performed using contrasts to evaluate the algal diet response. If any of the ANOVAs failed to be significant ($\alpha = 0.05$), then no contrasts were conducted.

Although survival and reproduction were the only two endpoints examined statistically, other endpoints were also observed during the conduct of the study. At the start of the test, ten eggs were placed in each of five replicates per diet. The replicates were examined throughout the study with observations being made on the following endpoints: larval hatch, larval mortality, pupal mortality, pupal emergence, number of live adults, and adult mortality before egg laying.

Results and Discussion

The data associated with the one-way ANOVAs used to examine the effects of diet on reproduction and survival are presented in Tables 3 and 4. Statistical examination of the results began with testing for the interaction between diet and the alfalfa suspension, as generated by linear contrast. The contrasts (Tables 3 and 4) were statistically significant for both reproduction (mean number of eggs/adult) and survival. This result indicates that the change in effects associated with the alfalfa- and nonalfalfa-containing diets does not follow the same pattern. The differences are reflected by the curves presented in Figs. 1 and 2.

Since interaction was observed, separate ANOVAs were conducted for the alfalfa- and nonalfalfa-containing diets for both reproduction and survival (Tables 5 to 8). The results indicated no significant differences ($\alpha = 0.05$) for reproduction in the nonalfalfa containing diets; however, survival within this group of diets was significantly affected. For the alfalfa-containing diets both reproduction and survival showed significant differences among diets. All statistically significant endpoints were observed (by linear contrast) to be diet dependent.

In the presence of interaction, interpretation of the alfalfa effect should be made with caution. The survival and reproduction data were interpreted to indicate a positive effect when the alfalfa suspension was included in the diet. Survival increased in all algal diets with the addition of alfalfa, and reproduction increased in the higher two algal diets (0.13 and 0.25 mg algae) when alfalfa was included. It can be further inferred from Figs. 1 and 2 that the positive alfalfa effects on survival and reproduction level off at the 0.13 mg algal diet.

Other observations taken during the study also support the 0.13 mg algae/0.2 mL alfalfa diet (Tables 9 and 10). The percent hatch was similar among diets ranging from 96 to 100% (based on starting with 50 eggs/diet). All egg hatching occurred within the first 24 to 48 h of the test. The 0.13 mg algae/0.2 mL alfalfa diet exhibited the lowest larval mortality (4%). Larval mor-

TABLE 3—*Analysis of variance table (ANOVA) and linear contrast statements used to examine the effect of diet on reproduction.*

DEPENDENT VARIABLE: REPRODUCTION

Source	DF	Sum of Squares	Mean Square	*F*-Value	Significance Level
Model	6	9 600.40	1600.07	6.29	0.0001
Error	209	53 167.56	254.39		
Corrected total	215	62 767.96			

Linear Contrast	DF	Sum of Squares	*F*-Value	*PR* > *F*
Total sum of squares	3	5299.36	6.94	0.0002
Main alfalfa effect	1	728.07	2.86	0.0922
Trend test	1	2540.06	9.98	0.0018
Lack of fit	1	14.18	0.06	0.8136

TEST FOR INTERACTION ON REPRODUCTION

SS interaction = 5299.36 − 728.07 = 4572.29
MS interaction = 4571.29/2 = 2285.65

$$F = \frac{\text{MS Interaction}}{\text{MSE}} = \frac{2285.65}{254.39} = 8.98\ (p > 0.01)$$

NOTE: DF = degrees of freedom; MSE = mean square error; SS = sum of squares; MS = mean square; F = Fisher.

TABLE 4—*Analysis of variance table (ANOVA) and linear contrast statements used to examine the effect of diet on survival.*

DEPENDENT VARIABLE: SURVIVAL

Source	DF	Sum of Squares	Mean Square	*F*-Value	Significance Level
Model	6	278.57	46.4	29.28	0.0001
Error	28	44.4	1.59		
Corrected total	34	322.97			

Linear Contrast	DF	Sum of Squares	*F*-Value	*PR* > *F*
Total sum of squares	3	198.50	41.73	0.0001
Main alfalfa effect	1	168.03	105.97	0.0001
Trend test	1	36.82	23.22	0.0001
Lack of fit	1	29.57	18.65	0.0002

TEST FOR INTERACTION ON SURVIVAL

SS interaction = 198.50 − 168.03 = 30.47
MS interaction = 30.47/2 = 15.23

$$F = \frac{\text{MS Interaction}}{\text{MSE}} = \frac{15.23}{1.59} = 9.58\ (p > 0.01)$$

NOTE: PR = probability.

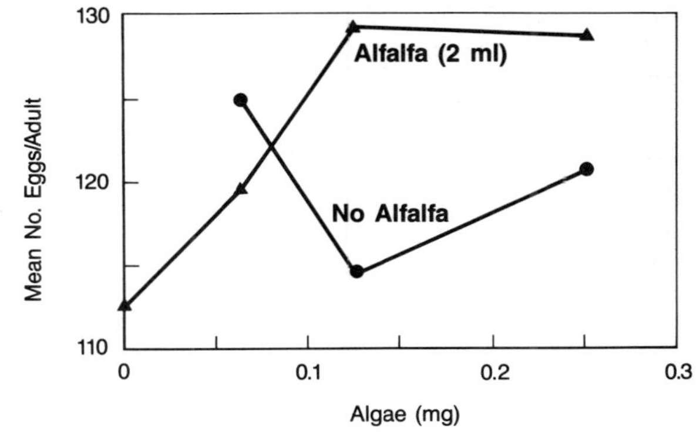

FIG. 1—*Presentation of the data illustrating the interaction between reproduction and diet.*

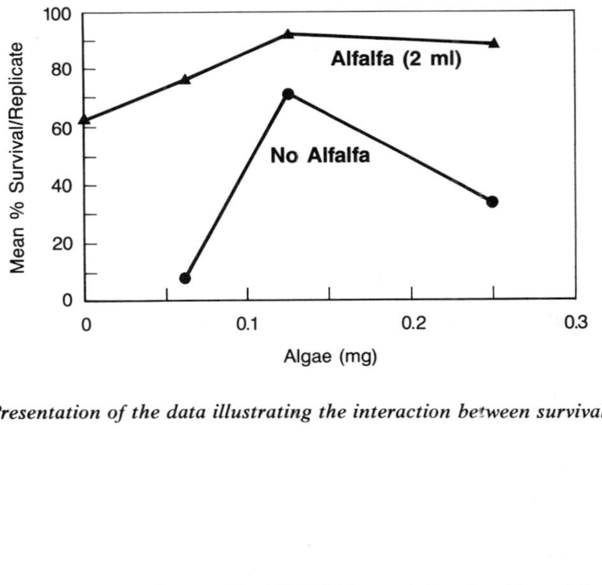

FIG. 2—*Presentation of the data illustrating the interaction between survival and diet.*

TABLE 5—*Analysis of variance table (ANOVA) examining the effect of the nonalfalfa containing diet on reproduction.*

Dependent Variable: Reproduction					
Source	DF	Sum of Squares	Mean Square	*F*-Value	Significance Level
Model	2	723.43	361.72	1.33	0.2722
Error	54	14 653.44	271.36		
Corrected total	56	153 676.87			

TABLE 6—*Analysis of variance table (ANOVA) examining the effect of the alfalfa containing diets on reproduction.*

		Dependent Variable: Reproduction			
Source	DF	Sum of Squares	Mean Square	*F*-Value	Significance Level
Model	3	7 161.45	2387.15	9.61	0.0001
Error	155	38 514.11	248.48		
Corrected total	158	45 675.56			

TABLE 7—*Analysis of variance table (ANOVA) examining the effect of the nonalfalfa containing diets on survival.*

		Dependent Variable: Survival			
Source	DF	Sum of Squares	Mean Square	*F*-Value	Significance Level
Model	3	103.6	51.8	29.88	0.0001
Error	12	20.8	1.73		
Corrected total	14	124.4			

TABLE 8—*Analysis of variance table (ANOVA) examining the effect of the alfalfa containing diets on survival.*

		Dependent Variable: Survival			
Source	DF	Sum of Squares	Mean Square	*F*-Value	Significance Level
Model	3	27.4	9.12	6.18	0.0054
Error	16	23.6	1.48		
Corrected total	19	51.0			

TABLE 9—*Survival and reproduction of* Tanytarsus dissimilis *cultured on seven different diets.*

Diet	Adult Survival, %	Mean No. of Eggs/Adult, ±S.D.
0.2 mL alfalfa	62.0 ± 11.0 (N = 5)	112.5 ± 21.3 (N = 31)
0.06 mg algae	8.0 ± 13.0 (N = 5)	125.0 ± 11.4 (N = 4)
0.13 mg algae	72.0 ± 13.0 (N = 5)	114.5 ± 17.6 (N = 36)
0.25 mg algae	34.0 ± 13.4 (N = 5)	120.8 ± 14.7 (N = 15)
0.06 mg algae/0.2 mL alfalfa	76.0 ± 16.7 (N = 5)	119.5 ± 14.8 (N = 39)
0.13 mg algae/0.2 mL alfalfa	92.0 ± 8.4 (N = 5)	129.3 ± 14.1 (N = 46)
0.25 mg algae/0.2 mL alfalfa	88.0 ± 11.0 (N = 5)	128.8 ± 13.4 (N = 44)

TABLE 10—*A summary of the data collected on the various endpoints in the* Tanytarsus dissimilis *nutrition study.*

Diet	No. of Eggs at Start	% Hatch	No. of Larvae Alive	% Larval Mortality	No. of Larvae Alive Before Pupation	% Pupal Mortality	No. of Pupae Alive	% Pupal Emergence	No. of Alive Adults	No. of Alive Adults Laying Eggs
0.2 mL alfalfa	50	98	49	20	38	16	32	84	32	31
0.06 mg algae	50	96	48	90	5	0	5	100	5	4
0.13 mg algae	50	100	50	20	40	5	38	95	38	36
0.25 mg algae	50	96	48	46	26	31	16	69	16	15
0.06 mg algae/0.2 mL alfalfa	50	100	50	14	44	9	40	91	40	39
0.13 mg algae/0.2 mL alfalfa	50	100	50	4	48	4	46	96	46	46
0.25 mg algae/0.2 mL alfalfa	50	100	50	8	46	2	45	98	45	44

TABLE 11—*A comparison of acute LC_{50} values (mg/L) between* Daphnia magna *(daphnid),* Tanytarsus dissimilis *(chironomid),* Pimephales promelas *(fathead minnow), and* Salmo gairdneri *(rainbow trout).*

Test Material	*D. magna*	*T. dissimilis*	*P. promelas*	*S. gairdneri*
Hexachloroethane [*13*]	8.1	1.7	1.5	...
1,2 dichlorobenzene [*14*]	2.4	11.8	...	1.5
1,4 dichlorobenzene [*14*]	11.0	13.0	4.0	1.1
Tetrachloroethylene [*15*]	17.7	30.8	15.7	5.2
Chromium [*16*]	6.4	59.9	43.1	69.0
Silver [*17,18*]	0.002	3.16	0.01	0.018
Cadmium [*19,20*]	0.009	0.004	0.60	0.001
Copper [*20,21*]	0.009	0.016	...	0.057
Zinz [*20,22*]	0.33	0.037	...	0.09
Lead [*20,23*]	0.612	0.258	...	1.1

tality in the remaining diets ranged from 8% (0.25 mg algae/0.2 mL alfalfa) to 90% (0.06 mg algae). Pupal mortality ranged from 0% in the 0.06 mg algae diet to 31% in the 0.25 mg algae diet. The majority of pupation occurred on Day 12 of the study. Adult midges began to emerge and lay egg masses on Day 13, and emergence and reproduction were completed on or by Day 14. The number of reproducing adults, collected from the original 50 eggs/diet, ranged from 4 (0.06 mg algae) to 46 (0.13 algae/0.2 mL alfalfa).

Throughout the study all dissolved oxygen measurements were >90% saturation. Temperature and pH measurements ranged from 19.2 to 20.2°C and 7.9 to 9.4, respectively. The higher pH measurements were associated with the 0.25 mg algae/0.2 mL alfalfa diet.

The most common invertebrate used and required for aquatic toxicity testing is *Daphnia magna.* A comparison of the sensitivity (acute LC_{50} values) of *D. magna, T. dissimilis,* and two freshwater fishes *(Salmo gairdneri* and *Pimephales promelas)* to selected organic compounds and metals is presented in Table 11. The data indicate that the chironomid was the least sensitive of the test organisms exposed to the various organics, with the exception of hexachloroethane. The acute data for metals did not show a clear sensitivity trend, and this might be attributed to variations in water quality variables (for example, hardness).

Although the chironomid appears to be as or less sensitive than the daphnid in its response to various toxicants, there is a benefit in including chironomids as part of our test battery of aquatic invertebrates. At present we are somewhat limited by emphasizing testing with one invertebrate species, *D. magna.* A recent U.S. EPA publication [*24*] concerned with deriving national water quality criteria includes a minimum data base which requires acute data on a benthic insect. The various states are becoming involved in generating their own water quality standards by concentrating on testing resident species and site-specific testing. The relative abundance and importance of chironomids in aquatic ecosystems supports the incorporation of a chironomid species into the battery of test organisms used for aquatic toxicity testing.

References

[*1*] Anderson, R. L. in *Aquatic Invertebrate Bioassays, STP 715,* American Society for Testing and Materials, Philadelphia, 1980, pp. 70–80.

[*2*] Coffman, W. P., "Chironomidae," *An Introduction to Aquatic Insects,* Kendall Hunt Publishing Company, Dubuque, Iowa, 1978, pp. 345–377.

[*3*] Wirth, W. W. and Stone, A., "Aquatic Diptera," *Aquatic Insects of California,* University of California Press, Berkeley, 1956, pp. 372–482.

[*4*] Johannsen, O. A., *Aquatic Diptera—Eggs, Larvae and Pupae of Aquatic Flies,* Entomological Reprint Specialists, Los Angeles, 1969, pp. 1–95.

[5] Johannsen, O. A., *New York State Museum Bulletin*, Vol. 86, 1905, pp. 7-352.
[6] Buikema, Jr., A. L., Geiger, J. G. and Lee, D. R. in *Aquatic Invertebrate Bioassays, STP 715*, American Society for Testing and Materials, Philadelphia, 1980, pp. 48-69.
[7] Cowgill, U. M., Emmel, H. W., Hopkins, D. L., Takahashi, I. T., and Parker, W. M., *International Revue der Gesamten Hydrobiologia*, Vol. 71, 1986, pp. 79-99.
[8] Nebeker, A. V., *Journal of the Kansas Entomological Society*, Vol. 46, No. 2, 1973, pp. 161-165.
[9] Cavanaugh, W. J. and Tilden, J. E., *Ecology*, Vol. 11, No. 2, 1930, pp. 281-287.
[10] Cowgill, U. M., Takahashi, I. T. and Applegath, S. L., *Journal of Environmental Toxicology and Chemistry*, Vol. 4, No. 3, 1985, pp. 415-422.
[11] Cowgill, U. M., States, S. J., and Marburger, J. E., *Environmental Pollution*, Vol. 22A, 1980, pp. 259-272.
[12] Winer, B. J., *Statistical Principles in Experimental Design*, McGraw Hill, New York, 1971.
[13] "Ambient Water Quality Criteria for Chlorinated Ethanes," EPA 440/5-80-039, U.S. Environmental Protection Agency, Washington, DC, 1980.
[14] "Ambient Water Quality Criteria for Dichlorobenzenes," EPA 440/5-80-039, U.S. Environmental Protection Agency, Washington, DC, 1980.
[15] "Ambient Water Quality Criteria for Tetrachloroethylene," EPA 440/5-80-073, U.S. Environmental Protection Agency, Washington, DC, 1980.
[16] "Ambient Water Quality Criteria for Chromium," EPA 440/5-80-035, U.S. Environmental Protection Agency, Washington, DC, 1980.
[17] "Ambient Water Quality Criteria for Silver," EPA 440/5-80-071, U.S. Environmental Protection Agency, Washington, DC, 1980.
[18] Lima, A. R., Curtis, C., Hammermeister, D. E., Call, D. J., and Felhaber, T. A., *Bulletin of Environmental Contamination and Toxicology*, Vol. 29, 1982, pp. 184-189.
[19] "Ambient Water Quality Criteria for Cadmium," EPA 440/5-80-025, U.S. Environmental Protection Agency, 1980.
[20] Anderson, R. L., Walbridge, C. T., and Fiandt, J. T., *Archives of Environmental Contamination and Toxicology*, Vol. 9, 1980, pp. 329-335.
[21] "Ambient Water Quality Criteria for Copper," EPA 440/5-80-036, U.S. Environmental Protection Agency, Washington, DC, 1980.
[22] "Ambient Water Quality Criteria for Zinc," EPA 440/5-80-079, U.S. Environmental Protection Agency, Washington, DC, 1980.
[23] "Ambient Water Quality Criteria for Lead," EPA 440/5-80-057, U.S. Environmental Protection Agency, Washington, DC, 1980.
[24] Stephen, C. E., Mount, D. J., Hansen, D. J., Gentile, J. H., Chapman, G. A., and Brungs, W. A., "Guidelines for Deriving Numerical National Water Quality Criteria for the Protection of Aquatic Organisms and Their Uses," PB85-227049, National Technical Information Service, Springfield, VA, 1985.

Stephen H. Ward[1]

The Requirements for a Balanced Medium in Toxicological Experiments Using *Mysidopsis bahia* with Special Reference to Calcium Carbonate

REFERENCE: Ward, S. H., **"The Requirements for a Balanced Medium in Toxicological Experiments Using *Mysidopsis bahia* with Special Reference to Calcium Carbonate,"** *Aquatic Toxicology and Hazard Assessment: 12th Volume, ASTM STP 1027,* U. M. Cowgill and L. R. Williams, Eds., American Society for Testing and Materials, Philadelphia, 1989, pp. 402–412.

ABSTRACT: A number of factors produce response differences in *Mysidopsis bahia* in toxicological experiments. In the experimental design of tests with *M. bahia,* the organism's requirements for an external source of certain ions should be considered because the absence or imbalance of these substances can affect their physiological requirements and, therefore, confound the response of *M. bahia* when tested against a toxicant. The purpose of this paper is to show the effects of two of these factors, buffering and ionic balance, and their interactions in culture and toxicity tests.

Also, because insufficient information is available about the response of *M. bahia* to calcium carbonate concentration, this was specifically studied. Mineral load can vary when performing toxicity tests, especially whole effluent tests, and may affect exoskeleton chemistry during ecdysis. Calcium, used as calcium carbonate in cuticle formation by all crustaceans including *M. bahia,* is derived directly from the external medium. Because of this, an important environmental factor to consider is the concentration of bicarbonate and how it may influence the uptake of calcium.

KEY WORDS: mysid, culture, reproduction, growth, calcium carbonate, cadmium, toxicity tests

Recently, the marine crustacean, *Mysidopsis bahia,* has been extensively used in toxicological experiments for the study and subsequent regulation of the discharge of point sources of pollutants into the nation's marine and estuarine waterways. If these tests are to provide quantifiable results that measure the toxicity of a pollutant, then it becomes increasingly clear that we must understand the needs of *M. bahia* for a balanced medium and that we must be able to distinguish a response due to toxicity from one due to the absence or imbalance of certain ions in the medium. The importance of ionized calcium along with an external source of bicarbonate (HCO_3) has been established for other crustaceans [*1,2*]. The exoskeleton of the mysid *M. bahia,* like that of other crustaceans, is heavily mineralized with calcium carbonate ($CaCO_3$).

Seawater is generally considered a stable medium with respect to the major ions found. A greater degree of variability is expected in estuarine systems. In toxicological experiments, as well as in culture, there could be occasions when the absence or imbalance of carbonate ($CaCO_3$) alkalinity can alter an acute response to a toxicant, or affect growth and reproduction in a life cycle test. This is especially true for crustaceans because of their need to shed their exoskeleton (ecdysis) during the molt cycle.

When studies reporting on the responses of *M. bahia* to a toxicant such as cadmium [*3,4*] are

[1] Aquatic biologist, U.S. Environmental Protection Agency, Region II, Edison, NJ 08837.

compared, they show that salinity and temperature affect the outcome of the toxicity test. There is also some indication in the literature of increased sensitivity during the molting cycle. In particular, Gentile et al. [*4*] reported work on *Mysidopsis bahia* and *Mysidopsis bigelowi* which showed problems with molting and reproduction. There has also been work done by Lee and Buikema [5] for *Daphnia pulex* which showed increased sensitivity during molting. On the other hand, experiments have not sufficiently addressed the importance of certain ions, especially $CaCO_3$, and how their absence or imbalance might affect the organism.

The purpose of this investigation is to demonstrate the effects on growth and reproduction when carbonate ($CaCO_3$) alkalinity is altered. In addition, mysids were exposed to cadmium in different concentrations of $CaCO_3$ to see if sensitivity to a toxicant would change at time of molt.

Methods and Materials

The mysid *Mysidopsis bahia*, used in these experiments, were obtained from in-house cultures as reported elsewhere [*6, 7*]. Cultures were maintained in 208-L glass aquaria at 25°C, 25‰ salinity, and a light cycle of 12-Light:12-Dark. Mysids were fed Reference *Artemia* II (RAC II), Guana Bara, Brazil (provided by U.S. Environmental Protection Agency, Narragansett, Rhode Island). Mysids in each aquarium were fed 500 mL of concentrated *Artemia* (24-h posthydrated) twice daily, following Ward [*7*]. Mysids were fed solely Reference *Artemia* II for four months prior to this study because nutritional quality has been established for a similar strain, Reference *Artemia* (RAC), by Bengtson et al. [*8*] for *M. bahia* used in cadmium chloride studies.

Growth and Reproduction Experiment

The seven-day survival, growth, and reproduction test developed by Lussier et al. [*9*] was used. This was modified with five replicates per treatment, ten mysids per Pyrex crystalline dish, and a test volume of 400 mL. Test and culture conditions were the same with the exception of $CaCO_3$ levels, which were varied for the treatment groups. The experiment was designed in this way to study the potential for $CaCO_3$ problems both when rearing *Mysidopsis bahia* in the laboratory and when testing. Twenty-four-hour-old mysids were first isolated in a 56-L capacity aquarium equipped with a recirculating system. These were held under the same conditions and offered the same feeding schedule as were animals in culture. The first group of mysids was used for the growth and reproduction experiment. A second group of juveniles was taken at the same time from the culture, and a growth test was started to study growth during the first six days postrelease from the female brood pouch. Mysids were removed with a wide mouth pipet and randomly placed in each replicate test vessel. Mysids were fed twice daily, 0.5 mL of concentrated *Artemia* (RAC II) 24-h posthydrated. The seawater was exchanged daily by siphon, and each dish was cleaned. All test vessels were covered during the experiment. The same procedures were used for the growth and reproduction experiment with the exception that the test was started with seven-day-old mysids. Physicochemical properties, including pH, dissolved oxygen ($\geq$60% for all treatments for both phases of the experiment), temperature, salinity, and carbonate ($CaCO_3$) alkalinity, were measured daily on each test vessel for both phases of this experiment.

The following procedures were used in order to derive a series of $CaCO_3$ concentrations in solution either by the removal or addition of carbonate. All carbonate treatments were made from the same batch of natural seawater. Seawater was continuously filtered through a carbon filter (24-h), then through a 0.45-μm membrane filter before use. Natural seawater (32‰ salinity) was placed in a 10-L borosilicate glass vessel and heated to 40°C with gentle aeration to drive off carbonate. This resulted in saline brines that were brought to either 60‰, 100‰, or 140‰ salinity. These brines were filtered through a 0.45-μm membrane filter and diluted back to 25‰ salinity with deionized water to achieve 70 mg/L, 50 mg/L, and 25 mg/L of $CaCO_3$, respectively. Then 150 mg/L and 200 mg/L $CaCO_3$ solutions were made by adding reagent grade sodium bicarbonate ($NaHCO_3$) to seawater (diluted to 25‰ salinity). The last treatment used was simply natural seawater diluted to 25‰ salinity resulting in a carbonate level of 100 mg/L. These procedures provided a series of approximately 25, 50, 75, 100, 150, and 200 mg/L of $CaCO_3$.

For both phases of this experiment, mean dry weight (mg) determinations of mysids were made. Mysids were rinsed with deionized water, placed on weighing boats, and dried in an oven at 80°C until constant weight was achieved ($\geq$48 h). Average dry weight per mysid was determined using a Mettler HL52 analytical balance. For the growth test, mysids were divided into groups of five and weights were determined for three-day-old and six-day-old mysids. For the rapid growth and reproduction test [*9*], sex determinations were first made using a dissecting microscope. For each replicate, females and males were weighed as two separate groups. If immature mysids were found, they were weighed with males. Sex and maturity were determined by observing the development of an oostegite brood pouch, the presence or absence of embryos within, presence of eggs in the oviduct, length of pleopods, and the presence or absence of male gonads.

One-way analysis of variance was used to test for differences in mean number of mature female mysids with embryos in the marsupium and mean dry weights of females, males, and juveniles of different ages. Arcsine transformation was applied to the data for female reproduction prior to analysis. A further analysis was made using Duncan's multiple range test to determine which $CaCO_3$ treatments were different from the natural seawater treatment (100 mg/L). Statistical analyses were made using PC ANOVA obtained from Human Systems Dynamics, Northridge, California.

Acute Toxicity to Cadmium

The 96-h acute toxicity test (static renewal) following standard test procedures [*10*] and ASTM Practice for Conducting Acute Toxicity Tests with Fishes, Macroinvertebrates, and Amphibians (E 729-80) was used. This test period was long enough to test the hypothesis that variation in response to cadmium could be caused by variation in carbonate ($CaCO_3$) alkalinity level during a critical growth phase (ecdysis). For the culture conditions employed in these experiments, it was verified that *M. bahia* completed a full molt cycle during the 96-h postrelease period. Each toxicity test consisted of a control and five concentrations in replicate with ten juvenile mysids (24-h old) randomly placed in each replicate. Pyrex crystalline dishes were employed as test vessels. Each test vessel was covered during the experiments. The test volume was 400 mL per replicate. All tests were performed under the same conditions as the culture and growth and reproduction experiments. The dissolved oxygen levels did not fall below 60% saturation for all tests. Mysids were fed once daily with concentrated *Artemia* (RAC II). A concentration of 0.1 mL per replicate was used.

Stock solutions of cadmium (1000 μg/L) were made daily from a 100-μg/mL total inorganic cadmium standard (2% HCl) obtained from Inorganic Ventures, Inc., Brick, New Jersey. Four $CaCO_3$ levels (25, 50, 100, 200 mg/L) were used in these experiments. Aliquots of the stock solution of cadmium were then added to the four treatment groups of $CaCO_3$ to achieve a concentration series of 10, 25, 50, 75, and 100 μg/L. To check the precision and accuracy of these methods, the 100-mg/L $CaCO_3$ treatment group was analyzed at the end of each test by atomic absorption spectrophotometry following U.S. EPA [*11*] Method 213.1 for total cadmium for each concentration used. The variation between nominal and measured levels was less than 5% in three different tests analyzed.

The acute toxicity data for mortality were analyzed following the probit method [*12*], using a statistical program probit (provided by U.S. Environmental Protection Agency, Cincinnati,

Ohio). Toxicity data were analyzed for 48, 72, and 96 h. The 95% confidence limits were compared within and between carbonate treatment groups and were considered significantly different if they did not overlap.

Buffer and Ionic Strength

Initial and final levels of total organic (TOC) and inorganic (TIC) carbon were measured for all experiments described above with a Beckman 915B carbon analyzer. Alkalinity was measured daily by sulfuric acid (0.02 *N*) titration with the pH end-point of 4.5 as milligrams of calcium carbonate ($CaCO_3$)/Liter following APHA Standard Methods [*13*]. Further, to assure that no significant changes in concentration of other major ions had occurred, analyses were made of samples of the above $CaCO_3$ concentration series for sodium, magnesium, potassium, and calcium by atomic absorption spectrophotometry, and for chloride by titrametric methods following U.S. Environmental Protection Agency [*11*]. With the exception of carbonate, neither the decarbonation procedure nor the addition of $NaHCO_3$ to form the carbonate series altered the ratio of these ions significantly. This ratio remained similar to that found in diluted (25‰ salinity) natural seawater used for rearing mysids in the lab.

Results

Calcium Carbonate Effects on Growth and Reproduction

Results of the dry weight measurements for the 1 to 6-day growth and the 7 to 15-day growth and reproduction tests for the different $CaCO_3$ treatments are given in Table 1. *Mysidopsis bahia* were only slightly influenced at three days of age by varying the carbonate levels; however, there was a distinct difference in dry weight per mysid by Day 6. The growth and reproduction test shows similar findings. For each treatment, the females were larger than the males. The male : female ratio was approximately 1 : 1 and the number of immature mysids found was similar within and among treatments. For the first phase of this experiment (one to six days), mortality was between 0 and 7% within and between treatments. During the second phase (7 to 15 days), no mortality occurred.

Figure 1 shows the relationship between total number of brooded and nonbrooded females found at the different carbonate levels during Phase 2 of this experiment. Among all mature females, for all treatments, and number of embryos in the marsupia varied between 12 and 18 and the stage of development was the same (Stage I, early embryos). A comparison of these data show that as you move away from 100 mg/L of $CaCO_3$ on both sides of the curve there is a significant increase of nonbrooded females.

Figure 2 shows the mean percentage of brooded females per $CaCO_3$ treatment and the standard deviation. For comparison purposes, it also presents the mean weight for the 15-day-old mysids for each treatment and shows how the increased availability of carbonate affects weight of the female. However, it is also interesting to note that a slight increase in weight at $CaCO_3$ levels above 100 mg/L does not result in an increase of brooded females. One-way analysis of variance showed a significant difference ($P \leq 0.01$) for number of brooded females when comparing $CaCO_3$ treatments. Duncan's multiple range test was applied and showed significant differences ($P \leq 0.01$) for 25, 150, and 200 mg/L of $CaCO_3$ as compared to 100 mg/L of $CaCO_3$.

Acute Effects When Exposed to Cadmium

Table 2 presents LC_{50}s from twelve 96-h acute toxicity tests at 48, 72, and 96 h. A comparison of the LC_{50} values and their confidence limits shows a significant difference between $CaCO_3$

TABLE 1—*Dry weight (mg) of mysids for the different $CaCO_3$ treatments and experiments reported as a mean and standard deviation.*

$CaCO_3$ (mg/L) nominal	Growth[a] 3 Day Old	6 Day Old
25	0.047 ± 0.005	0.096 ± 0.02[c]
50	0.050 ± 0.001	0.103 ± 0.01[c]
75	0.056 ± 0.003	0.110 ± 0.01
100	0.055 ± 0.002	0.107 ± 0.01
150	0.056 ± 0.010	0.116 ± 0.01[c]
200	0.059 ± 0.014	0.126 ± 0.01[c]

Reproduction and Growth[b]

$CaCO_3$ (mg/L) nominal	Females	*n*	Males and Immature	*n*
25	0.27 ± 0.04[c]	24	0.19 ± 0.06[c]	26(5)
50	0.25 ± 0.08[c]	23	0.21 ± 0.10[c]	27(4)
75	0.26 ± 0.07[c]	22	0.24 ± 0.05	28(5)
100	0.47 ± 0.07	24	0.39 ± 0.07	26(3)
150	0.46 ± 0.05	23	0.38 ± 0.08	27(5)
200	0.48 ± 0.10	25	0.36 ± 0.09	25(5)

100 mg/L $CaCO_3$ = natural seawater diluted.
NOTE: The numbers in parenthesis are the number of immature mysids found.
[a]Growth experiment for 1 through 6-day-old mysids.
[b]Reproduction and growth test for 7 through 15-day-old mysids.
[c]Significantly different from 100 mg/L $CaCO_3$ ($P \leq 0.05$).

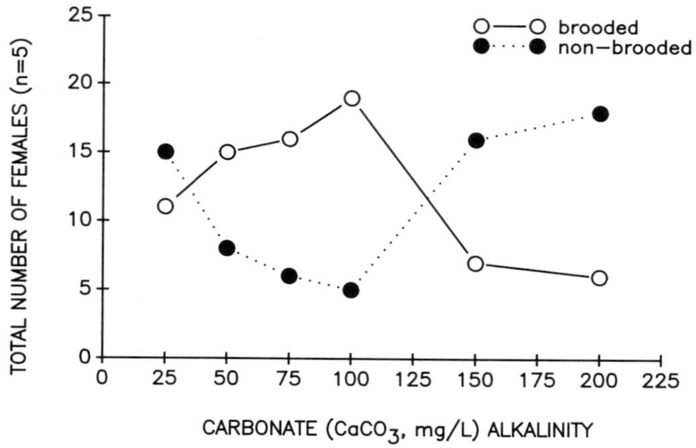

FIG. 1—*Total number of brooded and nonbrooded female* Mysidopsis bahia *as influenced by the different levels of $CaCO_3$.*

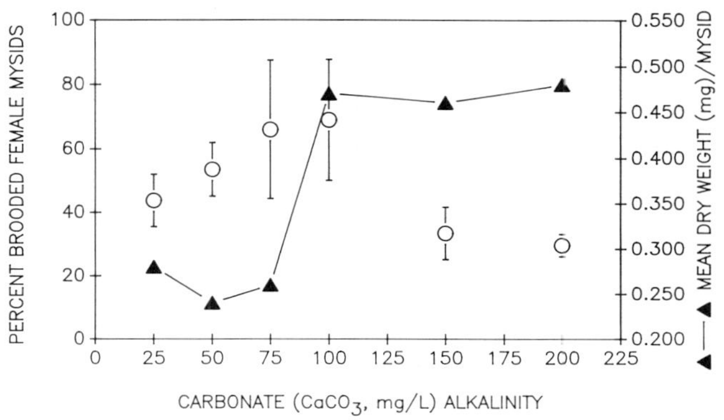

FIG. 2—*Mean percentage ±SD of brooded female* Mysidopsis bahia *and average dry weight (mg) as compared to the different levels of carbonate ($CaCO_3$) alkalinity.*

TABLE 2—*Acute toxicity for* Mysidopsis bahia *exposed to cadmium (μg/L) at different levels of carbonate ($CaCO_3$) alkalinity (mg/L).*

	LC_{50}, 95% Confidence Interval			
$CaCO_3$	48-h	72-h	96-h	*n*
25	52.75	32.29	21.68	3
	(41.37-67.47)	(24.59-40.41)	(16.14-27.37)	
	52.74	33.19	23.54	
	(41.82-66.17)	(25.26-41.60)	(17.78-29.78)	
	54.03	35.23	24.38	
	(41.62-71.72)	(26.88-44.20)	(18.66-30.31)	
50	88.47	42.71	30.73	2
	(71.32-133.25)	(33.28-53.24)	(24.17-37.36)	
	99.14	43.65	33.80	
	(73.34-195.87)	(34.97-52.50)	(27.06-40.41)	
100	87.40	42.53	31.74	4
	(65.89-149.75)	(33.69-51.81)	(25.07-38.44)	
	94.30	45.16	31.96	
	(75.32-153.13)	(35.03-57.04)	(25.65-38.15)	
	111.05	46.63	32.11	
	(78.63-273.67)	(37.73-55.69)	(25.97-38.11)	
	123.74	52.22	32.64	
	(76.63-552.75)	(43.88-59.99)	(25.76-39.57)	
200	31.13	21.53	17.93	3
	(24.03-38.47)	(16.76-26.57)	(14.25-22.05)	
	33.97	22.80	19.32	
	(26.71-41.33)	(18.15-27.60)	(15.25-23.76)	
	35.77	22.87	20.03	
	(28.11-43.59)	(18.07-27.90)	(15.79-24.61)	

NOTE: 100 mg/L $CaCO_3$ = natural seawater diluted. Salinity was 25 to 26‰ for all treatments and tests. $CaCO_3$ values shown are nominal.

treatments. In comparing the 95% confidence limits of 48, 72, and 96-h LC_{50}s for each treatment group, there is a significant increase in toxicity with carbonate imbalance. This factor is most dramatic at the 48-h time interval and corresponds well with the start of the molting process (ecdysis), a time when the mysid would be most susceptible to ionic imbalance. Figure 3 illustrates this variation and demonstrates that although the toxic effects of cadmium increased in parallel with increasing time, the rate of mortality is more pronounced for the two extreme carbonate treatments tested. The mysids in the 50 mg/L $CaCO_3$ treatment group responded similarly to the 100 mg/L treatment group.

Effects of Buffer and Ionic Strength

Table 3 shows the results of the analysis for pH, TIC, and TOC as well as the nominal and actual $CaCO_3$ concentrations for the growth (1 to 6 days), growth and reproduction, and acute toxicity tests. The measure of $CaCO_3$ in solution correlates well with available TIC for each of the carbonate treatment groups.

When TOC readings from high- or low-carbonate concentrations are compared with those from 100 mg/L, an increase in the release of organic carbon is observed in the former. This increase is dramatic for the well-buffered $CaCO_3$ concentrations, those with $NaHCO_3$ added. The lower carbonate concentration (25 mg/L $CaCO_3$) shows a lesser increase. The pH data show loss of buffer capacity for the lower concentrations of $CaCO_3$. Of the six different $CaCO_3$ levels tested, only the two groups with $NaHCO_3$ added showed a constant pH.

When compared to mysids held in 100 mg/L $CaCO_3$, animals in high- or low-carbonate concentrations either showed a loss in the female's ability to deposit eggs in the marsupium (empty) or the lamellae of the marsupium were not fully formed. Both high- and low-carbonate concentrations also produced greater sensitivity to cadmium during acute toxicity tests.

Discussion

The results of this study indicate that *Mysidopsis bahia* respond to the level of carbonate ($CaCO_3$) alkalinity in the medium even when salinity and temperature are held constant. The relationship between $CaCO_3$ level, growth and reproduction, and the results of the cadmium toxicity tests illustrate that physiological mechanisms associated with $CaCO_3$ level are closely connected to, or manifested during, certain phases of the life cycle for *M. bahia*.

In the laboratory, a standard practice is to increase the carbonate ($CaCO_3$) alkalinity level (especially for recirculating systems) to buffer against pH changes. In natural seawater, such as that used in these experiments, the pH normally approximates 8 and bicarbonate would be the dominant buffering ion. In recirculating systems, to counteract acidification caused by culture organisms and to neutralize hydrogen ion changes, Na_2CO_3 or $NaHCO_3$ is added. In a previous study [*7*], which subsequently led to this work, I sought to increase hatchability of *Artemia* fed to *M. bahia* by increasing the amount of carbonate in solution by the addition of Na_2CO_3 or $NaHCO_3$. This practice, however, often resulted in poor quality of hatch. In contrast, Trotman et al. [*14*] recently found that a deficiency of bicarbonate caused an abnormal emergence of *Artemia* prenauplii during incubation. The above studies indicate a sensitivity to carbonate/bicarbonate concentration that was previously little considered.

Although it was not the intent of this study to focus on the effects of carbonate ($CaCO_3$) alkalinity levels on long-term culturing practices, the question of acclimation should be considered. I have found that populations of *M. bahia* reared under the same conditions as stated above, with $NaHCO_3$ added, continue to decline over a three-month period when $CaCO_3$ levels exceed 150 mg/L. The number of juveniles per brooded female was similar to that in cultures without carbonate added. Thus, acclimation does not occur in these circumstances. While juveniles might have a different tolerance range for $CaCO_3$, this does not appear to be the case since

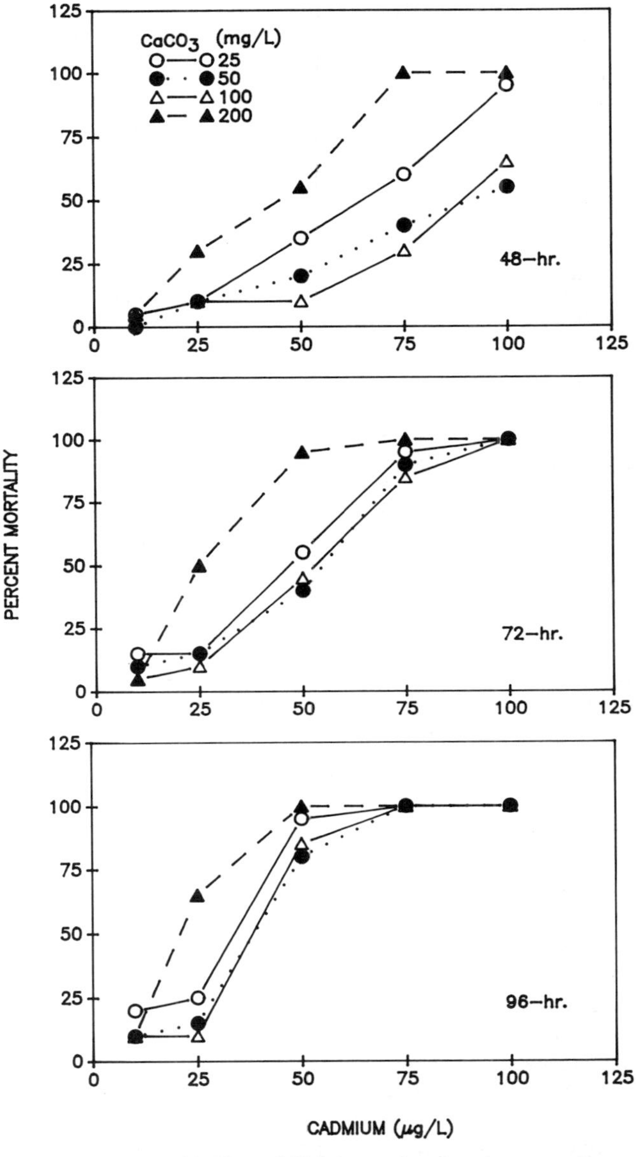

FIG. 3—*Toxicity of cadmium at 48, 72, and 96-h intervals, plotted as percentage mortality at the five concentrations of cadmium tested for the different $CaCO_3$ levels selected.*

mortality was not a key factor in the 1 to 6 day old growth test nor for the controls in the acute toxicity tests. It would appear, as shown in this study, that the female is affected at time of sexual maturity.

The results of the growth and reproduction test indicate that the concentration of $CaCO_3$ in the test medium can greatly influence growth and reproduction. Temperature and salinity are considered dominant environmental factors for determining rates of growth and reproduction.

TABLE 3—*A comparison of $CaCO_3$ (treatments), pH, total inorganic carbon (TIC), and total organic carbon (TOC) reported as a range over time.*

$CaCO_3$, mg/L				
Nominal	Measured	pH	TIC, mg/L	TOC, mg/L
		GROWTH[a]		
25	30–32	7.1–8.2	2.0– 2.5	2.0– 6.5
50	41–43	7.2–8.2	5.0– 5.5	2.5– 6.0
75	65–70	7.5–8.2	11.5–12.5	3.0– 6.0
100	85–90	7.6–8.0	15.5–18.5	3.0– 5.0
150	141–146	8.0–8.1	23.5–28.0	3.0–10.0
200	185–198	8.0–8.2	35.0–39.5	3.0–12.0
		REPRODUCTION[b]		
25	28–31	7.2–8.3	2.0– 3.0	2.0– 6.8
50	40–46	7.3–8.2	4.3– 6.6	3.0– 7.1
75	70–72	7.5–8.1	10.2–13.6	3.0– 5.4
100	85–96	7.7–8.1	14.2–19.3	3.0– 4.9
150	145–154	7.9–8.2	23.2–28.5	3.0– 8.0
200	199–212	8.1–8.2	35.0–37.0	3.0–11.0
		CADMIUM[c]		
25	19–25	7.0–8.0	2.0– 3.0	2.0– 3.8
50	41–47	7.2–8.1	5.0– 8.5	3.0– 3.8
100	77–86	7.5–8.0	17.4–19.5	2.8– 3.5
200	190–196	7.9–8.3	34.2–37.0	2.0– 6.5

NOTE: 100 mg/L $CaCO_3$ = natural seawater diluted. Salinity was 25 to 26‰ for all treatments and tests.
[a]Growth experiment for 1 through 6-day-old mysids.
[b]Reproduction test for 7 through 15-day-old mysids.
[c]96-h acute toxicity tests with mysids for cadmium.

Nevertheless, carbonate may provide a specific signal that has some threshold value which must be maintained for synchronization of the egg-laying process, the developing marsupium, and ecdysis in the female. Although this study did not focus on time of ecdysis for the growth and reproduction experiment, the female must first molt before extrusion of eggs from the ovaries into the brood pouch.

The formation of the new exoskeleton for *M. bahia*, as for other Crustacea, depends in part on the precipitation of carbonate from seawater and the incorporation of calcium and carbonate in the exoskeleton. Greenway [*1*] demonstrated that the absence of bicarbonate reduced the uptake of calcium by as much as 60% during certain periods of the molting cycle for crayfish. When inorganic carbon drops below the critical threshold as noted above, there is insufficient bicarbonate to satisfy the needs of the newly forming exoskeleton during molting.

The analysis of samples for inorganic carbon (principally HCO_3), as reported above, showed a range of concentrations. The 100 mg/L $CaCO_3$ seawater group had concentrations in the area of 15 to 20 mg/L inorganic carbon (carbonate). This range is similar to that found in Wickens' work (20 mg/L as inorganic carbon) [*15*]. Wickens identified problems with prawns in terms of growth and survival when inorganic carbon levels fell below 10 to 12 mg/L. My data also suggest that 15 to 20 mg/L inorganic carbon is a critical threshold range. Further analysis in this study showed that a 2 to 3 mg/L inorganic carbon concentration (15 mg/L $CaCO_3$) was acutely toxic to juvenile mysids. At this inorganic carbon level, 60% mortality occurred between 48 to 72 h. This correlated well with molting.

Compounds or effluents being tested, or excretion by organisms in toxicity tests and culture, may alter the pH, alter the buffering capacity of the test solution or culture (especially recirculating systems), and disrupt the hydrogen ion balance. Also, in toxicity tests a series of concentrations are used to derive the LC_{50} value. This technique, depending on the type of compound tested, may greatly change the balance of ions in solution. Such ion changes may induce different animal response because the animal's normal physiology is disrupted.

Mantel [*16*] reported that the uptake of water for crustaceans is greatly increased a few hours before molt and is at a maximum during the molting process. For the acute toxicity tests conducted here, it was established that *M. bahia* were more sensitive to cadmium during molting (ecdysis) when the concentration of $CaCO_3$ was low or high. Gentile et al. [*4*] showed problems with molting and reproduction for brooding female *M. bahia* and *M. bigelowi* exposed to cadmium and suggest that cadmium ions may interfere with the uptake of calcium from the external medium for the calcification of the exoskeleton. It has also been shown by Lee and Buikema [*5*] that survival is significantly shorter for *Daphnia pulex* for animals that molted during exposure to chromate.

A number of researchers have reported acute toxicity data for *M. bahia* exposed to cadmium or cadmium chloride at various temperatures and salinities. Nimmo et al. [*3*] identified a 96-h LC_{50} of 15.5 μg/L at 25 to 28°C, 10 to 17‰ salinity for juveniles. Roberts et al. [*17*] identified 96-h LC_{50}s of 17 and 29 μg/L at 22 ± 0.5°C, 20‰ salinity for adults. Gentile et al. [*4*] identified a 96-h LC_{50} of 110 μg/L at 21 ± 1°C, 30‰ salinity for juveniles. Bengtson et al. [*8*] identified 48-h LC_{50}s of 98.5 and 64.4 μg/L at 25 ± 1°C, 29 ± 1‰ salinity for adults with differences in response attributed to strain of *Artemia* fed to mysids. Comparing their findings with those reported here shows that some, but not all, values were similar, even in the cases where some had used adults instead of juveniles.

There have been several studies that have reported the importance of using reference toxicants to insure the reproducibility of toxicity tests, for example, Dorn et al. [*18*]. The increased sensitivity to cadmium shown in this study was caused by varying one important ion (carbonate), while differences in toxicity for the studies cited above were possibly caused by temperature and salinity differences. Such variation in response would indicate that for water quality-based toxics control using toxicity testing [*19*], a standard test would be desirable to insure reproducibility of the test.

Current test methods address two important factors, temperature and salinity, that could affect the outcome of the test for a specific organism. However, to better approximate toxic response, we must better understand not only all of the organism's requirements, but also what factors may alter sensitivity. Current methods allow the use of reconstituted formula and commercial brands of seasalts [*10*]. As shown by Bidwell and Spotte [*20*], commercial brands of seasalts can be extremely variable in the amount of HCO_3 they provide. This study suggests that when culturing and performing toxicity tests with *M. bahia* or other crustaceans, the medium in which they are grown or tested should have a controlled $CaCO_3$ level.

Conclusion

Altering the $CaCO_3$ concentration in natural seawater when testing and culturing with *Mysidopsis bahia* can affect growth and reproduction and increase sensitivity to a toxicant. The differences reported here are attributed to the carbonate concentration in the medium and can be of great importance as a first step in the understanding of a specific requirement of *M. bahia*. This study also suggests the limitations in culture and test practices of (1) using saline brines below a level of 50 mg/L $CaCO_3$ and (2) the addition of sodium carbonate. It is recommended that $CaCO_3$ be routinely measured when culturing and performing toxicity tests with *M. bahia*.

Acknowledgment

I wish to thank Kathleen I. Keating, Rutgers University, and Thomas J. Fikslin, U.S. EPA, Region II, for their constructive criticism and encouragement over the years. A special note of thanks goes to the personnel of the Inorganic Chemistry Section, Technical Support Branch, U.S. EPA, Region II, for their support in analyzing cadmium and mineral samples.

Disclaimer

Mention of commercial products does not imply endorsement by the U.S. Environmental Protection Agency.

References

[*1*] Greenway, P., *Journal of Experimental Biology*, Vol. 61, 1974, pp. 35–45.
[*2*] Wickins, J. F., *Aquaculture*, Vol. 41, 1984, pp. 49–60.
[*3*] Nimmo, D. R., Rigby, R. A., Bahner, L. H., and Sheppard, J. M., *Bulletin of Environmental Contamination and Toxicology*, Vol. 19, 1978, pp. 80–85.
[*4*] Gentile, S. M., Gentile, J. H., Walker, J., and Heltshe, J. F., *Hydrobiologia*, Vol. 93, 1982, pp. 195–204.
[*5*] Lee, D. R. and Buikema, A. L., *Journal of the Fisheries Research Board of Canada*, Vol. 36, 1979, pp. 1129–1133.
[*6*] Ward, S. H., *Progressive Fish-Culturist*, Vol. 46, 1984, pp. 170–175.
[*7*] Ward, S. H., *Progressive Fish-Culturist*, Vol. 49, 1987, pp. 29–33.
[*8*] Bengtson, D. A., Beck, A. D., Lussier, S. M., Migneault, D., and Olney, C. E. in *Ecotoxicological Testing for the Marine Environment*, Persoone, Jaspers, and Claus, Eds., State University of Ghent, and Institute for Marine Scientific Research, Bredene, Belgium, Vol. 2, 1984, pp. 399–418.
[*9*] Lussier, S. M., Kuhn, A., and Sewall, J. in "Users' Guide to the Conduct and Interpretation of Complex Effluent Toxicity Tests at Estuarine/Marine Sites," Contribution No. 796, S. C. Schimmel, Ed., Environmental Research Laboratory, U.S. Environmental Protection Agency, Narragansett, RI, 1986.
[*10*] Peltier, W. H. and Weber, C. I., "Methods for Measuring the Acute Toxicity of Effluents to Freshwater and Marine Organisms," EPNA 600/4-85/013, U.S. Environmental Protection Agency, Cincinnati, 1985.
[*11*] "Methods for Chemical Analysis of Water and Wastes," EPA 600/4-79-020, United States Environmental Protection Agency, Cincinnati, 1983.
[*12*] Stephan, C. E. in *Aquatic Toxicology and Hazard Evaluation, (First Conference), ASTM STP 634*, Mayer and Hamelink, Eds., American Society for Testing and Materials, Philadelphia, 1977, pp. 65–84.
[*13*] *Standard Methods for the Examination of Water and Wastewater*, 15th ed., American Public Health Association, Washington, DC, 1980.
[*14*] Trotman, C. N. A., Gieseg, S. P., Pirie, R. S., and Tate, W. P., *The Journal of Experimental Zoology*, Vol. 243, 1987, pp. 225–232.
[*15*] Wickins, J. F., *Aquaculture*, Vol. 9, 1976, pp. 19–37.
[*16*] Mantel, L. H., *The Biology of Crustacea*, Vol. 5, Academic Press, New York, NY, 1983.
[*17*] Roberts, M. H., Jr., Warinner, J. E., Tasi, C., Wright, D., and Cronin, L. E., *Archives of Environmental Contamination and Toxicology*, Vol. 11, 1982, pp. 681–692.
[*18*] Dorn, P. B., Rodgers, J. H., Jr., Jop, K. M., Raia, J. C., and Dickson, K. L., *Environmental Toxicology and Chemistry*, Vol. 6, 1987, pp. 435–444.
[*19*] "Technical Support Document for Water Quality-based Toxics Control," EPA 400/4-85-032, United States Environmental Protection Agency, Office of Water, Washington, DC, 1985.
[*20*] Bidwell, J. P. and Spotte, S., *Artificial Seawaters*, Jones and Bartlett Publishers, Inc., Boston, MA, 1985.

Philippe Léger,[1] *David A. Bengtson,*[2] *and Patrick Sorgeloos*[3]

Analytical Variation in the Determination of the Fatty Acid Composition of Standard Preparations of Brine Shrimp *Artemia:* an Interlaboratory Exercise

REFERENCE: Léger, P., Bengtson, D. A., and Sorgeloos, P., "**Analytical Variation in the Determination of the Fatty Acid Composition of Standard Preparations of Brine Shrimp *Artemia:* an Interlaboratory Exercise,**" *Aquatic Toxicology and Hazard Assessment: 12th Volume, ASTM STP 1027,* U. M. Cowgill and L. R. Williams, Eds., American Society for Testing and Materials, Philadelphia, 1989, pp. 413–423.

ABSTRACT: Brine shrimp, *Artemia* sp., nauplii are used throughout the world for the culture of larval aquatic organisms, especially fish and crustaceans. Recent studies have shown that the nutritional quality of *Artemia* can vary greatly from batch to batch and that *Artemia* fall basically into two categories: those that can only support survival, growth, and reproduction of freshwater organisms, and those that can also support survival, growth, and reproduction of saltwater organisms. The difference arises because marine organisms require certain essential fatty acids that freshwater organisms do not.

ASTM Standard E 1203 describes procedures for evaluation and use of *Artemia* nauplii as food for toxicity test organisms [Practice for Using Brine Shrimp Nauplii as Food for Test Animals in Aquatic Toxicology Tests (E 1203)]. One suggested procedure is the determination of the fatty acid composition of the *Artemia.* However, different laboratories around the world use different analytical methods for determination of lipid content and fatty acid composition and different ways of expressing the results. Until recently, most investigators reported the amount of each fatty acid as a percentage of the total amount of all fatty acids (the area percent method). Now, reports of absolute amounts of each fatty acid (mg fatty acid/g tissue) are increasingly seen. The results are sometimes expressed on a wet-weight basis, sometimes on a dry-weight basis.

An international interlaboratory exercise was conducted to investigate the variability associated with the preparation and analysis of samples and the reporting of fatty acid composition data for two samples of *Artemia* supplied to the laboratories by the Artemia Reference Center (ARC). Results reported to the ARC indicated that intralaboratory variability in determinations of total lipid content and fatty acid composition by the area-percent method was generally low, with a coefficient of variation (CV) around 5 to 7% on average. Intralaboratory variability in determination of fatty acid composition by the mg/g method was higher, with CVs greater than 10% in several cases. Interlaboratory variability was generally much greater, with CVs of >20% for total lipid content, 8 to 27% for composition of individual fatty acids determined by the area-percent method, and 12 to 49% for composition of individual fatty acids determined by the mg/g method. Thus, the precision of the data from any given laboratory is generally adequate, but the accuracy may not be. Nevertheless, in this case, data from each of the laboratories would support the decision to use these particular *Artemia* only for freshwater organisms.

KEY WORDS: brine shrimp, *Artemia,* fatty acids, interlaboratory exercise

[1]Research associate, Artemia Reference Center, State University of Ghent, Rozier 44, B-9000 Ghent, Belgium.

[2]Research assistant professor, Department of Zoology, University of Rhode Island, Kingston, RI 02881.

[3]Senior scientist, Belgian National Science Foundation, Artemia Reference Center, State University of Ghent, Rozier 44, B-9000 Ghent, Belgium.

Brine shrimp (*Artemia* sp.) nauplii are commonly used worldwide as live food for the larvae of many aquatic organisms cultured for toxicological research, as well as for commercial aquaculture. Brine shrimp cysts (dormant embryonic stage) are collected from salt production ponds in many parts of the world and, after cleaning and processing, can be incubated for 24 h in seawater to produce hatched nauplii.

The nutritional quality of the nauplii can vary, both among geographical strains and temporally or spatially within a single strain [*1–3*]. The variation is due primarily to the fatty acid composition of the nauplii [*4*], which in turn is dependent on the food conditions in the ponds where the cysts were produced. The nauplii tend to have high levels either of linolenic acid (18:3-omega 3) or of eicosapentaenoic acid (20:5-omega 3), but not both [*2*]. The variation is significant to culturists feeding the nauplii to predator organisms because freshwater organisms require linolenic acid in their diet and marine organisms require eicosapentaenoic acid [*5*].

The American Society for Testing and Materials recommends that each individual batch of *Artemia,* prior to its being used as a food for a particular species in toxicity testing, either be examined biochemically or be tested for its suitability as a food for that species [*6*]. If a particular batch of *Artemia* is nutritionally inadequate for a particular species used in toxicity tests, at least three negative outcomes can potentially occur: (a) the stock cultures may be unable to produce sufficient eggs or larvae to begin the test; (b) survival/growth/reproduction of the control organisms in the test may be so poor that valid test results are unobtainable, or (c) valid test results may produce LC_{50} values that are significantly lower than those for organisms of the same species fed nutritionally adequate *Artemia* [*7*]. Thus, it is very cost effective for toxicologists (and aquaculturists generally) to determine the fatty acid profile of a batch of *Artemia* to ensure that it is adequate for the species they are raising.

Recently, several techniques have been developed (see review in Ref *2*) to "enrich" *Artemia* nauplii, that is, to feed the nauplii various diets like algae or self-emulsifying concentrates in order to increase the levels of eicosapentaenoic acid and/or docosahexaenoic acid (22:6-omega 3) to improve the nutritional adequacy of the nauplii for marine organisms. During a workshop at the Second International Symposium on the Brine Shrimp, *Artemia,* a great discrepancy was noted in the maximum amount of enrichment that was attained and reported by different groups of scientists. Whether the discrepancy was due to differences in enrichment technique or to differences in analytical method was unclear [*8*]. In order to investigate the causes of the discrepancy, an international interlaboratory exercise was organized, to be conducted in two steps. In the first step, the participating laboratories would analyze the lipid contents and fatty acid profiles of two standard preparations of *Artemia:* (a) untreated *Artemia* cysts (for hatching and analysis of the nauplii by the participating laboratory) and (b) decapsulated cysts (for direct analysis without hatching). (Decapsulated cysts are those whose chorions have been chemically removed.) In the second step, the participants would all analyze the lipid content and fatty acid profile of nauplii after they had been hatched from the standard untreated cysts and enriched for 24 h with a reference enrichment product.

The purpose of this paper is to present the results of the first step of this interlaboratory exercise. The second step is in progress. In any case, most toxicologists, as well as aquaculturists generally, will most likely send samples of *Artemia* cysts to testing laboratories with instructions to determine the fatty acid profile. In the absence of any further instructions, the testing laboratories will most likely incubate the cysts under whatever conditions they deem appropriate to obtain the nauplii, extract the lipids by any of several procedures, further prepare the sample for chromatography, and obtain a fatty acid profile by means of gas chromatography. The report sent to the client will probably include: (a) data on the total lipid content [expressed either as per cent or mg lipid per g dry weight (or even wet weight) of sample]; and (b) either a qualitative fatty acid profile, amount of each detectable fatty acid methyl ester (FAME) expressed as a percentage of the total FAMEs (area percent method), or a quantitative profile, amount of each FAME expressed as mg FAME per g dry weight. The client must then evaluate these data and decide whether or not to purchase a large quantity of those particular *Artemia* cysts (sometimes at a cost of several thousands of dollars). A second purpose of this paper is to educate toxicologists and culturists on the variety of test results they can expect and to encourage them to specify test procedures to the testing laboratories to which they send their *Artemia* samples. Finally, it is hoped that this study will alert analytical laboratories to the needs of toxicologists and culturists so that the data provided to them will be maximally useful.

Materials and Methods

To each of the 20 laboratories that originally agreed to participate in the interlaboratory exercise, the following materials were sent: (a) dry, untreated *Artemia* cysts; (b) dry, decapsulated *Artemia* cysts (both of undisclosed geographical origin); and (c) the following written instructions:

> Since the first step of the interlaboratory exercise aims to assess interlaboratory variability of methodological and analytical procedures, no special instructions for Artemia hatching, sample preparation and chromatographic analysis are given. Each participating laboratory is asked to apply their own routine procedures for a) Artemia hatching: production of freshly hatched Instar I nauplii (only for the analysis of untreated Artemia cysts), b) sample preparation: including homogenization, lipid extraction, saponification, esterification, c) chromatographic analysis and data processing. The interlaboratory material consists of untreated Artemia cysts (for production and analysis of Instar I nauplii) and dry decapsulated cysts (for direct analysis without hatching). Enough material of both samples is supplied to carry out analyses in triplicate. For smooth processing of data of Step I and preparation of the next step it is asked to communicate following data and information: a) total lipid content of decapsulated cysts and freshly hatched nauplii from untreated cysts, expressed as mg total lipid per gram dry weight of decapsulated cysts or nauplii, b) fatty acid profile of decapsulated cysts and freshly hatched nauplii expressed in relative values (area percent composition: percent fatty acid ester of total fatty acid esters) and absolute values (mg fatty acid ester per gram dry weight of decapsulated cysts or nauplii); c) detailed outline of procedure followed for Artemia hatching (mentioning incubation temperature, salinity, duration, light and aeration conditions, etc.) sample preparation and chromatographic analysis (description of methods and materials used).

From the triplicate data on lipid content and fatty acid composition received from each laboratory, mean, standard deviation, and coefficient of variation (CV) were calculated for lipid content and for amounts of six selected fatty acids. The intralaboratory CV for lipid content for each lab was used in the calculation of a mean intralaboratory CV for all labs. The mean lipid content for each laboratory was used in the calculation of an overall mean and standard deviation for all labs, so that the standard deviation (and the associated interlaboratory CV) would represent only interlaboratory variability. The CVs for the six selected fatty acids from each laboratory were used to calculate a mean intralaboratory CV for fatty acid determinations for all labs. The mean content of a given fatty acid for each laboratory was used in the calculation of an overall mean and standard deviation (and interlaboratory CV) for that fatty acid.

Results

Results were received from only 9 of the 20 laboratories to which the samples were sent, and only 6 of those 9 laboratories provided all the data that was requested. Three laboratories (Labs 2, 4, and 9) were unwilling or unable to report the fatty acid profiles on a quantitative (mg/g) basis. One laboratory (Lab 4) presented the total lipid content of the *Artemia* nauplii on a wet weight basis only.

The information provided by the laboratories regarding the specific methods and equipment they used ranged from very detailed to minimal. *Artemia* cysts were incubated to effect hatching over a relatively narrow temperature range (25 to 30°C), but wide ranges of salinity (5 to 33 0/00) and duration of incubation (22 to 45 h) (Table 1). Lipids were usually extracted either by the method of Folch et al. [9] or by that of Bligh and Dyer [10], although one laboratory used a kit and one did not specify the method (Table 2).

The average intralaboratory variation in determination of lipid content was 4.9% (coefficient of variation) for decapsulated cysts and 7.1% for nauplii. The interlaboratory variation was much higher, however, being 28.5% for decapsulated cysts and 22.8% for nauplii. When the data were grouped according to the method of lipid extraction, no significant differences in lipid content determination were observed (ANOVA; $p > 0.05$) for those laboratories that used

TABLE 1—*Differences among the laboratories participating in the* Artemia *fatty acid interlaboratory exercise, with regard to methods of hatching* Artemia, *extracting lipids, and identifying particular fatty acids. Key:* — = *no information provided;* + = *peak identification was by reference standard.*

	Artemia Incubation Conditions				
Laboratory No.	Temperature, °C	Salinity, 0/00	Duration, h	Lipid Extraction Method	I.D. of Peaks by Reference Standards
1	...	...	...	Folch et al.	—
2	28	29.5	45	Folch et al.	—
3	25	30	28	Bligh and Dyer	+
4	25	...	24	Folch et al.	+
5	28	20	22	Reagent kit	+
6	28	35	24	Bligh and Dyer	+
7	26	5	26–27	...	—
8	28–30	33	24	Bligh and Dyer	+
9	...	...	24	Folch et al.	—

TABLE 2—*Lipid content (mg lipid/g dry weight) of decapsulated* Artemia *cysts and hatched* Artemia *nauplii. Data are given as mean ± standard deviation (s.d.), with coefficient of variation given in parentheses.*

	Lipid Content	
Laboratory No.	Decapsulated Cysts	Hatched Nauplii
1	147.5 ± 2.0 (1.4)	188.3 ± 8.3 (4.4)
2	165.5 ± 1.6 (1.0)	166.9 ± 2.7 (1.6)
3	40.2 ± 2.8 (7.0)	107.5 ± 16.1 (15.0)
4	157.3 ± 22.0 (14.0)	...
5	170.0 ± 1.5 (0.9)	185.0 ± 8.2 (4.4)
6	161.5 ± 15.1 (9.3)	212.7 ± 33.3 (15.7)
7	142.0 ± 7.6 (5.4)	141.3 ± 13.1 (9.3)
8	183.2 ± 5.0 (2.7)	199.4 ± 4.5 (2.3)
9	165 ± 4 (2.4)	238 ± 10 (4.2)
Average of intralaboratory CVs	(4.9)	(7.1)
Average of interlaboratory means	148.0 ± 42.2 (28.5)	179.9 ± 41.1 (22.8)

Bligh and Dyer extraction versus those that used Folch extraction. Laboratory 3 was an outlier from the other labs in determination of lipid content.

No two laboratories used exactly the same procedures for sample preparation (saponification and esterification) or for chromatography. Both the types and concentrations of reagents used for sample preparation differed from lab to lab. The labs also differed in their choices of gas chromatograph, column, column temperature, and carrier gas; however, flame ionization detectors were used by all those labs (2, 3, 5, 6, 7, 8) that specified the type of detector. Only Labs 3, 4, 5, 6, and 8 specified that reference standards were used to identify peaks.

Rather than presenting the total fatty acid profiles obtained from each laboratory, we present only the data for six major fatty acids that normally comprise about 75% of all the fatty acids in *Artemia.* Results are expressed both on an area percent basis for decapsulated cysts (Table 3) and nauplii (Table 4) and on a mg/g basis for decapsulated cysts (Table 5) and nauplii (Table 6).

The area percent data were less variable on both intralaboratory and interlaboratory levels than were the mg/g data. The intralaboratory variation in area percent determination of the six major fatty acids was less than 5% (mean of the individual CVs for each of the six acids) for all laboratories except Lab 7, which had mean CVs of 10.2% for nauplii and 22.1% for decapsulated cysts. Labs 3 and 6 had mean CVs of 14.3% and 8.3% for nauplii only. The interlaboratory variation in the area percent data for each of the six fatty acids, on the other hand, ranged from 8.3% to 26.3% for the individual fatty acids in the decapsulated cysts and from 11.2 to 27.2% for the individual fatty acids in the nauplii. The intralaboratory variation in mg/g determination of the six fatty acids was less than 5% in half the cases and greater than 10% in the other half. The interlaboratory variation in the mg/g data for each of the six fatty acids was quite high, ranging from 38.2 to 49.3% for the individual fatty acids in the decapsulated cysts and from 12.1 to 31.0% for the individual fatty acids in nauplii. A large part of the interlaboratory variation in the mg/g data was due to Laboratory 3, which again was an outlier from the other labs. It should be noted that Lab 3 was not an outlier with regard to the area percent data. The low levels (in mg/g) for all the individual fatty acids reported by Laboratory 3 undoubtedly follows from the low levels of total lipid content reported by that laboratory.

Discussion

It is by now well established that the fatty acid composition of *Artemia* is quite variable, both among geographical strains and within a given strain. An examination of data (published and unpublished) on fatty acid profiles of 142 samples of *Artemia* from around the world indicated that overall coefficient of variation for each of the six major fatty acids considered here ranged from 14.8% for 18:1-omega9 to 71.7% for both 18:3-omega3 and 20:5-omega3 [*2*]. All those data were based on the area percent method.

The results of the present study provided us with several lessons. First, even when certain analytical results were requested, they were not provided by the lab doing the analysis. Second, intralaboratory variability for determinations of lipid content and fatty acid profile (on an area percent basis) was small, indicating that precision of each lab's analyses was quite acceptable in almost all cases. Third, interlaboratory variability for those determinations was much larger than intralaboratory variability, indicating that accuracy may be more of a problem than precision. It should be pointed out that we do not know the truly "accurate" value for any of these determinations and each laboratory may have been quite accurate given its own methods and materials.

The problem for the culturist or toxicologist is to decide whether the laboratory that has analyzed his cysts has reported to him data that is accurate in relation to published information on fatty acid requirements of cultured organisms. For example, Leger et al. [*2*] commented that *Artemia* nauplii with levels of 20:5-omega3 of 4% or greater, by the area percent method, will

TABLE 3—*Fatty acid profile (area percent basis) for six major fatty acids in decapsulated cysts in the* Artemia *fatty acid interlaboratory exercise. Data for each FAME*[a] *are expressed as percentage of total FAMEs (mean ± s.d.), with coefficient of variation (CV) given in parentheses.*

Lab No.	FAME Percent Composition 16:0	16:1w7	18:1w9	18:2w6	18:3w3	20:5w3	Mean CV
1	13.48 ± 0.03 (0.2)	3.24 ± 0.01 (0.3)	20.88 ± 0.09 (0.4)	6.24 ± 0.05 (0.8)	28.10 ± 0.06 (0.2)	2.01 ± 0.01 (0.5)	(0.4)
2	12.50 ± 0.21 (1.7)	4.53 ± 0.06 (1.3)	25.57 ± 0.45 (1.8)	7.50 ± 0.17 (2.3)	27.20 ± 0.10 (0.4)	1.97 ± 0.06 (3.0)	(1.8)
3	12.60 ± 0.48 (3.8)	5.10 ± 0.21 (4.1)	31.76 ± 0.59 (1.9)	7.67 ± 0.04 (0.5)	21.60 ± 0.54 (2.5)	3.30 ± 0.38 (11.5)	(4.1)
4	14.01 ± 0.41 (2.9)	3.29 ± 0.16 (4.9)	23.31 ± 0.40 (1.7)	7.11 ± 0.04 (0.6)	33.13 ± 0.56 (1.7)	2.32 ± 0.08 (3.4)	(2.5)
5	15.38 ± 0.10 (0.7)	5.63 ± 0.05 (0.9)	29.96 ± 0.19 (0.6)	7.65 ± 0.03 (0.4)	33.08 ± 0.34 (1.0)	2.51 ± 0.02 (0.8)	(0.7)
6	13.58 ± 0.42 (3.1)	5.48 ± 0.44 (8.0)	26.40 ± 0.50 (1.9)	6.90 ± 0.25 (3.6)	30.28 ± 1.05 (3.5)	2.08 ± 0.10 (4.8)	(4.2)
7	10.70 ± 4.06 (37.9)	2.37 ± 1.44 (60.8)	34.00 ± 3.00 (8.8)	7.15 ± 0.26 (3.6)	33.67 ± 2.97 (8.8)	2.40 ± 0.30 (12.5)	(22.1)
8	15.00 ± 0.27 (1.8)	4.92 ± 0.03 (0.6)	35.12 ± 0.49 (1.4)	6.23 ± 0.47 (7.5)	21.66 ± 0.07 (0.3)	1.74 ± 0.02 (1.1)	(2.1)
9	14.80 ± 0.80 (5.4)	4.10 ± 0.20 (4.9)	20.40 ± 0.20 (1.0)	6.40 ± 0.10 (1.6)	26.60 ± 0.60 (2.3)	2.40 ± 0.20 (8.3)	(3.9)
Avg of interlab means	13.56 ± 1.47 (10.8)	4.30 ± 1.13 (26.3)	27.49 ± 5.49 (20.0)	6.98 ± 0.58 (8.3)	28.37 ± 4.64 (16.4)	2.30 ± 0.44 (19.1)	

[a] Fatty acid methyl ester.

TABLE 4—*Fatty acid profile (area percent basis) for six major fatty acids in newly hatched nauplii in the* Artemia *fatty acid interlaboratory exercise. Data for each FAME are expressed as percentage of total FAMEs (mean ± s.d.), with coefficient of variation (CV) given in parentheses.*

	FAME Percent Composition						
Lab No.	16:0	16:1w7	18:1w9	18:2w6	18:3w3	20:5w3	Mean CV
1	13.01 ± 0.04 (0.3)	3.15 ± 0.02 (0.6)	20.74 ± 0.17 (0.8)	6.25 ± 0.05 (0.8)	28.50 ± 0.06 (0.2)	2.12 ± 0.07 (3.3)	(1.0)
2	10.53 ± 0.21 (2.0)	4.83 ± 0.16 (3.3)	25.27 ± 0.25 (1.0)	8.80 ± 0.35 (4.0)	25.17 ± 0.58 (2.3)	2.27 ± 0.06 (2.6)	(2.5)
3	13.51 ± 1.08 (8.0)	5.96 ± 0.47 (7.9)	33.07 ± 2.96 (9.0)	6.57 ± 0.48 (7.3)	22.94 ± 2.71 (11.8)	2.19 ± 0.70 (32.0)	(12.7)
4	14.69 ± 0.15 (1.0)	3.44 ± 0.19 (5.5)	22.94 ± 0.61 (2.7)	6.84 ± 0.04 (0.6)	31.75 ± 0.11 (0.3)	2.24 ± 0.06 (2.7)	(2.1)
5	14.31 ± 0.03 (0.2)	5.53 ± 0.01 (0.2)	29.99 ± 0.01 (0.1)	7.81 ± 0.04 (0.5)	34.12 ± 0.06 (0.2)	2.55 ± 0.03 (1.2)	(0.4)
6	11.70 ± 2.42 (20.7)	5.48 ± 0.49 (8.9)	26.78 ± 0.32 (1.2)	6.98 ± 0.05 (0.7)	29.70 ± 1.17 (3.9)	2.00 ± 0.29 (14.5)	(8.3)
7	11.78 ± 2.28 (19.4)	2.46 ± 0.49 (19.9)	32.50 ± 1.31 (4.0)	6.75 ± 0.22 (3.3)	34.80 ± 0.96 (2.8)	2.53 ± 0.30 (11.9)	(10.2)
8	12.60 ± 0.27 (2.1)	4.31 ± 0.04 (0.9)	28.60 ± 0.40 (1.4)	6.81 ± 0.10 (1.5)	27.25 ± 0.18 (0.7)	2.36 ± 0.10 (4.2)	(1.8)
9	14.70 ± 0.70 (4.8)	4.20 ± 0.20 (4.8)	22.60 ± 0.40 (1.8)	5.20 ± 0.70 (13.5)	20.00 ± 0.30 (1.5)	3.80 ± 0.10 (2.6)	(4.8)
Avg of interlaboratory means	12.98 ± 1.46 (11.2)	4.37 ± 1.19 (27.2)	26.85 ± 4.29 (16.0)	6.89 ± 0.99 (14.4)	28.25 ± 4.98 (17.6)	2.45 ± 0.54 (22.0)	

TABLE 5—*Fatty acid profile (mg/g basis) for six major fatty acids in decapsulated cysts in the* Artemia *fatty acid interlaboratory exercise. Data for each FAME are expressed as mg of that FAME per g dry weight of tissue (mean ± s.d.), with coefficient of variation (CV) given in parentheses.*

Lab No.	Amount of FAME, mg/g						Mean CV
	16:0	16:1w7	18:1w9	18:2w6	18:3w3	20:5w3	
1	16.65 ± 0.18 (1.1)	4.00 ± 0.03 (0.8)	25.79 ± 0.34 (1.3)	7.71 ± 0.01 (0.1)	34.69 ± 0.33 (1.0)	2.48 ± 0.01 (0.4)	(0.8)
2	...	...	...	...	...	...	
3	3.02 ± 0.12 (4.0)	1.21 ± 0.04 (3.3)	7.53 ± 0.22 (2.9)	1.81 ± 0.02 (1.1)	5.12 ± 0.13 (2.5)	0.78 ± 0.08 (10.3)	(4.0)
4	...	...	...	...	...	...	
5	9.96 ± 0.32 (3.2)	3.76 ± 0.11 (2.9)	19.25 ± 0.52 (2.7)	5.00 ± 0.17 (3.4)	21.34 ± 0.83 (3.9)	1.62 ± 0.06 (3.7)	(3.3)
6	15.50 ± 2.51 (16.2)	6.25 ± 1.22 (19.5)	30.08 ± 3.76 (12.5)	7.88 ± 0.99 (12.6)	34.40 ± 4.14 (12.0)	2.35 ± 0.45 (19.1)	(15.3)
7	9.87 ± 3.75 (38.0)	2.18 ± 1.33 (61.0)	31.27 ± 2.75 (8.8)	6.58 ± 0.24 (3.6)	30.97 ± 2.77 (8.9)	2.21 ± 0.28 (12.7)	(22.2)
8	15.58 ± 0.27 (1.7)	5.11 ± 0.03 (0.6)	36.50 ± 0.51 (1.4)	6.48 ± 0.49 (7.6)	22.51 ± 0.07 (0.3)	1.80 ± 0.00 (0.00)	(1.9)
9	...	...	...	...	...	...	
Avg of interlab means	11.76 ± 5.21 (44.3)	3.75 ± 1.85 (49.3)	25.07 ± 10.37 (41.4)	5.91 ± 2.26 (38.2)	24.83 ± 11.26 (45.3)	1.87 ± 0.63 (33.7)	

TABLE 6—*Fatty acid profile (mg/g basis) for six major fatty acids in newly hatched nauplii in the* Artemia *fatty acid interlaboratory exercise. Data for each FAME are expressed as mg of that FAME per g dry weight of tissue (mean ± s.d.), with coefficient of variation given in parentheses.*

Lab No.	Amount of FAME, mg/g						Mean CV
	16:0	16:1w7	18:1w9	18:2w6	18:3w3	20:5w3	
1	20.07 ± 0.86 (4.3)	4.86 ± 0.25 (5.1)	31.99 ± 1.24 (3.9)	9.63 ± 0.35 (3.6)	43.96 ± 2.07 (4.7)	3.27 ± 0.24 (7.3)	(4.8)
2	...	...	...	...	...	...	
3	9.75 ± 0.36 (3.7)	4.28 ± 0.35 (8.2)	23.80 ± 0.42 (1.8)	4.78 ± 0.78 (16.3)	16.69 ± 3.55 (21.3)	1.62 ± 0.68 (42.0)	(15.6)
4	...	...	...	...	...	...	
5	12.63 ± 1.29 (10.2)	4.95 ± 0.49 (9.9)	26.27 ± 2.77 (10.5)	6.90 ± 0.71 (10.3)	29.90 ± 3.20 (10.7)	2.25 ± 0.21 (9.3)	(10.2)
6	14.63 ± 4.41 (30.1)	6.70 ± 0.83 (12.4)	33.08 ± 5.50 (16.6)	8.60 ± 1.34 (15.6)	36.58 ± 5.11 (14.0)	2.48 ± 0.69 (27.8)	(19.4)
7	10.84 ± 2.10 (19.4)	2.26 ± 0.45 (19.9)	29.93 ± 1.21 (4.0)	6.21 ± 0.20 (3.2)	32.03 ± 0.85 (2.7)	2.33 ± 0.28 (12.0)	(10.2)
8	13.60 ± 0.29 (2.1)	4.65 ± 0.04 (0.9)	30.49 ± 0.41 (1.3)	7.36 ± 0.11 (1.5)	29.42 ± 0.20 (0.7)	2.71 ± 0.11 (4.1)	(1.8)
9	...	...	...	...	...	...	
Avg of interlab means	13.55 ± 3.69 (27.2)	4.61 ± 1.43 (31.0)	29.18 ± 3.70 (12.7)	7.22 ± 1.77 (24.5)	31.37 ± 9.16 (29.2)	2.44 ± 0.54 (22.1)	

probably be adequate for marine organisms. Similarly, it is apparent from, although not explicitly stated in, the work of Watanabe et al. [*11,12*] that they consider levels of 20:5-omega3 of less than 4% (area percent method) to be inadequate for marine organisms. Data on adequate levels of 20:5-omega3 on a mg/g basis have only recently begun to appear in the literature [*13–15*] and indicate that about 4 mg/g of 20:5-omega3 is necessary for marine organisms and that higher levels (up to 35.2 mg/g reached through enrichment procedures) improve their survival and growth.

For the particular *Artemia* sample analyzed in this exercise, all the participating laboratories reported naupliar levels of 20:5-omega3, on an area percent basis, that were less than 4%. Therefore, any culturist who received the report on this sample from any of the nine participating laboratories would decide to use these nauplii for freshwater organisms but not for marine organisms. The same decision would be made based on the naupliar mg/g data. If the reporting of fatty acid data on a mg/g basis becomes more commonplace (as it should, because it is the truly quantitative information), one can appreciate that certain laboratories that, for whatever reason, underestimate the total lipid content of *Artemia* will underestimate the levels of individual fatty acids. Through the use of underestimated data, culturists may well choose not to use some batches of *Artemia* for marine organisms, when in fact it would be perfectly acceptable to do so.

The final message from this study is that both culturists and toxicologists will need to educate themselves about the nutritional requirements of their organisms so they can approach testing laboratories as knowledgable "consumers" of biochemical information. With their knowledge, they will be able to specify the kinds of information they need, to be sure they receive it, and to interpret it correctly. Furthermore, it is essential that analytical laboratories realize the critical need to verify the validity of their data (biological, chemical, instrumental variability) by intercalibration and data verification with specialized laboratories.

Acknowledgments

This research was supported in part by Cooperative Agreement CR-814721 between the U.S. Environmental Protection Agency and the University of Rhode Island and is Contribution Number X132 from the U.S. EPA Environmental Research Laboratory, Narragansett, RI. This research was also supported by the Belgian National Science Foundation through FKFO Project Artemia.

References

[*1*] Beck, A. D. and Bengtson, D. A. in *Aquatic Toxicology and Hazard Assessment (Fifth Conference), ASTM STP 766,* J. G. Pearson, R. B. Foster, and W. E. Bishop, Eds., American Society for Testing and Materials, Philadelphia, 1982, pp. 161–169.

[*2*] Léger, P., Bengtson, D. A., Simpson, K. L., and Sorgeloos, P., *Oceanography and Marine Biology Annual Review,* Vol. 24, 1986, pp. 521–623.

[*3*] Cowgill, U. M., Emmel, H. W., Boggs, G. U., Murphy, P. G., Gersich, F. M., Takahashi, I. T., and Bengtson, D. A. in *Artemia Research and its Applications,* Vol. 1, Universa Press, Wetteren, Belgium, pp. 173–188.

[*4*] Léger, P., Sorgeloos, P., Millamena, O. M., and Simpson, K. L., *Journal of Experimental Marine Biology and Ecology,* Vol. 93, 1985, pp. 71–82.

[*5*] Kanazawa, A., Teshima, S., and Ono, K., *Comparative Physiology and Biochemistry,* Vol. 63B, 1979, pp. 295–298.

[*6*] Practice for Using Brine Shrimp Nauplii as Food for Test Animals in Aquatic Toxicology Tests (E 1203), Vol. 11.04, *Annual Book of ASTM Standards,* American Society for Testing and Materials, Philadelphia, 1988.

[*7*] Bengtson, D. A., Beck, A. D., Lussier, S. M., Migneault, D., and Olney, C. E. in *Ecotoxicological Testing for the Marine Environment,* Vol. 2, State University of Ghent and Institute for Marine Scientific Research, Bredene, Belgium, 1984, pp. 399–416.

[*8*] Simpson, K. L. in *Artemia Research and its Applications*, Vol. 3, Universa Press, Wetteren, Belgium, pp. 515–516.
[*9*] Folch, J., Lees, M., and Sloane-Stanley, G. H., *Journal of Biological Chemistry*, Vol. 226, 1957, pp. 497–509.
[*10*] Bligh, E. G. and Dyer, W. J., *Canadian Journal of Biochemistry and Physiology*, Vol. 37, 1959, pp. 911–917.
[*11*] Watanabe, T., Oowa, F., Kitajima, C., and Fujita, S., *Bulletin of the Japanese Society of Scientific Fisheries*, Vol. 46, 1980, pp. 35–41.
[*12*] Watanabe, T., Ohta, M., Kitajima, C., and Fujita, S., *Bulletin of the Japanese Society of Scientific Fisheries*, Vol. 48, 1982, pp. 1775–1782.
[*13*] Léger, P., Bieber, G. F., and Sorgeloos, P., *Journal of the World Mariculture Society*, Vol. 16, 1985, pp. 354–367.
[*14*] Léger, P., Naessens-Foucquaert, E., and Sorgeloos, P. in *Artemia Research and its Applications*, Vol. 3, Universa Press, Wetteren, Belgium, 1987, pp. 411–424.
[*15*] Robin, J. H., LeMilinaire, C., and Stephan, G. in *Artemia Research and its Applications*, Vol. 3, Universa Press, Wetteren, Belgium, 1987, pp. 437–445.

Target Toxicant Analysis

Jeanette M. Van Emon[1]

Selected References Addressing the Development and Utilization of Immunoassay

REFERENCE: Van Emon, J. M., "**Selected References Addressing the Development and Utilization of Immunoassay,**" *Aquatic Toxicology and Hazard Assessment: 12th Volume, ASTM STP 1027,* U. M. Cowgill and L. R. Williams, Eds., American Society for Testing and Materials, Philadelphia, 1989, pp. 427–431.

ABSTRACT: A minisymposium on the uses of immunoassay techniques was organized by Dr. Bruce Hammork as a part of the 12th Aquatic Toxicology Symposium. Five papers were presented. This paper summarizes these activities and provides a pertinent bibliography.

KEY WORDS: references, bibliography

The field of immunoassay has made significant impacts in several areas of analytical chemistry. The technology is multidisciplinary and is based on components from immunology, enzymology, and synthetic, organic, clinical, and analytical chemistry. Due to this multifaceted nature of immunoassay, novices may be somewhat discouraged from utilizing the technology. Another hindrance to embracing immunoassay is the mistaken notion that immunoassay is synonymous with bioassay. Although antibodies are biologically derived reagents, they can be configured into extremely sensitive and quantitative assays. However, antibodies that are used in a quantitative assay may also be configured into a semiquantitative or qualitative assay by merely changing the assay conditions or format. The seemingly plethora of assay variations makes immunoassay a powerful and diverse analytical technology.

Another strength of the technology is that it can be used to analyze a wide variety of compounds. Abundant examples in the literature support the use of immunoassay for hormones, drugs, pesticides, microbial toxins, and environmental contaminants. Immunoassay is probably applicable to more classes of compounds than any other single analytical technique. In many instances, analyses can be performed on crude sample extracts obviating the need for extensive extraction or cleanup procedures. The technology does have limitations and is not particularly suitable for very small lipophilic compounds, or compounds that are water unstable since aqueous conditions are used to perform the assay.

Immunoassay development can be divided into several steps: hapten design and synthesis, hapten conjugation to carrier, immunization protocols for antibody production (either monoclonal or polyclonal antibodies), choice of indicator label, choice of assay format, data handling, and finally, the application of the specific immunoassay to solve a particular analytical need. The following bibliography encompasses this development scheme. Although this is not an exhaustive listing of applicable references, what is presented should aid in the understanding of assay development and utilization. This compendium of references is intended to encourage the continued use of immunoassay as well as stimulate assay development and implementation.

[1]Research molecular biologist, Exposure Assessment Research Division, Environmental Monitoring Systems Laboratory, Las Vegas, NV 89193-3478.

Bibliography

Immunoassay Development

Aguila, H. L., Pollock, R. R., Spira, G., and Scharff, M. D., "The Production of More Useful Monoclonal Antibodies. 2. The Use of Somatic-Cell Genetic and Recombinant-DNA Technology to Tailor-Make Monoclonal Antibodies," *Immunology Today,* Vol. 7, No. 12, 1986, pp. 380-383.

Avrameas, S., "Coupling of Enzymes to Proteins with Glutaraldehyde: Use of the Conjugates for the Detection of Antigens and Antibodies," *Immunochemistry,* Vol. 6, 1969, pp. 43-52.

Berson, S. A. and Yalow, R. S., "General Principles of Radioimmunoassay," *Clinica Chimica Acta,* Vol. 22, 1968, pp. 51-69.

Canellas, P. F. and Karu, A. E., "Statistical Package for Analysis of Competition ELISA Results," *Journal of Immunological Methods,* Vol. 47, 1981, pp. 375-385.

Delaage, M., de Veyrac, B., Morel, A., Cailla, H., and Drocourt, J. L., "Monoclonal Antibodies to Haptens," in *Monoclonal Antibodies and New Trends in Immunoassays,* C. A. Bizollon, Ed., Elsevier, Amsterdam, 1984, pp. 53-58.

Ehrlich, P. H., Moyle, W. R., Moustafa, Z. A., and Canfield, R. E., "Mixing Two Monoclonal Antibodies Yields Enhanced Affinity for Antigen," *Journal of Immunology,* Vol. 128, No. 6, 1982, pp. 2709-2713.

Eisen, H. N., *Immunology,* 2nd ed., Harper and Row, New York, 1980.

Ekins, R. P., "Current Concepts and Future Developments," in *Alternative Immunoassays,* W. P. Collins, Ed., John Wiley and Sons, New York, 1985, pp. 219-237.

Engvall, E., "Enzyme Immunoassay ELISA and EMIT," in *Methods Enzymology,* H. Van Vunakis and J. J. Langone, Eds., Academic Press, New York, Vol. 70, pp. 419-439.

Engvall, E., Jonsson, K., and Perlmann, P., "Enzyme-Linked Immunosorbent Assay II. Quantitative Assay of Protein Antigen, Immunoglobulin G, by Means of Enzyme-Labeled Antigen and Antibody-Coated Tubes," *Biochemica Biophysica Acta,* Vol. 251, 1971, pp. 427-434.

Engvall, E. and Perlmann, P., "Enzyme-Linked Immunosorbent Assay (ELISA) Quantitative Assay of Immunoglobulin G," *Immunochemistry,* Vol. 8, 1971, pp. 871-874.

Engvall, E. and Perlmann, P., "Enzyme-Linked Immunosorbent Assay, ELISA III. Quantitation of Specific Antibodies by Enzyme-Labeled Anti-Immunoglobulin in Antigen-Coated Tubes," *Journal of Immunology,* Vol. 109, No. 1, 1972, pp. 129-135.

Erlanger, B. F., "The Preparation of Antigenic Hapten-Carrier Conjugates: A Survey," in *Methods Enzymology,* H. Van Vunakis and J. J. Langone, Eds., Academic Press, New York, Vol. 70, 1980, pp. 85-104.

Fazekas de St. Groth, S. and Scheidegger, D., "Production of Monoclonal Antibodies: Strategy and Tactics," *Journal of Immunological Methods,* Vol. 35, 1980, pp. 1-21.

Fox, J. L., "Antibody Reagents Revolutionizing Immunology," *Chemistry and Engineering News,* 1 Jan. 1979, pp. 15-17.

French, D., Fischberg, E., Buhl, S., and Scharff, M.D., "The Production of More Useful Monoclonal Antibodies. 1. Modifications of the Basic Technology," *Immunology Today,* Vol. 7, No. 11, 1986, pp. 344-346.

Hammock, B. D. and Mumma, R. O., "Potential of Immunochemical Technology for Pesticide Analysis," in *Recent Advances in Pesticide Analytical Methodology,* J. Harvey, Jr. and G. Zweig, Eds., American Chemical Society Publications, Washington, DC, 1980, Chap. 18, pp. 321-352.

Hsu, I. C., Yolken, R. H., and Harris, C. C., "Ultrasensitive Enzymic Radioimmunoassay," in *Methods in Enzymology,* J. J. Langone and H. Van Vunakis, Eds., Academic Press, New York, 1981, Vol. 73, Part B, pp. 383-394.

Ishikawa, E., "Enzyme Immunoassay of Insulin by Fluorimetry of the Insulin-Glucoamylase Complex," *Journal of Biochemistry,* Vol. 73, 1973, pp. 1319-1332.

Kato, K., Hamaguchi, Y., Okawa, S., Ishikawa, E., Kobayashi, K., and Katunuma, N., "Use of Rabbit Antibody IgG Bound onto Plain and Aminoalkylsilyl Glass Surface for the Enzyme-Linked Sandwich Immunoassay," *Journal of Biochemistry,* Vol. 82, No. 1, 1977, pp. 261-266.

Kennedy, J. H., Kricka, L. J., and Wilding, P., "Protein-Protein Coupling Reactions and the Applications of Protein Conjugates," *Clinica Chimica Acta,* Vol. 70, 1976, pp. 1-31.

Kohler, G. and Milstein, C., "Continuous Cultures of Fused Cells Secreting Antibodies of Pre-Defined Specificities," *Nature,* Vol. 256, 1975, pp. 495-497.

Lane, D. and Koprowski, H., "Molecular Recognition and the Future of Monoclonal Antibodies," *Nature,* Vol. 296, 1982, pp. 200-202.

Landsteiner, K., *The Specificity of Serological Reaction,* Dover Publications, Inc., New York, 1962.

Litman, D. J., Hanlon, T. M., and Ullman, E. F., "Enzyme Channeling Immunoassay," *Analytical Biochemistry,* Vol. 106, 1980, pp. 223-225.

Littlefield, J. W., "Selection of Hybrids from Matings of Fibroblasts *in vitro* and their Presumed Recombinants," *Science,* Vol. 145, 1964, pp. 709–710.

Enzyme Immunoassay, Maggio, E. T. Ed., CRC Press, Boca Raton, FL, 1980.

Mattioli, C. A., Yagi, Y., and Pressman, D., "Production and Properties of Mouse Antihapten Antibodies," *Journal of Immunology,* Vol. 101, No. 5, 1968, pp. 939–948.

Miles, L. E. M. and Hales, C. N., "Labeled Antibodies and Immunological Assay Systems," *Nature,* Vol. 219, 1968, pp. 186–189.

Nakane, P. K. and Pierce, G. B., "Enzyme-Labeled Antibodies for the Light and Election Microscopic Localization of Tissue Antigens," *Journal of Cell Biology,* Vol. 33, 1967, pp. 307–318.

Parker, C. W., *Radioimmunoassay of Biologically Active Compounds,* Foundations of Immunology Series, Prentice-Hall, Inc., Englewood Cliffs, NJ, 1976.

Rodbard, D., "Statistical Aspects of Radioimmunoassays," in *Principles of Competitive Protein-Binding Assays,* W. D. Odell and W. M. Daughaday, Eds., J. B. Lippincott, Philadelphia, 1971, pp. 204–259.

Rubenstein, K. E., Schneider, R. S., and Ullman, E. F., "Homogeneous Enzyme Immunoassay a New Immunochemical Technique," *Biochemical and Biophysical Research Communication,* Vol. 47, No. 4, pp. 846–851.

Sevier, E. D., David, G. S., Martinis, J., Desmond, W. J., Bartholomew, and Wang, R., "Monoclonal Antibodies in Clinical Immunology," *Clinical Chemistry,* Vol. 27, No. 11, 1981, pp. 1797–1803.

Siddle, K., "Properties and Applications of Monoclonal Antibodies," in *Alternative Immunoassays,* W. P. Collins, Ed., John Wiley & Sons, New York, 1985, pp. 13–27.

Soini, E. and Hemmilla, I., "Fluoroimmunoassay: Present Status and Key Problems," *Clinical Chemistry,* Vol. 25, 1979, pp. 353–361.

Tijssen, P., *Practice and Theory of Enzyme Immunoassays,* Elsevier, New York, 1985.

Thompson, R. J., "Are Monoclonal Antibodies the End of Radioimmunoassay?," *Trends in Biochemical Science,* 1982, pp. 419–420.

Vallejo, R. P., Bogus, E. R., and Mumma, R. O., "Effects of Hapten Structure and Bridging Groups on Antisera Specificity in Parathion Immunoassay Development," *Journal of Agricultural and Food Chemistry,* Vol. 30, 1982, pp. 572–580.

Van Weemen, B. K. and Schuurs, A.H.W.M., "Immunoassay Using Antigen-Enzyme Conjugates," *FEBS Letters,* Vol. 15, No. 3, 1971, pp. 232–236.

Vinas, J., "Some Advantages and Problems Arising in the Substitution of Traditional Antisera by Monoclonal Antibodies," *Journal of Pure and Applied Chemistry,* Vol. 54, No. 4, 1985, pp. 577–586.

Voller, A., Bidwell, D. E., and Bartlett, A., "Microplate Enzyme Immunodiagnostics of Virus Infections," in *Manual of Clinical Immunology,* N. R. Rose and H. Friedman, Eds., American Society for Microbiology, Washington, DC, 1976, p. 506.

Wie, S. I. and Hammock, B. D., "Comparison of Coating and Immunizing Antigen Structure on the Sensitivity and Specificity of Immunoassays for Benzoylphenylurea Insecticides," *Journal of Agricultural Food Chemistry,* Vol. 32, 1984, pp. 1294–1301.

Wie, S. I., Sylwester, A. P., Wing, K. D., and Hammock, B. D., "Synthesis of Haptens and Potential Radioligands and Development of Antibodies to Insect Growth Regulators Diflubenzuron and BAY SIR 8514," *Journal of Agricultural Food Chemistry,* Vol. 30, 1982, pp. 943–948.

Yalow, R. S., "A Probe for the Fine Structure of Biologic Systems," *Science 200,* 1978, pp. 1236–1245.

Immunoassay Utilization

Brimfield, A. A., Lenz, D. E., Graham, C., and Hunter, K. W., "Mouse Monoclonal Antibodies Against Paraoxon: Potential Reagents for Immunoassay with Constant Immunochemical Characteristics," *Journal of Agricultural and Food Chemistry,* Vol. 33, No. 6, 1985, pp. 1237–1242.

Bushway, R. J., Savage, S. A., Lekousi, S. J., and Ferguson, B. S., "Determination of Atrazine Residues in Water and Soil by Enzyme Immunoassay," *Bulletin Environmental Contamination and Toxicology,* Vol. 40, 1988, pp. 647–654.

Chait, E. M. and Ebersole, R. C., "Clinical Analysis, A Perspective on Chromatographic and Immunoassay Technology," *Analytical Chemistry,* Vol. 53, No. 6, pp. 682A–692A.

Cheung, P. Y. K., Gee, S. J., and Hammock, B. D., "Pesticide Immunoassay as a Biotechnology," in *The Impact of Chemistry on Biotechnology: Multidisciplinary Discussions,* M. P. Phillips, S. P. Shoemaker, R. D. Middlekauff, and R. M. Ottenbrite, Eds., American Chemical Society, Washington, DC, 1988, Chapter 18, Vol. 362, pp. 217–229.

Cook, C. E., Ballentine, N. H., Seltzman, T. B., and Tallent, C. R., "Warfarin Enantiomer Disposition: Determination by Stereoselective Radioimmunoassay," *Journal of Pharmacology and Experimental Therapeutics,* Vol. 210, No. 3, 1979, pp. 391–398.

Ercegovich, C. D., "Analysis of Pesticide Residues: Immunological Techniques," in *Pesticide Identification at the Residue Level,* R. F. Gould, Ed., *Advances in Chemistry Series,* Vol. 104, American Chemical Society, 1971, pp. 162-177.

Ercegovich, C. D., "Relationship of Pesticides to Immune Responses," *Federal Proceedings,* Vol. 32, No. 9, 1973, pp. 2010-2016.

Ercegovich, C. D., Vallejo, R. P., Gettig, R. R., Woods, L., Bogus, E. R., and Mumma, R. O., "Development of a Radioimmunoassay for Parathion," *Journal of Agricultural and Food Chemistry,* Vol. 29, 1981, pp. 559-563.

Fleeker, J., "Two Enzyme Immunoassays to Screen for 2,4-Dichlorophenoxyacetic Acid in Water," *Journal of the Association of Official Analytical Chemists,* Vol. 70, No. 5, 1987, pp. 874-878.

Groopman, J. D., Trudel, L. J., Donahus, P. R., Marshak-Rothstein, A., and Wogan, G. N., "High Affinity Monoclonal Antibodies for Aflatoxins and Their Application to Solid-Phase Immunoassays," *Proceedings of the National Academy of Science,* Vol. 81, 1984, pp. 7728-7731.

Hammock, B. D., Gee, S. J., Cheung, P. Y. K., Miyamoto, T., Goodrow, M. H., Van Emon, J. M., and Seiber, J. N., "Utility of Immunoassay in Pesticide Trace Analysis," in *Pesticide Science and Biotechnology,* Proceedings of the 6th International Congress of Pesticide Chemistry, R. Greenhalgh and T. R. Roberts, Eds., Blackwell Scientific Publications, Oxford, 1987, pp. 309-316.

Haugen, A., Groopman, J. D., Hsu, I. C., Goodrich, G. R., Wogan, G. N., and Harris, C. C., "Monoclonal Antibody to Aflatoxin B_1-Modified DNA Detected by Enzyme Immunoassay," *Proceedings of the National Academy of Science,* Vol. 78, No. 7, 1981, pp. 4124-4127.

Hemingway, R. J., "Improved Cost-Effective Approaches to Pesticide Residue Analysis," *Pure & Applied Chemistry,* Vol. 56, No. 8, 1984, pp. 1131-1152.

Huber, S. J., "Improved Solid-Phase Enzyme Immunoassay Systems in the ppt Range for Atrazine in Fresh Water," *Chemosphere,* Vol. 14, 1985, pp. 1795-1803.

Huber, S. J. and Hock, B., "A Solid-Phase Enzyme Immunoassay for Quantitative Determination of the Herbicide Terbutryn," *Journal of Plant Disease and Protection,* Vol. 92, No. 2, 1985, pp. 147-157.

Hunter, K. W. and Lenz, D. E., "Detection and Quantification of the Organophosphate Insecticide Paraoxon by Competitive Inhibition Enzyme Immunoassay," *Life Sciences,* Vol. 30, 1982, pp. 335-361.

Kelley, M. M., Zahnow, E. W., Petersen, W. C., and Toy, S. T., "Chlorsulfuran Determination in Soil Extracts by Enzyme Immunoassay," *Journal of Agricultural Food Chemistry,* Vol. 33, 1985, pp. 962-965.

Kennel, S. J., Jason, C., Albro, P. W., Mason, G., and Safe, S. H., "Monoclonal Antibodies to Chlorinated Dibenzo-*p*-dioxins," *Toxicology and Applied Pharmacology,* Vol. 82, 1986, pp. 2256-2263.

Langone, J. J. and Van Vunakis, H., "Radioimmunoassay for Dieldrin and Aldrin," *Research Communications in Chemical Pathology and Pharmacology,* Vol. 10, No. 1, 1975, pp. 163-172.

Lukens, H. R. and Williams, C. B., "Fluorescence Immunoassay Technique for Detecting Organic Environmental Contaminants," *Environmental Science, & Technology,* Vol. 11, No. 3, 1977, pp. 292-297.

Mount, M. E. and Failla, D. L., "Production of Antibodies and Development of Enzyme Immunoassay for Determination of Monensin in Biological Samples," *Journal of the Association of Official Analytical Chemists,* Vol. 70, No. 2, 1987, pp. 201-205.

Mumma, R. O. and Brady, J. F., "Immunological Assays for Agrochemicals" in *Pesticide Science and Biotechnology,* Proceedings of the 6th International Congress of Pesticide Chemistry, R. Greenhalgh and T. R. Roberts, Eds. Blackwell Scientific Publications, Oxford, 1987, pp. 341-348.

Newsome, W. H. and Shields, J. B., "Radioimmunoassay of PCBs in Milk and Blood," *International Journal of Environmental Analytical Chemistry,* Vol. 10, pp. 295-304.

Newsome, W. H. and Shields, J. B., "A Radioimmunoassay for Benomyl and Methyl 2-Benzimidazolecarbamate on Food Crops," *Journal of Agricultural and Food Chemistry,* Vol. 29, No. 2, 1981, pp. 220-222.

Newsome, W. H., "An Enzyme-Linked Immunosorbent Assay for Metalaxyl in Foods," *Journal of Agricultural and Food Chemistry,* Vol. 33, No. 3, 1985, pp. 528-530.

Newsome, W. H., "Development of an Enzyme-Linked Immunosorbent Assay for Triadimefon in Foods," *Bulletin of Environmental Contamination and Toxicology,* Vol. 36, 1986, pp. 9-14.

Newsome, W. H., "Determination of Iprodione in Foods by ELISA," in *"Pesticide Science and Biotechnology,"* Proceedings of the 6th International Congress of Pesticide Chemistry, R. Greenhalgh and T. R. Roberts, Eds., Blackwell Scientific Publications, Oxford, 1987, pp. 349-352.

Numazawa, N., Haryu, A., Kurosaka, K., and Nambara, T., "Picogram Order Enzyme Immunoassay of Oestradiol," *FEBS Letters,* Vol. 79, 1977, pp. 396-398.

Rinder, D. F. and Fleeker, J. R., "A Radioimmunoassay for 2,4-Dichlorophenoxy acetic Acid and 2,4,5-Trichlorophenoxyacetic Acid in Surface Water," *Bulletin of Environmental Contamination and Toxicology,* Vol. 26, 1981, pp. 375-380.

Roberts, T. R., "Current Trends in Pesticide Residue Analysis," *Trends in Analytical Chemistry,* Vol. 4, No. 1, 1985, pp. 3-7.

Schwalbe, M., Dorn, E., and Beyermann, K., "Enzyme Immunoassay and Fluoroimmunoassay for the Herbicide Diclofop-methyl," *Journal of Agricultural and Food Chemistry,* Vol. 32, 1984, pp. 734–741.

Uotila, M., Ruoslahti, E., and Enguall, E., "Two-Site Sandwich Enzyme Immunoassay with Monoclonal Antibodies to Human Alpha-Fetoprotein," *Journal of Immunological Methods,* Vol. 42, 1981, pp. 11–15.

Van Emon, J., Hammock, B. and Seiber, J. N., "Enzyme-Linked Immunosorbent Assay for Paraquat and Its Application to Exposure Analysis," *Analytical Chemistry,* Vol. 58, 1986, pp. 1866–1873.

Van Emon, J. M., Seiber, J. N., and Hammock, B. D., "Applications of Immunoassay to Paraquat and Other Pesticides," in *Bioregulators for Pest Control,* P. A. Hedin, Ed., American Chemical Society, Washington, DC, Vol. 276, 1985, pp. 307–316.

Van Emon, J., Seiber, J., and Hammock, B., "Application of an Enzyme-Linked Immunosorbent Assay (ELISA) to Determine Paraquat Residues in Milk, Beef, and Potatoes," *Bulletin of Environmental Contamination and Toxicology,* Vol. 39, 1987, pp. 490–497.

Wie, S. I., Andrews, R. E., Hammock, B. D., Faust, R. M., and Bulla, L. A., "Enzyme-Linked Immunosorbent Assays for Detection and Quantitation of the Entomocidal Parasporal Crystalline Protein of *Bacillus thuringiensis* subspp. *kurstaki* and *israelensis,*" *Applied and Environmental Microbiology,* Vol. 43, No. 4, pp. 891–894.

Wie, S. I. and Hammock, B. D., "Development of Enzyme-Linked Immunosorbent Assays for Residue Analysis of Diflubenzuron and BAY SIR 8514," *Journal of Agricultural and Food Chemistry,* Vol. 30, 1982, pp. 949–957.

Wie, S. I. and Hammock, B. D., "The Use of Enzyme-Linked Immunosorbent Assay (ELISA) for the Determination of Triton X Nonionic Detergents," *Analytical Biochemistry,* Vol. 125, 1982, pp. 168–176.

Wild, C. P., Umbenhauer, D., Chapot, B., and Montesano, R., "Monitoring of Individual Human Exposure to Aflatoxins (AF) and N-Nitrosamines (NNO) by Immunoassays," *Journal of Cellular Biochemistry,* Vol. 30, 1986, pp. 171–179.

Wing, K. D., Hammock, B. D., and Wustner, D. A., "Stereoselectivity of a Radioimmunoassay for the Insecticide S-Bioallethrin," *Journal of Agricultural and Food Chemistry,* Vol. 26, 1979, pp. 1328–1333.

Author Index

T

V

W

Y

Subject Index

C

D

E

F

N

O

P

V

W

X

Z